21st Century Manufacturing

Paul Kenneth Wright
University of California at Berkeley

PRENTICE HALL
Upper Saddle River, NJ 07458

Library of Congress Cataloging-in-Publication Data

Wright, Paul Kenneth.
21st century manufacturing / Paul Kenneth Wright.—1st ed.
p. cm.
Includes bibliographical references and index.
ISBN 0-13-095601-5
1. Electronic industries. 2. Electronic apparatus and appliances—Design and construction. 3. Production engineering. I. Title.

TK7836.W75 2001
658.5—dc21 00-039181

Vice president and editorial director: **Marcia Horton**
Acquisitions editor: **Laura Curless**
Editorial assistant: **Lauri Friedman/Dolores Mars**
Production editor: **Arik Ohnstad, Carlisle Communications**
Managing editor: **David A. George**
Executive managing editor: **Vince O'Brien**
Art director: **Jayne Conte**
Cover design: **Bruce Kenselaar**
Marketing manager: **Danny Hoyt**
Manufacturing buyer: **Dawn Murrin**
Manufacturing manager: **Trudy Pisciotti**
Assistant vice president of production and manufacturing: **David W. Riccardi**

Cover photograph by Dan Chapman; pictured items courtesy of Alan Kramer and colleagues at ST Microelectronics, Inc.

Printed in the United States of America

10 9 8 7 6 5 4 3 2

ISBN 0-13-095601-5

Prentice-Hall International (UK) Limited, *London*
Prentice-Hall of Australia Pty. Limited, *Sydney*
Prentice-Hall Canada Inc., *Toronto*
Prentice-Hall Hispanoamericana, S.A., *Mexico*
Prentice-Hall of India Private Limited, *New Delhi*
Prentice-Hall of Japan, Inc., *Tokyo*
Pearson Education Asia Pte. Ltd., *Singapore*
Editora Prentice-Hall do Brasil, Ltda., *Rio de Janeiro*

Contents

Preface

This is a book that deals with today's technologies and the future of *manufacturing.* It includes details of the product design process, rapid prototyping, and a survey of manufacturing techniques relevant to today's production of consumer electronics or electromechanical devices. Biotechnology has been added because of the substantial future career opportunities in this field of manufacturing. The book also aims to provide a balanced view for the *management of technology.*

WHAT WILL 21ST CENTURY MANUFACTURING LOOK LIKE?

Within our imaginations, we probably all share a similar futuristic vision of electronic commerce, product design, and automated manufacturing.

Quite certainly the Internet and the World Wide Web of the 21st century will be vastly enriched. Using virtual reality and a haptic interface, a future consumer might "reach into" a computer and feel the virtual texture of a sweater that they want to mail-order. Quite certainly, keyboards will disappear: thus, in a voice-activated conversation with a virtual salesagent, the consumer might negotiate batch size (in many cases as low as one), size, color, and price, and then arrange for overnight fabrication and immediate delivery of a fully customized product. Somewhere else, clothing designers will already have sent beautifully rendered computer graphics images to fully automated factories. These images will sit quietly—waiting to be customized to an incoming order. And when the order comes, sophisticated machine tools and robots will spring to life automatically and smoothly fabricate the product for that specific consumer of the 21st century. The words "mass customization" are being used today for such a scenario.

At the beginning of the 21st century, *electronic commerce, product design, and manufacturing* are now global enterprises, increasingly integrated by the World Wide Web. Reliable electronic infrastructures and prompt customer delivery mechanisms mean that design services and manufacturing plants can be installed in any country. Any country? Perhaps any planet. By the 22nd century, surely someone will be exploiting as-yet-unknown minerals on a remote planet. These will be partially processed on the spot and subsequently converted to consumer products for people living throughout our solar system and even beyond. The Website <**Mars-manufacturing.com**> might be worth reserving now.

This is a realistic vision. One that is perhaps rooted in the television documentaries over the past two decades showing welding robots on the automobile lines in Detroit. Today's exponential growth of the Internet and the World Wide Web seems to further expand our personal boundaries, with visions of access to a wide variety of services, including opportunities for online shopping and custom designing. Our natural curiosity about the future then extrapolates today's capabilities to more Hollywood-esque images of design studios and automated manufacturing systems. These might be distributed throughout our solar system and guided from the mission control deck of a "Starship Enterprise."

THE ECONOMIC CONTEXT FOR 21ST CENTURY MANUFACTURING

With this future in mind, what should be included in a college level manufacturing course? What do future students need to know? What is exciting?

Some economic issues must be mentioned before answering the above questions. New constraints have been forced upon all manufacturers in the last 10 years or so. Being knowledgeable and efficient in the basic processing methods is still very important but not sufficient. Introducing new automation and robotic systems to reduce factory-floor labor costs is also important but not sufficient.

Many of these new pressures on all manufacturers have been the result of international competition. At the same time, consumers have been made more aware of their choices. Here is a quote from *The Economist* magazine that emphasizes the power of consumer choice:

> Suppose one had walked into a video shop a decade ago looking for Betamax tapes. Sony's Betamax was the better standard, almost everyone agreed: but the VHS had the marketing muscle, and customers fell into line. They wanted three walls of films to choose from, not one.

In the final analysis, if a manufacturing company is going to be successful in the 21st century, being good at just "the technology" is not enough to survive. A company must be alert to change; it must offer its customers the most innovative product at the best price and the best all-around service.

WHY DID I WRITE THIS BOOK?

The University of Birmingham in England was like any other leading engineering school in the 1970s. We studied the "physics" of individual manufacturing processes in great depth. My thesis discovered new methods for measuring temperatures very close to a cutting tool edge and correlating them with wear patterns when machining aerospace alloys. Later as a postdoctoral student at Cambridge University, my colleagues and I made movies through transparent sapphire cutting tools and studied the friction at the interface between the tool and the flowing chip. Actually it was great fun. So, not really knowing any better, these were the topics I lectured on in my first years as a professor. However, especially after I moved to Berkeley and Silicon Valley around 1990, these one-by-one studies of individual processes (whether for metals or semiconductors) seemed an inadequate preparation for students who were going to work for Intel, Hewlett Packard, IBM, and—more recently—dot.com start-ups. Today, although these students graduate and go off to manufacture the next generation of semiconductors, computers, disc drives, and all manner of peripherals and consumer products, their day-to-day careers involve designing, prototyping, and fabricating these electromechanical products rather than just refining one of the physical processes in great depth.

It thus seemed that a more global view of manufacturing was needed for students going into product development and probably management. This book emerged from that perception. Thus Chapters 1 and 2 begin with a review of the history of manufacturing, its present state, the need for integration, and a summary of some basic principles. These first two chapters cover ground that can also be found in the other excellent and comprehensive texts (listed in the Bibliography of Chapter 1) that focus on the general field of manufacturing.

Moving into Chapter 3, a different approach from these other texts has been adopted. Speaking generally, other manufacturing-oriented textbooks begin with a review of material properties and then mechanics (if they are targeted at mechanical engineers) or basic electronics (if they are targeted at electrical engineers). They continue with a comprehensive description of many manufacturing processes and then conclude with some manufacturing system issues that tie the whole landscape together. However, this previous approach has some limitations for today's students. The evidence indicates that they will probably start off their careers in the technology of manufacturing, but after only a few years they will become "managers of technology."

For these future managers, the word "manufacturing" will mean much more than the basic fabrication technology. It will involve market analysis, design, production planning, fabrication (including outsourcing), distribution and sales, customer service, and, finally, being agile enough to reconfigure the factory for the next product "six months down the road." Of course one could argue that this has always been the case: but now, the pace of change is so dramatic and being first to market is so critical that there is a much greater obligation for faculty to train students for this environment.

Therefore, the new approach beginning with Chapter 3 guides students through a *product development cycle*. The goal is to embed each fabrication process in its appropriate place in the whole activity of *manufacturing in the large*.

WHO MIGHT BENEFIT FROM THIS BOOK?

The audience that has been kept in mind is a class consisting of both engineering and business students, who are interested in a survey of manufacturing processes and their strategic consequences for business and the international economy. The course has been taught for a number of years at Berkeley, but the emphasis changes somewhat according to whether it is a junior/senior course or a first-year graduate course. The level also influences the topic chosen for the semester-long CAD/CAM project outlined in the Appendix. In the last few years the course has also been part of a management of technology program.

The analytical material is easy to digest without an extensive background in stress analysis, electronics, or biochemistry. The rationalizations for this level of treatment are that:

1. The ideas try to move beyond the basic science in each field to the strategic issues such as time-to-market.
2. On most campuses there are several subsequent graduate courses that do go into the detailed engineering issues in each domain.
3. There is a bibliography of research articles and books for the future specialist.
4. There is always the hope that other audiences, outside the academic community, might get something out of this book if it is written in a more conversational style rather than jam-packed with equations.

The first few chapters thus serve as a readable survey of the current economic factors before moving into Chapters 3 through 9, which have more technical content. The analysis of each basic process in chapters 3 through 9 is then presented in the context of business. While the central sections of these chapters focus on analysis, the market issues and the management context issues are discussed at the beginning and end of each chapter.

An especially valuable way of dealing with the new approach in a semester-long class has been to place emphasis on two activities:

1. Group projects in CAD/CAM, where students design, prototype, and fabricate a new product, including its marketing plan
2. Factory tours that support the understanding of integrated manufacturing, after which students, again in groups, write up a case study on the company, its business model, and future growth

Chapter 10 considers future management issues in more detail. It contains more open-ended topics that often come up in class discussions. For example, we may wonder about the more frightening side of automation and technology: will these future factories create inhumane relationships between machines and society, as depicted in Carel Kapek's famous play of the 1920s, *Rossum's Universal Robots?* Many people in the world today may feel the same as the Luddites—an informal protest group in the late 18th century that opposed the loss of craft skills during the first industrial revolution (1770–1820). Whether locked to a word processing terminal or an assembly line, many of today's jobs are still soulless. Or perhaps worse, unemployment in several European countries is widespread. With such pressing social issues, can we really justify fully automated factories? A further concern revolves around ecological issues. Not only are these advanced manufacturing processes energy-hungry, but they can often result in dangerous chemical by-products. How does one country create manufacturing systems that are ecologically friendly and yet efficient enough to compete against those of other countries that may have less strict environmental laws?

In summary, the "old manufacturing mentality" (certainly pre-1980) was mostly focused on getting products through machines and out the door to the loading dock. This had several weaknesses. In particular, it relied on a distant marketing organization to make the link to the customer. This is not so today, and this book focuses on "manufacturing in the large" and associated "business issues." Throughout the next century, manufacturing will be much more than machining metals, etching wafers, assembling computers, or controlling bioreactors. Manufacturing will be an integral part of an extended social enterprise. Today, it drives the "global economy"; probably in the future, it will drive a "solar system economy."

OUTLINE: A JOURNEY ALONG THE PRODUCT DEVELOPMENT PATH

The following subjects and chapters are organized as a journey along the product development path with emphasis on the fabrication techniques.

The following figure is a summary of this approach, using one of today's cell phones or handheld computers as a metaphor for the fabrication techniques needed.

Chapter 1: Manufacturing: art, technology, science, and business
Chapter 2: Manufacturing analysis: some basic questions for a start-up company
Chapter 3: Product design, computer aided design (CAD), and solid modeling
Chapter 4: Solid freeform fabrication (SFF) and rapid prototyping
Chapter 5: Semiconductor manufacturing
Chapter 6: Computer manufacturing
Chapter 7: Metal-products manufacturing
Chapter 8: Plastics-products manufacturing and system assembly
Chapter 9: Biotechnology
Chapter 10: Conclusions

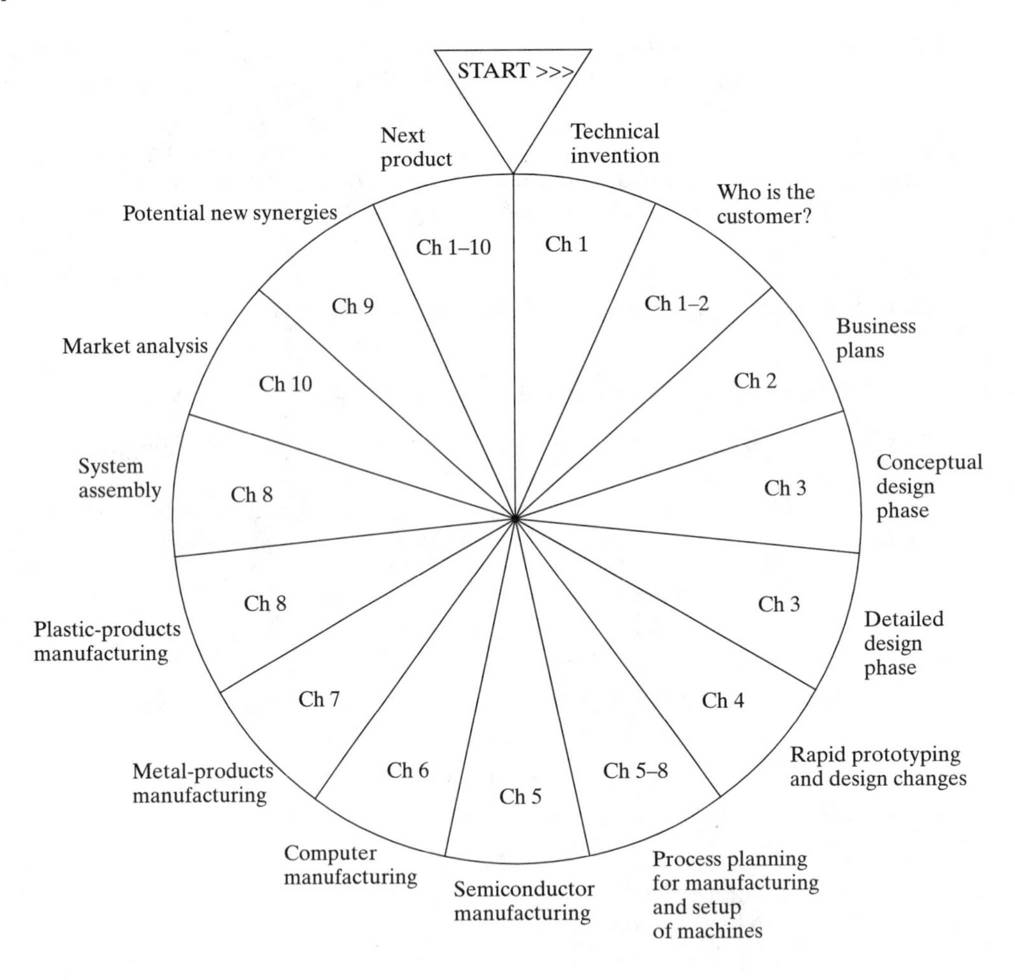
START >>>
Technical invention
Ch 1
Who is the customer?
Ch 1–2
Business plans
Ch 2
Conceptual design phase
Ch 3
Detailed design phase
Ch 3
Rapid prototyping and design changes
Ch 4
Process planning for manufacturing and setup of machines
Ch 5–8
Semiconductor manufacturing
Ch 5
Computer manufacturing
Ch 6
Metal-products manufacturing
Ch 7
Plastic-products manufacturing
Ch 8
System assembly
Ch 8
Market analysis
Ch 10
Potential new synergies
Ch 9
Next product
Ch 1–10

Acknowledgments

I have been fortunate in having many talented and congenial colleagues during the 30 years in which I have been actively interested in "manufacturing." Many ideas in this book have evolved from these close friendships, coauthorships, joint projects, informal discussions at various conferences, and many interactions with students. Warm thanks are extended to all these colleagues and students. If any of you feel that I have misrepresented you, or forgotten a citation to your work, please let me know at <**pwright@robocop.me.berkeley.edu**>. There is certain to be a revision of this book, and corrections can easily be made.

Looking back on my career in manufacturing, I recall my thesis work at the University of Birmingham, England, carried out with Dr. Edward Moor Trent. Sadly, he passed away in the spring of 1999 after a long decline, but it was possible to honor his great influence on my life by joining him as a coauthor in the Fourth Edition of his original book, *Metal Cutting*. Thanks are also due to Professor G. Rowe, Dr. D. Milner, Dr. T. Childs, Dr. R. Lorenz, Dr. P. Dearnley, and Mr. E. Smart, who were colleagues in the "machining research group" at Birmingham.

At the Cavendish Laboratory, Cambridge University, England, I would also like to thank my colleagues in the "transparent sapphire tool group": Professor D. Tabor, Dr. N. Gane, Dr. J. Williams, Dr. D. Doyle, and Dr. J. G. Horne.

At the University of Auckland in New Zealand, thanks are due to Mr. P. D. Smith, Professor R. F. Meyer, the late Professor J. H. Percy, Dr. J. L. Robinson, Dr. P. S. Jackson, Dr. A. W. Wolfenden, Professor G. Arndt, and Mr. George Moltzchianiwisky. I also thank Mr. L. S. Aiken, Mr. W. Beasley, Dr. John Meikle, and Mr. Peter Connor at the "old" Department of Scientific and Industrial Research in New Zealand. The late Professor P. L. B Oxley, Professor E. J. A. Armarego, Professor R. H. Brown, and Dr. M. G. Stevenson in Australia are thanked for their interactions.

At Carnegie Mellon University in Pittsburgh, Pennsylvania, I warmly acknowledge Professors Raj Reddy, Angel Jordan, and William Sirignano—who together inspired the growth of the Robotics Institute and manufacturing research—and the following colleagues: the late Dr. J. L. Swedlow, Dr. D. A. Bourne, Dr. C. C. Hayes,

Dr. R. S. Rao, Dr. M. Fox, Dr. A. Bagchi, Dr. A. J. Holzer, Dr. J. G. Chow, Dr. S. C. Y. Lu, Dr. D. W. Yen, Dr. C. King, Mr. H. Kulluk, Dr. M. R. Cutkosky, Dr. F. B. Prinz, Dr. S. Finger, Dr. M. Nagurka, Dr. P. Khosla, Dr. L. Weiss, Dr. R. Sturges. Many of these colleagues are now successful professors and researchers around the United States.

At the Courant Institute of Mathematical Sciences at New York University, acknowledgments are due to Professor J. T. Schwartz, Professor K. Perlin, Mr. I. Greenfeld, Mr. F. B. Hansen, Mr. L. Pavlakos, Dr. J. Hong, and Dr. X. Tan. These colleagues launched the open-architecture manufacturing research, and I thank them for their inspiration and teamwork.

At the University of California, Berkeley, Professor David Dornfeld, Professor Carlo Sequin, Dr. Frank Wang, Dr. Sung Ahn, Dr. Chuck Smith, Professor Paul Sheng, Professor Robert Brodersen, Professor Jan Rabaey, Professor David Hodges, Professor Robert Cole, and Professor Ken Goldberg are my day-to-day research colleagues. At the same time, Professors Erich Thomsen, Joseph Frisch, and the late Shiro Kobayashi were the founders of manufacturing-related research at Berkeley, and I thank them for their interest in this newer approach to manufacturing. Also, for their friendship and support, I thank Professor David Bogy, Professor Hami Kazerooni, Professor Paul Gray, Professor Masayoshi Tomizuka, Professor Kyriakos Komvopoulos, Professor Dan Mote, Professor Karl Hedrick, Professor Shankar Sastry, and Professor John Canny. Dr. Sara Okamura was a patient guide for the chapter on Biotechnology (Chapter 9).

For their detailed assistance in the preparation of the manuscript I thank Bonita Korpi, William Chui, Lisa Madigan, Andrew Lipman, and Zachary Katz. Dan Chapman has played an invaluable role in preparing art work and taking care of many details.

Thanks are also due to my graduate students from the Integrated Manufacturing Laboratory and Design Studio at Berkeley. They include Dr. Sanjay Sarma, now a Professor at M.I.T.; Dr. Steven Schofield, now at Siemens; Dr. James Stori, now a Professor at the University of Illinois, Urbana-Champaign; V. Sundararajan; Ganping Sun; Shad Roundy; Jaeho Kim; John Michael Brock; Juan Plancarte; Ryan Inouye; Matt Mueller; Louis Marchetti; Dan Odell; Kiyoshi Urabe; Kevin Chan; Fadel Hamed; Damien Crochet; Naomi Molly An; Mark McKenzie; Louis Marchetti; Roshan DeSouza; Balaji Kannan; Michael Montero; Ashish Mohole; and Socrates Gomez. Also the students that work with Professor Carlo Sequin are thanked: Jordan Smith, Jane Yen, Jainling Wang, and Sara McMains.

In addition, Berkeley has been the temporary home for many visitors who have also been contributors to the ideas in this book. I warmly thank Dr. Bruce Kramer from the National Science Foundation, Professor Jami Shah from Arizona State University, and Professor Jay Gunasekera from Ohio University.

From around the world I acknowledge the following inspiring friends in manufacturing: Professor Richard DeVor and Professor Shiv Kapoor at the University of Illinois, Urbana-Champaign; Professor B. F. von Turkovich at the University of Vermont; Professors R. Komanduri and J. Mize, now at Oklahoma State University; Professor J. T. Black at Auburn University; Professor S. Kalpakjian at the Illinois Institute of Technology; Professor H. Voelcker at Cornell University; Professor F. Ling at the

University of Texas, Austin; Dr. O. Richmond, Dr. M. Devenpeck, and Dr. E. Appleby from Alcoa Research Laboratories; Professors N. P. Suh and D. Hardt at M.I.T.; Professors T. Kurfess, S. Liang, S. Melkote, and J. Colton at Georgia Tech.; Professors S. K. Gupta and D. Nau at the University of Maryland; Professor W. Regli at Drexel; Professors S. Ramalingam and Kim Stelson at Minnesota; Dr. R. Woods at Kaiser Aluminum; Professors Y. Koren, G. Ulsoy, D. Dutta, and J. Stein at the University of Michigan; Professors K. Weinmann and J. Sutherland at Michigan Technological University; Professor K. P. Rajurkar at the University of Nebraska; Professor J. Tlusty at the University of Florida; Professors A. Lavine and D. Yang at UCLA; Professors I. Jawahir, A. T. Male, and O. Dillon at the University of Kentucky; Professor Tony Woo at the University of Washington; Professor S. Settles, now at the University of Southern California; Professor W. DeVries at Iowa State University; Professors M. DeVries, J. Bollinger, and R. Gadh at the University of Wisconsin; Professor D. Williams at Loughborough University, England; Professor M. Gregory and colleagues at Cambridge University, England; Professor Nabil Gindy, Dr. T. Ratchev, and colleagues at Nottingham University, England; Professor M. Elbestawi and his group at McMaster University, Canada; Professor Y. Altintas at the University of British Columbia, Canada; and the many colleagues at NSF, Boeing, and Ford/Visteon mentioned below.

Several years ago, Cincinnati Milacron was a great source of support for my work, and I acknowledge Mr. R. Messinger, Dr. R. Kegg, and Mr. L. Burnett. Similarly, for the support and interactions from Westinghouse I thank Tom Murrin (now at Duquesne University), Jose Isasi, Gary Schatz, and Jerry Colyer.

Collaboration with Sandia National Laboratories has been invaluable, and I thank Mr. L. Tallerico, Dr. R. Stoltz, Mr. A. Hazleton, and Dr. R. Hillaire for their support.

I am also grateful to colleagues at The Boeing Company, Dr. Donald Sandstrom in particular, for their funding and for new insights on high-speed machining.

Funding from the Ford Motor (and particularly the Visteon organization) is gratefully acknowledged, and I particularly thank Mr. Charles Szuluk and Dr. Shuh Liou for their long-term interest in integrated manufacturing. I also thank Mr. Craig Muhlhauser, Mr. Ashok Goyal, Mr. Glenn Warren, and Dr. Eugene Greenstein.

Support from the National Science Foundation has been invaluable in developing some of the fundamental ideas. Many colleagues have served at NSF during the course of the research: Dr. B. Kramer, Dr. C. Astill, Dr. B von Turkovich, Dr. W. Spurgeon, Dr. J. Meyer, Dr. T. Woo, Dr. S. Settles, Dr. J. Lee, Dr. M. DeVries, Dr. W. DeVries, Dr. L. Martin-Vega, Dr. C. Srinivasan, Dr. B. Chern, Dr. C. Wardle, Dr. M. Foster, and Dr. G. Hazelrigg have all played a role in this work. Support from DARPA, AFOSR, and ONR is also gratefully acknowledged as is the personal interest of Dr. W. Isler, Dr. E. Mettala, and Dr. J. Sheridan.

Thanks are due to the staff at Prentice Hall, Carlisle Communications, and the several colleagues who deserve a double mention for reviewing the early drafts of the book: Professor Costas Spanos, Professor Richard Devor, Professor Angel Jordan, Professor Serope Kalpakjian, Professor J. Black, Professor Klaus Weinmann, Professor Robert Cole, Professor Kenneth Goldberg, Professor M. Elbestawi, Professor Caroline Hayes, Professor Mark Cutkosky, Professor Mikell

Groover, Professor Kim Stelson, Dr. Calvin Kin, Mr. Randy Erickson, and Professor David Hodges, who first encouraged this combination of manufacturing and management of technology.

My wife Terry has contributed in innumerable ways to the writing, especially by spending many hours discussing the content and scope. My brother Derek and my late parents are implicitly part of the book. The loving company of my three sons Samuel, Joseph, and Thomas and my two stepdaughters Jessi and Jen has made the writing a lighthearted task. Nevertheless, I want to thank you all—my family, my students, and my colleagues—for "putting up with me" during the recent months in which I know I have been hyper-preoccupied.

Paul Wright,
Berkeley, CA
Spring 2000

CHAPTER

1 MANUFACTURING: ART, TECHNOLOGY, SCIENCE, AND BUSINESS

1.1 INTRODUCTION: WHAT IS "MANUFACTURING"?

The word has Latin roots: *manu,* meaning by hand, joined to *facere,* meaning to make. The dictionary definition is "*Making of articles by physical labor or machinery, especially on a large scale.*" Even this simple definition shows a significant historical trend. For hundreds of years, manufacturing was done by physical labor, in which a person with hand tools used craft skills to make objects. Since the industrial revolution 200 years ago, machinery has played an increasing role, as summarized in the second column of Figure 1.1. Also, the models for manufacturing processes have become better understood. And in more recent decades, computer aided design and manufacturing (CAD/CAM) and new concepts in quality assurance (QA) have been introduced to improve efficiency in production. It is expected that the 21st century will bring even better process models, more exacting control, and increased integration.

During the early part of the 20th century, the words *large scale*—used above in the dictionary definition—were synonymous with the mass production of Henry Ford. Most people would agree that the present trends created by the Internet have now set the stage for an even larger scale or global approach to manufacturing. We can expect to see global networks of information and distributed manufacturing enterprises. The 20th century concept of a monolithic organization clinging to one centralized corporate ethos may fade. The new culture may well be smaller, more agile corporations that can spring up for specific purposes, exist while the market sustains the new product, and then gracefully disband as the market changes. The Internet is certainly providing the infrastructure for these more flexible and informal ways of creating new enterprises that respond to people with a naturally entrepreneurial spirit. In Chapter 1, the goals are to set the stage for these broad views of manufacturing and a new era of global change.

Manufacturing: Past, Present, and Future			
Early 18th Century	19th Century	20th Century	21st Century
A person with an anvil and hammer Poorly understood process Craftspeople Cottage industry	Steam-powered machinery Improved understanding of processes Factory conditions in cities	Computer aided design, planning, and manufacturing Limited process models using closed loop control Increased factory automation	Systemwide networks and information Robust processes and intelligent control Global enterprises and virtual manufacturing corporations

Figure 1.1 Four centuries of manufacturing leading to 21st century manufacturing.

1.2 THE ART OF MANUFACTURING (FROM 20,000 B.C. TO 1770 A.D.)

In the most general sense, manufacturing is central to existence and survival. Historians connect the beginning of the last European Ice Age, approximately 20,000 years ago, to a period in which "*technology took an extra spurt*" (Pfeiffer, 1986). Cro-Magnons retreated southward from the glacial ice that, more or less, reached what are now the northern London suburbs. They manufactured rough pelts for warmth, simple tools for hunting, and crude implements for cooking. This general period of prehistory around 20,000 B.C. to 10,000 B.C. is referred to as the *Stone Age.* The availability of simple manufacturing tools and methods around the period of 10,000 B.C. also created the environment for community living, rather than an opportunistic, nomadic-tribe mentality. Such communities set the stage, at that time, for the *agrarian revolution.*

Manufacturing must have then evolved from these arts and crafts roots with occasional similar spurts prompted by climate, famine, or war. For example, the accidental discovery that natural copper ore, mixed with natural tin ore, would produce a weapon much more durable than stone replaced the Stone Age with the Bronze Age. Archaeologists believe that bronze weapons, drinking vessels, and other ornaments were made in Thailand, Korea, and other Eastern civilizations as early as 5000 B.C. At a similar time, in the Western civilizations, the evidence suggests that tin was mined in the Cornwall area of England. The two contemporaneous societies of Egypt and Mesopotamia appeared on the historical scene around 3000 B.C. While the historical roots of these cultures appear hazy, they were blessed with sophisticated artisans (Thomsen and Thomsen, 1974). Their early arts and crafts skills were then passed on to the Greeks and Romans, thereby setting the stage for European manufacturing methods. These grew very slowly indeed until the Iron Age and, finally, the industrial revolution of the 17th and 18th centuries.

One example of these early arts and crafts skills was the lost-wax casting process. It was discovered by both the Egyptians and the Koreans around the period 5000 B.C. to 3000 B.C. In the process, an artist carves and creates a wax model—say of a small statue. Sand or clay is then packed around this wax model. Next, the wax model is melted out through a small hole in the bottom to leave a hollow core. The small hole is plugged, and then liquid metal is poured into a wider hole at the top of the hollow cavity. After the metal freezes and sets, the casting is broken out of the sand. Some hand finishing, deburring, and polishing render the desired art object. Later chapters in the book will describe modern rapid-prototyping shops, connected to the Internet, producing small batches of trial-run computer casings for AT&T, Silicon Graphics, and IBM. These are high-tech operations by anyone's standards. Ironically, however, this lost-wax process remains one of the basic processes that is widely used in prototyping.

If the roots of forging and casting are with the Egyptian and Korean artisans, what about slightly more complex processes such as turning and milling? Bronze drinking vessels extracted by archaeologists from the tombs in Thebes, Egypt, show the characteristic turned rings on their bases as if made on a crude lathe. (As described later, a lathe is a turning machine tool, predominantly used today to change the diameter of a bar of steel.) The manufacturing date is estimated to be before 26 B.C., because Thebes was sacked in that year (Armarego and Brown, 1969). In the British Museum and the Natural History Museum in New York, many art objects show these characteristic turned circles from early machining operations.

Even the word *lathe* has romantic roots. It derives from the word *lath,* related to the description for a flexible stick or slender tree branch used to spin the bar as described below. Early lathes were operated by two people: one holding the tool, the other turning the bar being machined. Sooner or later someone figured out (probably one of the exhausted turning guys) that one could rig up a crude system something like an old-fashioned sewing machine treadle. A rope was wrapped and looped around the free end of the bar being machined. One end was tied to the turner's foot, rather like a stirrup; the other end was tied to the end of a springy stick or tree branch (the lath) that was nailed up into the roof rafters. As the turner raised his foot up and down, the motion rotated the bar back and forth, and the lath functioned as a return spring for the rope. Obviously this was a relatively crude process from a modern day view of achievable precision! But from the word *lath* comes today's word *lathe.* And in Britain, the word "turner" is frequently used instead of the American word "machinist" for the lathe operator.

This introduction to manufacturing from an artistic point of view brings up the first thoughts on design for manufacturability (DFM) (see Bralla, 1998). It must be clear from the above descriptions of open-die forging, casting, and machining that there is always a trade-off between the complexity of the original design and how easily it is made. It was certainly clear to the original artisans. In any natural history museum showing European art, one can see many functional items such as cooking pots, ordinary tools, eating implements that are rather dull looking: no fancy fleurs-de-lis or insets, no beautifully rounded corners. By contrast, exotic jewelry and necklaces do contain these fanciful additions. The most decorative items are the handles and scabbards of swords. These were obviously the most important objects to even an average soldier's heart, and they were willing to pay relatively large sums of money to the artisan to create beauty as well as functionality. Asian cultures had different ways of demonstrating wealth or societal position, where simplicity was synonymous with beauty. Nevertheless, the very best materials and refined structures were employed.

An economic analysis of design for manufacturability should always keep in mind the ultimate customer. An overly fanciful, nearly impossible to fabricate sword scabbard (or the 21st century equivalent) may be exactly what the customer wants and is happy to pay for. But not all customers. Walmart, Kmart, and McDonald's show that the greatest wealth is to be obtained from the mass markets where aesthetics and highest quality materials are compromised in favor of low cost. Any new enterprise embarking on the design, planning, and fabrication of a new product should therefore begin with the market analysis. How much time and money go into each step of design, planning, and fabrication is a recurring theme of this book. Without the best case reading of the marketplace—to analyze which group of consumers is being targeted, how many items will be sold, and at what margin—no amount of fabulous technology will win.

The brief case study at the end of this chapter expresses the same opinion. The article refers to "*the next bench syndrome*" coined at Hewlett-Packard. The idea is that, in the past, engineering designers would create devices to impress their engineering colleague seated at the next bench, rather than the ultimate consumer. Today, the evidence is that HP products have improved, now that its designers have redirected their efforts to become more customer oriented. The article also mentions the early (1993–1994 era) prototypes of pen-input computers. Some readers may recall how bulky and slow these were. But today, designers and manufacturers understand what consumers genuinely want from mobile, palm-size, pen-input devices: for example, the Palm Pilot and similar products have now become well established, useful consumer products.

This section is entitled "The Art of Manufacturing," and it introduces the important link between design and manufacturing (DFM). The relationship between art, design, and manufacturing is complex. The word *art* is derived from the Latin *ars,* meaning skill. Thus, especially before the industrial revolution (1770–1820), new products were designed and manufactured by artists and craftspeople: their hand skills were predominant. By contrast, in the modern era, new products are most likely to be designed and manufactured by mathematically trained engineers. Today, some degree of intuition, and trial and error, is still needed on the factory floor to operate machinery and to set up other equipment effectively, but throughout the 21st century, the role of the craftsperson or artisan will fade away.

Does this mean that art will no longer play a role in design and manufacturing? The answer is "probably not," because art involves more than just a hand skill itself. Most scholars of art describe the concept of *aesthetic experience* that elevates a basic skill into the artistic realm. It is observed that the most successful artists—in any field such as music, dance, literature, painting, architecture, or sculpture—communicate an aesthetic experience to their audience. Communication of this aesthetic experience to the consumer will always be key for the "design artist" or "manufacturing artist" no matter how mathematically sophisticated and high-tech these fields eventually become. As this book moves on to the technology and business of manufacturing, it is suggested that new students in the field keep this concept of aesthetic experience in mind.

1.3 THE TECHNOLOGY OF MANUFACTURING: FROM THE 1770s TO THE 1970s

The first watershed that changed manufacturing from a purely artistic or at least artisan type endeavor must surely have been the industrial revolution in England, which took place in the approximate period 1770 to 1820. The most gifted historians have come up with no single reason for this revolution (Plumb, 1965; Wood, 1963). It was a combination of technological, economic, and political factors, as follows:

1. A rapid increase in the day-to-day health and living conditions of people, hence increased population for marketing purposes and the supply of a labor force for the expanding factories.
2. Access to large markets, not only in England and the rest of Europe but also in Asia and Africa, as explorers opened up new colonies and global markets. Also, historians point out that even though England had lost the American War of Independence, there nevertheless remained a huge market for goods in the rapidly expanding United States.
3. A long period of social and political stability in Britain. This provided the stage for a more entrepreneurial mood in business and commerce.
4. New techniques in banking and the handling of credit. Added to this were faster communications and reliable methods for handling mail and business documents.
5. Many successive years of successful commerce, which caused capital to accumulate and interest rates to fall. Available and cheap capital favored business. Both large-scale operations and smaller middle-class businesses were formed, which then added to a general "gold rush" type fervor around London and the industrial cities north of England.
6. For sure, the industrial revolution could not have taken off without the steam engine. (In exactly the same way, the current information age could not have taken off without the invention of the transistor and microprocessor 200 years later.) Thomas Newcomen built one of the first steam engines in 1712, but it was James Watt's improved engine designs that made steam power usable by industry: in particular, his patent for a separate condenser was granted in 1769. This steam-powered machinery that was thus set in motion during the industrial revolution paved the way for massive increases in productivity in all fields. The historians (Plumb, 1965; Wood, 1963) provide many examples, in iron, in textiles, and in machinery manufacturing. For example, a rapid series of inventions took cotton spinning out of the house and into the factory. Arkwright's water frame (1769), Hargreaves's multiplied spinning wheel (or jenny) (1770), and Crompton's mule (1779) enormously increased the amount of thread one person could spin. And it took little time to apply the steam engine to these industries such as cotton spinning. The first steam-powered machines were developed in 1785, and in the space of 15 years or so, the transition from cottage industry to factory life was complete. This naturally increased the demand for cotton, which could not have occurred without another invention

by Eli Whitney. Records show that in April 1793, he built the first cotton gin, which revolutionized and rapidly accelerated the output of cotton in Georgia and the southern states.

Historians and economists emphasize strongly, however, that the new technology alone was not enough to account for the dramatic expansion in productivity and commerce. In fact, the historians point out that even steam technology was not a brand new idea. Evidence indicates that the ancient Greeks played with steam-powered toys and that the ancient Egyptians used steam-powered temple doors. Also, in the 15th century, the evidence indicates that China had already developed a rather sophisticated set of technological ideas that included steelmaking, gunpowder, and the ability to drill for natural gas. None of these previous cultures capitalized on such technologies to launch an industrial revolution. All six factors above, added together between 1770 and 1820, were needed. It is thus interesting to review the list in the context of manufacturing growth in the 21st century. In many respects, while today's technology is more related to electronics and telecommunications, to maintain growth, the social and economic drivers must remain much the same.

Once the Industrial Revolution was well under way, beyond 1820, it gave way to a more sustained period of consolidation in both Europe and the United States. The rise and consolidation of the machine tool industry, for example, was important during the period from 1840 to 1910. Increased standardization, improved precision, and more powerful machines provided a base for many other metal-product type industries. These secondary industries could expand only because of the availability of reliable machine tools. *Indeed even today, the machine tool industry is a key building block for industrial society, since it provides the base upon which other industries perform their production.* Rosenberg (1976) has written a comprehensive and engaging review of the origins of the machine tool industry and its crucial supporting role for secondary industries such as the gun making industry. Here are some typical extracts:

> Throughout the whole of the first half of the nineteenth century and culminating perhaps with the completion of Samuel Colt's armory in Hartford in 1855, the making of firearms occupied a position of decisive importance in the development of specialized precision machinery . . . it is clear that both Eli Whitney and Simeon North employed crude milling machines in their musket producing enterprises in the second decade of the nineteenth century as did John D. Hall in the Harper's Ferry Armory . . . the design of the plain milling machine was stabilized in the 1850s and rapidly assumed a prominent place in all the metal trades.

This interaction between gun making and machinery invention created the important manufacturing idea of *interchangeable parts.* Prior to this concept, each gun was handcrafted and fitted together as a unique item. This was because the dozens of subcomponents were machined with no quality control of size or geometry. By contrast, using the interchangeable parts concept, the individual subcomponents were produced with strict uniformity. In this case, any combination of them would fit together nicely. It also meant that assembly could proceed using relatively unskilled labor. Eli Whitney is often credited with the "invention" of the interchangeable parts idea. However, many historians modulate this view. It is more likely that these new manufacturing methods were developed and refined over a period of many years by the craftspeople in several New England armories, all of whom were struggling with the same problems of quality control and delivery time (see Rosenberg, 1976). At the same time, in France, LeBlanc developed similar methods for interchangeable parts.

How did these new methods affect customer satisfaction? Some historical accounts point out an initial anxiety from the customers, who were government agencies involved in war (accounts indicate that Thomas Jefferson gave Whitney one of the earliest contracts to make muskets). This was because the interchangeable parts concept required careful machinery setup and exacting quality control. It meant that the initial deliveries were probably slower than usual because of this added setup time. By contrast, the accounts then indicate that the customers were subsequently astonished by how quickly Whitney and other armories could mass-produce large batches. These customers were doubly happy in the event that they needed a spare part. Previously, such a repair might require the hand polishing and adjustments skills of a "fitter and turner." However, with interchangeable parts, it was just a question of buying a replacement and being fairly confident it would be the same size as the original subcomponent.

F. W. Taylor wrote *Principles of Scientific Management* in 1911, much of which was based on his factory experiences in the period 1895–1911, first at the Midvale Steel Company and then at the Bethlehem Steel Corporation. Taylor was an extraordinarily successful man in many areas. First, he co-invented, with Maunsel White, high-speed steel cutting tools that allowed a four times increase in cutting speed in the basic production processes of turning, drilling, and milling. Second, Taylor carefully analyzed individual manufacturing processes such as metal machining and tried to bring them under closer control. The Taylor equation that relates cutting speed to tool life is still used today. This work sprang naturally from the interchangeable parts concept, but for Taylor, even more systematic measurement was the main goal. Third, when he turned his attention to factory organization, he created order out of chaos. He quantified manual labor tasks by breaking them down into substeps. These smaller steps were then more efficiently organized. The goals were to shorten each subtask and get the overall task done more quickly. Such *time-and-motion* studies were so effective for industrial organization at that time that they were soon to be used by all the larger, emerging industrial corporations.

Consequently, *mass production*—usually attributed to Henry Ford—was the natural culmination of the interchangeable parts idea and Taylor's careful methods for dissecting and optimizing industrial tasks. By 1912, the first automobiles were beginning to roll off well-organized production lines (Rosenberg, 1976). The First and Second World Wars further increased the need for speed and efficiency. Weapons produced with great efficiency in the United States were crucial to the success of the Allied forces in Europe.

The knowledge gained from these efficient production methods meant that after the end of the Second World War in 1945–1946, the United States had a world monopoly, especially in comparison with the rest of the world, which was devastated from the war and would need many years to rebuild. Not only did the United States have detailed knowledge of basic manufacturing processes, but it was also very skilled in operations research (OR)—meaning the logistics of how to organize large agricultural and manufacturing enterprises. This was further fueled by the expanding electronics industry. The first *numerically controlled* (NC) machine tools were invented and refined over the period from 1951 to 1955. This period also saw the beginnings of *computer aided manufacturing* (CAM).

In summary, from the end of the Second World War, throughout the 1960s, and into the early 1970s, the United States enjoyed more than 25 years of unparalleled wealth. Excellence in manufacturing was one of the key components to this wealth.

1.4 A SCIENCE OF MANUFACTURING: THE 1980s TO THE PRESENT

1.4.1 Overview: Engineering Science and Organizational Science

It is perhaps human nature to "fall asleep at the wheel" when we are successful. Despite the commercial prowess of the United States in 1970, and despite the early promise of the new ideas in computer aided manufacturing (CAM), many manufacturing operations in the United States were left vulnerable to the new Japanese efficiency and quality assurance (QA) methods. These began to make a very noticeable impact by the mid-1970s.

Consumer items such as VCRs, microwave ovens, televisions, and cameras were the first to be taken over by Japanese manufacturers (such as Matsushita and Sony) and, subsequently, by other Pacific Rim countries. Furthermore, given the reluctance of the "Big Three" U.S. automobile manufacturers to change their designs to reflect increased gas prices in the 1970s, and a general demand from consumers for more reliable vehicles, it was not long before Toyota, Honda, and Nissan were stealing away a worrisome chunk of the U.S. car markets.

Beginning around 1980, how did the United States respond to these challenges? At first, the responses were emotional and a little derogatory. Magazine articles of the early 1980s alleged that our ever-cunning competitors in Europe and Japan were at worst "dumping" steel, autos, and memory chips at below real market costs, merely to penetrate the U.S. market. Or perhaps, more mildly, these new competitors were successful only because of cheaper labor costs. Not surprisingly then, the first rational U.S. response was to heavily invest in *robotics* and unattended *flexible manufacturing systems* (FMS) in order to reduce factory floor labor costs. Taken together, robotics and unattended flexible manufacturing systems can be defined as *computer integrated manufacturing* (CIM).

By the mid-1980s, this investment in CIM did begin to show considerable promise. Nevertheless, as emphasized in the preface, to compete in manufacturing, no amount of fabulous technology alone can win the day. To make a true turnaround in manufacturing excellence, these investments in robotics and FMS needed to be executed in the context of *total quality management* (TQM).

The next two subsections review more details of these issues under two headings:

- *Engineering Science,* defined for this book as the hardware and software of CIM
- *Organizational Science,* defined for this book as the management and TQM issues

It is thus likely that we are now in a historical period, beginning around 1980, in which certain ideas are solidifying that will become the cornerstones of manufacturing analysis. To call them a science may be an exaggeration, but it is useful to nonetheless set the stage for a handful of irrefutable concepts that will stay with manufacturing analysis in the future. These ideas build upon each other, as shown in Figure 1.2.

1.4.2 Engineering Science

The initial visionary work on the application of computers to manufacturing was done by a handful of people including Harrington (1973), Merchant (1980), and Bjorke (1979). They created the idea of computer integrated manufacturing (CIM) as the way to automate, optimize, and integrate the operations of the total system of manufacturing. During the 1980s, CIM naturally expanded to include the use of robotics and artificial intelligence (AI) techniques (see Wright and Bourne, 1988).

The three circles in Figure 1.3 show the early emphasis from the CIM era of Harrington, Merchant, and Bjorke. It includes the basic physics of each process (such

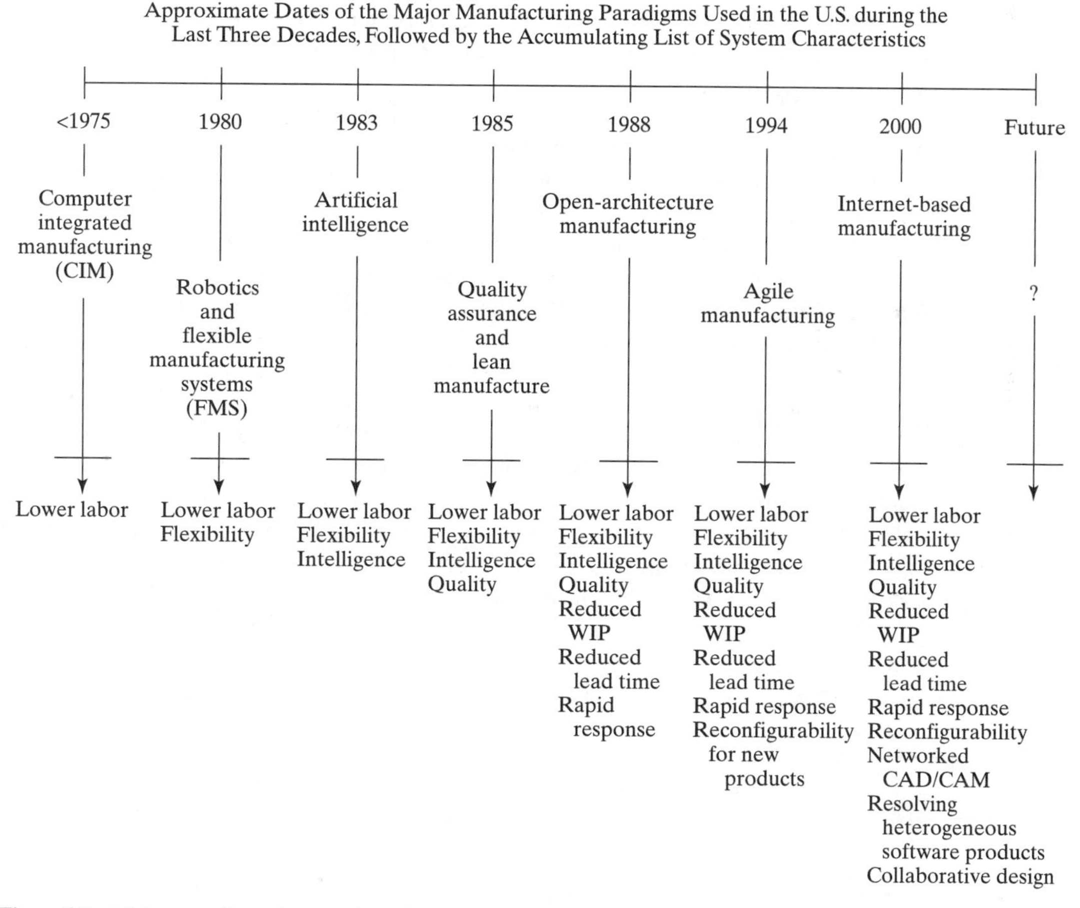

Figure 1.2 Major paradigms in manufacturing.

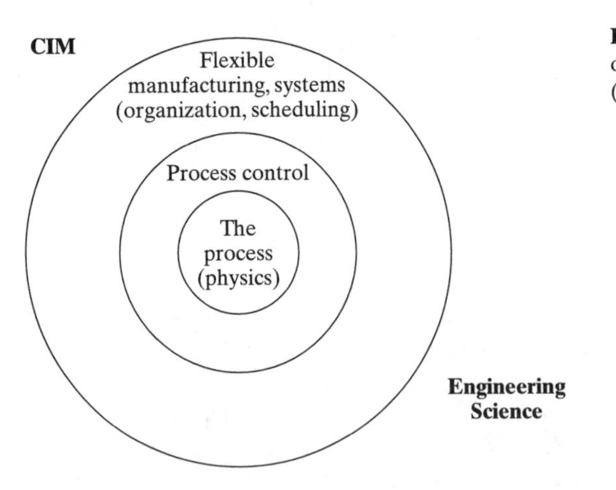

Figure 1.3 Engineering science aspects of Computer Integrated Manufacturing (CIM).

as machining, welding, or semiconductor manufacturing), control issues (such as servo-control of robots and processing machinery), and flexible manufacturing system (FMS) scheduling (e.g., for the production of machined components or IC wafers).

Does Figure 1.3 constitute an engineering science, where the word *science* may be defined in a dictionary as "systematically formulated knowledge"? The key issue is whether mathematical formalism and rigorous proofs can be developed. The pro-science data include the following observations.

First, in the inner circle, there is the physics of the basic processes in materials processing and semiconductors. These processes have, at their deepest roots, topics such as dislocation theory for the basic understanding of plasticity and lattice physics for the basic understanding of the way in which transistors work. At the same time, as metals are deformed in plastic deformation processes such as machining and forging, there are now some very standardized methods (such as the finite element analysis method) that can be universally applied to predict forces across a wide range of individual processes. Similarly, whether ICs are made by NMOS, CMOS, or BiPolar (see Chapter 5), the fundamentals of lithography and doping and the like stay the same. Such observations are a genuine basis for scientific methods and principles that can be widely applied across several manufacturing steps.

Second, in the next circle, there is now a well-established body of knowledge in control theory that prescribes stability, settling time, and accuracy in machines used for manufacturing. Combined with the standard kinematic analyses for linkages, cams and drive mechanisms, and friction, another body of scientific knowledge has been established for this part of manufacturing, primarily concerned with machinery control. Especially as integrated circuits get smaller and smaller, precise machine motions of the lithography patterns are crucial to the success of the whole industry. And here again a body of scientific knowledge exists for optics, materials science, and related issues in solid mechanics.

Third, in the outer circle, the scheduling of a flexible manufacturing system is shown. This involves the analytical areas of discrete event simulation, statistical modeling, optimization, and queuing theory. These are the cornerstones of many Industrial Engineering and Operations Research Departments. In recent years, the AI community has added some additional science to this area, referred to as constraint based reasoning. In summary, the mathematics behind these scheduling issues is now well established in its own right as a science and has very important applications to the scheduling of semiconductor plants where wafers must be economically moved in and out of lithography machines and ovens (Leachman and Hodges, 1996). Despite the more engineering-science approaches to manufacturing already described, they also needed the more organizational methods reviewed next to really make a full impact.

1.4.3 Organizational Science

Ironically, many of the new philosophies involving total quality management were invented by U.S. industrial engineers such as W. E. Deming. Also, historians point out that although the first industrial robot was patented in the United States in 1961, its most enthusiastic users were first in Japan (Engelberger, 1980). Quite simply, the new competitors from Japan and other Pacific Rim countries merely took the "best ideas that were out there" and then applied them with absolute dedication and perfection.

For example, Taiichi Ono championed the Toyota Production System, which reduced work in progress by *pulling* products through a flexible manufacturing system (FMS) rather than *pushing* unnecessary amounts of subcomponents into an already log-jammed system. *Just in time* (JIT) manufacturing is often used to describe this method of operation. *Lean manufacturing* is another associated phrase emphasizing a focus on reduced work-in-progress (WIP) and inventory.

At the same time a new approach to quality control was pioneered by Toyota. In the "old" definition of quality control (QC), a component was measured, *after* it had been made, to see if the processing operation(s) had created the dimensions specified by the designer. Then, if the component did not meet the specified dimensions, it was rejected.

By contrast, today, the "new" Toyota methods focus on measurements *during* the production line activities. Therefore the focus changes: rather than measuring at the end of the line and rejecting parts, measurements are done along the way. Also, machines are adjusted to prevent faulty parts from occurring in the first place. In summary, this has come to be called in-process quality control—or total quality management (TQM). And it puts responsibility on the individual worker and/or machine rather than "punting" problems downstream to be eventually uncovered by inspectors (Cole, 1999). TQM is thus added as a new circle in Figure 1.4, which enlarges the previous Figure 1.3 with organizational and business issues.

As implied in Figure 1.4, *concurrent engineering* (CE)—also known as concurrent design—is a topic that is closely related to TQM. Concurrent engineering also became important during the late 1980s, because too many U.S. companies indulged in *over-the-wall manufacturing.* This catchphrase can be explained as follows: The evidence was widespread in many companies that designers did their work in a social vacuum. Beautiful CAD images were rendered on high-end, graphics-oriented work

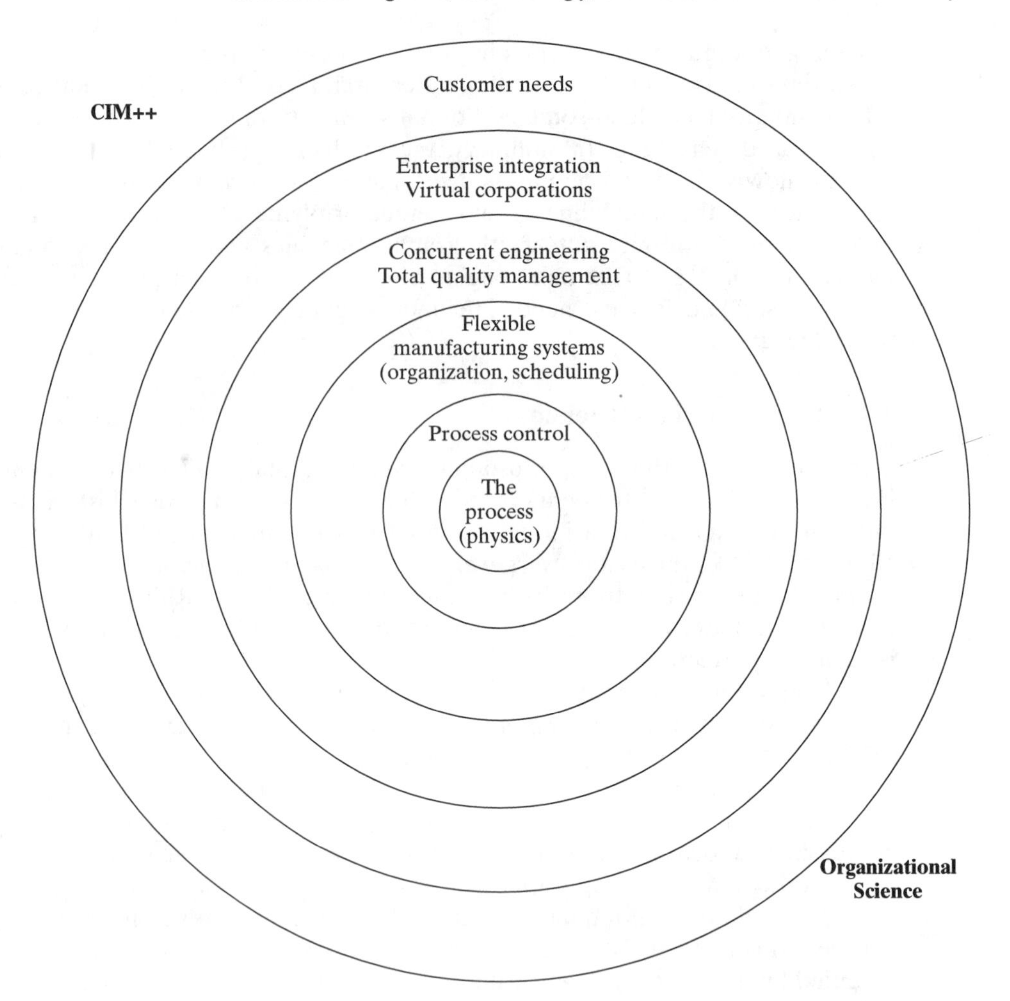

Figure 1.4 Organizational science aspects of computer integrated manufacturing, with more focus on the customer rather than the technology itself (shown earlier in Figure 1.3).

stations, but these images had to be reinterpreted for the specific operation of robots and machine tools on the factory floor. And during this translation, many ambiguities and errors arose, causing long delays between design and manufacturing.

Some of the reasons for the over-the-wall manufacturing of previous decades actually go back to the autocratic F. W. Taylor. He believed that only the design engineers were intelligent enough to make the decisions for production. He asserted that the manufacturing engineers should stay out of the decision loop and just do what they were told. This also seems to have had an influence on the way in which university courses were organized for many decades and the way in which pay scales and responsibilities were divided up in most factories. In general, the designers were university trained; the manufacturers, trade-school trained. By the 1980s, this compartmentalization was not as helpful. *Taylorism* is used today with an unpleasant tone. It creates gulfs between designers and manufacturing engineers, breaks down communication, and creates time delays in a manufacturing system.

Sadly, it took U.S. manufacturers several years to honestly acknowledge that an increased focus on concurrent engineering and total quality management was the only way out of the mess. Here is a quote from "The Quality Wars" by Jeremy Main:

> All of the Big Three started off on more or less the same footing, took different paths to get out of the crisis, but then all ended up doing essentially the same thing. They have found no substitute for TQM.

Luckily for the general economy of the United States, it was not all bad news in the 1980s. The rise of the computer industry led to enormous growth in both hardware and software. And despite a somewhat roller-coaster behavior, the manufacturing of semiconductors in the United States continued with increasing health (Macher et al., 1998). Hardware, operating systems, and software continued to be the forte of U.S. companies, thanks to the creative venture capitalists and the computer culture of Silicon Valley in particular. At the same time, the biotechnology and pharmaceutical industry boomed during the 1980s.

By the late 1980s the organizational sciences of TQM, JIT, CE, and lean manufacturing, combined with the engineering sciences of CIM, all began to create an important improvement in U.S. manufacturing (Schonberger, 1998; Macher et al., 1998). And this set the stage for the economic growth of the 1990s, as described in the next section.

1.5 THE BUSINESS OF MANUFACTURING

Ayres and Miller (1983) provide the succinct definition of manufacturing as the "confluence of the supply elements (such as the new computer technologies) and the demand elements (the consumer requirements of delivery, quality, and variety)." This definition perhaps needs some minor clarification. It relates to the natural "push" of new technologies onto the general marketplace on the one hand. Today, for example, new chips, faster computers, and faster modems are in constant development and being announced almost on a daily basis. On the other hand, there is a hungry "pull" from the marketplace. For example, users want to download programs faster from the Internet and run more lifelike graphics with their video games.

In summary, Ayres and Miller observe that at any time in technological progress, there is a confluence of these "push-pull factors" that stimulates more efficient methods of design and manufacturing on the one hand, and more demanding consumers on the other.

What do these demanding consumers of the 21st century want? Ayres and Miller's definition states that they want delivery, quality, and variety. In more colloquial language, they want pizza, eyeglasses, and their vacation photographs in "one hour or less or their money back." Even in more industrial settings, large computer design companies such as Sun and IBM make similar demands on their manufacturing oriented subsuppliers such as Solectron—a fast growing company in the assembly of printed circuit boards (PCBs).

Thus, in the 1990s, the best companies extended concurrent engineering and TQM to a higher level. This meant a "seamless" connection, all the way from factory floor manufacturing to the desires of the consumer. While this may seem obvious and sensible today, the "old" (certainly pre-1980) factory mentality was mostly focused on getting products out the door and leaving things to a distant marketing organization to make the link to the customer. This is not so today, and this section of Chapter 1 focuses on business issues and manufacturing-in-the-large.

These broader views are shown on the right of Figure 1.2. *Open-architecture manufacturing* and *agile manufacturing* were thus new paradigms that permeated the 1990s. These emphasized quickly reconfigurable enterprises that could respond to the new customer demands of "delivery, quality, and variety" (Greenfeld et al., 1989; Goldman et al., 1995; Anderson, 1997).

By the mid-1990s, Internet-based manufacturing was the natural extension of these paradigms, emphasizing the sharing of design and manufacturing services on the Internet (Smith and Wright, 1996).

The availability of the Internet, videoconferencing, and relatively convenient air travel seem to pave the way for increased global commerce. Large business organizations can be split up but then orchestrated over several continents—perhaps to take advantage of excellent design teams in one country and low-cost, efficient manufacturing teams in another. But in fact, for a variety of cultural and economic reasons, industrial growth has always been dependent on situations where "large businesses are distributed." This was just as true in the year 1770 when cotton from Georgia in the United States was shipped to Bradford in England for manufacturing into garments and then exported to an expanding population throughout the increasingly global British Empire. It was still true in the year 1970, just before the creation of the Internet: product design in the United States and the use of cheaper "offshore manufacturing" was a standard practice. In the 21st century, with the World Wide Web and videoconferencing, there is the *potential* for much faster exploitation of advanced design studios in one location and cheap labor in another. Nevertheless, clear communications—first, between the customer and the designer, and second, between the designer and the manufacturer—remain vital for realizing this potential and obtaining fast time to market. In later chapters of the book, examples will show that those companies that beat their competitors in launching the next chip, cell phone, or any consumer product will usually gain the most profit (see Ulrich and Eppinger, 1995).

Enterprise integration thus appears in the fifth circle of Figure 1.4. This term is actually the idea of concurrent engineering carried to a much larger scale and covering the whole corporation. The key requirement is the integration of all the divisions of a manufacturing-in-the-large enterprise. To reiterate, before 1980, Taylorism created competitiveness rather than cooperation between these various divisions (Cole, 1999). The more 21st century approach must involve the breaking down of barriers between people and subdivisions of an organization so that the whole of the enterprise can share problems openly, work toward shared goals, define shared productivity measures, and then share the dividends equally. Time to market will then benefit from this integrated design and manufacturing approach. This is one central message of this book.

Beyond such intercorporation trust comes the possibility of agreements with outside corporations. These agreements might spring up for a temporary period to suit the commercial opportunity at hand. This more ephemeral version of the old style monolithic business is called the *virtual corporation.* Nishimura (1999) argues that a successful 21st century virtual corporation must continue to rely on the core competency skills of each player, but at the same time, each participant must become more experienced in partnering skills.

Thurow (1999) goes further and argues that "cannibalization is the challenge for old business firms." It means that older well-recognized companies must now fragment into smaller business divisions. These will interact tightly for certain business ventures but then disband when their usefulness is over.

Open-architecture manufacturing, agile manufacturing, Internet-based manufacturing, and the virtual corporation all sound exciting. However, it does not take much imagination to look at Figure 1.2 and realize that a new buzzword or phrase will arrive soon. The reader is left to fill in the question mark. Perhaps the most important thing, emphasized in Figure 1.2, is that each era builds upon the previous one, and that under no circumstances should the organizational sciences built around total quality management be forgotten. New engineering science technologies, such as the Web, offer new ways of creating products and services, but efficiency and in-process quality control in basic manufacturing will always be mandatory.

1.6 SUMMARY

By reviewing the art, technology, science, and business aspects of manufacturing, it can be concluded that the activity of manufacturing is much more than machining metals or etching wafers: manufacturing is an extended social enterprise. In the last 250 years, people have been dramatically changed by the advances in manufacturing. Society has moved from an agrarian society, to handcrafts in cottage industries, to the operation of machinery in factories, to computer automation/robotics (and all its associated software writing and maintenance), and finally to telemanufacturing by modem and the Web.

Gifted philosophers such as Marx and Maslow have noted that people actually prefer to work rather than do nothing. But they want to get recognition for their labors beyond a paycheck. In the early transitions described in Section 1.3, up until the 1950s, craftsmanship often lost out to mass production and the dehumanization of work. Today, by and large, people are not inclined to work in dangerous factory situations or sit in a sea of cubicles carrying out monotonous word processing tasks just for the paycheck.

As the futurist Naisbitt says, people want "high-tech high-touch," meaning all the modern conveniences of life with a softer approach. Thus, once people have enough money, they strive to re-create their jobs, to make them more interesting, or to reeducate themselves for a more intellectually rewarding job. In today's corporations, this generally means moving off the factory floor. Initially, a person's reeducation might lead to a position in machinery diagnostics and repair or in the organization of production. In time, such a position might grow into general management, personnel, and business oriented decision making. It is likely that for several more decades, a *combination of people and partial automation* solutions will be seen on the factory floor. Today, the cost-effective solution is to use mechanized equipment for, say, moving pallets of printed circuit boards through a reflow solder bath but to concurrently use human labor for inspection, monitoring, rework, and the occasional corrective action.

Despite this partial-automation/partial-human situation, the long-term trend is to invest in sophisticated capital equipment that can work completely unattended by humans. This has always been the stated goal of computer integrated manufacturing (Harrington, 1973; Merchant, 1980).

This leaves the people to work with knowledge issues. The trends in both Figures 1.1 and 1.2 from left to right emphasize this change from Taylor's "hired hands" to "*knowledge workers*"—a term first coined by Peter Drucker in the 1940s. For many industries, there is also a shift in balance from *capital-intensive machinery* to *software and corporate knowledge.* Many top managers are being forced to rethink the way their organization functions. Indeed the role of "management" in and of itself is being reevaluated. This is especially true in newer start-up companies where the culture is informal and youth oriented.

Drucker (1999) reexamines the foundations of management within this new context. He argues that management policy within a firm should focus on "customer values and customer decisions on the distribution of their disposable income." This is consistent with the themes at the beginning of Chapter 2 and throughout this book. Without a clear answer to the question "Who is the customer?" product development, design, prototyping, and fabrication may be misguided.

In the 21st century, providing an environment that promotes creativity and flexibility will continue to be the social trend—a rather different emphasis than that of the early "time-and-motion studies" at the beginning of the 20th century! Furthermore, in contrast to working for one company for a lifetime, new graduates see themselves as *free agents,* namely, gaining more skills by moving from one company to another every one to three years (see Jacoby, 1999; Cappelli, 1999).

Given these trends, this introductory Chapter 1 ends with the question, "Will there be manufacturing, and will people work in the year 2100?"

The answer is probably "No" to anything that looks like manual labor, but "Yes" to collective enterprises where people design, plan, and install automation equipment and make things for consumers. And probably, those consumers (in the outer circle of Figure 1.4) will need or want pretty much the same things they have always needed or wanted since before the Greeks and the Romans: good health, nice food, happy relationships, attractive clothes, safe and comfortable housing, as-fast-as-possible transportation, and gizmos for entertainment.

We might telecommute and telemanufacture: one day we might, as admired on "Star Trek," even teletransport—but the human soul will probably stay pretty earthy and basic.

1.7 REFERENCES

Anderson, D. M. 1997. *Agile product development for mass customization.* Chicago: Irwin Publishing.

Armarego, E. J. A., and R. H. Brown. 1969. *The machining of metals.* Englewood Cliffs, N.J.: Prentice-Hall.

Ayres, R. U., and S. M. Miller. 1983. *Robotics: applications and social implications.* Cambridge, MA: Ballinger Press.

Bjorke, O. 1979. Computer aided part manufacturing. *Computers in Industry* 1, no. 1: 3–9.

Bralla, J. G., Ed. 1998. *Design for manufacturability handbook,* 2nd ed.New York: McGraw-Hill.

Cappelli, P. 1999. Career jobs are dead. *California Management Review* 42, no. 1: 146–167.

Cole, R. E. 1999. *Managing quality fads: How american business learned to play the quality game.* New York and Oxford: Oxford University Press.

Drucker, P. F. 1999. *Management challenges for the 21st century.* New York: HarperCollins Publishers.

Engelberger, J. F. 1980. *Robotics in practice.* New York: Amacom Press.

Goldman, S., R. Nagel, and K. Preiss. 1995. *Agile competitors and virtual organizations.* New York: Van Nostrand Reinhold.

Greenfeld, I., F. B. Hansen, and P. K. Wright. 1989. Self-sustaining open system machine tools. In *Transactions of the 17th North American Manufacturing Research Institution,* 304–310.

Harrington, J. 1973. *Computer integrated manufacturing.* New York: Industrial Press.

Jacoby, S. M. 1999. Are career jobs headed for extinction? *California Management Review* 42, no. 1: 123–145.

Leachman, R. C., and D. A. Hodges. 1996. Benchmarking semiconductor manufacturing. *IEEE Transactions on Semiconductor Manufacturing* 9, no. 2: 158–169.

Macher, J. T., D. C. Mowery, and D. H. Hodges. 1998. Reversal of fortune? The recovery of the U.S. semiconductor industry. *California Management Review* 41, no. 1: 107–136. Berkeley: University of California, Haas School of Business.

Merchant, M. E. 1980. The factory of the future—technological aspects. *Towards the Factory of the Future,* PED-Vol. 1, 71–82. New York: American Society of Mechanical Engineers.

Nishimura, K. 1999. Opening address. In *Proceedings of the 27th North American Manufacturing Research Conference.* Berkeley, CA.

Pfeiffer, J. E. 1986. Cro-magnon hunters were really us: working out strategies for survival. *Smithsonian Magazine,* 75–84.

Plumb, J. H. 1965. *England in the eighteenth century.* Middlesex, England: Penguin Books.

Rosenberg, N. 1976. *Perspectives on technology.* Cambridge, England: Cambridge University Press.

Schonberger, R. 1998. *World class manufacturing: The next decade.* New York: Free Press.

Smith. C. S., and P. K. Wright. 1996. CyberCut: A World Wide Web based design to fabrication tool. *Journal of Manufacturing Systems* 15, no. 6: 432–442.

Taylor, F. W. 1911. *Principles of scientific management.* New York: Harper and Bros.

Thomsen, E. G., and H. H. Thomsen. 1974. Early wire drawing through dies. *Transactions of the ASME, Journal of Engineering for Industry* 96, Series B, no. 4: 1216–1224. Also see Thomsen, E. G. Tracing the roots of manufacturing technology: A monogram of early manufacturing techniques. *Journal of Manufacturing Processes.* Dearborn, Mich.: SME.

Thurow, L. 1999. Building wealth. *The Atlantic Monthly* 283, no. 6: 57–69.

Ulrich, K. T., and S. D. Eppinger. 1995. *Product design and development.* New York: McGraw-Hill.

Wood, A. 1963. *Nineteenth century Britain.* London: Logmans.

Wright, P. K., and D. A. Bourne. 1988. *Manufacturing intelligence.* Reading, MA: Addison Wesley.

1.8 BIBLIOGRAPHY

1.8.1 Technical

Compton, W. D. 1997. *Engineering management.* Upper Saddle River, N.J.: Prentice-Hall.

Cook, N. H. 1966. *Manufacturing analysis.* Reading, MA: Addison Wesley.

DeGarmo, E. P., J. T. Black, and R. A. Kohser, 1997. *Materials and processes in manufacturing,* 8th ed. New York: Prentice Hall.

Groover, M. P. 1996. *Fundamentals of modern manufacturing.* Upper Saddle River, NJ: Prentice-Hall.

Jaeger, R. C. 1988. *Introduction to microelectronic fabrication.* Reading, MA: Addison Wesley Modular Series on Solid State Devices.

Kalpakjian, S. 1997, *Manufacturing processes for engineering materials.* 3rd ed. Menlow Park, CA: Addison Wesley Longman.

Koenig, D. T. 1987. *Manufacturing engineering: Principles for optimization.* Washington, New York, and London: Hemisphere Publishing Corporation.

Pressman, R. S., and J. E. Williams. 1977. *Numerical control and computer-aided manufacturing.* New York: Wiley and Sons.

Schey, J. A. 1999. *Introduction to manufacturing processes.* New York: McGraw-Hill.

Womak, J. P., D. T. Jones, and D. Roos. 1991. *The machine that changed the world.* New York: Harper Perennial.

1.8.2 Social

Sale, K. 1996. *Rebels against the future: the Luddites and their war on the industrial revolution.* Reading, MA: Addison Wesley.

1.8.3 Recommended Subscriptions

The Economist, **<www.economist.com>,** 25 St. James St., London SW1A 1HG. This often includes special "pull-out sections" on "high technology": for example, see the June 20, 1998, copy that contains "Manufacturing" and the June 26, 1999, copy that contains "Business and the Internet."

Fast Company, **<www.fastcompany.com>,** 77 North Washington St., Boston, MA, 02114-1927.

The Red Herring, **<www.redherring.com>,** Redwood City, CA, Flipside Communications.

Scientific American, **<http://www.sciam.com>,** 415 Madison Ave., New York, NY, 10017-1111.

Wired **<www.wired.com>,** 520 3rd St., 3rd Floor, San Francisco, CA, 94107-1815.

1.9 CASE STUDY: "THE NEXT BENCH SYNDROME"

Many of the chapters in the book contain a case study that attempts to combine an engineering view of a situation or a product with the management context. Ideally, this combination gives a balanced approach for the management of technology. Some key points that may be learned in this first introductory case study include:

- Product design and prototype manufacturing should involve as much engineering creativity as possible. But!—along the way, always ask some tough, consumer-oriented questions. A sample list follows:
  - Which group of consumers is going to buy this product?
  - Is it at the right price point for this group?
  - Does it have "shelf appeal" among equally priced products?
  - Will consumers enjoy using the product and spread the word to friends?
  - Will customers return to buy the next revision of the product because they have come to appreciate its aesthetic qualities as well as its functional ones?
  - In the 21st century, these *customer needs* will remain as an all-embracing topic—shown in the outer circle of Figure 1.4.

The text below is abstracted from "Tech-Driven Products Drive Buyers Away," written by Glenn Gow in the *San Francisco Chronicle,* March 1995.

> Technology companies are usually great innovators. Most of their new ideas come from engineers. But when engineers (alone) use their ideas to drive new product planning, companies risk failure. The Lisa computer from Apple was an engineering-driven failure, as were most (early) pen-based computers and many computer-aided software engineering packages.
>
> Hewlett-Packard (HP) used to suffer from engineer-driven production so often they developed a name for it: "next-bench syndrome." An engineer working on a new product idea would turn to the engineer on the next bench and ask him what he thought. The first engineer, then, was building a product for the engineer on the next bench.
>
> HP has since developed some very ingenious ways to truly understand the needs of their customers. While the next-bench syndrome may not be completely eliminated, HP has grown significantly in several areas (printers, Unix systems, systems management software, etc.) . . . by demonstrating the value of customer input to the engineering team. To help marketing gain a better understanding of customer needs, HP created customer focus groups, with the engineering team attending the focus groups.

1.10 REVIEW MATERIAL

1. In a spreadsheet with four columns, list the main attributes of manufacturing through four centuries, 18th to 21st, under the headings of equipment, process, and people.

2. Beginning with James Watt's invention of a separate condenser for the steam engine in 1769, list the six factors that historians usually identify that then led to the first industrial revolution between 1770 and 1820. In addition, for each factor, write a sentence or two about the same needs in today's information age revolution, beginning with the transistor in 1947, the first IC in 1958, and the first microprocessor in 1971.
3. Define in short bullets of 25 to 50 words (a) the next bench syndrome, (b) interchangeable parts, and (c) design for manufacturability/assembly (DFM/A).
4. List in a table format five or six reasons why the United States was "asleep at the wheel during the early 1970s," soon leading to losses in competitiveness against Honda/Toyota/Sony. In a second column, list next to each entry some of the organizational science approaches to manufacturing promoted especially by Toyota.
5. List in a table format the six or seven "major manufacturing paradigms" in the last three decades.

CHAPTER 2

Manufacturing Analysis: Some Basic Questions for a Start-Up Company

2.1 INTRODUCTION: WWW.START-UP-COMPANY.COM

Imagine that you and a group of friends are launching <**www.start-up-company.com**> or spinning off a smaller company from within a major corporation. Since this book is about manufacturing, it will be assumed that the company will develop, fabricate, and sell a new product, rather than be a service organization or a consulting group.

This book is built around the idea that your company will be brainstorming a technical idea, analyzing the market, developing a business plan, creating a conceptual product, fabricating a prototype, executing detailed designs, overseeing manufacturing, and then launching the product for sale. Figure 2.1 shows more details of these steps, arranged in a clockwise order. Beginning at the top of Figure 2.1, some of the very first survival questions that must be asked are:

- Who is the **customer?** Specifically, who is going to buy this product?
- How much will the product **cost** to manufacture? Specifically, what will be the start-up, overhead, operating, and payroll costs associated with the product? What will be the annual sales volume of the product? What will be the profit margin?
- What level of **quality** is needed for the identified group of consumers?
- What is the **delivery** time? Specifically, what is the time-to-market for the first sales income of a new product? Will another company get to market quicker?
- How fast can the next product line be delivered to ensure **flexibility?**
- What are the management of technology issues that will ensure **long-term growth** of the company? And is there a **barrier to entry** to hostile competition?

This chapter of the book contains six main sections that address these questions.

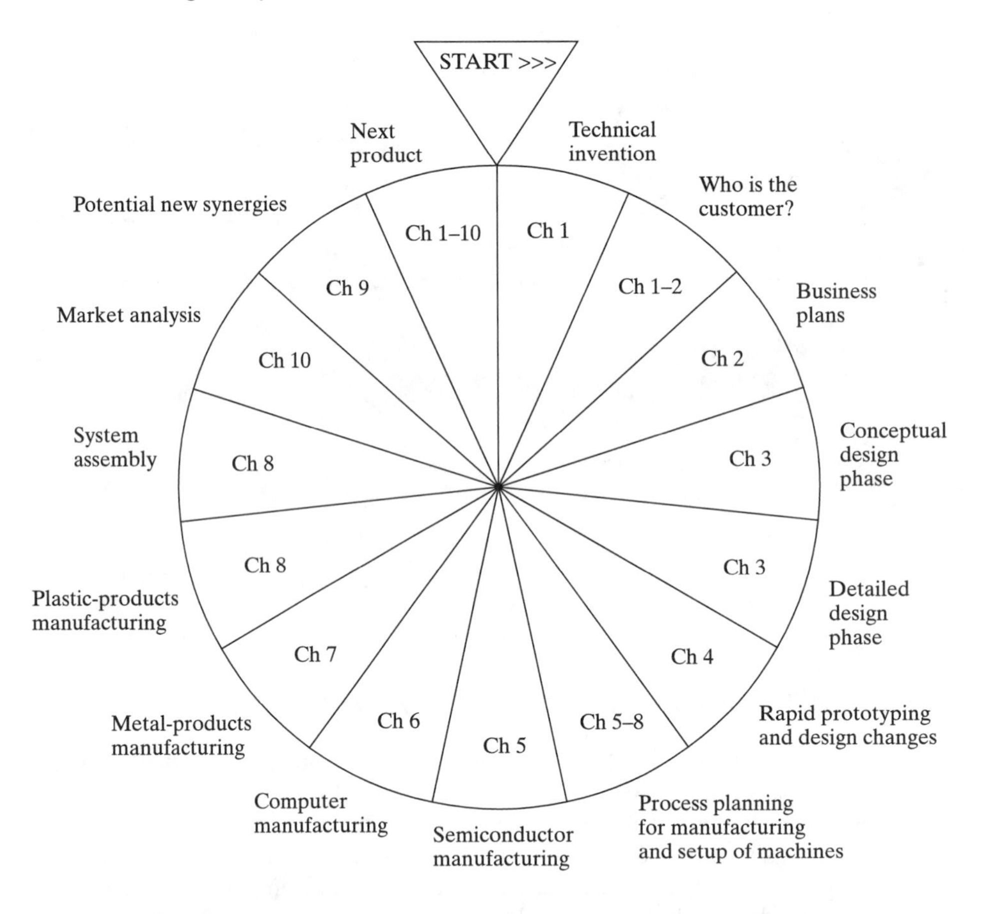

Figure 2.1 Steps in the product development and fabrication cycle. The chart moves clockwise from analyzing "Who is the customer?" to business plans, to design, to prototyping, to different types of fabrication, to sales. The content of Chapter 2 and the order of the other chapters in the book are approximately organized around this chart.

2.2 QUESTION 1: WHO IS THE CUSTOMER?

To establish the correct market niche for the product, an inevitable trade-off will occur between the four central factors of cost, quality, delivery, and flexibility (CQDF). These issues are discussed in detail in this chapter, but first, let's be a little entertaining. Consider a spectrum of possible customers for the products that will be made by <**www.start-up-company.com**>.

First, assume the customer is one of the U.S. national laboratories, and the new company is going to make a device that will go into a nuclear weapon. In this case, no matter how much it costs, or how long it takes to deliver, it has to be of the highest integrity. No compromises on safety or reliability can be made. High costs and long delivery times are likely.

Second, assume the customer is the aerospace industry, and your new company will be one of many subcomponent suppliers. The emphasis on safety and reliability will still be paramount, but some eye to cost will begin to be raised. Boeing knows that the European Airbus is courting its customers and that Japanese manufacturing companies are entering the commercial aircraft business.

Third, assume that the customer is a major automobile producer, and again, the new company will be a subcomponent supplier. Reliability will still be of some concern, in this era of 50,000-mile bumper-to-bumper warranties, but obviously cost competitiveness will now be a bigger issue. Some compromise between quality (see Section 2.3) and cost must occur.

Fourth, if the intended customer is Harrods of London or Nordstroms or The Sharper Image, the consumer products that your company plans to supply must be attractive and well priced but not as reliable as a nuclear weapon! Finally, if the consumer product is destined for Kmart, high-volume, low-cost, and adequate reliability are the market forces behind the design and manufacturing decisions.

2.2.1 Market Adoption Graphs

A key challenge for a new company, especially in "high tech," is as follows:

- The engineering founders of the new company will almost certainly want to be creative and build something new and exciting.

However, if the product and the company are going to be successful in the long run:

- The company must focus on who, or which group of consumers, will be the first real market adopters. This provides and maintains the serious cash flow needed to grow the company.

Measured over a long period of time, most products go through different stages of research and development, initial acceptance in the marketplace, sustained growth, maturity, and possible decline. The diagrams on the next two pages should be interpreted in a qualitative rather than a quantitative way, but they are based on trends that are seen in other publications on product development (Moore, 1995; Poppel and Toole, 1995).

These articles usually trace just one product as follows: (1) the new product is adopted slowly into the market at first; (2) if it is successful, a period of rapid growth then occurs; (3) much later on in life, the market becomes well established and even saturated; (4) finally, the market might fade as new products take over the role of the original product.

For this book, several products have been placed along the graph in Figure 2.2. This confuses the issue a little because all these products grow at different rates and their gross incomes in the market are substantially different. But for product acceptance in general, as measured by a concept of *market adoption,* a graph that combines many products is quite useful. (When speaking in public about the graph, the audience will usually "politely laugh" with the observation that buggy whips would be way down in the bottom right corner of the graph and products related to Dolly the

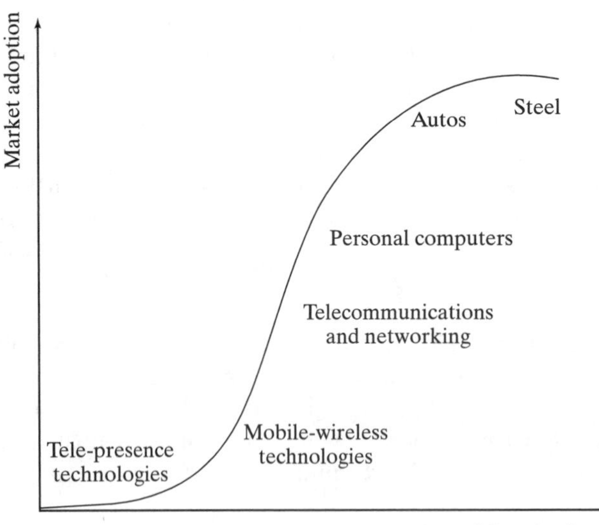

Figure 2.2 The market adoption curve, modified to include several products.

cloned sheep would be way down in the bottom left corner.) Whether or not one of these technologies will climb the market adoption curve depends on the following:

- The cost benefit to the consumer
- The robustness and usefulness of the technology to the consumer
- Complementary uses for the consumer (in the case of consumer electronics this may well mean other applications or software that can run on the new device)
- Whether or not the device falls in line with industry standards

Such projects bring out the difficulties associated with launching radically new products. Geoffrey Moore (1995) introduced the phrase *crossing the chasm* to emphasize this (Figure 2.3). In the early stages of a product's life, there will always be some measure of a market. There is a small group of consumers who love technology enough to buy a new product, no matter how useful it is, perhaps just for amusement or to be able to show off to their friends that they have the latest "cool" thing on the market. But this early market of impressionable "technology nuts" or "technophiles" only lasts so long. The real growth of the product depends on the first bullet above, namely, the cost benefit to the *average* consumer. This begs the question, How can a product survive across this chasm between the early enthusiasts and the real market? Geoffrey Moore goes on to use a "head bowling pin" analogy for capturing this phase of *stable* growth. As an example, he chooses the now-familiar personal pager. Apparently, medical doctors were the first *real market adopters* of the pager, and once the general public saw how useful they were, the product "crossed the chasm" and accelerated rapidly up the S-shaped curve in Figure 2.2.

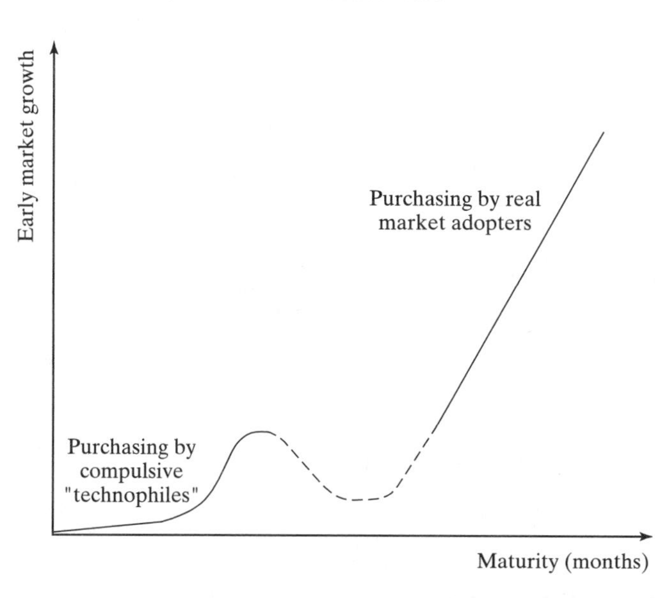

Figure 2.3 The concept of "crossing the chasm" (from Geoffrey Moore, 1995).

2.2.2 Inevitable Trade-Offs between Cost, Quality, Delivery, and Flexibility (CQDF)

This brief and informal introduction illustrates a wide spectrum of market opportunities with a wide variety of quality levels and consumer choices. Thus, inevitably, any company must face the rising costs (C) of adding higher quality (Q), faster delivery (D), or having a more flexible (F) manufacturing and supply line (Figure 2.4).

Despite the rising costs of quality, delivery, and flexibility, today's customers are more informed and they expect more than they used to. For example, they expect a Pentium chip to perform perfectly and yet be competitively priced. So how do manufacturers respond to this? At first glance, manufacturing seems to require too many trade-offs to simultaneously achieve high quality and fast delivery in small lot sizes, while still maintaining low cost. For example:

- Supercomputers and high-end, graphics-oriented UNIX machines cannot be fabricated quickly and in small lot sizes and then sold at CompUSA in a price range suitable for families and college students.
- A Lamborghini or a Ferrari cannot be fabricated quickly and in small lot sizes and then sold at the price of a Honda Accord or a Ford Taurus.

However, on closer inspection, the marketplace does set up a compromise situation between the engineering constraints and the broader economic goals. This compromise involves a reliable prediction of market size and type, supported by design and fabrication systems that can appropriately address each market sector. In this way, luxury automobiles will always be more expensive than economy cars, but in the luxury car sector, the industry leaders will be those who refined their design and manufacturing skills the best to give a high quality-to-cost ratio. Consumers are,

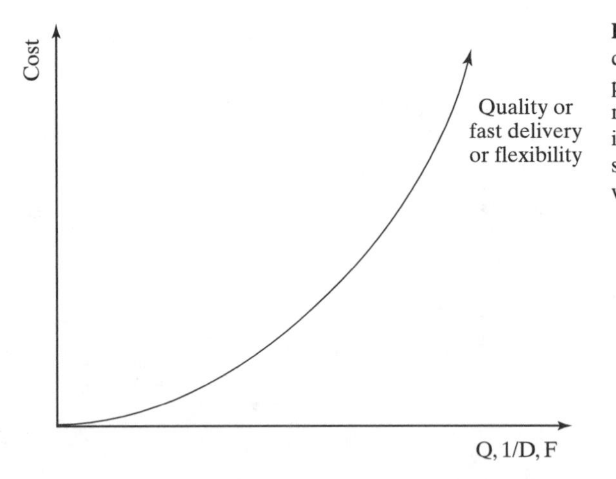

Figure 2.4 Schematic graph of rising costs for adding more quality to the product, faster delivery (schematically measured on the x axis by the shortness in days or weeks to deliver), and more system flexibility (CQDF). Also compare with Figure 2.10.

of course, highly influenced by cost, which must be clearly related to the benefits that accrue. The relationship between cost and quality will thus be examined in the next two main sections. The relationship between cost, delivery, and flexibility will follow.

2.3 QUESTION 2: HOW MUCH WILL THE PRODUCT COST TO MANUFACTURE (*C*)?

2.3.1 General Overview to Calculating the Manufactured Cost of a Product

Assume that N = the number of devices sold over the life of a particular product. The cost of each product is, in the simplest terms:

The cost of an individual product, $C = (D/N + T/N) + (M + L + P) + O$ (2.1)

In this general equation the terms are:

- D = design and development costs. This includes conceptual design, detailed design, and prototyping.
- T = tooling costs. This particularly includes costs for production dies.

These first two costs are amortized over the number of products made and therefore divided by the number made, N. The next three costs are consumed by every product, and finally there are the overhead costs.

- M = material costs per product.
- L = labor costs per product for operating machines, assembly, inspection, and packaging.
- P = production costs per product associated with machine utilization, including loan payments for machinery.
- O = overhead costs. This might include rental space for offices and factory, computer networking installation and servicing, phone installations and

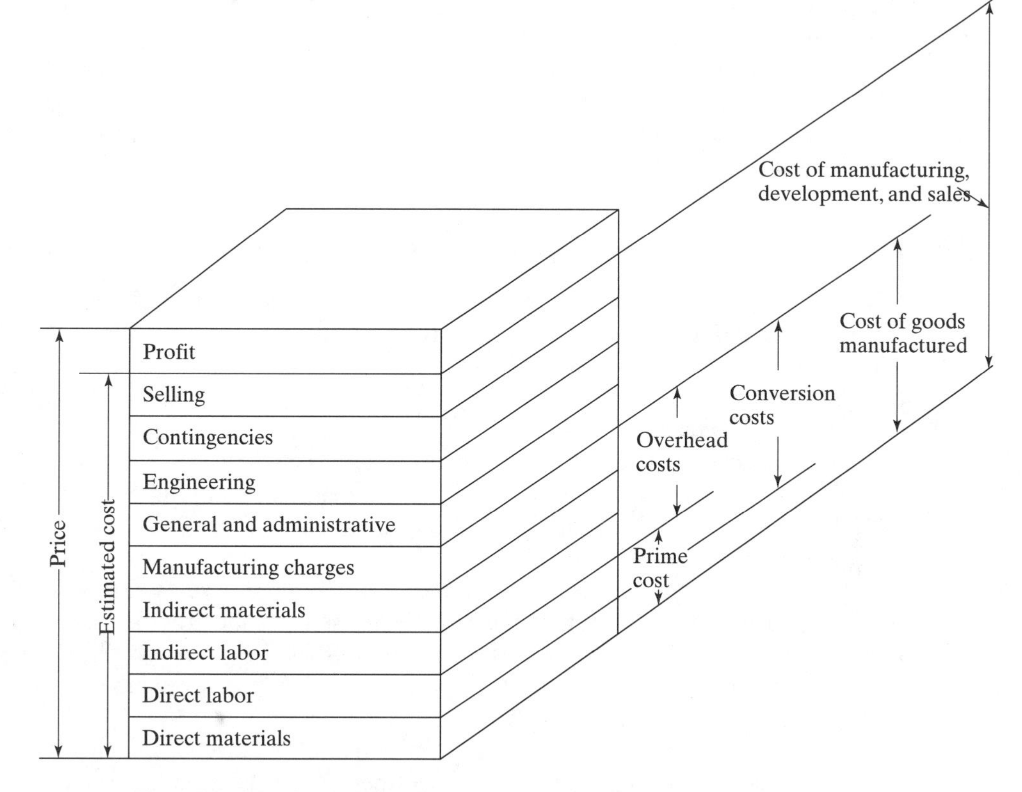

Figure 2.5 Cost breakdowns for manufacturing, similar to Equation 2.1—not to scale (courtesy of Ostwald, 1988).

monthly costs, electrical and other physical services, general advertising, and general support staff. This cost is more of a flat rate for all activities and must eventually be allocated per component. Reviewing this list of potential overheads, it is not surprising that many small companies build the first prototypes in the garage or basement of one of the founders.

Another view of the above equation is shown in Figure 2.5 from Ostwald (1988). It provides an overview—not to scale—of all the costs involved in creating, manufacturing, and selling a product. At the bottom of the chart, the prime costs are for the direct labor and materials. *Manufacturing overhead* is then shown, which accounts for the labor costs to run the overall factory, consumable materials such as fixtures, and other charges for maintenance and the like. On top of this are many other overhead type charges for running the organization. Included in the engineering costs are the *design and development costs* for a particular product in Equation 2.1.

Time-to-market is a phrase that will be used in this book to measure the time between (a) the first moment an engineer starts to bill his or her time to the company at the "conceptual product" stage in Figure 2.1 and (b) the moment the product is sold to the first customer and some revenue occurs. Note in Figure 2.1 that the time-to-market extends all the way around the diagram to the "9 o'clock position." Colloquially speaking, there are many opportunities to run up huge debts and "go broke" before selling anything. Section 2.5 on delivery therefore examines the time-to-market issue in more detail.

2.3.2 Specific Costs for Individual Manufacturing Processes

Given these introductory remarks, an interesting question now arises: Are some manufacturing processes cheaper to use than others?

If so, since the designers and the production planners have a wide variety of possible methods to choose from, why not choose one that gets the cheapest and most predictable results?

The book now reviews the manufacturing processes of the "mechanical world" and discusses the constraints that arise. For a reader unfamiliar with mechanical manufacturing processes, a brief taxonomy is given in Table 2.1. More details are included in Chapters 4, 7, and 8.

The Manufacturing Advisory Service at <**cybercut.berkeley.edu**> shows illustrations and related information. The first entry is the solid freeform (SFF) family that appeared after 1987 with the introduction of SLA. The other families have been categorized in that way by the Unit Manufacturing Process Research Committee of the National Academy (Finnie, 1995).

The analysis that follows is based on personal experiences in the metal processing industries and could probably be refined to suit other domains of manufacturing such as semiconductors. Also, the texts by Kalpakjian (1997) and Schey (1999) cover similar material in their last few chapters. Some of their diagrams and concepts are integrated (with acknowledgment) into this text. Other work being expanded at the time of writing is a "process selector" by Esawi and Ashby (1998).

2.3.3 Manufacturing Advisory Service (MAS) at <cybercut.berkeley.edu>

Picture any of the standard mechanical components in a lawn mower, washing machine, or even a simple can opener.

- Should the mechanical component be completely machined from a solid block?
- Should it be near-net-shape cast or forged and then finish machined?
- Should it be welded or riveted together from several pieces of standard stock?

These are only three of many possible manufacturing routes. The route to choose usually depends on a rather complex interaction between guiding subprinciples of manufacturing process selection shown in Table 2.2.

TABLE 2.1 A Brief Taxonomy of Mechanical Manufacturing Processes (Courtesy of Finnie, 1995)

Family	Processes	Brief explanation of processes
1. Solid free-form fabrication (SFF)—also could be categorized as layered manufacturing	1. Stereolithography 2. Selective laser sintering (SLS) 3. Fused deposition modeling (FDM)	1. Stereolithography (SLA) uses a laser to photocure liquid polymers. 2. Selective laser sintering (SLS) uses a laser to fuse powdered metal. 3. Fused deposition modeling extrudes hot plastic through a nozzle. Like "mini-toothpaste, hot extrusion," it builds up a model.
2. Mass-change processes	1. Drilling 2. Milling 3. Turning 4. Grinding 5. EDM and ECM	These processes remove shapes from a solid block, called the stock. 1. A simple drill from the hardware store creates holes of different depth and diameter. 2. A milling cutter has a flat end and can cut on its sides. It can carve out flat pockets in a block to make an "ashtray." 3. Turning is done on a lathe. The stock is round. The turning tool passes up and down the rotating stock removing layers. "A round bar can become a sculpted chair leg." 4. Grinding/polishing use abrasives to remove thin layers of metal to greater accuracy than processes 1–4. 5. Electrodischarge machining uses electric arc: Electrochemical machining uses charged chemicals to remove fine layers.
3. Phase-change processes	1. Casting 2. Injection molding of plastics (could include FDM)	1. In casting, molten metal is poured into a hollow cavity in sand, initially created from a mold. 2. Injection molding "shoots" hot liquefied plastic into a mold.
4. Structure-change processes	1. Coatings 2. Surface alloying 3. Induced residual stresses	1. Hard surface coatings can be deposited chemically or physically on softer or tougher substrates—chrome plating is an example. 2/3. Alloying or shot blasting toughens surfaces.
5. Deformation processes	1. Rolling 2. Sheet drawing 3. Extrusion 4. Forging	1. Slabs can be rolled down to strip as thin as everyday "aluminum kitchen foil." 2. Such sheets can be cut and bent into office furniture, filing cabinets, or soup cans. 3. Like large-scale hot toothpaste, extrusions of different cross sections can be made if the die (the hole at the end) is a premade shape—using milling or EDM. 4. Hot or cold forging involves "slamming" metal into a die cavity. The metal stock plastically deforms to the desired shape.
6. Consolidation processes	1. Powder metals 2. Composites 3. Welding/brazing	1. Powdered metal is formed in a die and then sintered to give full strength. 2. Layers of different carbon fiber sheets are an example of composite materials. 3. Welding involves local melting and "mini casting together" of adjoining plates. Brazing uses solid-state bonding between a filler metal and the two plates.

2.3.4 Batch Size

When considering these subprinciples for a suggested manufacturing method, it makes sense to begin with the criteria that make the most impact on the costs. Usually batch size, strength, geometry, and tolerance are the four most significant factors.

TABLE 2.2 Subprinciples of Manufacturing Process Selection

General principles driven by designer	Implied considerations for the manufacturing process (mechanical manufacturing processes)
1. Cost	Cost is driven by all the principles below. It is reviewed in the text in relation to all parameters.
2. Batch size	Only the SFF, rapid-prototyping processes, machining, and possibly casting are suitable for "just one" component. For a structurally useful product, only CNC machining and casting are realistic, but FDM can provide an alternative for low strength ABS-plastic parts. For two to five components, CNC machining is likely. As batch size increases, short-run, plastic injection molding or casting is considered. The cost of the die or mold is always a key factor.
3. Strength and weight—related to material choice	A need for high strength drives the choice toward metal processing over plastic molding. Even higher strengths/performance will force the choice toward forging or machining over casting. A need for light weight may drive the choice to plastic, aluminum, or titanium. In general, costs will increase when high strength and performance are required. Materials such as titanium are very costly.
4. Geometry	Wide, thin cross sections will drive choice toward blow molding for plastic parts or sheet-metal forming for metal parts. "Chunky" cross sections drive the choice to casting/forging/milling operations. "Cylindrical" cross sections drive the choice to turning. In general, parts with complex geometries will cause high manufacturing costs. In small batches, the CNC programming and execution times will be long. In large batches, dies are expensive to create and operate.
5. Tolerances	Tolerances tighter than +/− 50 microns (0.002 inch) will begin to drive the choice to the machining/grinding/polishing family of processes. Grinding and polishing are expensive operations—if possible the designer should avoid such finishing costs.
6. Product life	A desired long service life will probably drive the choice to metal processing over plastics. Design constraints on fatigue properties may require special tooling and/or lapping processes. Longer desired service life will almost certainly increase costs because of more or better materials used, improved surface finish, and more careful design optimizations.
7. Lead time	To deliver the part to the designer quickly, production planners hope to use standard processes and tooling. Weird designs might require special tools and lengthy hand assembly and finishing. Special tools and fixtures will rapidly drive up the costs and delivery time. Any process that requires a die or mold will be more expensive and will take longer than machining or SFF.
8. Design for assembly (DFA) issues	The way in which an individual part is mated or fixed to another one in an assembly is also important. Welding, riveting, and bolting costs are high, and these processes are often poorly controlled. Thus, cost reductions may result from new assembly or single-piece manufacturing methods.

2.3.4.1 Batch Size of 1

If only one component is needed, then one of the rapid prototyping methods such as stereolithography (SLA), selective laser sintering (SLS), fused deposition modeling (FDM), CNC machining, or casting is the obvious manufacturing choice. The first three listed are collectively known as solid freeform fabrication (SFF). Note that SLA produces a low strength photopolymer model, not a structural part. SLS can create low strength metal parts, and FDM can create low strength ABS plastic parts. The SFF and CNC machining processes are economic for *one-off* components because they do not require the time to make an expensive die or mold prior to production. Casting is also possible for large, complex single components that cannot feasibly be made by machining. However, an expensive mold of wax or wood is needed. This creates the shaped cavity in sand before the molten metal is poured in.

Actually, if the batch size is as low as only one or two components and if the part geometry is simple, a skilled craftsperson might even use a manual milling machine or lathe because the programming time for a CNC machine might not be worthwhile. However, if the part geometry is complex, it is worth the time spent to program the CNC machine even for one-off components. The reason for this is reasonably subtle: on a manual machine, if an error is made near to the point when the part is almost complete, "all is lost"—not only the piece of work material but all the time that was invested up until that point. But on a programmable machine, if an error is made toward the end, there may be a scrapped workpiece but all the geometrical programming steps that were invested up until that point are stored in the computer. This of course begs the question, How is part complexity measured? One answer might be that part complexity increases with the number of lines of CNC code needed to machine the part.

For a batch size of one, if the desired part has a complex geometry, SFF rather than machining processes will be used. An object that resembles a doughnut will be easy to make by the SFF methods but almost impossible to make by CNC machining unless it can be made in two halves and joined along its equator. Speaking very generally, the more complex the shape, the more likely SFF processes will be used. However, component strength is a vital consideration. CNC machining and casting are the only viable alternatives if high strength is needed. Selective laser sintering (SLS) can produce a metal part, but it will be weaker than one produced by machining. Fused deposition modeling can produce a plastic part with reasonable strength—but, again, one that is weaker than a part from plastic injection molding.

2.3.4.2 Batch Size[1] 2 to 10

If only a few components, say 2 to 10, are needed, then CNC machining is today the most likely choice, unless the geometry is highly complex and sculptured, in which case a batch of SLS or FDM components might be realistic and cost effective.

2.3.4.3 Batch Size 10 to 500

A batch of 10 to 500 might well be done by CNC machining. This batch size is beginning to enter the realm of a *production shop* rather than a *custom prototyping shop* (Figure 2.6). However, manual transfer of parts between machines will be quite likely since this batch size is not sufficient to warrant investments in automation.

It should also be noted that if the customer is increasing its order from just one to this higher batch size, it might be best to "backtrack" and make a good stereolithog-

[1]These numbers for batch size are very approximate and depend on several factors including part complexity. The ranges in these next few pages deliberately overlap.

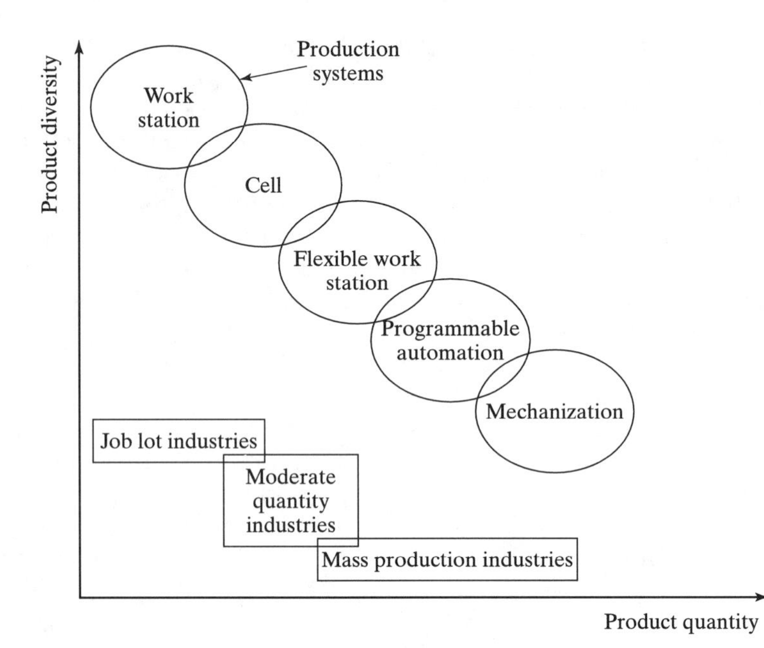

Figure 2.6 Trends from manual, to CNC, to reprogrammable systems (FMS), and to "harder automation" (courtesy of Ostwald, 1988).

raphy pattern that will then be used for castings. This option will work if the desired tolerances are within the capability of both the SLA master and the subsequent casting process, rather than machining.

2.3.4.4 Batch Size 100 to 10,000

CNC machines, arranged in larger complexes called *flexible manufacturing systems* (FMS), will be favored as batch size grows. Perhaps robots or automated guided vehicles (AGVs) will be used to move parts from one machine to another. Efficiency will be very dependent on the communication software needed to orchestrate the system. However, batch sizes of several thousands will begin to warrant the fabrication of an expensive die that can rapidly punch out products by a cold forging or stamping process. While processes that require a premade die or mold are rarely, if ever, used for one-off or short batch runs, the cost of the die can be amortized over these larger runs and the *cost of the tooling die per component* decreases (parameter T in Equation 2.1). Sands (1970) presents a comprehensive analysis for different forming processes showing at which batch size the use of a die becomes efficient. Die costs and manufacturing system costs increase from left to right in Figure 2.6. This cost factor places an important responsibility on the designer. In an ideal situation the newly designed component will be made on existing factory floor machinery, readily leading to an "off-the-shelf" automation solution. In the best case, existing fixtures and even some parts of existing dies will also be reused.

2.3.4.5 Batch Size 5,000 to millions

As batch size increases, automation plays a bigger role. However, for extremely large batch sizes, it might even be economic to revert to noncomputerized machines. Speaking colloquially, this batch size moves into the realm of "ketchup in bottles," where fixed conveyor lines pump out the same product day in, day out. This is often referred to as fixed or *hard automation,* literally because "hard stops" are fixed in place with wrenches. These hard stops establish the positions where components rest in place while being filled or labeled. Some basic computer control and sensors are needed to keep things on track, but reprogramming will not be needed.

2.3.5 Material Choice

The material that the designer chooses for the part will be influenced by weight considerations, cost factors, and desired strength. This desired strength of the finished object is obviously a key factor. Even though metals are generally stronger than plastics, injection molded and thermoformed plastics are preferred for ordinary consumer products such as household appliances, consumer electronics, and many automotive products.

In general manufacturing costs for plastics are lower than those for comparable metal products. This is because plastic forming requires much lower forces than metal forming, and so the machinery is cheaper, the dies are less complex, and the labor costs are often lower. However, critical components such as transmission gears need to be made from closed die forging blanks to obtain a well-distributed grain structure. Finish machining completes the critical gear tooth involute profiles. This issue of basic strength is obviously related to the primary material that the object is to be made from—not only its basic data-book strength but also its purity, heat treatment, and in-process characteristics. The latter include the work-hardening properties of metals and the shrinkage characteristics of plastics. This is also an important moment in the text to reemphasize that solid freeform (SFF) prototyping techniques such as stereolithography create plastic components from a photo-curable liquid. This material from the SLA bath is by no means as structurally sound as standard plastics such as ABS and polystyrene. FDM can create ABS parts of reasonable strength, but not with the same structural integrity as injection molded ABS.

2.3.6 Part Geometry

The product's geometry embodies the aesthetic qualities and functional properties of the part, but it also restricts the selection of suitable manufacturing processes. Figure 2.7 is taken from Schey (1999) to show how one aspect of overall part geometry drives process choice. To quickly understand this graph, begin by noting that the cold rolling process nearest the x axis produces a flat strip that is at the extremes of wide and thin. In fact the strips are often several feet wide and still only a few thousandths of an inch thick. No other process can match rolling for such dimensions. Cold rolling is thus one of the starting points for a large range of subsidiary processes such as sheet forming and stamping, which then produce automobile body panels, office furniture, and even the humble soup can.

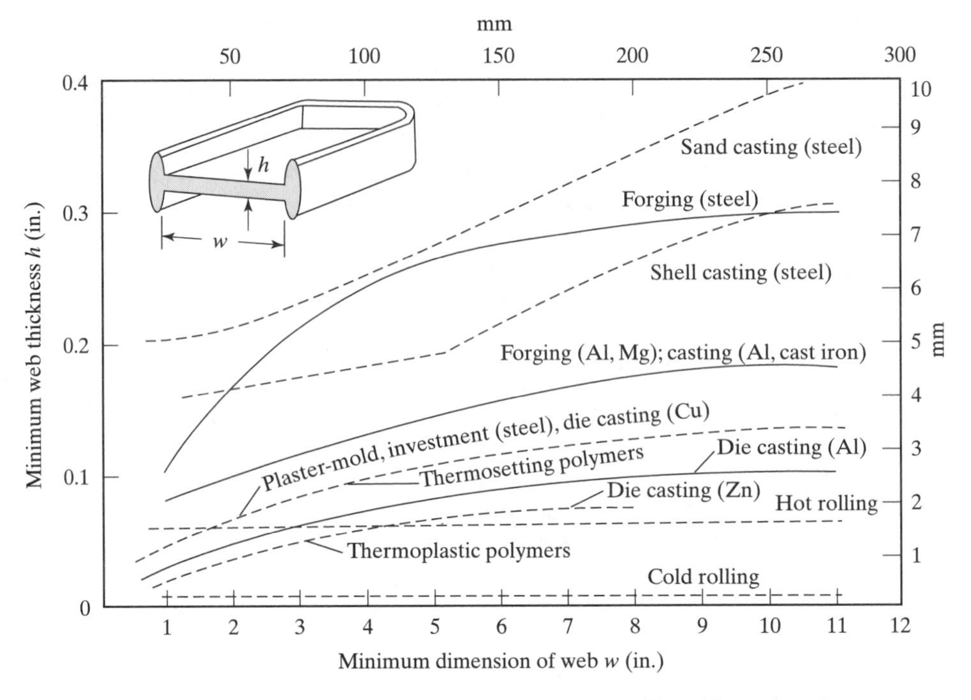

Figure 2.7 Process capabilities related to part geometry. Very thin sections favor rolling and thermoforming; "chunky" sections favor machining and injection molding (from *Introduction to Manufacturing Processes* by J. A. Schey, © 1987. Reprinted with permission of the McGraw-Hill Companies).

The thermoforming of plastic sheets is slightly above cold rolling in the graph. This also creates sections that are relatively thin, and thus it competes with cold rolled metal products for many common items that require less structural rigidity. The middle part of the graph relates to processes that create more "chunky" looking parts of greater thickness (the y axis in the figure). Finally, note that the mold making procedures in sand casting prevent it from being selected if one of the dimensions is less than 5 millimeters (0.2 inch).

2.3.7 Accuracy, Tolerances, and Fidelity between CAD and CAM

In all fabrication processes—semiconductors, plastics, metals, textiles, or otherwise—the physical limitations of each process have a major impact on the achievable accuracy. Each processing operation comes with a bounding envelope of performance that is constrained by the physical and/or chemical processes that, during fabrication, are imposed on the original work material. This begs the following question: How much fidelity will there be between (a) the specified CAD geometry, tolerances, and desired strength and (b) the final physical object that is manufactured? In the best case scenario, the CAD geometry will be perfectly translated into the fabricated geometry. Also, the properties of the original piece of work material stock will be either unchanged or possibly work-hardened into an even more preferred state.

TABLE 2.3 Routine Accuracies for Mechanical Processes (One "Thou" Approximately = 25 Microns)

Process	Accuracy microns	Accuracy inches
Hot, open die forging	+/− 1250 microns	+/− 0.05 inch
Hot, closed die forging	+/− 500 microns	+/− 0.02 inch
Investment casting	+/− 75–250 microns	+/− 0.003–0.01 inch
Cold, closed die forging	+/− 50–125 microns	+/− 0.002–0.005 inch
Machining	+/− 25–125 microns	+/− 0.001–0.005 inch
Electrodischarge machining	+/− 12.5 microns	+/− 0.0005 inch
Lapping and polishing	+/− 0.25 microns	+/− 0.00001 inch

In the worst case situation, a poorly controlled process will damage a perfectly good work material. Examples of this were widespread in the early days of welding, where heat-affected zones reduced the fracture toughness of materials. Controlling this envelope for each process is quite complex and relies on a number of factors, which include:

- The properties of the work materials that are being formed/machined/deposited
- The properties of the tooling/masking/forming media
- The characteristics of the basic processing machinery and its control structure
- The number of parameters in the physics or chemistry of the process
- Sensitivity of the process to external disturbances such as dirt, friction, and humidity

Table 2.3 and Figure 2.8 convey the typical tolerances that can be obtained.

Note that even within one particular process there can be subtle differences in performance, resulting in a range of tolerance. The darkest bars in the center of each process are the normally anticipated values. This range is given the name *natural tolerance* (NT) of the process and is crucially important in both design and manufacturing work.

It cannot be emphasized enough that the cost of manufacturing, and the subsequent cost of any consumer product, is related to the designer's selection of part accuracy and dimensional tolerance.

Once the design and its related tolerances reach a factory floor, the manufacturers will be obliged to choose processes that deliver the accuracy and NT implicit in the decisions made by the designer. Quite clearly, costs will rise rapidly if the designer has been overdemanding or just thoughtless. Poor design decisions could result in the obligatory choice of an inherently expensive manufacturing process.

The next concept to emphasize is that of *process chains* within a particular family of manufacturing processes. Examples of these are also shown on the Website <**cybercut.berkeley.edu**>. In general, several processes are used sequentially to gradually achieve a highly accurate, smooth surface. A common chain in mechanical manufacturing is to start with a flame-cut plate, a casting, or a forging to obtain the

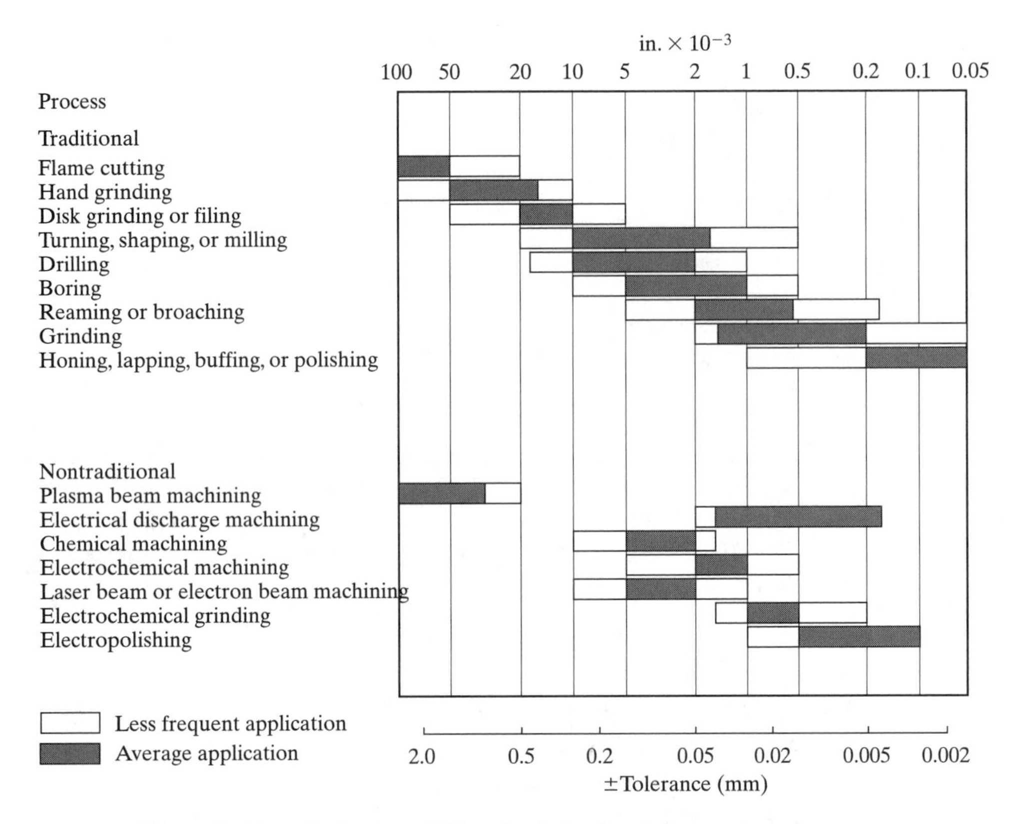

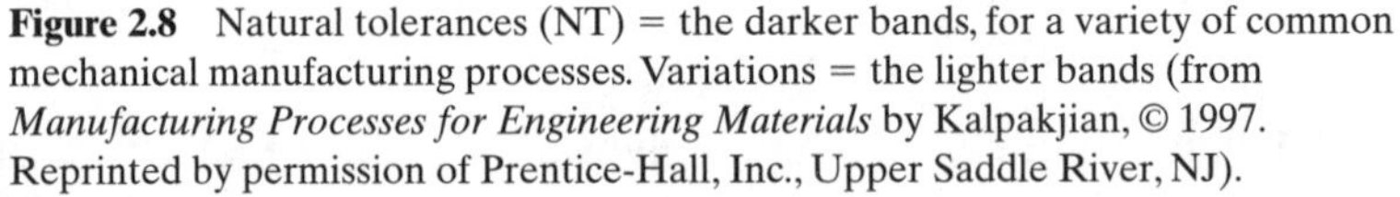

Figure 2.8 Natural tolerances (NT) = the darker bands, for a variety of common mechanical manufacturing processes. Variations = the lighter bands (from *Manufacturing Processes for Engineering Materials* by Kalpakjian, © 1997. Reprinted by permission of Prentice-Hall, Inc., Upper Saddle River, NJ).

bulk shape. Flame cutting could then be followed by a series of machining operations to obtain further accuracy. These can then be followed by grinding and polishing if high accuracy and finish are desired by the designer.

In Figure 2.8, the NTs of flame cutting, machining, and grinding are shown, moving across from left to right with finer accuracy. Several points should be made:

- The designer should realize that these process chains exist, as summarized in the simple diagram of Figure 2.9.
- Each additional process is needed after a certain *transitional tolerance.* If the designer is unaware of these transitions, unnecessary finishing costs may be created, as shown in Figure 2.10. The other side of this coin is that manufacturing costs can be saved if the designer is willing to loosen desired tolerances.
- The manufacturing quality assurance at one step in the process chain must be carefully executed before moving on to the next process. If a "parent" process is "ended too early," the next "child" process may have too much or an impossible amount of work to do. (Imagine cleaning a rusty garden tool; heavy

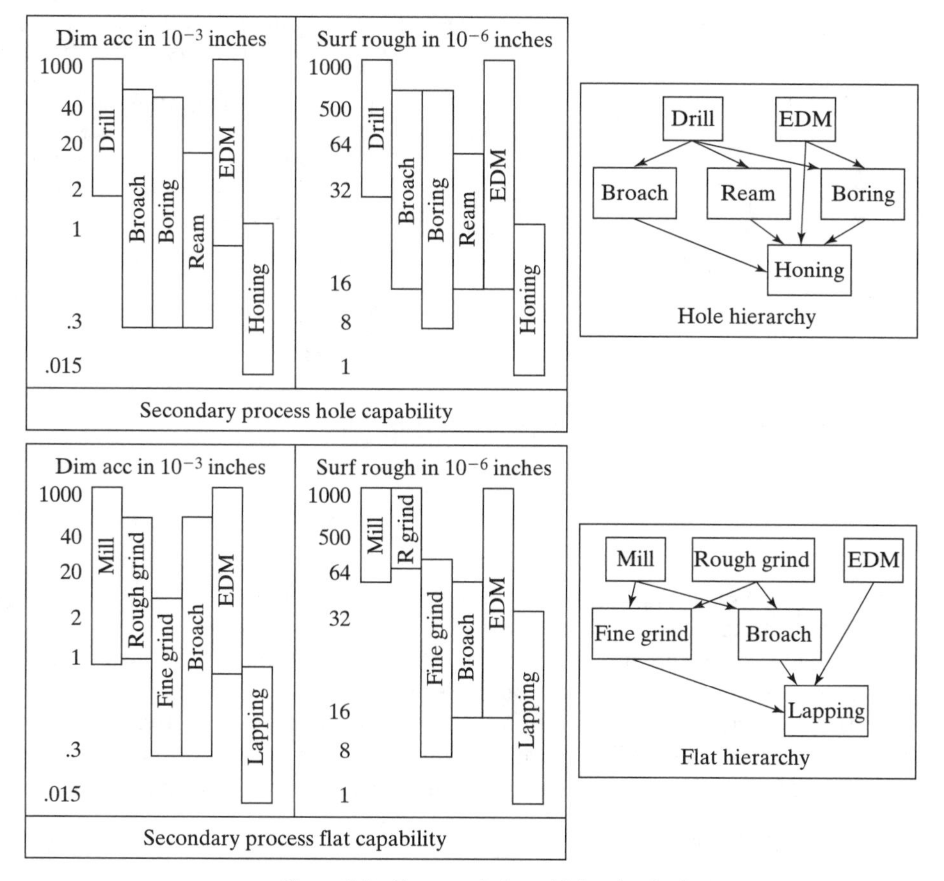

Figure 2.9 Process chains with levels of tolerance.

grinding or heavy abrasive papers are needed before moving on to the final polishing steps.)

2.3.8 Product Life Expectancy

Recall that part strength is listed as the third criterion in Table 2.2. It is related to the design geometry, tolerances, material selected, and chosen manufacturing method. These factors also have a *coupled* influence on the long-term *in-service life.* Aerospace and structural engineers are probably the designers who are most concerned with these long-term properties. Hertzberg (1996) and Dowling (1993) describe the fatigue properties of metals and polymers. The influences of material composition and local-geometry effects are also described. A fatigue failure always begins at a stress concentration. A sharp corner, a small hole, a rapid transition in diameter are examples of danger zones for crack initiation. Designers in such fields will specify high integrity grades of steel and aluminum, will choose processes like forging and forming (rather than casting) to maintain a homogeneous grain structure, and will specify additional final finishing operations such as grinding and lapping. These

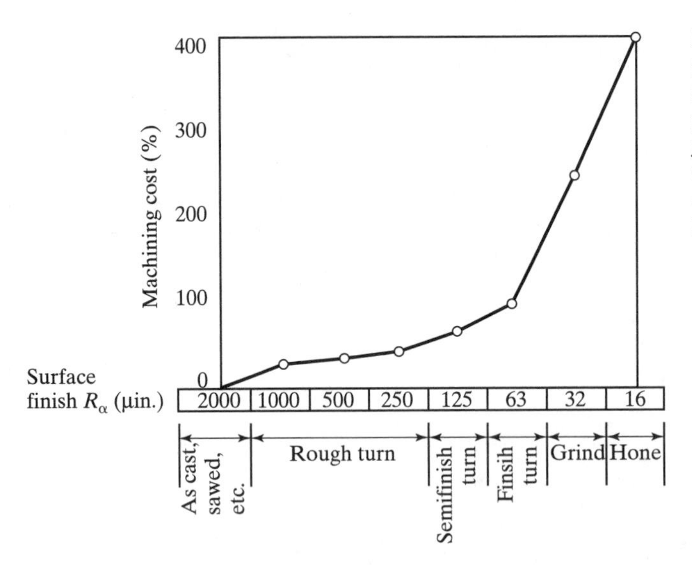

Figure 2.10 Finishing costs increase as a part moves from a rough casting, to a finish-machined part, to fine-honed final product (from *Manufacturing Processes for Engineering Materials* by Kalpakjian, © 1997. Reprinted by permission of Prentice-Hall, Inc., Upper Saddle River, NJ).

additional operations lead to very smooth surfaces that give dramatically improved long-term fatigue life.

Figure 2.10 illustrates the costs of these additional fine finishing operations. The additional grind and hone operations add 400% more cost over the as-forged, or as-cast, surfaces. Even in comparison with turning on a lathe, they add 200 to 300% more cost. It is not surprising that carefully manufactured aircraft components, or the surface of a production quality plastic injection mold, are so very expensive.

2.3.9 Lead Time

Lead time is defined for this book as "the number of weeks between the release of detailed CAD files to the fabrication facility and the actual production of the part." It is a small subcomponent of the total time-to-market. This broader topic will be reviewed in greater depth in Section 2.5. For this overview, the important point is that lead time is very dependent on the designer's decisions, which then have direct implications on the choice of manufacturing process. The desired batch size, part geometry, and accuracy are the main factors. As a benchmark, a small batch of medium complexity metal parts with +/−50 microns (+/− 0.002 inch) accuracy can be obtained from a production machine shop with a two- to three-week turnaround time, obviously depending on normal business conditions.

However, several weeks of lead time will be experienced as soon as a serious mold or die is needed. For the processes like forging, sheet metal forming, and high-volume plastic injection molding, the die making involves *many* extra steps. During die design, factors such as springback for metals and shrinkage for plastics need to be incorporated. Since the deformation stresses that build up during manufacturing are high, the die designer also has to create supporting blocks and pressure plates. The designer will also need to consider parting planes and the draft angles that give slight tapers to any vertical walls: these are needed to ensure that the part can be ejected after forming. Unfortunately, perfect analytical models do not exist yet for predicting the precise amounts of springback or the best draft angle. This usually means that handcrafting is needed in the production of the first die. Subsequent trial-and-error adjustments and iterations to the die surfaces are always needed.

The above paragraph still pertains only to one machine and one process. Several months of lead time are needed to set up the large-scale FMS systems and high-volume batches indicated at the bottom right of Figure 2.6. These contain a large number of manufacturing processes, linked together and scheduled to make complex subassemblies. And of course as product complexity and scope increase, the lead time increases proportionally. At the extreme, for a completely new model of aircraft or automobile, the lead time from design to first product will run into years rather than months.

2.3.10 Cost Factors Especially Related to Adjoining Parts

The following example shows how design and manufacturing keep changing to suit a rather complex interaction between (a) the availability of innovative manufacturing techniques and (b) new economic conditions. In other words, recalling Ayres and Miller's (1983) quotation in Chapter 1, "*CIM is the confluence of the supply elements and the demand elements.*"

Twenty or even ten years ago, it would not have seemed reasonable to machine very large structures from a solid monolithic slab. However, innovative machining programs at Boeing Aircraft are proceeding in that direction. Inside the ceiling of the plane, structural members that resemble giant coat hangers are spaced across the plane at intervals to give it torsional stability. Today, most are made from many conjoined pieces. This arrangement is shown in the upper photograph of Figure 2.11. However, newer designs are favoring machining from one very large solid slab, as shown in the lower photograph. This eliminates costly and unpredictable joining operations in the factory.

Thomas (1994) has observed that such manufacturing innovations *flowing back* into the design phase must be the new way of organizing the relationship between design and manufacturing. Using Ayres and Miller's definition of CIM we can observe that the new innovations, or the new *supply elements,* include:

- Improved cutting tool technology and an understanding of how to control the accuracy of very high speed machining processes
- The availability of stiffer machine tools and very high speed spindles
- More homogeneous microstructures that give uniformity in large forging slabs
- The ability to carry out comprehensive testing and show that these one-piece structures are at least if not more reliable than multiple-piece structures.

Meanwhile the new *demand elements* include:

- Escalating costs of joining and riveting operations, which can only be partially automated. Specifically these operations often require manual fixturing of the workpieces.
- A preference for eliminating multiple fabrication steps, which always demand more setup, fixturing, documentation, and quality assurance.
- General pressures on the whole airline industry, since deregulation, to cut costs and yet improve the safety and the integrity of the aircraft.

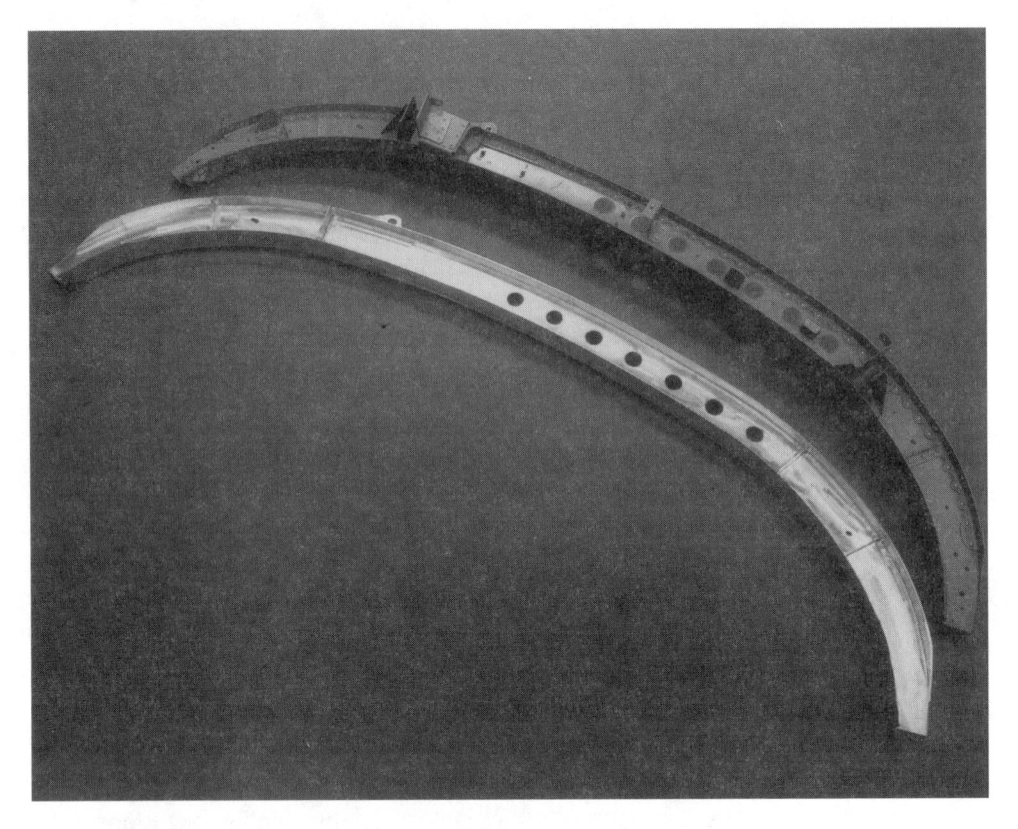

Figure 2.11 Integrated product and process design allows this aerospace component to be completely machined from the solid as shown in the lower photograph (courtesy of Dr. Donald Sandstrom, The Boeing Company).

These trends introduce a great deal of complexity into the design and manufacturing process, but on the other hand, creative companies can exploit them to their advantage. The conclusion to be drawn is that *no single component should be analyzed and optimized in isolation.* There will always be something that can be improved, simplified, or made cheaper if design and manufacturing are viewed from a slightly wider system perspective.

2.3.11 Analyzing Costs in Terms of the Profit Potential

Hewlett-Packard's return map (RM) is another method for analyzing design and manufacturing costs. However, it focuses not just on cost but on these survival questions:

- How much profit, ΔP, will be made at any given time?
- How long, T_b, will it take to make any profit?

Figure 2.12 plots the costs or revenues against logarithmic time expelled. The key curves on the chart (modeled on House and Price, 1991) are:

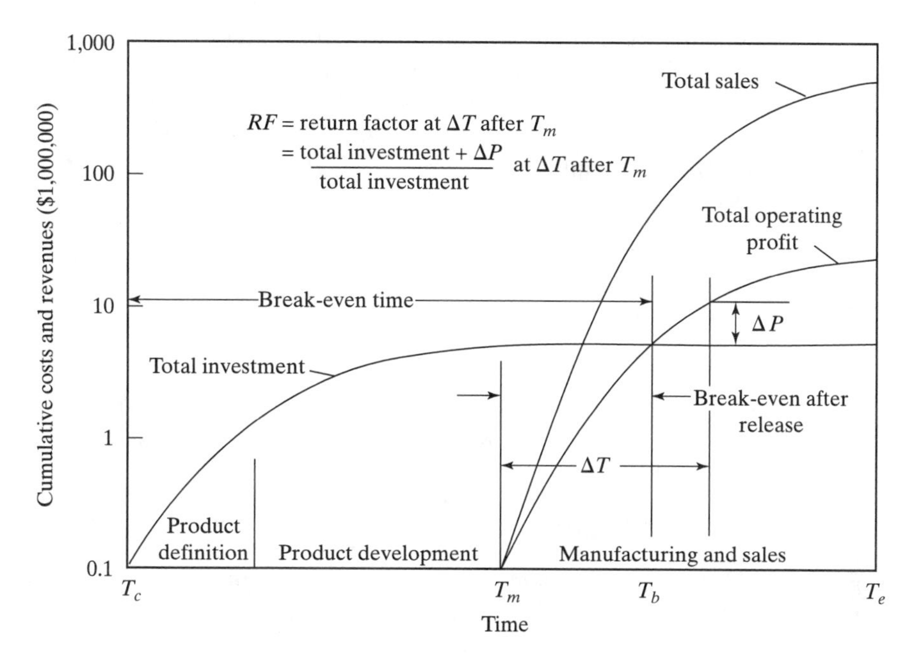

Figure 2.12 Hewlett-Packard's return map (diagram based on House and Price, 1991; Magrab, 1997).

- The total investment of dollars starting from the first instant (T_c) that engineers start dreaming up the project (see the top of Figure 2.1 at the beginning of the chapter).
- The total sales that begin as soon as possible after the first product is manufactured and sold, T_m. Note that setting up and debugging the manufacturing line generate no sales.
- The total profit that starts to be gained at T_b.

The key points on the time axis are:

- T_c—the project initiation point. This is followed by product definition, product development, process planning for manufacturing, setting up machines, debugging the assembly line, and first launch into manufactured product around point T_m.
- T_m—the point where real manufacturing begins and products get sold.
- T_b—the *break-even time* from the very beginning of the "conceptual product definition" to the point where a positive profit occurs ($T_b - T_c$). Note that the chart also shows the *break-even-after-release time,* which measures ($T_b - T_m$) and focuses more on manufacturing productivity. Obviously, fast production and high volumes of product are desirable. The goal is to quickly amortize all the development costs.
- ΔT and ΔP—any arbitrary point (ΔT, ΔP) beyond the break-even-after-release time (T_m) where Hewlett-Packard's return factor (RF) is calculated. The RF is

calculated by dividing total profits by total investments. The goal is to maximize *RF* in the shortest time after T_m. More importantly, it is also possible to measure an *RF′* from the break-even time, T_c. In terms of profitability for the whole company, break-even time is more crucial than break-even-after-release time. New companies with limited cash flow should focus more on the former measure.

What is the income stream from the product? The following definitions are often used:

- Sales price = estimated sales price of one unit from company to distributor (not retail)
- **Net sales** = individual sales price × number of products sold
- Cumulative net sales = integrated net sales over several consecutive years

What are the costs of being in business and producing that particular product? The following definitions are often used:

- Unit cost = prime manufacturing and related manufacturing overhead costs of a single unit of product (see the cost of goods manufactured on the right of Figure 2.5)
- **Cost of the product** = unit cost × number of products sold
- Development costs = conceptual and detailed design + launch + support
- Marketing costs = a percentage of net sales (Magrab, 1997, uses 13%)
- Other promotional and running costs = a percentage of net sales (Magrab, 1997, uses 8%)

What is the potential profit or loss? The following definitions are often used:

- **Gross margin = net sales − cost of product**
- Percentage gross margin = gross margin / net sales × 100%
- Pretax profit = gross margin − development costs − marketing costs − other
- Cumulative profit = integrated profits (or losses) on a year-by-year basis

Table 2.4 has been reproduced from Magrab (1997) to show some specific figures. In that example, the first two years have no sales. However, the design and development costs are running up all the time showing a bottom line, *temporary loss of $1.6 million.*

This particular illustration shows that by the year 2005, the product makes an impressive profit. But the risks of the first two to three years cannot be emphasized enough. And what if the customer does not like the product when it is released to the market? What if the development time is too long and another company launches a similar product first? Or a better product a few weeks later? The risks of a company are far too evident here.

Also it is useful to ask, Where will the 1.6 million come from? Obviously from a loan of some kind (new company) or a strategic investment (larger, existing company). At what effective interest rate? 8%? 10%? 12%? What other products might

TABLE 2.4 An Example of Magrab's Baseline Hypothetical Profit Model. (Reprinted with permission from *Integrated Product and Process Design* by E. B. Magrab. Copyright CRC Press, Boca Raton, Florida.)

	Year								
	1997	1998	1999*	2000	2001	2002	2003	2004	2005
	j = 1	j = 2	j = 3	j = 4	j = 5	j = 6	j = 7	j = 8	j = 9
A Sales price			$65.90	$65.90	$67.90	$67.90	$67.90	$67.90	$67.90
B Number of units sold			100,000	250,000	300,000	350,000	250,000	200,000	150,000
C Net sales [=AB]			$6,590,000	$16,475,000	$20,370,000	$23,765,000	$16,975,000	$13,580,000	$10,185,000
D Cumulative net sales [=SUM C(j)]			$6,590,000	$23,065,000	$43,435,000	$67,200,000	$84,175,000	$97,755,00	$107,940,000
E Unit cost (target)			$34.00	$33.50	$33.00	$33.00	$33.50	$34.00	$34.50
F Cost of product sold [=BE]			$3,400,000	$8,375,000	$9,900,000	$11,550,000	$8,375,000	$6,800,000	$5,175,000
G Gross margin ($) [=C−F]			$3,190,000	$8,100,000	$10,470,000	$12,215,000	$8,600,000	$6,780,000	$5,010,000
H % gross margin [=100G/C]			48.41%	49.17%	51.40%	51.40%	50.66%	49.93%	49.19%
I Development cost	$800,000	$800,000	$400,000	$50,000	$50,000	$50,000	$50,000	$50,000	$50,000
J Marketing (13% net sales) [=0.13C]			$856,700	$2,141,750	$2,648,100	$3,089,450	$2,206,750	$1,765,400	$1,324,050
K Other (8% of net sales) [=0.08C]			$527,200	$1,318,000	$1,629,600	$1,901,200	$1,358,000	$1,086,400	$814,800
L Total operating expense [=I+J+K]	$800,000	$800,000	$1,783,900	$3,509,750	$4,327,700	$5,040,650	$3,614,750	$2,901,800	$2,188,850
M Pretax profit [=G−L]	($800,000)	($800,000)	$1,406,100	$4,590,250	$6,142,300	$7,174,350	$4,985,250	$3,878,200	$2,821,150
N % profit [=100M/C]			21.34%	27.86%	30.15%	30.19%	29.37%	28.56%	27.70%
O Cumulative profit [=SUM M(j)]	($800,000)	($1,600,000)	($193,900)	$4,396,350	$10,538,650	$17,713,000	$22,698,250	$26,576,450	$29,397,600

*Product enters market midyear.

be launched by <**www.start-up-company.com**> that would make less money overall but involve a much lower risk than $1.6 million? What other projects might a large existing company sponsor? Would another project be more central to the company mission?

These questions are really beyond the scope of the present book. *Economics* by Parkin (1990) or *Engineering Economy* by Thuesen, Fabrycky, and Thuesen (1971) have many chapters devoted to such issues as the economic analysis of alternatives.

2.4 QUESTION 3: HOW MUCH QUALITY (*Q*)?

2.4.1 Introduction: Process Quality versus Organizational Quality

What is quality? How much quality does a product need? Can quality be measured numerically, and/or does it also involve aesthetic issues? In particular what is the "cost of quality"? Also, is there a "cost of not enough quality"?

To begin to answer these questions, a definition of quality is needed. In these next few sections, several definitions will be given.

- The first definition considers a parameter such as the measured diameter of a manufactured shaft in relation to the desired diameter requested by the designer. This is one aspect of *process quality.*
- The second definition considers more global measures of a company's overall quality. This is more related to *organizational quality or total quality management* (TQM). In Chapter 1, it was emphasized that U.S. engineers, particularly W. E. Deming, were the early advocates of TQM but Japanese companies, such as Toyota (Ohno, 1988), were the first to passionately apply them. Fortunately, by 1982, books such as Tom Peters's *In Search of Excellence* made U.S. manufacturers realize what they had to do to restore U.S. manufacturing competitiveness: (a) quality assurance on the factory floor, (b) TQM throughout the complete organization, and (c) leaner management hierarchies. Garvin (1987) has described eight aspects of TQM, and they are also evaluated by the Malcolm Baldrige Award and the ISO 9000 system. Cole's (1999) recent book describes the use of such procedures within a "learning organization." These will be reviewed later in this chapter.

2.4.2 Process Quality on the Factory Floor: Quantitative Measurements Using Statistical Quality Control (SQC)

Process quality is directly related to the physics of a manufacturing process, specifically, its inherent accuracy and how well it is controlled.

Imagine a group of friends in a British pub playing darts. Each player is given three darts. The goal is to hit the bull's-eye with each throw. Each player takes a turn, and then the round resumes. After an hour of pleasant drinking and playing, how many bull's-eyes does each player score? And what type of clustering occurs around the bull's-eye?

- Player One is very experienced and scores many bull's-eyes. In addition, even when the bull's-eye is not hit, the darts are clustered symmetrically around it in a small circle 50 millimeters (2 inches) in diameter.
- Player Two is less experienced and scores some bull's-eyes. However, the darts are scattered all over the board in a much larger circle of diameter 325 millimeters (13 inches). This is about the diameter of the scoring circle of a standard dart board.
- Player Three has never played before. No bull's-eyes are scored, and to great laughter, the board is often missed altogether and the darts ricochet onto the floor.
- Player Four has a strange style. All the darts are grouped together on the left-hand side of the scoring board (this is the number eleven zone on a standard dart board). No bull's-eyes are scored. However, the darts are consistently grouped in a 2-inch diameter circle close to the legal edge of the scoring target. The other players wonder why Player Four just can't pull them all over to the bull's-eye.
- Player Five is quite good. At the beginning of the evening, many bull's-eyes create a score ahead of Player Two. But too much beer is consumed, and by the end of the evening Player Five is much worse than Player Two.
- Player Six is in principle better than Player Two, but this is a player who is easily distracted by other people in the pub. There are a great number of bull's-eyes and on average a better clustering than Player Two. However, quite often, a dart goes way off target. So far off, in fact, that the errors are more dangerous than those of Player Three.

The obvious point of these entertaining thought experiments is that manufacturing processes are subject to the same problems.

In the semiconductor industry, many process steps occur: they involve lithography machines, dry-etching machines, diffusion chambers, and vapor deposition machines. All are subject to the inherent behavior seen in the dart players.

In the machine tool industry, imagine now that a circular shaft is being machined on six different lathes. The shaft might be going into the central axis of a lawn mower. There is a target dimension of 25 mm or 1 inch. The shafts are measured as they come off the six machines by an automatic touch sensor:

- Machine One is very accurate and repeatable. All the shaft diameters are clustered around the 25 mm or 1 inch target. The spread is 50 microns (0.002 inch). Machine One is delivering 25 mm shafts with +/−25 microns in their diameter.
- Machine Two is less accurate. The shaft diameters have a much bigger spread, as much as 500 microns (0.02 inch) to give a 25 mm diameter (micron +/−250). As with the first two dart players, Machine Two is less accurate than Machine One. Perhaps Machine Two can be used for some rough cutting on cylindrical

parts where accuracy is not critical. Or, more importantly, the SQC Quality Assurance Department can recommend machine maintenance to improve the machine.

- Machine Three has hopeless accuracy. Some parts are so far off the 25 mm or 1 inch target that the Quality Assurance Department stops any kind of production on the machine and begins serious maintenance work. Perhaps an actuator or lead screw is damaged, and occasionally, the machine sticks in place—way off from its desired settings.
- Here is an important question: What is the difference between accuracy and precision? Machine Four, like Player Four, demonstrates this difference when compared with Machine One, or Player One. The results of Machine Four demonstrate great precision. However, the precision is demonstrated in the wrong place. Something is wrong with the machine's ability to locate an accurate location. Perhaps a fixture slips right at the beginning of a batch run. The first shaft is incorrect right from the beginning, but all shaft diameters cluster around that incorrect location.
- Machine Five starts off well, but tools wear out (on a lathe) or the alignment drifts (on a lithography machine) and the process deteriorates. The SQC team must recognize the deteriorating factor and fix it.
- Machine Six is quite good overall, but occasionally a really poor part is produced. Perhaps this is a machine with a controller error, which shows a short circuit and causes major errors from time to time.

For this example, the data on the dimensions of the shafts would be monitored and overseen by a Quality Assurance Department. The results would be statistically analyzed and stored in extensive computer databases.

These statistical quality control (SQC) databases are the key to maintaining the highest levels of quality assurance. They provide the information for careful machine adjustments, machine-maintenance scheduling, timely reporting of errors or drifting behavior, and machine diagnostics. Recommendations on scheduled maintenance for a particular machine can also be tied into the factory scheduling.

Such quality assurance can also include the *Pokayoke* approach (Ohno, 1988; Black, 1991). *Pokayoke* in Japanese simply means "defect free." It can be applied to machinery design where some additional devices are added to a machine to prevent an operator from making a mistake like loading a bar in the wrong way around.

Finally, the quality assurance methods might include the formal techniques of *Taguchi*. *Taguchi* methods focus on the types of noise in a manufactured product and then proceed to reduce their occurrence by documented statistical means. In addition, *Taguchi* methods document the lost time that the consumer of the product consumes, getting the product up and running and/or getting it repaired at some later date. All such problems are then traced back through the factory to the source of the noise and allocated a cost function.

This is not "rocket science." Much of it is commonsense quality assurance. In fact, it is no different from carefully maintaining an automobile: checking the oil, following the recommended maintenance schedule, fixing problems before they lead to a major breakdown, scheduling the maintenance when life is not too hectic.

2.4.3 "Specification Limits" versus "Process Control (PC) Limits"

It is important to understand two definitions that are at first glance similar but that are "on different sides of the CAD/CAM fence":

- The *specification limits* set by the designer (often called "the specs" for short)
- The *process control limits* that are inherent to the manufacturing process being used

Within the "specs" there is an important definition:

- *Tolerance:* which is related to the designer's needs. The tolerance might be equal on each side of the "target dimension." This is called a "bilateral tolerance spec." Note that there are many cases where the tolerance will deliberately not be symmetrical about the mean. Often shafts will need to fit and rotate in a central bearing. Thus the shaft cannot be too big without jamming. But it could be a little smaller without a problem. In the example above perhaps the tolerance would be written 25 mm (+0/−50 microns).

Within the process control limits there are other important definitions:

- *The mean value:* such as the mean value of all the cylindrical shafts made. For reference, the value of (x) at the mean value is assigned μ_x. In Figure 2.13, 13 mm is the mean value.
- *The variance* around this mean value. This is the same as the natural tolerance (NT) for the process (not the designer's tolerance). In standard SQC monitoring, a value of {NT= 6σ = $+/- 3\sigma$} is commonly used. It represents each side of the Gaussian or bell-shaped normal distribution curve shown in Figure 2.13. The ranges of diameter starting from the left tail might be [12.950 to 12.955], [12.955 to 12.960], and so forth. Next, the histogram plots the number of shafts in each band all across to the right side tail.

If the company accepts all the shafts of diameter within ($6\sigma = +/- 3\sigma$), then the rejection rate will be 27 out of 10,000 parts. If the company accepted ($12\sigma = +/- 6\sigma$), then the rejection rate would be 2 out of 1 billion parts.

The desired accuracy—the specs—is summarized in the lower part of Figure 2.13: it is the range given by the *upper specification limit* minus the *lower specification limit,* namely (USL − LSL). This might simply be the range of acceptable diameters of the metal shaft to be used for the central axle of a lawn mower. The ideal diameter and allowable (USL−LSL) will be set by the designer based on functional considerations.

Next, it is common sense to choose a manufacturing process that has the right capability. Put in other words, its accuracy and NT should match the designer's demands. Ideally, the designer's value of (USL−LSL) should be wider than the achievable $+/-3\sigma$ (or NT = 6σ) of the manufacturing process.

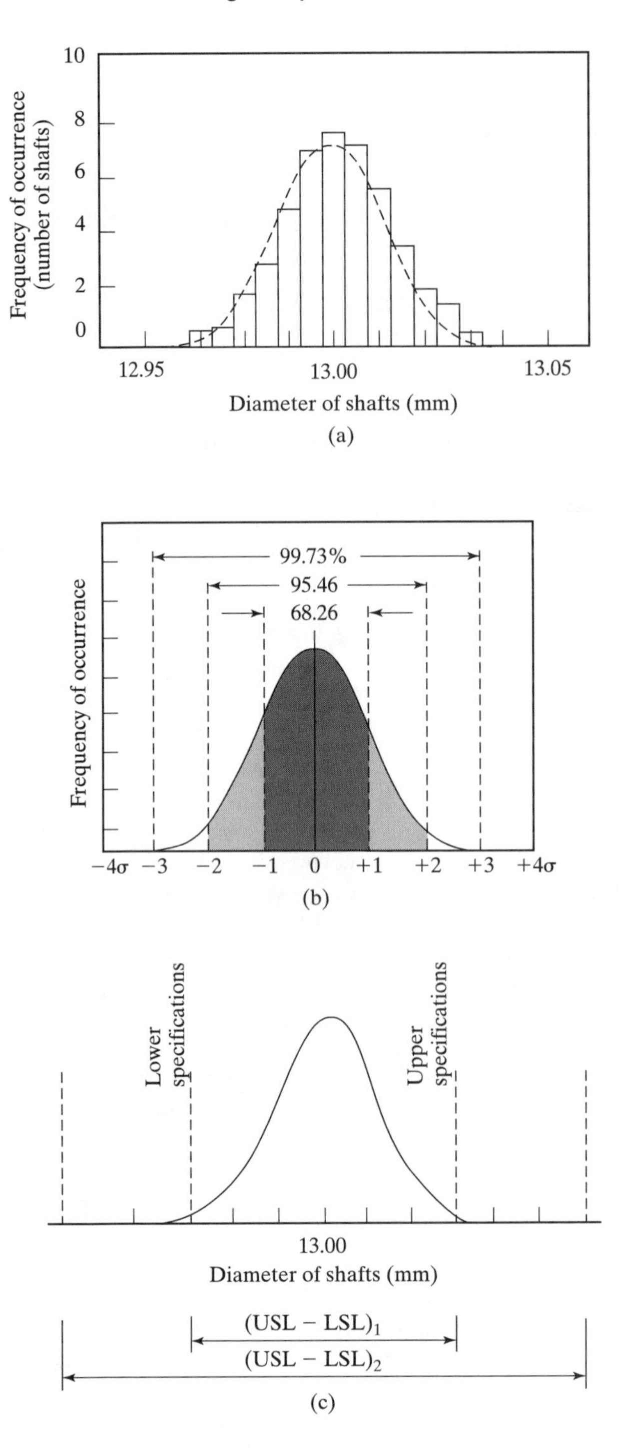

Figure 2.13 The upper diagram (a) is from the manufacturing *process* itself showing data clustered around a mean shaft diameter of 13 mm. In standard statistical quality control (SQC) work the Gaussian or bell-shaped normal curve is assumed, as shown in (b), meaning that 99.73% of the manufactured parts will lie inside the range $6\sigma = +/- 3\sigma$. The lower diagram (c) represents the *designer's* desired range. The upper specification limits (USL) and lower specification limits (LSL) are shown. In one case, $(USL-LSL)_1$ is the same as the manufacturing tolerance: this represents $C_p = 1$ and is the minimum acceptable condition but not really desirable because, with manufacturing capability of $+/- 3\ \sigma$, some parts are certain to fall outside the designer's specification. In the second case, $(USL-LSL)_2$ is twice as large meaning that $C_p = 2$. This is more desirable. Note, however, if C_p gets too large, then the chosen manufacturing process may be "too good" for the rather loose constraints set by the designer. (diagrams adapted from Kalpakjian, 1997, and DeVor et al., 1992)

Speaking colloquially, to build a precision telescope with tight tolerances and small values of (USL−LSL), nobody would go into their basement and use an old hacksaw and blunt drills that have a large value of NT. Surprisingly though, many manufacturers do struggle with the wrong manufacturing process or worn-out tools to try and satisfy a designer's needs. Likewise, some designers specify values of (USL−LSL) that might be "too good" for the needs of the part being designed: in this scenario, to achieve the tolerances, the manufacturer may have to follow standard milling or turning operations with a costly finishing process such as precision grinding or even polishing.

2.4.3.1 *Process Capability Index,* C_p

The process capability divides the range between the upper and lower specification limits (USL−LSL) by the width of the bell-shaped curve. Standard SQC uses $+/-3\sigma$ (namely, +/−3 standard deviations). This measure is known as C_p:

$$C_p = \frac{\text{USL} - \text{LSL}}{6\sigma_X} \tag{2.2}$$

The minimum acceptable value for C_p is considered to be 1, but as can be seen in Figure 2.13 a value between 1 and 2 is more desirable.

2.4.3.2 *Process Capability Index,* C_{pk}

The preceding discussions basically assume that the mean value of the manufacturing process coincides with the desired size of the part set by the designer. In other words, it assumes that a $+/- 3\sigma$ "viewing window" on the manufacturing process coincides with a (USL−LSL) "viewing window" from the designer.

However, what if this is not the case? Recall Dart Player Four: the darts are all tightly spaced in a small "viewing window," but the center of the window has drifted way off target. Machine Four's shafts on the extreme left of the tail will soon be in error. Following the example given by DeVor, Chang, and Sutherland (1992), suppose the ideal size of a shaft, as set by a designer, is 145 millimeters. Additionally, suppose that although a chosen lathe operation is more than capable of giving the desired SQC constraint of $+/- 3\sigma$, the tool post on the lathe has been distorted and all the components have shifted in size, meaning that the mean value is 130 millimeters not 145 millimeters. To account for such "drift" of the mean value it is now common for manufacturers to use a capability index referred to as C_{pk}. The value of C_{pk} relates the actual process mean to the nominal value of the specification.

For specifications in which the designer sets a desired value and then sets the same +/− tolerance on each side (recall this is called a bilateral spec), C_{pk} is defined in the following manner. First, it is necessary to determine the relationship between the process mean μ_X and the specification limits in units of standard deviations:

$$Z_{\text{USL}} = \frac{\text{USL} - \mu_X}{\sigma_X}, \quad Z_{\text{LSL}} = \frac{\text{LSL} - \mu_X}{\sigma_X} \tag{2.3}$$

The minimum of these two values is selected. The reader might pause for a moment and think about why this is the case. The answer is as follows: if the shaded curve in Figure 2.14 "drifts" to either end of the set range, it is desirable to look at the "worst case." The schematic figure shows the manufacturing process drifting to the left-hand side, and so manufactured parts in the left side of the tail will "go out of spec" first. The analysis must therefore consider how close the Gaussian curve is to the left side, LSL = 100 mm.

$$Z_{\min} = \min\left|Z_{\text{USL}}, \text{or}(-Z_{\text{LSL}})\right|. \tag{2.4}$$

The C_{pk} index is then found by dividing this minimum value by 3. The division is by 3 because this represents one side of the bell-shaped curve or the distance between the mean value, μ_X, and either LSL or USL depending on which way the "drift" occurs:

$$C_{pk} = \frac{Z_{\min}}{3} \tag{2.5}$$

In manufacturing, C_{pk} should be ≥ 1.00 for the process capability to be acceptable. Again following the example of DeVor, Chang, and Sutherland (1992), consider that the process mean has drifted and is located at 130, somewhat away from the nominal of 145, with a standard deviation of $\sigma_X = 10$. The calculations proceed as follows:

$$\begin{aligned}
Z_{\text{USL}} &= \frac{190 - 130}{10} \\
&= 6 \\
Z_{\text{LSL}} &= \frac{100 - 130}{10} \\
&= 3 \\
Z_{\min} &= \min[|6, \text{or}(-(-3))|] \\
&= 3 \\
C_{pk} &= \tfrac{3}{3} \\
&= 1.00
\end{aligned}$$

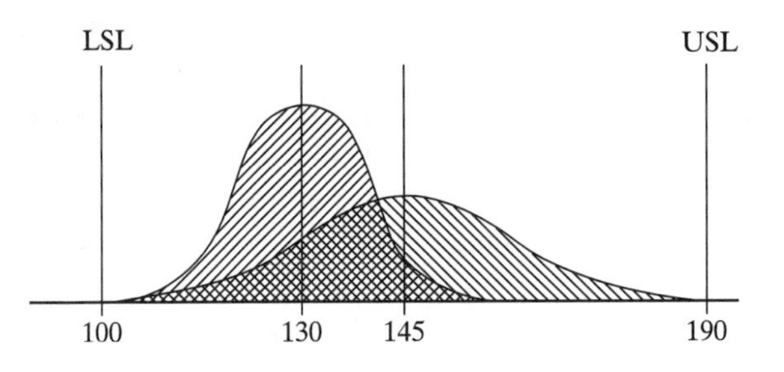

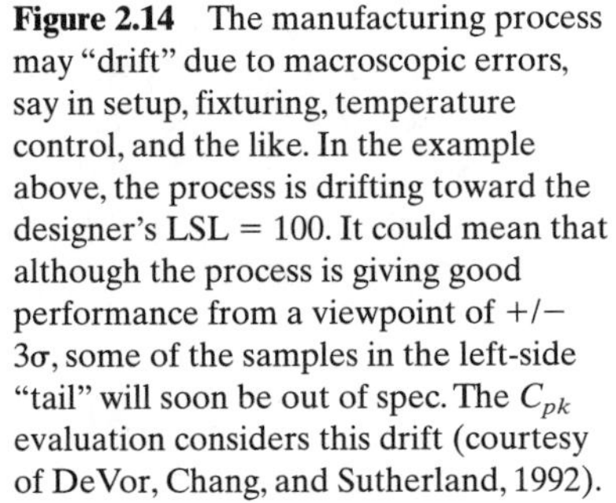

Figure 2.14 The manufacturing process may "drift" due to macroscopic errors, say in setup, fixturing, temperature control, and the like. In the example above, the process is drifting toward the designer's LSL = 100. It could mean that although the process is giving good performance from a viewpoint of +/− 3σ, some of the samples in the left-side "tail" will soon be out of spec. The C_{pk} evaluation considers this drift (courtesy of DeVor, Chang, and Sutherland, 1992).

However, if the process mean were recentered at the nominal of 145, then:

$$
\begin{aligned}
Z_{\mathrm{USL}} &= \frac{190 - 145}{10} \\
&= 4.5 \\
Z_{\mathrm{LSL}} &= \frac{100 - 145}{10} \\
&= -4.5 \\
Z_{\min} &= \min[|4.5, \text{or}(-(-4.5))|] \\
&= 4.5 \\
C_{pk} &= \frac{4.5}{3} \\
&= 1.50
\end{aligned}
$$

The example shows that by recentering the process, the value of C_{pk} would be increased by 50%.

2.4.4 Motorola's 6 Sigma Program

Six sigma quality is a phrase made famous by Motorola once it decided to refocus on quality in the late 1970s and early 1980s. It is a quality assurance program that has the goal of reducing the defective parts in a batch to as low as 3.4 parts per million.

Of more academic interest is the precise way in which Motorola implements this quality standard, which is under some scrutiny from several statisticians (Tadikamalia, 1994). A rigorous interpretation of 6 sigma really translates to 2 defects per billion parts made. The brief explanation is as follows, and it all boils down to whether the process is allowed to stay centered or not on the desired mean value (Figure 2.15).

First, consider the production of a million components using a manufacturing process that is centered on the mean (i.e., the target) value. The area under the normal curve can be calculated for various sigma bands. If we consider the two vertical lines that can be drawn at +/− 4.5 sigma each side of center, then 6.8 components per million will lie in the tails of such a curve, with 3.4 on each side. Clearly, this is a much tighter tolerance than was allowed in the +/− 3 sigma bands shown in Figure 2.13.

What happens if the process drifts off-center? Let us say that the center of the normal curve drifts by 1.5 sigma, this time to the right. If this second curve is still viewed through a window that is centered on the original origin and is still +/− 4.5 sigma wide, there will be virtually no defects in the left-side tail but a rather large number in the right-side tail: 1,350 to be exact.

However, if this shifted curve is viewed through a much wider window (a window that is +/− 6 sigma wide, centered on the original origin), the analysis returns to a more favorable view with only very few data in the tails. In this case these tails will contain 3.4 parts per million, probably all in the right-side tail. Actually,

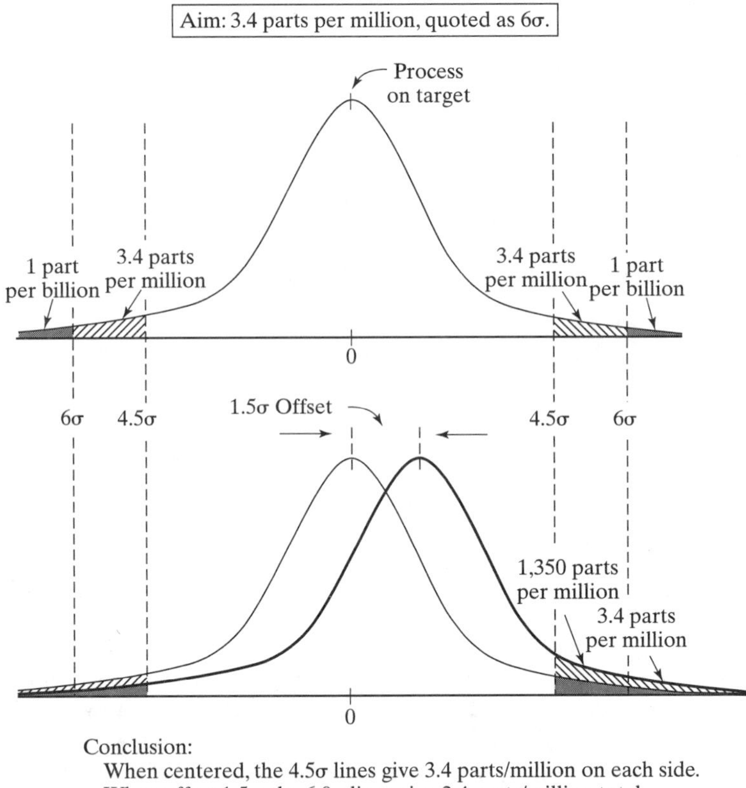

Conclusion:
When centered, the 4.5σ lines give 3.4 parts/million on each side.
When offset 1.5σ, the 6.0σ lines give 3.4 parts/million total.

Figure 2.15 Motorola's 6-sigma quality assurance method.

there is an infinite combination of "*m*-sigma offset plus *n*-sigma viewing window" ways of staying at a quality performance of 3.4 defects per million.

The positive outcome of all this is that Motorola is rightly known for excellent products with less than *3.4 defective parts per million.* It is this quantitative number that should be focused on as the benchmark rather than the rigorous definition of 6 sigma. However, the statisticians do have a point when they wonder why Motorola seems to allow a monitoring scheme in which some of its manufacturing processes might be allowed to drift off-center.

2.4.5 Summary on Process Quality

The best industry practice from a statistical point of view is to keep careful track of both the process *mean* and the process *variance.*

The process variance can be improved by (a) creating *quality circles* (groups of engineers who work diligently on a machine performance to improve process physics) and (b) investing in new capital equipment that is more precisely controlled. Both these solutions are quite expensive and time-consuming. Again it relates to the dart players: Player Two has to gain more experience, put in more training time, and try to be as good as Player One. Machine Two has to be studied and modified in certain aspects. Or it has to be sold and replaced with a better one that can deliver the desired accuracy.

The process mean, however, can often be addressed more cheaply by monitoring the output continuously and using the statistics to keep the mean of each batch centered on the target value. Measured errors in the process mean or target value can often be traced to a fixed offset. This might be related to a misoriented fixture or lithography mask in semiconductor manufacturing, or to a worn cutting tool that has not been adjusted from one batch run to the next.

As one might expect, *both* the mean and the variance can be thrown off in some cases. Perhaps a wide variance in the hardness of an incoming work material from an unreliable subvendor will cause scattered results plus tool wear, which will quickly move the mean value as well. Thus in conclusion there is no quick fix to obtaining 3.4 defects per million parts. But companies that want to stay in front obviously have to be part of this goal, refining their techniques in a way that really improves their products at an appropriate level of cost.

2.4.6 The "Bigger Picture"—Organizational Quality

The Motorola view is that quality assurance touches all aspects of product realization, not just the factory floor itself. When analyzed "in the large" manufacturing becomes a complex art form. Full-scale industrial design of both the product itself *and* the production processes that will fabricate the product relies on a huge team of people ranging from classical mechanical and electrical engineers, to marketing experts, to venture capitalists, to industrial psychologists, to advertising executives. Not only does manufacturing in the large encompass the complete assembly line for a Sony Walkman, or the much larger assembly line for a Ford Mustang, or the gigantic assembly hangers for a Boeing 777, but it also encompasses market analysis and consumers' reaction.

Manufacturing in the large can be effective only in the context of rigorous quality assurance within a "learning organization." Cole (1999) distinguishes *organizational learning* from *individual learning* by contrasting two social styles. Individual learning can be applied to one person or to one factory unit, but the key thing is that there are walls around the unit, almost related to a kind of protectionism tinged with paranoia. The old-style British Trade Union attitude seems to capture this the most: craftspeople jealously guarding their secret techniques, and factory units operating for their own bottom line and not worrying about what comes next. One industrial case study included a metal-extrusion unit that manufactured bar stock that was geometrically correct but so nonuniformly tempered that the downstream machine shop could not meet production schedules. In the old days the harried machine shop had to take care of this problem alone, while the extrusion unit celebrated record production.

With organizational learning, a problem such as the one above becomes everyone's problem, including the sales force. This cooperative attitude toward quality was a very important change in the way U.S. companies began to operate during the period after 1980. It became known as total quality management (TQM).

Today, quality assurance is the favored phrase, as discussed in Chapter 10. Many group seminars and books have become available to teach the social styles for these new ways of doing business in which everyone is a learner and people are encouraged to reveal rather than hide the problems that are occurring in the organization (Senge et al., 1994).

Today, the phrase *TQM,* in and of itself, is viewed with a small amount of suspicion (Cole, 1999). For too long it was used as lip service, ignoring the real need for improved and/or controlled quality using more formal statistical methods. In the worst situations, TQM was the "warm and fuzzy" *qualitative* approach to quality that logic-oriented engineers and MBAs can get grumpy and restless about, since it seems to be common sense. Nevertheless, quality assurance—meaning the careful analysis of process quality and cross-division quality in an organization—is now mandatory for success in modern manufacturing.

2.4.7 Definition of Quality at the TQM Level

At the TQM level, the general term *quality* can be measured in many different ways (see Cole, 1999; Garvin, 1987). The eight below are from Garvin's work. Rather than summarize a dry list of characteristics, imagine going shopping this weekend for a car or computer. The bullets below the generic category show the kinds of topics that fall into that subcategory:

1. **Performance** is a measure of basic issues that can be quantified and ranked:
   - Car: horsepower, top speed, acceleration, weight, miles per gallon
   - Computer: processor speed, amount of RAM, amount of hard disk space, screen size
2. **Features** are secondary aspects of performance[2]:
   - Car: moon-roof, leather seats, designer wheel rims, cup holders
   - Computer: CD player, graphics chip, high-speed modem
3. **Conformance** is a measure of how well the product fits operational and safety standards:
   - Car: emission standards, air-bag requirements, miles per gallon
   - Computer: operating system standards, I/O port standards, shielding standards
4. **Reliability** is concerned with the frequency of breakdowns or failures:
   - Car: consumer reports, the J. D. Powers quality survey on faults and breakdowns
   - Computer: mean time between failures, system crash frequency, disk drive reliability

[2]Note the "gray line" between performance and features. Twenty-years ago cup holders were certainly "features." Today, advertisers on television seem to regard the number of cup holders in a minivan as a performance measure.

5. **Durability** is linked to reliability but more concerned with long-term life:
   - Car: life of tires, miles before a recommended major part change (e.g., timing belts)
   - Computer: long-term life expectancy
6. **Serviceability** relates to frequency and ease of repair:
   - Car: frequency of oil changes, other servicing schedules, ease and cost of service work
   - Computer: ease and cost of upgrades, accessibility of major parts
7. **Aesthetics** relates to how a product looks, feels, sounds, tastes, and smells:
   - Car: Porsche versus minivan—enough said!
   - Computer: cream-colored cubes versus the iMacs
8. **Perceived quality** is concerned with the built-over-time reputation:
   - Car: despite the dramatic improvements in the U.S. companies, Toyota still wins
   - Computer: while consumers might be swayed by price point, larger companies will prefer to buy name-brand products from Sun, IBM, HP. "Intel inside" is important.

2.4.8 The Malcolm Baldrige Award and the ISO 9000 Scheme

Two well-known schemes have now emerged for evaluating the TQM ability of a particular company. Both awards bring enhanced marketability and recognition.

- The Malcolm Baldrige National Quality Award presented by the U.S. Commerce Department to recognize U.S. companies that excel in quality management and quality achievement
- The ISO 9000 certification of the International Organization for Standardization, whose objective it is to promote the development of quality standards, testing, and certification

The criteria for the awards are somewhat different (Table 2.5), but they both emphasize the creation of a "learning organization" (Cole, 1999).

2.4.9 A Case Study on Organizational Quality

Some notes on a visit to the Daihatsu Motor Corporation in Osaka, Japan, are now introduced, not to promote Daihatsu in any particular way but to illustrate how a focus on quality assurance has helped the company "swim with much bigger fish" and establish a market niche in the extraordinarily competitive, global automobile market. Daihatsu has extensively relied on the analysis of "What is quality?" and has now established a very clear view of who its customer is. It is especially conscious of establishing its place in the minicar and minitruck market. To do this, it matches its sought-after customer needs to the size, comfort, and fuel efficiency of the vehicle. Thus its objective function is optimizing cost, safety, and fuel efficiency for this

TABLE 2.5 Similarities and Differences between the Malcolm Baldrige National Quality Award and ISO 9000. (From G. Hutchins, *ISO 9000: A Comprehensive Guide to Registration, Audit Guideline, and Successful Certification,* Oliver Wight Publications, Inc., Essex Junction, VT. Copyright (c) 1993 by Oliver Wight Publications, Inc. reprinted with permission of John Wiley & Sons, Inc.)

Baldrige	ISO 9000
U.S. based	Global
Highest level of quality	Highest common denominator criteria
"World-class" quality	Doable and attainable quality
Advanced TQM award	First step in the TQM journey
Systems-oriented	Systems-oriented
Broad quality criteria covered	Version ISO 9001 generically covers Baldrige criteria
Focus on control, participation, and improvement	Focus on control
Exclusive, only two winners per category	Inclusive, all can become registered
Quality criteria higher and more demanding, stressing customer satisfaction, quantifiable results, and continuous improvement	Quality criteria generic; customer satisfaction and continuous improvement not emphasized

limited market sector. Daihatsu minitrucks are ubiquitous in the small commercial alleyways of Tokyo, making small volume deliveries to shops and the like. On the congested commuter roads of Osaka, Kyoto, and Tokyo, its minicars are also very evident. Younger first-time buyers also seem to represent a fair share of Daihatsu's customers.

During the visit, Daihatsu crash tested a car and emphasized that it is possible to make a high-quality yet inexpensive product. Its quality movement begins with the study of the intended market. It then leads into integrated design and manufacturing for the establishment of a low cost product with high quality. For example, its quality assurance studies showed that customers still valued safety above all else, despite the need for a relatively low cost vehicle. Thus Daihatasu continued to emphasize safety issues in its design, and later safety was particularly emphasized in its marketing. As a specific example, while good design practices were used to minimize the number of weld points in the body (thus reducing cost), no compromise was made to structural safety of the chassis's crumple zone.

Another observation from all such studies of the automobile industry (whether in Osaka, Detroit, or Coventry) is that the right type of automation increases overall quality. This is especially true in the welding and bulk vehicle assembly lines where the work is heavy and requires good alignment. In the future, engineers will still be striving to automate as many operations as possible and move into other more exacting areas of vehicle assembly. An interesting finding from studies by Xerox and by Boothroyd and Dewhurst (see Chapter 8) is that to push automation to the limit and create the best quality, any peg-in-hole-like assembly insertions and so forth should be vertical. From an "integrated manufacturing," or TQM, viewpoint it is thus useful to take a very broad view and even consider the redesign of key elements of the engine, transmission, or body just to promote vertical assembly.

2.4.10 A Colloquial Conclusion

The word *quality* is used on an everyday basis. However, this section of the book has shown that it can be interpreted in many ways.

First, it is useful to conclude by reiterating the *quantitative* methods for quality assurance based on statistical quality control (SQC) and the process capabilities of C_p and C_{pk}. These are based on the $+/-3\sigma$ process variances but now could include Motorola's 6σ analysis (see DeVor, Chang, and Sutherland, 1992).

Second, however, in TQM, many *qualitative* issues have been raised, and these must be interpreted with caution. For example, one of Garvin's measures is *aesthetics.* As individuals we all know what we mean by quality and aesthetics when it comes to choosing cars, clothing, music, food, wine, and which movie to see. Right? The following *options* indicate possible *preferences:*

- True or false: Dinner at a celebrity restaurant in New York City at $100 per person is of higher quality than dinner at the local burger joint at $10 per person.
- True or false: A $300 polo shirt by Georgio Armani is of higher quality than a $30 polo shirt by Land's End, which is, in turn, of higher quality than a $13 polo shirt from the local open-air market.

Whether the answer is true, false, or maybe depends on an *objective function* that is strongly dependent on *context* (see Hazelrigg, 1996, for a review). This context is dependent on the specific needs and specific circumstances of the moment.

One of these circumstances is strongly tied to how much disposable income each person has. However, even people with phenomenal unlimited wealth are unlikely to wear the $300 shirt and pick the $100 per person dinner on their way to a "night out with the kids at the baseball stadium."

Furthermore, the context is not merely that of disposable income. A $300 shirt might be scorned by a wealthy backwoodsman as a product for fashion victims but highly desired by a similarly wealthy socialite in a big city who wants to make the best impression at the opening night of the opera.

"Art is in the eye of the beholder." Any type of product can be made from a large range of materials. And the product can be sold with a wide variety of marketing strategies and objective functions. The savvy manufacturing-in-the-large organization must therefore try as hard as possible to "walk a mile in the shoes" of its major consumer group and then design a product with the most appealing quality-to-cost price point.

2.5 QUESTION 4: HOW FAST CAN THE PRODUCT BE DELIVERED (*D*)?

In the 21st century, time-to-market is key. Today's consumers demand "instant gratification." Vacation photographs, reading glasses, and pizza can be delivered in an hour or less. Why not manufactured parts?

All evidence shows that companies like Intel are so successful because they get the new chip into the marketplace first and consequently get the lion's share of the market profits. Those companies that enter the market later avoid the risk. But they make less profit.

In any manufacturing endeavor, one of the greatest challenges is to design a quality (Q) product at the right cost (C) *and* make it quickly with fast delivery (D). In the semester-long projects described in the Appendix, student groups start with a fascinating design in the early part of a semester. It is always the case that many compromises are made by the end of the semester when the design really has to be manufactured in piece parts and assembled. But this is a lesson for life. All product developments in the real world involve a compromise or a "balancing act" between the factors of designed-in quality (Q), manufactured cost (C), how fast the product can be delivered (D), and how flexible (F) the enterprise is.

A key challenge that always arises for both practicing engineers and student groups is: At what level do such compromises get made?

- During the conceptual design phase?
- During the detailed engineering phase?
- During the prototyping phase?
- During process planning for actual production and manufacture?
- After the product has reached the market and customers give feedback?
- During all of the above?

The evidence now seems pretty convincing that there should be several feedback loops. However, the earlier any problems can be pinpointed and eliminated, the faster the time-to-market will be. Thus, there is a great deal to be gained from doing conceptual design well. By contrast, waiting for customer feedback is probably very risky.

2.5.1 Time to a Finished Conceptual Design

To speed up the conceptual design phase, the team should contain a wide variety of representatives from all over the company. A general observation in the auto industry is that one high-level product drawing spins off into many hundred associated, ancillary manufacturing and assembly tools. Therefore, there is an enormous payoff in both cost and time if conceptual designs are accurate in the first place in determining "who is the customer." The goal is to not keep backtracking but do the market analysis well. Just a quick analysis of Figure 2.16 shows how long it took to launch common products. This is from Ulrich and Eppinger (1995), a relatively recent and excellent monograph on *product design and development.* It presents many details on the conceptual design strategies that lead to fast and accurate concepts.

The best situations are those where the market analysis and conceptual designs are correct in the first place. In this best-case scenario, several months or even a year or two later, many consumers will desire and be able to afford the final manufactured product when it arrives on the shelf or in the showroom. The very successful Mazda Miata's conceptual design team even had car insurance advisers on their panel of experts. They helped decide on engine power, chassis stability, and overall cost. In

	Stanley Tools Jobmaster screwdriver	Rollerblade Bravoblade in-line skates	Hewlett-Packard DeskJet 500 printer	Chrysler Concorde automobile	Boeing 777 airplane
Annual production volume	100,000 units/year	100,000 units/year	1.5 million units/year	250,000 units/year	50 units/year
Sales lifetime	40 years	3 years	3 years	6 years	30 years
Sales price	$3	$200	$365	$19,000	$130 million
Number of unique parts (part numbers)	3 parts	35 parts	200 parts	10,000 parts	130,000 parts
Development time	1 year	2 years	1.5 years	3.5 years	4.5 years
Internal development team (peak size)	3 people	5 people	100 people	850 people	6,800 people
External development team (peak size)	3 people	10 people	100 people	1,400 people	10,000 people
Development cost	$150,000	$750,000	$50 million	$1 billion	$3 billion
Production investment	$150,000	$1 million	$25 million	$600 million	$3 billion

Attributes of five products and their associated development efforts. All figures are approximate, based on publicly available information and company sources.

Figure 2.16 Product development times in row five for common products (from *Product Design and Development* by Karl T. Ulrich and Steven D. Eppinger, © 1994. Reprinted with permission of the McGraw-Hill Companies).

this way, the younger driver market that was being targeted could eventually afford the insurance rates.

2.5.2 Time to a Finished Detail Design

New techniques to speed up the detail design phase include design for assembly (DFA) software tools (Boothroyd and Dewhurst, 1999). Such techniques have led to many successes at Compaq computer and at Chrysler for its Intrepid and Neon lines. For example, Nissan's president, Yoshifumi Tsuji, was quoted in the October 29, 1994, issue of the *Economist* with the following praise for Chrysler: "Where we would have five parts to make a component, the Neon has three. Where we would use five bolts, the Neon body side was designed so cleverly, it needs only three." The general goals of the DFA software are to drastically reduce the number of subcomponents used in assemblies, to avoid screws and attachments that require complex hand-operated tools, and to streamline design shapes so that plastic molds are cheaper to make. Chapter 8 deals with DFA in detail.

2.5.3 Time to a Finished Prototype

When the designers have finished their conceptual designs, have considered the above DFM/A issues, and are at the first iteration of their detail designs, it is often useful to obtain a prototype of the component(s). A prototype is defined in the dictionary as "*The original thing, in relation to any copy, imitation, representation, later specimen, or improved form*" (Webster's, 1999).

In the VLSI world this could well mean going to the Metal Oxide Semiconductor Implementation Service (MOSIS). MOSIS (2000) was created approximately 15 years ago and is now a well-established brokering service currently located at the University of Southern California. Clients use standardized circuit layout tools and then submit their designs over the Internet in the electronic data interchange format (originally the CalTech Interchange Format [CIF]). After some checking, the chips are sent to fabrication services that guarantee manufacturability based on EDIF descriptions. A prototyped chip is returned within 6 to 10 weeks. Chips and other components can then be assembled onto printed circuit boards (PCBs) by custom houses.

In the mechanical world, something that is just for looks is more a *model* and might even be made as a paper model, a foam-core model, or a crude wooden carving of the final imagined object. Simple models allow the design group to share a common view. A physical prototype "structures the design process and coordinates sub-suppliers" (Kamath and Liker, 1994).

Several levels of sophistication are then available beyond the simple model-making step. A more substantial *prototyping technique* is needed if the designers want something that looks better, or if the prototype is going to be used as the first positive mold in a casting process. In such cases it is generally better to make the prototype by stereolithography, selective laser sintering, fused deposition modeling, or machining. Several prototyping methods are described in Chapter 4 (Weiss et al., 1990; Ashley, 1991; Au and Wright, 1993; DTM, 1993; Jacobs, 1992; Weiss and Prinz, 1995; Weiss et al., 1997; Jacobs, 1997; Sachs et al., 1998; Weiss and Prinz, 1998).

It is also useful to make *small batches of prototypes,* in the range of 10 to 500 or so, from high-strength plastic prototypes. These plastic prototypes can be injected into relatively cheap aluminum cast or machined molds. In this scenario, the CAD/CAM team orders an aluminum *prototyping mold* before scaling up to a steel *production mold* for the final injection-molding process. The final production molds for high-volume batch runs of, say, kitchenware products, toys, or automobile components need to be wear resistant and stable. Made from high-strength steel and polished to perfection, such molds could cost $100,000 or more. Figure 2.17 is a summary of product development thus far. Note that each stage builds upon the previous one. A theme of the case study at the end of this chapter is that the CAD/CAM software should gracefully and unambiguously make the transitions from one step to the next.

2.5.4 Time to a Finished Process Plan and the First Production Run

When mechanical designers have finished the CAD designs and *prototypes* for a specific product, they want to see their *real components* manufactured quickly and with fidelity. Process planning is the "bridge" from the rendered "virtual object" on the CAD screen to the machined "physical object" leaving the machine shop and on its way back to the designer.

Process planning involves several steps including (a) recognizing the features that the designer created, (b) analyzing how the features overlap and intersect, (c) mapping the geometry of these features to the capabilities and geometries of the "downstream" manufacturing machines, (d) selecting appropriate fixtures and associated setup rou-

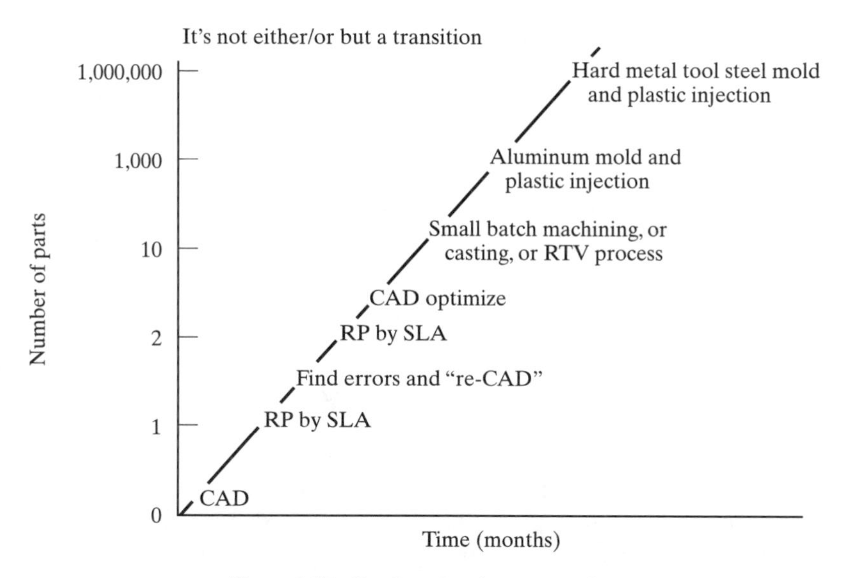

Figure 2.17 Product development cycle.

tines for the processing, (e) specifying the running parameters of the machinery, (f) detailing the in-process and post-process inspection routines, and (g) providing a quality assurance report that ties all the information together along with the part itself.

Even when the specific processes have been decided upon, there is still the selection of which machines will be used and how the parts will be routed through a flexible manufacturing system (FMS). This is the general domain of shop-level process selection and production planning. The constraints that are introduced consider machine availability, raw material availability, and customer delivery times. Some of the techniques needed to schedule the parts flow through the various machines are described by Bourne and Fox (1983), using constraint-directed reasoning; Cho and Wysk (1993), using scheduling algorithms; and Adiga (1995) and Kamath, Pratt, and Mize (1995), using object-oriented programming techniques.

These techniques address the scheduling of an "orchestra" of milling centers, drilling machines, and lathes for metal production; *or* a series of lithography, deposition, and etching steps for IC wafer fabrication.

At a more detailed single-machine level, process planning needs to be done at each individual machine. In today's factories the precise combination of which hole needs to be drilled first and so on is still pretty much the domain of mature, skilled machine operators who have been promoted to CNC programmers and who write the set up sheet, or the traveler, that goes along with all the machine tool programs. The text by Wysk and associates (1998) provides a comprehensive review of the above topics.

2.5.5 Time to the First Customer

Simulation software is available to view the passage of parts through the factory on different machines. Such simulation results speed up machinery setup and debugging time. Also, if fixtures, dies, and tools can be reused from previous models, time is reduced. At this point, the first customer is on the horizon. Round-the-clock setup and debugging work is almost certainly needed.

2.6 QUESTION 5: HOW MUCH FLEXIBILITY (*F*)?

In addition to production systems that fabricate very high *quality* products, at low *cost,* and with ultrarapid *delivery,* many strategic planners and economists point to the need for *flexibility.*

Publications from the United States and Europe (Agile, 1991; Black, 1993; Cole, 1999; Greis and Kasarda, 1997; Anderson, 1997; Kramer, 1998) specifically refer to the need for "agile manufacturing" systems that focus on "improving flexibility and concurrence in all facets of the production process, and integrating differing units of production across a firm, or among firms, through integrated software and communication systems" (Agile, 1991).

Publications from Japan (Yoshio, 1994; Ohsono, 1995) express a similar view, and the more recent J. D. Powers comparative surveys on automobiles indicate that "now that others are closing the quality gap, the Japanese have to compete in other areas" (see Rechtin, 1994; and the annual J. D. Powers report series). Emphasis is thus placed on these *combined* factors of quality, cost, delivery, and flexibility (QCDF). The ability to react to smaller lot sizes and the quest for ultrarapid delivery are major concerns, culminating in the possibility of a three-day car (Iwata et al., 1990).

In an ideal situation, once the various market sectors have been established, production will settle into a groove and be constantly refined and improved but with no major upheavals. Unfortunately, in recent years, manufacturers have not been able to rely on long periods of uninterrupted production because events in the world economy have forced rapid changes in consumer demand and the range of consumer preferences.

Henry Ford's favorite aphorism—that his customers could have any color of car they wanted as long as it was black—is in sharp contrast to today's range of consumer preferences. This has led to the proposal by some academics that manufacturing can be built for "customized mass production." This sounds nice on first hearing. However, for products like automobiles, the degree of customization can go only so far for a given batch size and price point. Only hyperwealthy CEOs and movie stars can get precise customization in products like automobiles.

Nevertheless, an ability to be prepared for any sudden market shifts is becoming more of an issue. As new equipment is purchased, manufacturing companies must decide between hardware that is dedicated to only a few tasks and is thus relatively inexpensive, and more costly but more versatile equipment that might perform unforeseen tasks in the future. The methodologies for analyzing capital expenditures, returns-on-investment (ROI), and depreciations are given in many texts (see Parkin, 1992).

These can be used to analyze the ROI for new machinery that has been identified as useful and is therefore about to be purchased. However, since today's market trends are so uncertain, such analyses do not help to predict the specific systems to install in the first place. The hope is that some of the engineering solutions presented later in this book will provide much more flexible machinery for only a modest increase in cost (Greenfeld et al., 1989). In this way, the investment dilemma might be less critical.

The preceding discussions emphasize that flexibility is a main challenge for the continued growth of a new company. The main question is: Can a design and fabrication system that is first set up to respond to one market sector be quickly reconfigured to respond to the needs of another market sector, or even another product, and be just as efficient?

Today, the answer to this question is "Probably not." For example, if a machine shop is well equipped with lathes but has no vertical boring machines, there will be a natural limit on achievable tolerances. It is unlikely that it will be able to suddenly jump from truck transmissions to helicopter transmissions. And even in the reverse scenario, if a shop has dedicated itself to precision boring, it is unlikely that the equipment and the craftspeople will be able to be quickly redeployed in a cost-effective manner to routine production procedures and less demanding tolerances; their competitive advantage would be lost. These same comparisons can be made for semiconductor manufacturing. Manufacturers who are currently focusing on the high-volume production of memory chips will not readily switch to application-specific devices or vice versa. The general conclusion may be drawn that today's manufacturing tools—specifically machine tools, robots, and manufacturing systems—are still too dedicated to specific market sectors and are not flexible enough.

This general need for flexible, reconfigurable manufacturing systems was of course a key aspect of CIM in its original conception. Merchant (1980) led a number of industry forecasts between 1969 and 1971 that refined the details and needs of the CIM philosophy. However, these forecasts overestimated the rate at which flexible manufacturing systems and related technology would be absorbed into factories. During the 1970s and 1980s, machines exchanged "handshakes" when tasks were completed. If these tasks were completed properly and on time, then a flexible manufacturing system (FMS) continued to operate satisfactorily. However, if the machines went seriously out of bounds, then the communications broke down and too frequent human intervention was needed to make the FMS efficient. During this era, the experiences of several research and development groups showed that the inadequacy of cell communication software was probably the key impediment to the industrial acceptance of CIM (Harrington, 1973; Merchant, 1980; Bjorke, 1979). Of interest was that by the late 1980s, the review articles on CIM were advocating much smaller FMSs of only three or four machines as the most efficient way of utilizing the cell concept. All these trends suggested more sophisticated computer- and sensor-based techniques at the factory floor, as described later.

2.6.1 Design for Flexibility (Reuse)

Design for flexibility in the automobile industry can pay off in a big way if there is some reusability of fixture families. The automated assembly lines where the frames, doors, and chassis are assembled with robots and welded together are obviously intensely expensive. These are usually two-story-high lines as big as many football fields where robots, fixtures, and alignment cradles bring the body components together for welding and assembly. The intense cost of these lines is hard to picture without a visit to a standard automobile plant. The key issue is to maximize the use and reuse of these fixturing lines. If designers were allowed total freedom, each vehicle in the family might require special tooling. This would not allow cost-effective manufacturing. As mentioned earlier, this factor places an important responsibility on the designer. In an ideal situation the newly designed component will be made on existing factory-floor machinery, readily leading to an "off-the-shelf" automation solution. In the best case, existing fixtures and even some parts of existing dies will also be reused.

Some companies, those with smaller batch sizes, might use a *mixed* production line. As one views such a line, several body styles go by: perhaps the mix is as simple as regular sedans and station wagons, but the mix can often be stretched beyond this to different cars of more or less the same size. With good design for multiple usability, many of the hard fixtures and robots can be used for all the differing vehicles in the family.

Also, with good cooperation between manufacturing and design, the existing robots and fixtures might even be able to "upwardly constrain the vehicle design space" for *future* vehicles. Therefore, viewed across several years and more than one family of vehicles, automation costs are relatively lower per individual vehicle.

2.6.2 Concluding Remarks: Design Aesthetics versus Manufacturing

Just to keep a proper perspective at the close of this subsection, it must be emphasized that design for manufacturability (or flexibility) has to be prudently applied with the perceived end user constantly in mind. The Japanese articles concerned with TQM (Yoshio, 1994; Iwata et al., 1990; see Hauser and Clausing, 1988) increasingly emphasize the more qualitative aspects of aesthetics as one of their next thrusts, even though this is much more difficult to measure as a design objective.

At one extreme, a component that is destined to be buried deep in a car, a washing machine, or a furnace does not need to look good. DFM and DFA methods can be applied at every step in Figure 2.1.

At the other extreme, there will always be a market for high-quality expensive products such as the $300 polo shirt discussed in Section 2.4.10, a new Jaguar, or an expensive Bang and Olfson music system. In these cases the buyer is actively seeking style and luxury. Therefore, the design-for-manufacturability engineers cannot have it all their own way, or a car might end up looking like a rectangular box on four wheels: cheap, it's true, but hard to sell.

To conclude with personal observations, it is clear from visits to Tokyo and Kyoto that wealthy Japanese people prefer a Mercedes Benz, a BMW, a Jaguar, or the large Toyotas and Accuras. Not many American cars are seen on the streets, even in the financial districts and embassy areas of Tokyo. One does see the occasional Cadillac, some fully loaded Jeeps, and some of the newer Ford Mustangs, but not many.

The "Big Three" U.S. companies complain that the reason for this is that Japan imposes trade barriers on U.S. vehicles. However, perhaps the real reason is that a wealthy Japanese businessman wants a car with *brando*. This is a Japanese phrase meaning "brand appeal." So, perhaps U.S. cars in Japan just don't have enough *brando* at the present time. After all, Japan is swamped with other U.S. products that do have *brando*: Levi's 501, McDonald's, Hollywood action movies, and CDs by American musicians. And, inevitably, the Starbuck's Coffee shops in Tokyo are swamped.

The moral of the story is that product design and product manufacturing have art and irrational emotions lurking in their corners that design and manufacturing engineers should not ignore. A product that can evoke the *aesthetic experience,* discussed in Chapter 1 in the section called "The Art of Manufacturing," will always capture some of the market.

2.7 MANAGEMENT OF TECHNOLOGY

Chapter 1 reviewed the art, technology, science, and business of manufacturing. During the last 100 years design and manufacturing have clearly moved from an *art/technology* endeavor to a *business/science* endeavor. Emphatically, in this new business/science environment, being gifted in just the science and technology is not enough to win the day. Specifically, U.S. manufacturing in the late 1980s was clearly in a panic. Basic industries such as semiconductors, automobiles, consumer electronics, and machine tools were all losing out to international competitors.

Today, things do seem a lot better on all fronts. In "How the World Sees Us" the *New York Times* boldly stated: "Our technology—computerized weapon systems, medical scanners, the Internet—sets the standard to which developing countries aspire."[3]

No single miracle has happened, but steady progress has occurred in:

- Creativity in design and manufacturing
- Quality assurance and control of basic manufacturing methods
- Downsizing companies to be more efficient
- Global commerce
- Internet commerce

The fact that some of our international competitors did not fare too well and struggled with unfavorable currency exchange rates in the late-1990s also made it easier for U.S. companies to compete.

At the time of this writing, the challenge for the future is to keep this real growth going. For a few years, U.S. manufacturers were chasing their Pacific Rim and European competitors and even enjoying the athletic-event psychology of "coming from behind." Staying up with the front-runners or, better still, being ahead requires an extra degree of creativity and commitment.

Similar to the above list, perhaps some specific areas for continued creativity include:

- Further exploiting global markets through Internet commerce.

[3] *The New York Times Magazine,* June 8, 1997, 37–83.

- Focusing on complex systems, specifically developing CAD/CAM techniques for the electro/mechanical/biological products that are on the horizon.
- Continuing with the time-to-market awareness, balanced by aesthetic creativity.
- Continually striving for the 6-sigma quality goal.
- Creating an organization that can cope and even thrive on change. For example, all industries—from semiconductors, to machine shops, to steel mills—are being told to be more environmentally conscious. Being so, and yet still being competitive with other countries that might care much less about issues such as pollution or industrial safety, challenges manufacturers to be especially creative and efficient.
- Responding proactively to government funding opportunities, public policy, and federal regulations that impact all industries at some level. For example, a project such as the original Internet was launched more than 25 years ago by the Defense Advanced Research Project Agency (DARPA), and it has continued to be nurtured by the National Science Foundation (NSF). The MOSIS (2000) story is similar. It can thus be reasonably argued that much of the wealth of Silicon Valley in northern California and Route 128 near Boston had its birthplace in projects such as these. Companies that are willing to understand the constraints and then cosponsor this type of federally funded research can benefit greatly. However, reaching consensus can often be frustrating and time-consuming, as seen recently in the regulations surrounding the telecommunications industry. (On another constraining note in biotechnology, the FDA-supervised drug trials do impose a long period of time between initial research and development (R&D) investments and the marketplace. Perhaps for this reason, many small biotechnology start-ups are bought out by the deep-pocket pharmaceutical companies.)
- The initial time-to-market of a new product and the ongoing delivery time of an established product are ongoing themes of this book. The integration issues that will be discussed around design, planning, and fabrication are obvious areas to focus on technically. In particular, new hardware and software environments allow the connections between design, planning, and fabrication to be simplified. Particular benefits include the reduction of the time taken to obtain the first prototype of a designed object, whether it is a chip or a computer casing. New techniques and standards for distributed software systems also provide a more information-rich dialogue between the design function and the manufacturing function. The ability to rapidly obtain an initial prototype allows designers to assess the aesthetic aspects of a design. It also allows a preliminary analysis of how a single object in a subassembly will interact with mating components, and finally allows some preliminary decisions to be made on the future manufacturing methods for the component. The benefits of obtaining an initial physical prototype are seen to embrace both the component itself and the way in which the component will be produced. The ability to evaluate both the product *and* the process by which it will be made is an essential concept in concurrent or simultaneous engineering.

Although there will be many new trends and unexpected disturbances, one basic business goal will remain constant: the winners will be those who design, plan, and manufacture a high-quality product at the right price and get it to market first. To address this need, Chapter 2 has considered some general principles of manufacturing analysis in the context of four parameters: quality, cost, delivery, and flexibility (QCDF). Generic approaches for quality assurance methods were also reviewed. The chapter also discussed some guiding subprinciples for process selection in mechanical manufacturing.

2.8 REFERENCES

ACIS. 1996. *ACIS technical overview, ACIS geometric modeler, programming manual.* ACIS 3D Toolkits. Boulder, CO: Spatial Technology Inc.

Adiga, S. 1993. *Object oriented software for manufacturing systems.* London: Chapman & Hall.

Agile Manufacturing Enterprise Forum. November 1991. *An industry led view of 21st century manufacturing enterprise strategy,* 2 vols. Bethlehem, PA: Lehigh University.

Anderson, D. M. 1997. *Agile product development for mass customization.* Burr Ridge, ILL: Irwin Publishers.

Ashley, S. April 1991. Rapid prototyping systems. *Mechanical Engineering Magazine,* ASME, 34–43.

Au, S., and P. K. Wright. 1993. A comparative study of rapid prototyping technology. In *Intelligent concurrent design: Fundamentals, methodology, modeling and practice,* ASME DE 66: 73–82.

Ayres, R. U., and S. M. Miller. 1983. *Robotics: Applications and social implications.* Cambridge, MA: Ballinger Press.

Bjorke, O. 1979. Computer aided part manufacturing. *Computers in Industry* 1, no. 1: 3–9.

Black, J. T. 1991. *The design of a factory with a future.* New York: McGraw-Hill.

Boothroyd, G., and P. Dewhurst. 1999. *DFMA Software,* on CD from the company, or contact <**www.dfma.com**>.

Bourne, D. A., and M. S. Fox. 1983. Autonomous manufacturing: Automating the job shop. *Computer Magazine,* IEEE, 17, no. 9.

Chang, T. 1990. *Expert process planning for manufacturing.* Reading, MA: Addison-Wesley.

Cho, H., and R. A. Wysk. 1993. A robust adaptive scheduler for an intelligent workstation controller. *International Journal of Production Research* 31, no. 4: 771–789.

Cole, R. E. 1999. *Managing quality fads: How American business learned to play the quality game.* New York and Oxford: Oxford University Press.

Cutkosky, M. R., and J. M. Tenenbaum. 1990. A methodology and computational framework for concurrent product and process design. *Mechanism and Machine Theory* 25, no. 3: 365–381.

DeVor, R. E., T. H. Chang, and J. W. Sutherland. 1992. *Statistical quality control.* New York: MacMillan Publishing Co.

Dewhurst, P., and G. Boothroyd. 1988. Early cost estimating in product design. *Journal of Manufacturing Systems* 7, no. 3: 183–191.

Dowling, N. E. 1993. *Mechanical behavior of materials.* Upper Saddle River, NJ: Prentice-Hall.

DTM Corporation. 1993. Selective laser sintering. *Product Information Bulletin* 1, no 1.

Esawi, A. M. K., and M. F. Ashby. 1998a. Cost-based ranking for manufacturing process selection. In *Proceedings of the Second International Conference on Integrated Design and Manufacturing Mechanical Engineering (IDMME 1998).* Compiègne, France 4: 1001–1008.

Esawi, A. M. K., and M. F. Ashby. 1998b. The development and use of a software tool for selecting manufacturing processes at the early stages of design. In *Proceedings of the Third Biennial World Conference on Integrated Design and Process Technology (IDPT).* Berlin, Germany 3: 210–217.

Finnie, I. (Chair of Committee). 1995. *Unit manufacturing processes.* Washington, D.C: National Academy Press.

Garvin, D. A. 1987. Competing on the eight dimensions of quality. *Harvard Business Review* (November–December): 101–109.

Greenfeld, I., F. B. Hansen, and P. K. Wright. 1989. Self-sustaining, open-system machine tools. In *Proceedings of the 17th North American Manufacturing Research Institution* 17: 281–292.

Greis, N. P., and J. D. Kasarda. 1997. Enterprise logistics in the information era. *California Management Review* 39, no. 3: 55–78.

Harrington, J. 1973. *Computer integrated manufacturing.* New York: Industrial Press.

Hauser, J. R., and D. Clausing. 1988. The house of quality. *Harvard Business Review* (May–June): 63–73.

Hazelrigg, G. A. 1996. *Systems engineering: An approach to information-based design.* Upper Saddle River, N.J.: Prentice-Hall International Series in Industrial and Systems Engineering.

Hertzberg, R. W. 1996. *Deformation and fracture mechanics of engineering materials.* New York: John Wiley.

House, C. H., and R. L. Price. 1991. The return map: Tracking product teams. *Harvard Business Review* (January–February): 92–100.

Inouye, R., and P. K. Wright. 1999. Design rules and technology guides for Web-based manufacturing. The *Design Engineering Technical Conference (DETC) on Computer Integrated Engineering,* Paper Number DETC'99/CIE-9082, Las Vegas.

Iwata, M., A. Makashima, A. Otani, J. Nakane, S. Kurosu, and T. Takahashi. 1990. *Manufacturing 21 report: The future of Japanese manufacturing.* Wheeling, IL: Association of Manufacturing Excellence.

Jacobs, P. F. 1992. *Rapid prototyping and manufacturing: Fundamentals of stereolithography.* Dearborn, MI: Society of Manufacturing Engineers Press.

Jacobs, P. F. 1996. *Stereolithography and other RP&M technologies.* Dearborn, MI: Society of Manufacturing Engineers Press.

Jones, R., S. Mitchell, and S. Newman, 1993. Feature based systems for the design and manufacture of sculptured products. *International Journal of Production Research* 31, no. 6: 1441–1452.

Kalpakjian, S. 1997. *Manufacturing processes for engineering materials.* Menlo Park, CA: Addison Wesley. (See in particular Chapter 15).

Kamath, M., J. Pratt, and J. Mize, 1995. A comprehensive modeling and analysis environment for manufacturing systems. In *Proceedings of the 4th Industrial Engineering Research Conference,* 759–768. Also see **http://www.okstate.edu/cocim.**

Kamath, R. R., and J. K. Liker, 1994. A second look at Japanese product development. *Harvard Business Review* (Reprint Number 94605).

Kochan, D. 1993. *Solid freeform fabrication.* Amsterdam: Elsevier Press.

Kramer, B. M. 1998. *Proceedings of the 1998 NSF Grantees Design and Manufacturing Conference.* Monterrey, Mexico. See annual volumes of this conference. Arlington, VA: National Science Foundation.

Machining Data Handbook. 1980. 3d ed. 2 vols. Cincinnati, OH: Institute of Advanced Manufacturing Systems (IAMS).

Magrab, E. B. 1997. *Integrated product and process design and development.* Boca Raton and New York: CRC Press.

Mead, C., and L. Conway. 1980. *Introduction to VLSI systems.* Reading, MA: Addison Wesley.

Merchant, M. E. 1980. The factory of the future—Technological aspects. In *Towards the Factory of the Future.* PED 1: 71–82 New York: American Society of Mechanical Engineers.

Moore, G. A. 1995. *Inside the tornado.* New York: Harper Business.

MOSIS. 2000. *University of Southern California's Information Sciences Institute—The MOSIS VLSI Fabrication Service,* **http://www.isi.edu/mosis/**

NSF. 1993. *Agile Manufacturing Initiative,* program solicitation, National Science Foundation's Directorate for Engineering.

Ohno, T. 1988. *Toyota production system: Beyond large scale production.* Portland, OR: Productivity Press.

Ohsono, T. 1995. *Charting Japanese industry.* London and New York: Cassell Publishers.

Ostwald, P. F. 1988. *American Machinist cost estimator,* 4th ed. Cleveland, OH: Penton Education Division.

Parkin, M. 1990. *Economics.* Reading, MA: Addison Wesley.

Peters, T. J., and R. H. Waterman. 1982. *In search of excellence: Lessons from America's best-run companies.* New York: Harper and Row.

Poppel, H. L., and M. Toole. 1995. The bleeding edge of information technology. *The Red Herring Magazine,* (June): 82–86.

Pressman, R. S., and J. E. Williams. 1977. *Numerical control and computer-aided manufacturing.* New York: Wiley and Sons.

Queenen, A. 1979. *The modular approach to productivity in flexible manufacturing systems.* Technical Report 79–824. Dearborn, MI: The Society of Manufacturing Engineers.

Rechtin, M. 1994. "Europeans roar to no. 2 in CSI. *Automotive News,* (July 11): 59.

Sachs, E., M. Cima, J. Bredt, A. Curodeau, T. Fan, and D. Brancazio. 1992. CAD casting: direct fabrication of ceramic shells and cores by three dimensional printing. *ASME Manufacturing Review* 5, no. 2.

Sands, R. L. 1971. *Selection of processes for the manufacture of small components, in competitive methods of forming.* (103-112). ISI Publication Number 138. London: The Iron and Steel Institute.

Schey, J. A. 1999. *Introduction to manufacturing processes.* New York: McGraw-Hill.

Senge, P., A. Kleiner, C. Roberts, R. Ross, and B. Smith. 1994. *The fifth discipline fieldbook.* New York: Doubleday Books.

Shah, J. J. 1991. Assessment of features technology. *Computer Aided Design* 23, no. 5: 331–343.

Smith, G. W. 1973. *Engineering economy: Analysis of capital expenditures.* Ames: Iowa State University Press.

Suh, N. P. 1990. *Principles of design.* New York: Oxford University Press.

Sungertekin, U. A., and H. B. Voelcker, 1986. Graphic simulation and automatic verification of machining programs. In *Proceedings of the 1986 IEEE Conference on Robotics and Automation,* 156–165.

Tadikamalia, P. R. 1994. The confusion over six-sigma quality. *Quality Progress,* 83–85.

Thomas, R. J. 1994. *What machines can't do.* Berkeley, Los Angeles, London: University of California Press. See in particular Chapter 7, "The Politics and Aesthetics of Manufacturing," 246–258.

Thuesen, H. G., W. J. Fabrycky, and G. J. Thuesen. 1971. *Engineering economy.* Englewood Cliffs, N.J.: Prentice-Hall.

Toye, C., M. R. Cutkosky, L. J. Leifer, M. Tenenbaum, and J. Glicksman. 1994. SHARE: A methodology and environment for collaborative product development. *International Journal of Intelligent Systems* 3, no. 2: 129–153.

Ulrich, K. T., and S. D. Eppinger. 1995. *Product design and development.* New York: McGraw-Hill.

Urabe, K., and P. K. Wright. 1997. Parting planes and parting directions in a CAD/CAM system for plastic injection molding. 1997 ASME Design for Manufacturing Symposium. The Design Engineering Technical Conferences, Sacramento, CA,

Wang, F.-C., and P. K. Wright. 1998a. Web-based design tools for a networked manufacturing service. Paper presented at the 1998 ASME Design Technical Conference, Atlanta, GA. September 13–16.

Wang, F.-C., and P. K. Wright. 1998b. Internet-based design and manufacturing on an open architecture machining center. Paper presented at the 1998 *Japan–USA Symposium on Flexible Automation,* Ohtsu, Japan, July 13–15.

Wang, F-C., and P. K. Wright. 1998. Collaborative design: A case study on InfoPad, a wireless, networked computer. Anaheim, CA: The International Mechanical Engineering Congress and Exposition.

Webster's New World College Dictionary. 1999. Foster City, CA: IDG Books Worldwide.

Weiss, L., E. Gursoz, F. B. Prinz, P. Fussel, S. Mahalingam, and E. Patrick. 1990. A rapid tool manufacturing system based on stereolithography and thermal spraying. *ASME Manufacturing Review* 3, no. 1.

Weiss, L. E., and F. B. Prinz. 1995. Shape deposition processing. Paper presented at the NSF Workshop II on Design Methodologies for Solid Freeform Fabrication, Pittsburgh, PA, June 5–6.

Weiss, L. E., R. Merz, F. B. Prinz. G. Neplotnik, P. Padmanabhan, L. Schultz, and K. Ramaswami. 1997. Shape deposition manufacturing of heterogeneous structures. *SME Journal of Manufacturing Systems* 16: 239–248.

Weiss, L. E., and F. B. Prinz, 1998. Novel applications and implementations of shape deposition manufacturing. Naval Research Conference.

Wright, P. K., D. A. Bourne, J. P. Colyer, G. S. Schatz, and J. A. E. Isasi. 1982. A flexible manufacturing cell for swaging. *Mechanical Engineering,* 76–83.

Wright, P. K., and D. A. Bourne. 1988. *Manufacturing intelligence.* Reading, MA: Addison Wesley.

Wysk, R. A., T. C. Chang, and H. P. Wang. 1998. *Computer aided manufacturing,* 2d ed. Prentice-Hall.

Yoshio, T. 1994. Japan's competitiveness in industrial technology. *Journal of Japanese Trade and Industry* 13, no. 4: 8–10.

2.9 BIBLIOGRAPHY

Barr, A., and E. Feigenbaum. 1981. *The handbook of artificial intelligence,* 1–3. Los Altos, CA: Heuristech Press, Stanford, and William Kaufmann Inc.

Boothroyd, G., P. Dewhurst, and W. Knight, 1994. *Product design for manufacture and assembly.* New York: M. Dekker.

Chanan, S. 1994. *Concurrent engineering: Concepts, implementation and practice,* 1st ed. Edited by Chanan S. Syan and Unny Menon. London, New York: Chapman & Hall.

Chang, T. 1990. *Expert process planning for manufacturing.* Reading, MA: Addison-Wesley.

Cook, N. H. 1966. Manufacturing economics. In *manufacturing analysis,* 148–159, Reading, MA: Addison Wesley.

Corbett, J. 1991. *Design for manufacture: Strategies, principles, and techniques.* Edited by John Corbett et al. Addison-Wesley Series in Manufacturing Systems. Reading, MA: Addison-Wesley.

DeGarmo, E. P., J. T. Black, and R. A. Kohser. 1997. *Materials and processes in manufacturing,* 8th ed. Upper Saddle River, NJ: Prentice-Hall.

El Wakil, S. D. 1998. *Processes and design for manufacturing.* Boston: PWS Publishers.

Groover. M. P. 1996. *Fundamentals of modern manufacturing.* Prentice-Hall.

Koren, Y., F. Jovane, and G. Pritschow. 1998. *Open architecture control systems: Summary of global activity.* Institute for Industrial Technologies and Automation.

Peterson, I. 1997. Fine lines for chips. *Science News* (8 November): 302–303.

2.10 CASE STUDY

2.10.1 TouchChip™: A Fingerprint Recognition Device

This case study presents the product development process of a novel fingerprint recognition device. Some 250 prototypes were produced as evaluation kits for marketing purposes. Some key points that may be learned in the case study include:

- A "naked printed circuit board" is not the best marketing device.
- A single prototype in SLA, SLS, or FDM is only the starting point for a short prototype run (250) and then for real production (many thousands).
- A clear focus on "who is the customer" significantly influences the required batch size decision. This in turn determines the best manufacturing processes and also has an effect on the CAD tool that is chosen to create the design.
- If the conceptual design phase and the fabrication phase are concurrently considered at the beginning of the project, time-to-market is improved.
- The aluminum tooling for a new product was created with an integrated design-to-machining system. The "clean interfaces" between various modules for CAD, process planning, and fabrication also created more reliable and faster time-to-market. The system's front end was a set of design tools that encapsulated machining knowledge for checking manufacturability. This valid design was passed to an automated process planning module through a clean, unambiguous interchange format that eliminated unnecessary personal conversations between the designers and process planners. The process plan was then sent to an open-architecture three-axis CNC milling machine, which executed standard G & M codes for machining.

2.10.2 "Who Is the Customer?": Manufacturing Analysis and Conceptual Design

The TouchChip™ is a small sensor manufactured by ST Microelectronics. It uses techniques based on capacitance variation for fingerprint recognition. Other fingerprint recognition units on the market use optical devices, but this is a "silicon solution." The TouchChip can be used for a variety of applications that might include door entry, passwords for computers, replacing PINs for bank cards, and credit card verification.

The project began with the "naked" chip-on-board device shown in Figure 2.18. However, it was quickly realized that potential customers—typically computer peripheral makers—would be more impressed if a good prototype of a packaged unit could be created. Therefore, the goal was to create a small batch of product prototypes (about 250 assemblies) as evaluation kits for potential customers in the recent COMDEX Trade Show at Las Vegas. The entire project encompassed conceptual product design through prototyping to final production and packaging. The final product is shown here, and the manufacturing analysis follows.

2.10.2.1 Electronic Components

The device shown is ST Microelectronics's TouchChip STFP2015-50. It has a 20 mm × 15 mm surface with a 384 × 256 sensor array with approximately 100,000 individual sensors. The TouchChip utilizes a feedback capacitive sensing scheme. The sensing cell detects the distance between the skin and the sensor and identifies the presence of ridges and valleys. An array of sensing cells is used to sample the fingerprint pattern. The chip architecture and schematic diagram for the sensing cell are shown in Figure 2.19. The chip is manufactured in a 0.7 micron two metal-layer CMOS manufacturing process. A special coating protects the device from contamination, scratching, and electrostatic discharge. The completed chip also has a standard connection through external cable to the serial ports of computers (Table 2.6).

2.10.2.2 Mechanical Package

The fabricated mechanical package holds the above electronic components (Figures 2.20 and 2.21). It also supports the user's index finger, when its tip is resting and slightly pushing on the sensing window. The conceptual prototype is thus a mouselike enclosure

Figure 2.18 TouchChip sensor and electronics.

TABLE 2.6 Details of the Device

Features	
Sensor surface	20 mm × 15 mm
Module dimension	35 mm × 35 mm × 5 mm
Sensor array	384 × 256
Capture rate	10 frames/sec
Data interface	8 bit parallel data interface
Resolution	500+ DPI
Pixel	50-micron
Power	<200 mW @ 5 Volts
Integrated submodules	A-to-D converter, biases, on-board oscillator, and crystal

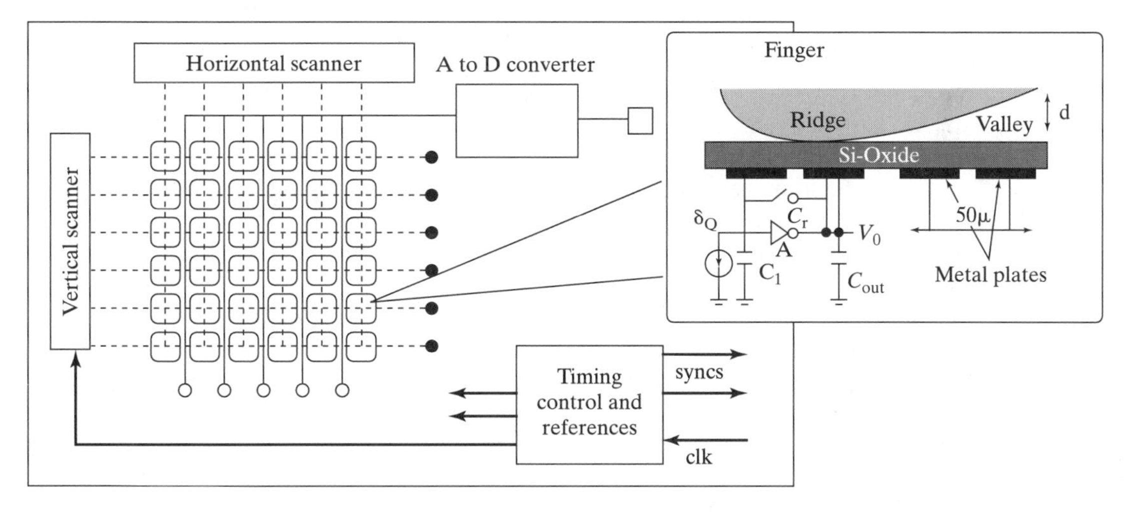

Figure 2.19 The sensor.

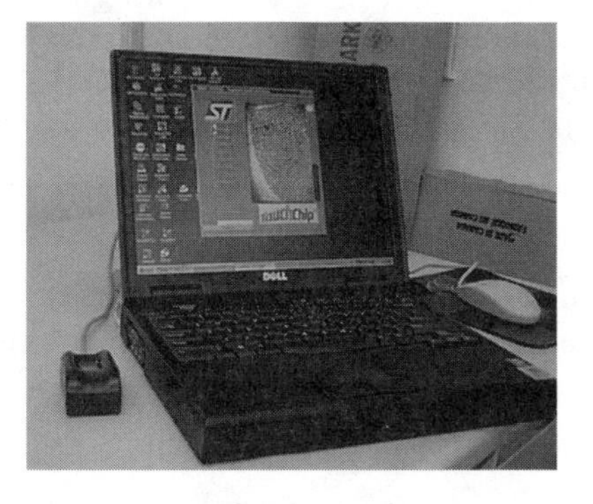

Figure 2.20 Integrated unit as a password entry for a standard machine.

Figure 2.21 Mechanical enclosure.

with standard cable connection to a computer. The top surface of the enclosure has to be smooth and ergonomic for the finger to sit comfortably on the sensing array.

2.10.3 Detail Design

At the "detail design" stage, a CAD representation (in *SDRC-IDEAS*) was created to check the fit of the electronics and the basic form factor. The top lid was designed to package the PCBs, single- and double-board versions, and to provide a comfortable support for the user's finger. The base provided additional space for connection ribbon cables and a power cable. Both the top lid and the base were designed to be fastened together.

2.10.4 Prototyping

The next step was to decide among the different rapid prototyping methods. It was important to fabricate a prototype that was strong enough to be structurally safe when assembled and handled during the evaluation period. At the same time, relatively high tolerances were needed to allow the PCBs to be securely fastened to the top lid. The strength, tolerance, and surface finish achieved by the SLA process seemed to meet these needs and was thus chosen for prototyping.

An informal group of "human users" was asked to evaluate the ergonomic aspects of the upper surface. The user group recommended that a flat slope be replaced with a more contoured slope that would mirror the inside surface of the index finger.

To create this more comfortable contour for finger positioning, an important decision was made. It was decided to switch to a *Web-based, freeform design tool* (Wang and Wright, 1998a, 1998b) to create the freeform shape on the top surface. The key issue was again time-to-market.

First, it was important to have a design tool that would produce the freeform surface. Second, and equally important, the Web-based, freeform design tool incorporated a *design rule checker* for "downstream mold machining" (Urabe and Wright, 1997). This software assisted the designer to create both the part and a feasible mold. By anticipating the milling process early in the design process, possible errors were eliminated, therefore accelerating the overall design-to-manufacturing time (Table 2.7 and Figure 2.22). The Web-based freeform design was rendered with the ACIS 3D tool kit (ACIS, 1996). A second prototype was then fabricated to evaluate the ergonomic freeform surface (Figure 2.23).

2.10.5 Aluminum Tooling Design

After the second prototype was evaluated and modified as shown in Figure 2.24, the shape was fixed and ready for the aluminum tooling stage for the injection molding process. Again, the Web-based, freeform design tool, plus the ACIS solid modeling tool kit, was selected for the complete mold design. The mold halves for the top part of the casing are show in Figure 2.25.

While designing these mold halves, some specific molding characteristics were considered as well. For instance, an enlargement factor of 1.014 was applied to the

TABLE 2.7 Manufacturing Analysis for the Development of the TouchChip Fingerprint Device

Questions	Answer	Criteria and comments
Who is the customer?	Original equipment manufacturer (OEM) of biometric security devices for computer applications	1. A "pre-mass-production" (OEM) product prototype as an "evaluation kit" for potential customers only 2. A potential market need for such a device but the first step is to demonstrate the product to computer makers and similar equipment makers
Conceptual design?	1. Small and ergonomic 2. Simple computer interfacing	1. Much smaller and cheaper than the current optical-based fingerprint devices on the market 2. Easy integration with current computer peripherals
What is the cost? (a) batch size?	200–300 assemblies	Product prototype as an "evaluation kit"
What is the cost? (b) accuracy and tolerances?	Injection molding into aluminum tooling and ABS plastic	1. Have a direct access to an injection molding service 2. Have a direct access to CyberCut machining service
How much quality?	Not the final production version—delivery time should dominate over perfection	Production quantity (200–300) and "prototype only" decisions merit soft aluminum tooling instead of hard-steel tooling to save machining time and cost. "Quality" will wait until a larger batch run is needed
Delivery time?	9 weeks	The production prototype will be demonstrated at the Fall COMDEX computer trade show
Flexibility?	Need eventually to switch to steel molds	CAD/CAM software allows ramp-up to more production

design to compensate for the shrinkage factor of the ABS material when it cooled down after the molding process. A "draft angle" of 1 degree was also applied to the side walls of the molds to allow for easy part-mold separation. Subsequently, injection molds were fabricated on a three-axis CNC milling machine.

2.10.6 Mold Machining and Injection Molding of 250 Casings

The final mold cavity designs were sent directly to the CyberCut machining service over the Internet. CyberCut is an experimental research platform for networked manufacturing. It provides a "pipeline" in which the features generated from the Web-based design tools can be analyzed for process planning, tool-path planning, and the creation of the machine codes (G&M codes) for a three-axis CNC milling machine. CyberCut also provides a detailed manufacturability check on the designs as they are created (see Wang and Wright, 1998a, 1998b).

The completed mold cavities were sent to Metalcast Engineering in Oakland, California, to complete the plastic injection process. The mold cavities were sand-blasted to improve the surface texture of the final product. Backing plates were machined and attached to the back of "cope" mold halves for adding the spruce, runners and gates, and the ejecting mechanism (see Urabe and Wright, 1997).

A total of 250 casing parts were produced. The TouchChip was mounted to the top casing and a daughter board complete with cable connection was connected to

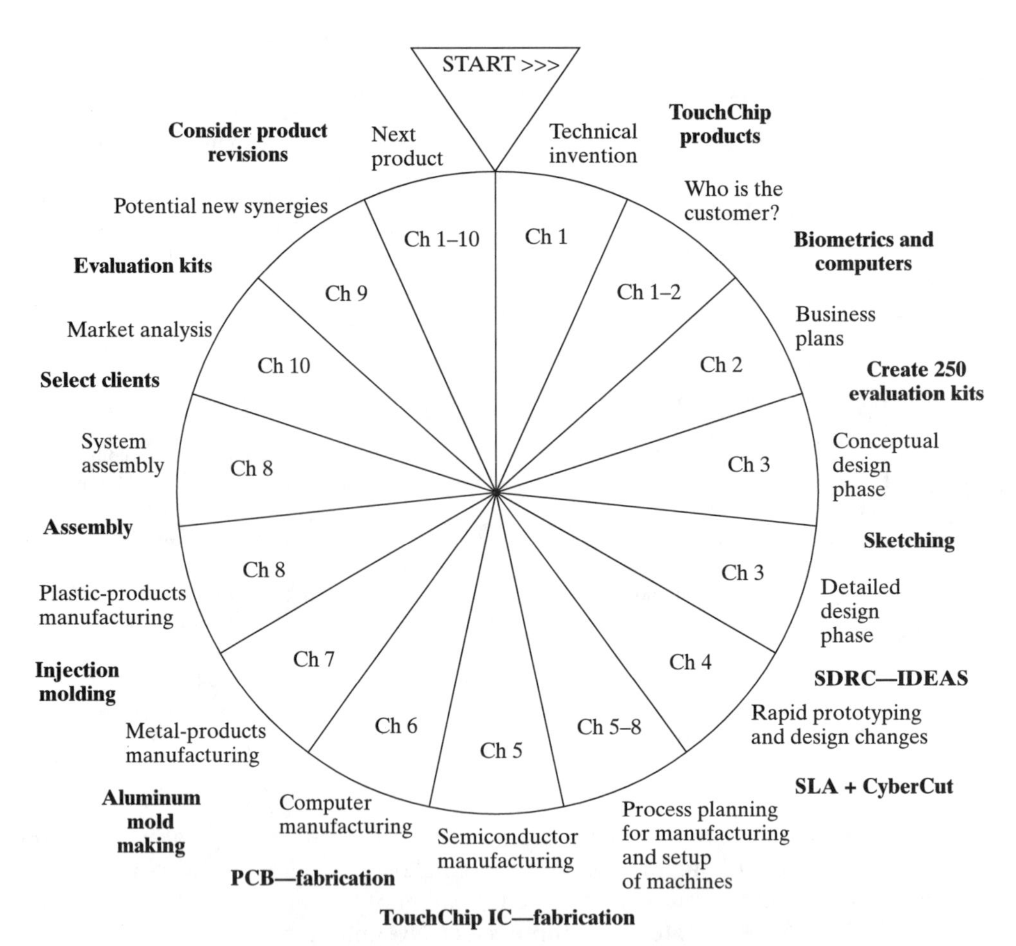

Figure 2.22 Figure 2.1 reproduced with TouchChip steps.

Figure 2.23 Stereolithography prototype.

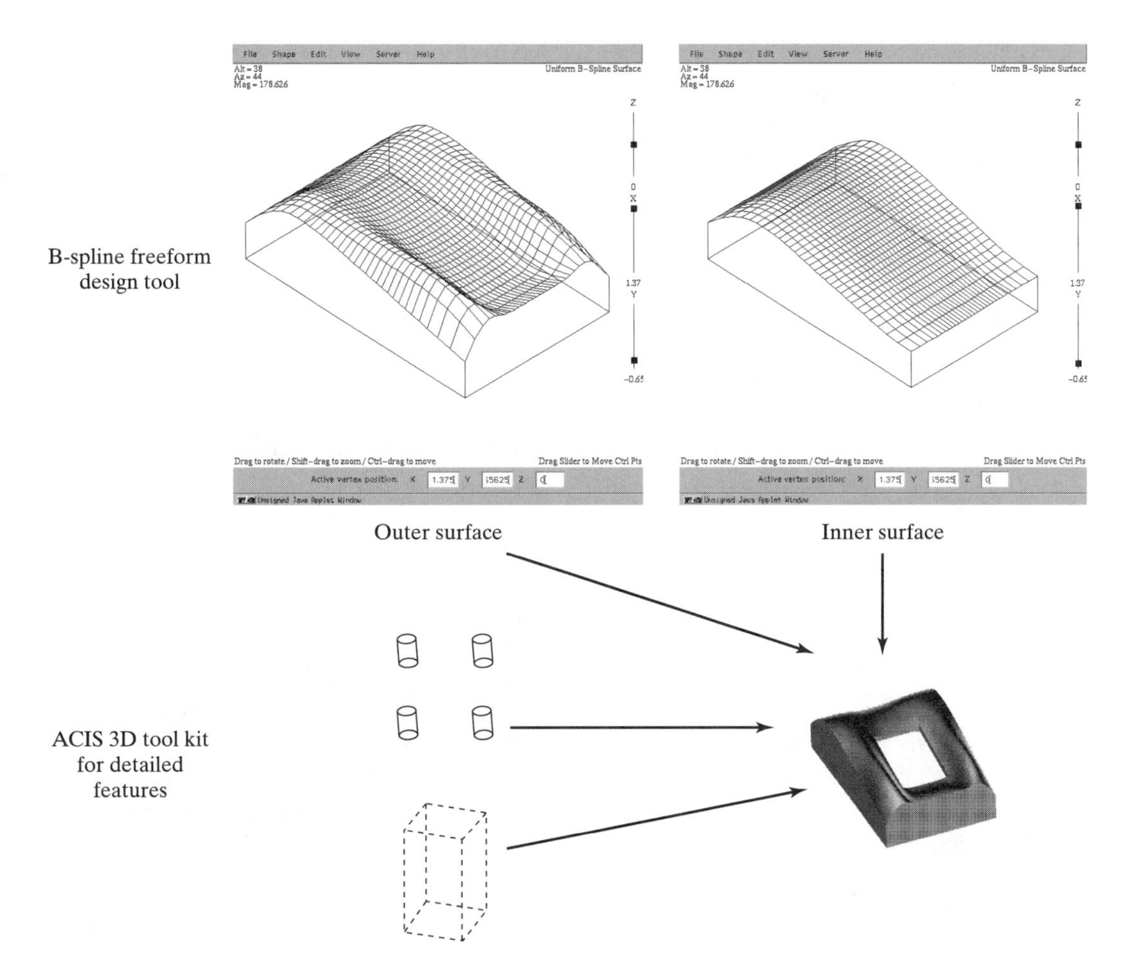

Figure 2.24 Second prototype designed using a Web-based freeform design tool and ACIS tool kit.

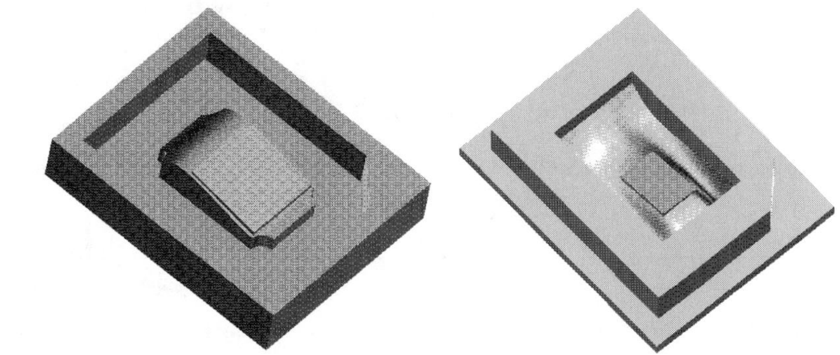

Figure 2.25 Mold design for a freeform top casing.

the TouchChip and finally assembled to the bottom casing. Figures 2.19 through 2.24 show the device and its connections.

2.10.7 Time Analysis

Product development was thus divided into four main stages: the prototyping stage (2.5 weeks), the tooling design stage (3 weeks), the production stage (1.5 weeks), and the postprocessing stage (1.5 weeks). Details of the development processes at each stage are shown in Figure 2.26.

The tooling design and machining of the mold was the most time-consuming aspect of production with a total of about 150 worker-hours. This task included the design of the molds, design review meetings, design iterations, and the time consumed to perform the process planning to guarantee the fabrication of the molds on a three-axis milling machine. The time consumed during this stage was mainly attributed to a combination of the design iterations of the injection molds and the process planning of the mold with a freeform surface.

2.10.8 Conclusions from the Case Study

1. The integrated CAD/CAM environment called CyberCut allowed a series of prototypes to be prepared in quick succession. When the design was converted into a hard aluminum mold, the design team had confidence that the resulting batch of 200 prototypes would be correct.
2. The "clean interfaces" between different stages of the product development process ensured that the final mold was dimensionally correct for the final assembly processes (of PCBs, displays, etc.) so that mold rework was eliminated.
3. Specifically, improving the design and mold machining process, and predicting the suitability of the final design of the mold, enabled a smooth "hand off" to the injection molding company (Metalcast Engineering).
4. Elements shown on the time line could be completed in parallel because of the confidence in the decisions in the design. This reduced the overall elapsed time of the project and avoided the need for expensive rework.

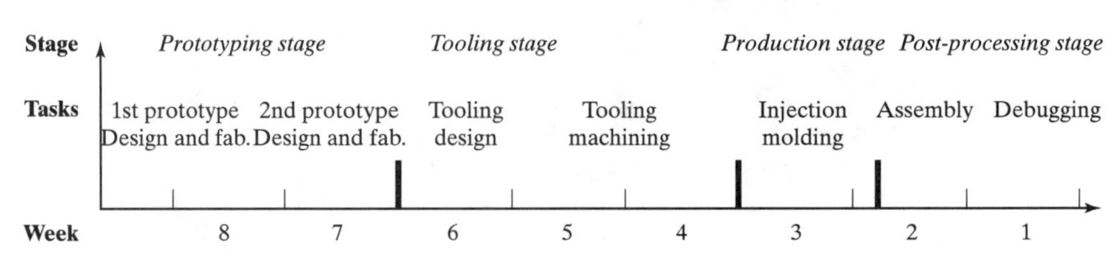

Figure 2.26 Time-to-market in nine weeks.

2.11 INTERACTIVE FURTHER WORK

2.11.1 Background

A manufacturing advisory service can be found at the following URL: <**http://cybercut.berkeley.edu**>. However, "don't try this at home" without a Java-enabled browser! Anything over a Pentium 100 should be satisfactory. For a slower machine, read the instructions for work-arounds. The goal is to provide an automatic way of negotiating the choices in Table 2.2 concerned with process selection.

2.11.2 Tutorial

Introductory tutorials have been set up for use at <**http://cybercut.berkeley.edu/mas2/html/tutorials.html**>. With a big enough screen, put the Applet on one side and the tutorial on the other.

Follow the steps in the tutorial, the key aspects of which are:

- Click on **Process Search.**
- Click on one of the attributes listed in the **Facet** column.
- Specify a value for this attribute underneath the **Facet** column. Note: Some Facets are selected by clicking on the choice beneath Facet List and selecting one of the options that appears. Other Facets are defined through numerical entry and/or the slider bar. Do not use commas (e.g., a large batch size is 30000 not 30,000).
- Go through the 11 Facets using the **Get Info** button for definitions.
- Keep an eye on the **p-rank** column, making sure that tolerances are not over-specified resulting in all zero rankings.
- After **Process Search,** next click on **Material Search.**
- Repeat the same steps for **Material Search.**
- Then, finally, click the **Results Survey** button to combine the two searches.

2.11.3 Some Other Notes

The MAS is not a "Black Box Tool" that blindly chugs numbers. It's more of an "Exploratory Tool." So, if a process seems acceptable but has a zero rank, click on it to find out why it is being rejected. Then in the **f-rank** column, it will show which facets have a zero ranking. Go back to these facets and explore how new readings on the slider bar or choices give a nonzero rank. A rank of unity is the best.

2.11.4 Specific Work

Consider three domestic or industrial products and go through the selection procedure to find out how each product would be made. Bear in mind, given the limits of the project to date, that it should be a *single* item or a component in an assembly rather than the entire multiple-piece assembly (see **Facet descriptions**).

2.12 REVIEW MATERIAL

1. During quality assurance of a drilled hole in an aircraft component, the following hole diameters were measured by an inspector. The hole was originally toleranced (i.e., on the drawing) at 10.000 mm +0.500/−0.000.

1	10.042	11	10.022
2	10.029	12	10.030
3	10.031	13	10.012
4	10.029	14	10.048
5	10.011	15	10.099
6	10.054	16	10.015
7	10.025	17	10.023
8	10.031	18	10.034
9	10.047	19	10.042
10	10.016	20	10.053

- What is the upper specification limit (USL) and lower specification limit (LSL)?
- What is the natural tolerance (NT) for the drilling process used to make these holes? This question first involves calculating the standard deviation, σ.
- Calculate the process capability, C_p.
- Calculate the C_{pk} for the process.
- What does this say about the *real* process capability?
- Using the data above, define the upper control limit (UCL) and lower control limit (LCL). Do any of the holes fall outside of the control limits?

2. The new company, **www. start-up-company.com,** plans to market a plastic car for under $39.95, otherwise it will not sell. The tool steel die costs $500,000 to make. Refer to Figure 2.17 to recall the steps needed to get to this die. Next:

- Plot a graph below showing how this die cost gets amortized over the number of cars sold.
- If the company only had to worry about the die cost, what is a reasonable number of cars to sell?
- If it factors in some of the other costs represented in the graph, make another "ballpark" estimate of what is a reasonable number of cars to sell.
- Using a short, 50-word description, what are the total development costs that **www. start-up-company.com** needs to take into account? Add this to the manufacturing costs above, and then make an estimate of how many cars must be sold. Note: This is not an easy task, nor is there a specific number, but it makes a project group think about how to "price-point" a product (refer to the Hewlett- Packard return map).

CHAPTER 3

Product Design, Computer Aided Design (CAD), and Solid Modeling

3.1 INTRODUCTION

Chapter 3 discusses the general methods of design in engineering. The specific focus is on different methods of computer aided design (CAD). Since CAD tools are constantly being improved by the rapid advances in computer power, the state of the art will change in the mere few months it takes to get this book into print. However, the general principles of CAD and solid modeling should remain relatively unchanged.

Chapter 4 will follow with the creation of an initial *prototype* and the subsequent manufacturing of the first *real part* from the prototype. It may seem an unusual juxtaposition to talk about the geometric operations in feature-based design and the metallurgy of casting in adjacent chapters. Certainly no other books on manufacturing do so. The goals of doing this are to emphasize the *integration* between the various divisions of a manufacturing enterprise, rather than having them split up with the negative consequences of Taylorism described in Chapter 2.

In recent years, CAD design tools have been developed that can be explicitly linked in an electronic sense to rapid prototyping methods and full production. Chapter 4 will describe how to "tessellate" the CAD models shown here in Chapter 3 and then link these data structures to the rastering movements of a laser beam in a stereolithography (SLA) process. Many of these newer prototyping methods did not appear until 1987. Therefore, any manufacturing company that wants to be innovative should be constantly aware of developments in this general field that links CAD to prototyping and then to CAM. In fact, it is also much easier today to link such CAD databases to traditional processes such as machining. Even lost-wax casting—a process that dates back more than 5,000 years to the ancient Koreans and Egyptians—is still in wide use because the links from design to mold making and investment casting have been radically improved by CAD/CAM methods.

3.2 IS THERE A DEFINITION OF DESIGN?

Here are some common definitions of design that are useful but do show a rather wide range of activities. Note that this list begins with a very high level artistic view of the design world. The next definition is also high level but more engineering related. The third definition is concerned with analyzing the inevitable trade-offs that occur as a designer begins to calculate the dimensions of the components in a specific design. The fourth definition is more utilitarian and concerns drawing the details. The next four main sections review these four levels:

- *Art related and high-level:* "Design in any of its forms should be functional, based on a wedding of art and engineering" (W. A. Gropius, founder of the Bauhaus movement).
- *Engineering related and high-level:* "Design is the process of creating a product (hardware, software, or a system) that has not existed heretofore" (Suh, 1990).
- *Engineering related and at the analytical level:* "Design is a decision making process" (Hazelrigg, 1996).
- *Detailed design:* "Design is to make original plans, sketches, patterns, etc." (Webster's Dictionary).

As a result of this range of possibilities, the design community continues to argue passionately about the definition of the word *design*. And since the definition of design can have such a broad scope, each engineering school around the world seems to have a slightly different focus in its teaching methods.

Some engineering schools emphasize the creative aspects of design. This might mean building models or prototypes and thinking about the markets for the next-and-better consumer product. Other schools are more analytical and proceed with detailed VLSI and circuit design for electrical engineers, or traditional stress analysis and fatigue analysis of gears, shafts, and bearings for mechanical engineers.

3.3 THE ARTISTIC, CREATIVE, OR CONCEPTUAL PHASE OF DESIGN

Outside the engineering schools of a university, the word *design* has an even broader scope and includes residential living spaces and art objects. To the conceptual designers of automobiles, living spaces, many consumer products, and clothing, the activity of design is an art form. It conceptualizes creative images that are aesthetically pleasing.

The *artistic,* or *creative,* or *conceptual* phase of design is a multidisciplinary activity that considers many factors. Chapter 2 discussed these factors, which include consumer needs and desires, potential markets, manufacturing costs, the desired quality level, time-to-market, flexibility, and long-term growth.

This "high-level" creative phase of design involves a conceptual mapping between consumer demands and physical objects. Designers at this level prefer to be unconstrained. When interviewed, they say they still prefer pencil sketching on large sheets of paper and argue that it liberates the more creative right side of the brain.

Creative design should be a team effort, and the whiteboard or pencil and paper still seem to be the best tools. In the Appendix it is thus recommended that the design groups create their first initial concepts with pencils on art paper that is 550 × 700 mm (22 × 28 inches) in size. It provides a fluid exchange of ideas and in addition facilitates the round-table team discussion. Once the design team has its vision of the new device, specific design tools such as CAD can then come into play. Also, specific analysis tools can be used for decision making among alternatives and possible optimization of a design.

Computer aided design (CAD) programs have been developed in several formats and with several strategies in mind. Regardless of the strategy, traditional CAD programs have been difficult to leverage as *conceptual design* tools. It is a common experience that using a standard CAD system too early can actually extend the overall design time! The old cliché applies: "If you only have a hammer everything looks like a nail." Since all standard CAD programs have limits, structured statements, and standard software operations, the creative process will be straitjacketed if new ideas have to conform to the program's logic. Conceptual design should be a process that is characterized by brainstorming, market identification, and aesthetics. It requires a fluid approach, and this fluidity may not fit the CAD program's logic.

For the conceptual design of new cars and trucks, some of the newer, very expensive *computer aided styling* tools do encourage the artistic spirit (see Sequin's work at **http://www.cs.berkeley.edu/~sequin/CAFFE/cyberbuild.html**). However, by contrast, the majority of the day-to-day CAD packages (ProEngineer, SDRC, AutoCAD, SolidWorks, etc.) fall short of the fluidity needed for full creative work.

3.4 THE HIGH-LEVEL ENGINEERING PHASE OF DESIGN

3.4.1 Quality Function Deployment (QFD)

QFD is a well-known acronym for quality function deployment (see Hauser and Clausing, 1988). It focuses on the "quality versus cost" aspects of the broader CQDF issues in Chapter 2. A team of marketing specialists begins the process of QFD. Customer groups are subjected to lengthy surveys that ask about their general preferences, likes, and dislikes. These surveys are next transformed into perceived product qualities, then into design features, and finally into manufactured product characteristics (Figure 3.1a). The reader is referred to Compton's book (1997) for a more detailed analysis, but the basic steps are now reviewed.

A list of desirable characteristics is first drawn up by the survey team. For the polo shirt example in Chapter 2, this might be material type, fit, styling, color range, and cost. An individual person being surveyed is then asked to assign a numerical value between 1 and 10 to each of the characteristics, based on the importance to that customer. "Best-in-class" competing shirts from other manufacturers can also be evaluated in the survey form. The survey team then collates the results from many respondents and identifies the most important customer requirements for the product. These requirements are then ranked.

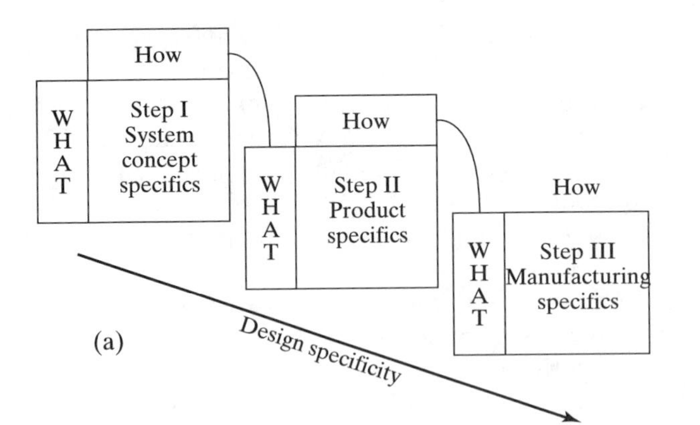

(b)

Customer requirements (Student)	Evaluation of requirements: According to customer ranking (1–10)	Evaluation of requirements: According to best-in-class competitor (1–10)	Relative numerical ranking of requirements	Rank ordering
Easy to read	9	9	18	2
Easy to understand	10	9	19	1
Easy to carry/handle	6	7	13	5
Good aesthetics	4	8	12	5
Content relevant to course	9	6	15	3
Reasonable price	7	7	14	4
Good resale value	5	5	10	8
Freedom from errors	5	6	11	7

Figure 3.1 The mapping process in QFD and Compton's (1997) example for a textbook (from *Engineering Management* by Compton, © 1997. Reprinted by permission of Prentice-Hall, Inc., Upper Saddle River, NJ).

Figure 3.1b is Compton's hypothetical example for a textbook. This ranks the topics from "easy to understand" as number 1, down to "good resale value" as number 8. In another matrix, these customer requirements such as "easy to understand" can then be translated into specific product requirements such as "worked examples," "organization of material," "number of diagrams," and so on. Through a series of similar matrices the manufacturing actions needed to achieve the critical product characteristics are finally identified and launched into the process planning and production departments. For Compton's textbook example, the production department would select layouts, artistic style, and even the weight and dimensions of the book to reflect the desires of the customers.

One can also imagine QFD applied to automobile design and fabrication. Suppose a consumer group—typical of Volvo owners—values crash resistance above all else. In the "product specifics" matrix (center of Figure 3.1a), the "what" down the left side might be "avoid crushing a door from a side impact," and the "how" along the top might be "an attainable stress level." In the "manufacturing specifics" the necessary sheet metal thickness and grade of steel might then emerge along the top. Sheet thickness and grade will also influence the final cost of the product in the manufacturing table.

There is a drawback to the QFD method for the design of mass-produced products such as standard automobiles. QFD assumes (a) the existence of distinct market sectors and (b) that everyone inside each sector has the same ordered set of preferences. QFD *for a group* breaks down if the preferences of the group members are not the same (see Hazelrigg, 1996). For example, the group of consumers who value crash resistance above all else might well lean toward the marketing messages of Volvo. But if that group of people does not like the style, price, or color selections, the crash-resistance characteristic may not be high enough to boost sales. Having said this, QFD has been used successfully in situations where the customer group is small or where there is a limited range of product characteristics (Hauser and Clausing, 1988).

3.4.2 Axiomatic Design

An alternative to QFD has been developed by Suh (1990) and colleagues. It also begins with customer needs. Three steps follow: identifying functional requirements (FR), design parameters (DP), and process variables (PV). In this regard, the mapping shown in Figure 3.2 below is philosophically the same as Figure 3.1a for QFD.

The idea is to map the qualitative desires of the customer to more concrete engineering terms. Functional requirements (FR) might well be the performance issues described for computers (amount of RAM and so forth) and automobiles (acceleration and the like) in the TQM discussion in Section 2.4.7. Design parameters (DP) might then be the specific design parameters for a computer chip or a car engine needed to attain the FR. Process variables (PV) would then be the semiconductor or production-line technologies needed to create the design of the chip or the automobile engine.

To make these mappings converge as the design team moves across Figure 3.2, Suh and colleagues propose two axioms:

1. Maintain the independence of the functional requirements (FRs).
2. Minimize the information content of the design.

The first axiom aims to create a design where the functional elements are decoupled. For example, Suh analyzes a variety of designs for an injection molding machine (see

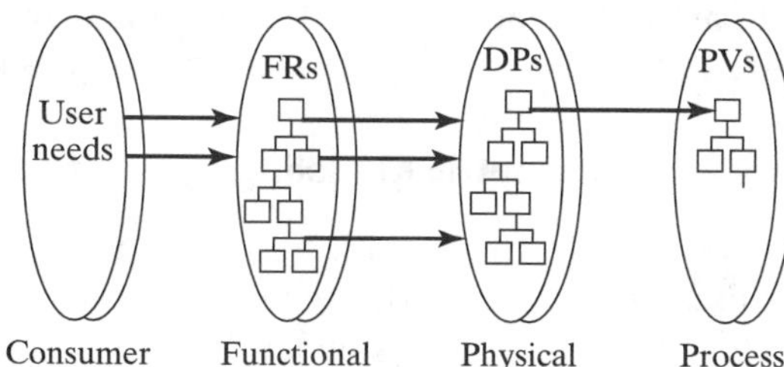

Functional

Process

Figure 3.2 The framework for axiomatic design (from *The Principles of Design* by Nam P. Suh, © 1990 by Oxford University Press, Inc.).

Suh, 1990, 72–78). In some designs, three functional requirements (FRs)—the melting rate of the plastic, the flow rate, and the pressure rise in the extruder—are all affected by just one design parameter (DP), the rotational speed of the screw (the reader might glance ahead to Chapter 8 to see the screw-machine operation). This coupled design is criticized because it will be impossible to regulate any of the three functional requirements (FRs) independently. As a result, alternative designs are explored where individual functional requirements (FRs) can be controlled independently of a screw mechanism.

The mathematical way to create such independence is to set up a "design matrix," **A,** where (FR) = **A** (DP). The matrix elements define the nature of the relationship between each of the functional requirements (FR_i) and each design parameter (DP_j). The individual elements of the matrix are given by

$$\mathbf{A}_{ij} = \frac{\delta FR_i}{\delta DP_j} \tag{3.1}$$

A "decoupled" product design will be achieved if **A** is a diagonal matrix. In other words, **A** should be a square matrix, and its nonzero elements should appear only on the main diagonal. There should be zeros elsewhere to exclude any potential coupling. Various designs can thus be analyzed. The aim is to arrive at a situation where any specific function of the device is related *to one and only one* design parameter. Speaking colloquially, if one were to "tweak" that design parameter, it should only influence that one functional requirement and not cascade into other functions of the device.

Intuitively, the second axiom is quickly grasped. It advises the designer to create simple subcomponents and devices. In fact this is also one of the key ideas in the Boothroyd and Dewhurst (1999) DFM/A software described in Section 3.5.2. They advocate simplifying the shape of individual components in an assembly and simplifying the fit of one component with another, for example, using press fits where possible rather than screws. This is further described in Chapter 8 with reference to the redesign of IBM's ProPrinter.

3.5 THE ANALYTICAL PHASE OF DESIGN

Soon after the conceptual design and the high-level analysis are in place, detail designs should be pinned down. Although these can be done with simple "desktop CAD/drawing systems," most of today's CAD programs encourage their users to add software modules to their basic CAD package. These include, but are not limited to, (a) constraint-based design and parametric modeling, (b) design for manufacture/assembly/environment (DFM/A/E) scoring, and (c) finite element analysis (FEA).

3.5.1 Constraint-Based Design and Parametric Modeling

Rather than build up a model with specific dimensions, it is often useful to create constraints between certain features or lines. Consider the simple cover plate to a regular domestic light switch. Using the U.S. standard dimensions as an example, the outer dimensions of the plate are usually 112 × 68 millimeters (4.5 × 2.72 inches). But in fact, these dimensions do not have to be constrained; for example, in toy stores one can find light switch covers where the outer shape is a popular cartoon character. However, in the United States, there is a central rectangular slot for the actual light switch plus two holes for the screws that attach the plate to a standard switch box. The dimensions of the central slot and the positions of the screw holes must always be constrained to fit the switch box. In summary, an artistic redesign of a light switch cover can thus be wildly imaginative on the outside contours—but it must be strictly constrained on these inner, more engineering-like dimensions.

Another example from Shah and Mantyla (1995) is shown in Figure 3.3:

- Line 3 is parallel to line 5.
- Line 2 is a circular arc, tangent to lines 1 and 3.
- Line 4 is oriented at α to line 3.
- Line 1 is horizontal and length *b*.
- Line 5 is perpendicular to 1 and length *a*.

A more sophisticated development of such constraints is parametric modeling. This procedure allows fast scaling of an object into unique variations from one original.

Consider a simple aluminum can for soft drinks (Figure 3.4). Rather than specify that the dimensions are 120 millimeters (4.75 inches) high and 62 millimeters (2.5 inches) in diameter the ratio 4.75/2.5 = 1.9 might be used. The height and diameter are then described in proportion to each other without specific dimensions being used. Then the model can be scaled up or down without having to numerically respecify each of the new dimensions.

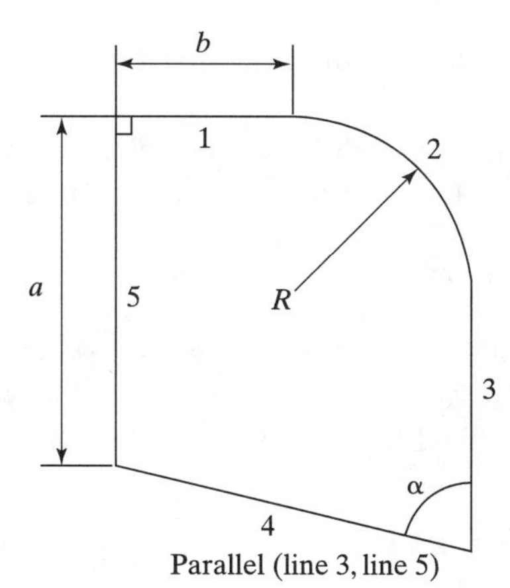

- Line 3 is parallel to line 5
- Line 2 is a circular arc, tangent to lines 1 and 3
- Line 4 is oriented at α to line 3
- Line 1 is horizontal and length *b*
- Line 5 is perpendicular to 1 and length *a*

Figure 3.3 Example parametric definitions (from *Parameter and Feature-Based CAD/CAM,* Jami Shah and M. Mantyla, © 1995. Reprinted by permission of John Wiley & Sons, Inc.).

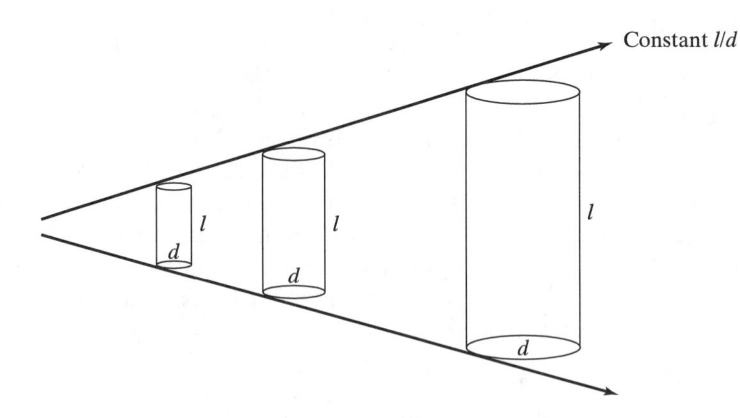

Figure 3.4 Parametric scaling of a soft drink can.

3.5.2 Design for Assembly, Manufacturing, and the Environment (DFA/M/E)

DFA/M/E software packages give manufacturing and environmental feedback to the designer or design team. Ideally, designs can be modified early in the design process to improve manufacturability and decrease environmental impact. New techniques to analyze a particular design for its manufacturability include the design for manufacturing and assembly (DMFA) software tools by Boothroyd and Dewhurst (1999). Theirs is a commercialized product that designers can access by CD or the Web. A suite of tools is available that contain, for example, DFM software for machining, DFM software for sheet metalworking, DFE software to assess environmental impacts, and their best-known DFA module for evaluating assembly.

The DFA for assembly module involves two key ideas:

- The quality of individual subcomponents must be high. Also, their number must be reduced as much as feasible.
- Assembly operations must be as simple as possible. For example, factory layouts should be orderly, the shape of individual components should be simple, design features should simplify the assembly of one component with another, and assembly operations should not fight gravity.

Their DFM for machining module "Machining for Windows" assists a designer with the following issues: developing operation and process plans, obtaining cost estimates at the earliest stages of conceptual design, developing quotations, and planning for production.

3.5.3 Analysis and Decision-Based Design

Finite element analysis (FEA) allows the optimization of material use and performance in a quantitative and automatic manner, both of which are fundamental to the detail design process. In many cases this detailed engineering analysis provides technical innovations, but it rarely influences the original high-level concept.

Despite the apparent rigor of FEA calculations, they should be interpreted with caution. In particular, the common "safety factors" that have been developed over many decades for design work should still be applied. This is because there is uncertainty in the engineering materials that are used today, and also in the boundary conditions that are used for the finite element model. Siddall's (1970) simple diagram succinctly captures this concept of "uncertainty" as it relates to decision-based engineering design (Hazelrigg, 1996). In general the designer does not have the luxury of measured, or a priori known, values of stress due to loading; instead the designer can only calculate stress based on a probability function. Similarly, given today's steelmaking and other production methods, the designer does not know precise values for the yield strength and similar properties; again there is a probability function. The diagram shows the great dangers that can arise when the "tails" of the two probability functions overlap in the "failure zone."

A key question arises: What does a designer do to address the "failure zone"? The answer critically depends on the preferences, or values, of the designer. For routine consumer products the goal of a design is to make money, and more is better (Hazelrigg, 1996).

With "making money" as the main objective, the designer might choose material properties in such a way that "an occasional failure is acceptable"; for example, 3.4 parts per million might be acceptable (Figure 2.15). In such circumstances, the consumer might be temporarily disappointed by a product failure. But if a fast, courteous warranty procedure is put in place by the company, then in the end the design utility, **u,** will be acceptable. In other words, a consumer product will not be "overdesigned and overcostly" from choosing material properties that are "oversafe" so that no failures will occur.

By sharp contrast, for the aerospace and aircraft industry, a designer should reevaluate the goal of making money and balance it with "stability and reliability." In the extreme case of a nuclear weapon, mentioned at the beginning of Chapter 2, society demands "reliability" rather than "cost effectiveness" as the main design goal. In Figure 3.5, this means that the two curves should move apart on the x axis leaving no possibility for a failure zone.

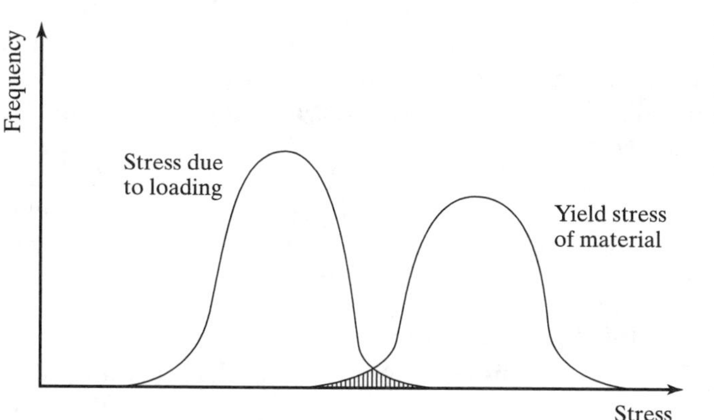

Figure 3.5 A frequency distribution for predicting the strength of a design. It shows the inevitable spread in material properties on the right and loading stresses on the left (courtesy of James Siddall, 1970).

With these thoughts in mind the designer can now begin to create a specific design with specific dimensions.

3.6 THE DETAILED PHASE OF DESIGN

In the "old days," the creative designers would pass their drawings down to the drafting office, where pencil and paper were used to document the detailed, formal drawings with appropriate tolerances. With the advent of CAD stations, it was obvious that much of this pencil-and-paper work could be replaced with pull-down computer menus and the drag-and-drop positioning from the mouse. The detailed CAD methods that are available fall into several categories, including wire frame methods and solid-geometry techniques. Chapter 3 presents some tutorials on these basic methods. CAD drawings are created step by step to show the construction techniques. However, reading these tutorials, or buying the book *Teach Yourself AutoCAD,* is no substitute for going to a CAD workstation, logging in, and persevering through the tedious little commands that eventually create a nice isometric view and the three principal orthogonal views of a fairly standard object. Detailed design using CAD is like driving a car: no amount of tutorials in high school drivers' education classes can prepare you for the actual experience of cornering, even at 10 mph.

For *detail design* (as opposed to conceptual design) the utility of traditional CAD programs as a communication tool is undeniable. Drawings remain consistent, are repeatable, and conform to standards. Designers and draftspersons can compare component designs to check for clearance, flushness, perpendicularity, and other fit-related qualities. Although this process is often tedious, it is a great improvement over manual methods. CAD has also been useful in articulating precise details so that designers can better understand their implications during each design iteration.

Most contemporary CAD tools have both wire frame and solid modeling capabilities. Solid modeling provides a sense of form usually lacking in wire frames. Solid modeling is a more intuitive system, and the process of constructing a model also provides insights into the manufacture (particularly the machining) of parts and assemblies. Solid modeling by both constructive solid geometry (CSG) and destructive solid geometry (DSG) is described later in this chapter.

3.7 THREE TUTORIALS: AN OVERVIEW

The object pictured in Figure 3.6 is an example of a prototype of a joystick for the user of a virtual reality (VR) environment. The joystick has a somewhat heavy mass as its base. As the mass is moved in three-dimensional space, accelerometers in the base adjust the virtual environment appropriately. Information is transferred to the computer via a parallel port.

With the exception of the handle, the device is fabricated from components with rather simple geometries. The joystick will now be designed with three different CAD methods:

- Wire frame using the AutoCAD package (Section 3.8)
- Constructive solid geometry (CSG) using the AutoCAD package (Section 3.10)
- Destructive solid geometry (DSG) using the SolidWorks package (Section 3.11)

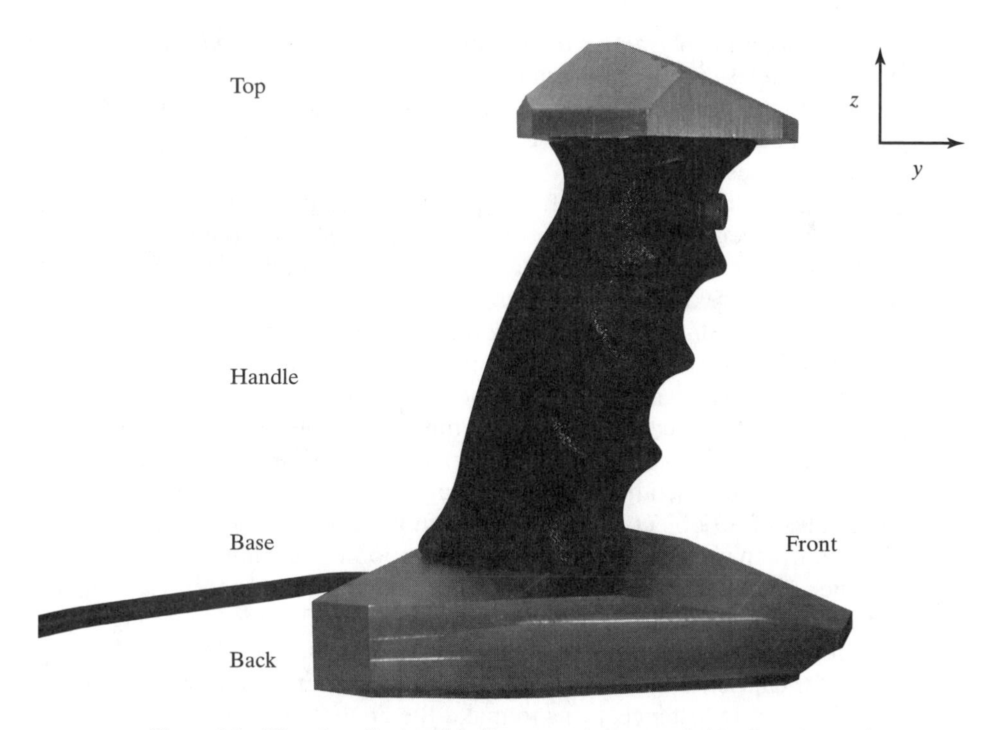

Figure 3.6 Virtual reality joystick (from a student group led by Ryan Inouye).

The three tutorials show the characteristics of each CAD technique. In addition the case study at the end of the chapter shows the features of the SDRC package and some aspects of parametric based design.[1]

3.8 FIRST TUTORIAL: WIRE FRAME CONSTRUCTION

Simple wire frame CAD systems use basic mathematical and computer graphic technologies. Wire frame programs begin by allowing users to choose points from a local or relative reference frame (the choice of the local frame is often arbitrary). These points are then mapped onto a global reference frame. Finally, lines are drawn between points.

The final image is therefore a connection of lines that may be hard to view clearly. For example, (a) lines that should be hidden, perhaps representing the back of an object, will remain visible during construction, and (b) no helpful shading of the front face will be possible. A quick glance ahead to the figures depicting wire frame constructions shows these difficulties with visualization.

[1]The use of AutoCAD, SolidWorks, and SDRC is not intended to be an endorsement of these products or a deliberate exclusion of the other CAD products listed in the URLs at the end of the chapter. Furthermore, all CAD packages are capable of wire frame, CSG, and DSG. Even the cheaper systems can do some parametric design. A variety of CAD systems was chosen for the chapter to deliberately show some product mix rather than choosing just one.

Although *wire frame* models lack the advantages of *solid* representations, they are nevertheless useful. The inner workings of wire frame programs are also simple to understand and thus adapt for special user-generated purposes. Also, the computer's calculations mirror those commonly found in linear algebra, dynamics, and robotics classes.

The fact that most wire frame programs require less powerful computers has until very recently been the most compelling reason for their use. However, the inevitable progress in semiconductor design and manufacturing (see Chapter 5 and associated figures) and the availability of cheaper, more powerful machines now bring solid modeling tools to the average desktop machine.

For illustrative purposes, consider the base of the joystick. By selecting a number of desired lengths for the base, a series of lines can be developed to outline the object. In CAD constructions of symmetrical objects it is quite common to draw only half of the object and then *mirror* it across a central plane. This general approach will be used here. The advantage is that the overall time taken is less, and both sides of an object are identical from a symmetry viewpoint. Figure 3.7 is the right half, plan view of the base. The halfway point of the line segment that represents the back of the base will be the mirroring plane origin, **a.** Therefore, point **a** is at coordinate $x = 0, y = 0, z = 0$, or more simply (0,0,0). The line beginning at point **a** was drawn with the *line* command. The line travels from the middle of the back bottom edge of the base to the end of that edge. The line then travels toward the front of the base for a short distance, turns to make the angle along the right-side bottom edge of the base to **r,** and finally moves across the front bottom edge.

Figures 3.7 and 3.8 provide the results of some of these initial suggestions on how big the base of the object should be. Figure 3.7 is the top view of the external edges of half of the base. Figure 3.8 is an isometric view of the same lines. It provides insights into the three-dimensional nature of the lines.

The viewpoint in Figure 3.8 was accomplished with the *vpoint* command in AutoCAD. The setting for the viewpoint is (1,-1,1), or positive in the x direction, neg-

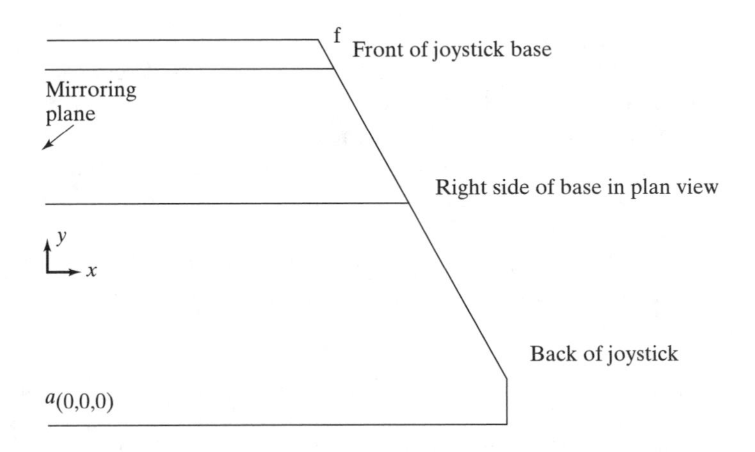

Figure 3.7 Top view of one-half of the base. The line segment contains point **a** in its center.

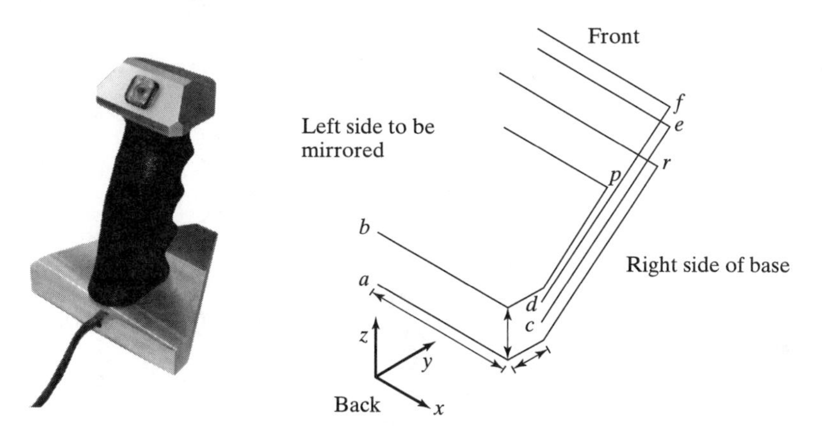

Figure 3.8 Isometric view of one-half of the base (viewing from the back of the joystick where the wires can be seen in the earlier photograph).

ative in the y direction, and positive in the z direction, with each axis held to 1:1 proportions (i.e., isometric). While making drawings, it is a good idea to experiment with the overall view by changing the viewpoint proportions to see the effects of foreshortening.

The line starting at point **b** was drawn in a similar manner according to the dimensions for the top surface of the object. All points on the **b** line have a constant, z dimension above the points on the **a** line. The lines originating at **c** and **d** follow the same x-y profile. Each of these lines has its own, constant z coordinate.

Points **a, b, c,** and **d** were specified with Cartesian coordinates. The choice of coordinates was dictated by the desired object size and ease of calculation. The *points* at **c** and **d** were first determined with Cartesian coordinates. However, the *lines* that originate at **c** and **d,** going along the angled portion and the front edge, have been determined using polar coordinates. Points **e** and **f** were determined by a fixed distance and an angle from points **c** and **d.** Temporary construction lines establish all the points for the shape **df** and **ce,** but are later removed. They are there temporarily to validate the correct shape of the slanting surface along **ref.** The outline of the base exterior begins to become clear in Figure 3.9. The *mirror* command was used to duplicate the objects across the y-z plane at $x = 0$. Figure 3.9 contains some overhangs and extraneous line segments. The overhangs can be trimmed with the *trim* command to leave segments **ce, df, ij,** and **gh** as the required edges. This is done by first clicking on the excess segments **jl, hk, fm,** and **en** as the objects to trim. The *erase* command then leaves segments **ce, df, gh, ij, ej,** and **hf.**

The outline of the object is revealed by constructing lines between points **o** and **h, h** and **j, j** and **q, p** and **f, f** and **e,** and **e** and **r,** all with the *line* command. For example, lines along the z direction (between **e** and **f, j** and **h,** ...) at the intersection of planes complete the general outline of the object. Figure 3.10 presents the results thus far.

With the general outline of the object described, the generation of inner details and cavities becomes the next significant step.

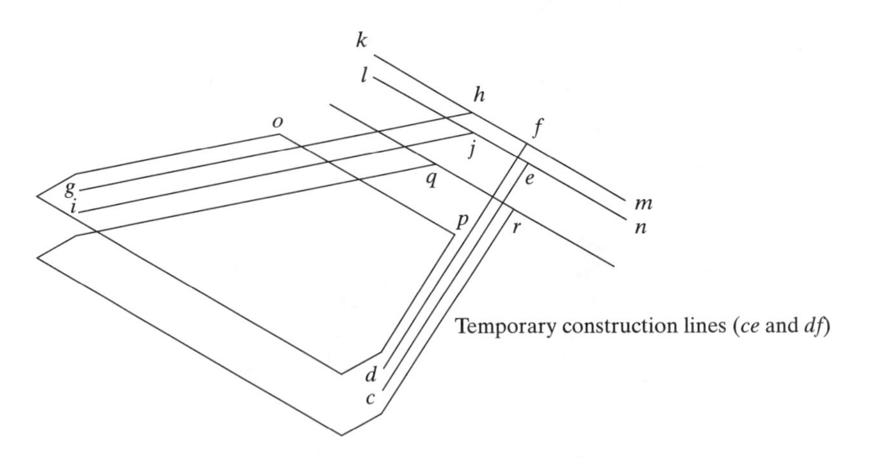

Figure 3.9 Outline of the base exterior before the trimming process.

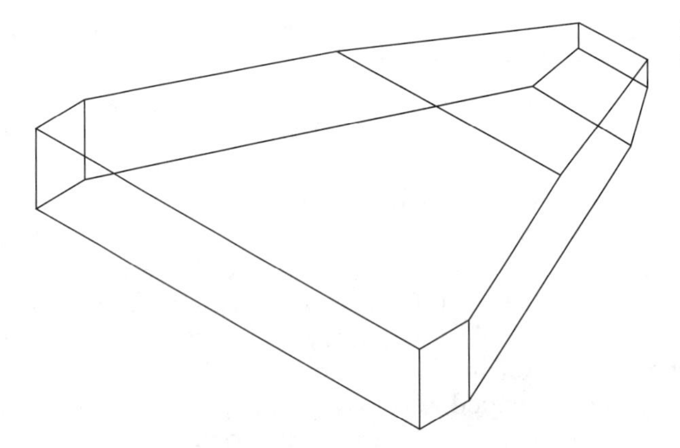

Figure 3.10 Outline of the base exterior after the trimming process.

A good place to start is the shallow rounded pocket in the bottom face, because many consumer electronics products, including telephones, have a cover plate on the bottom. This is more like a shallow recess that the cover plate will eventually nestle into. The cover plate will close over the larger internal cavity that houses the printed circuit boards and wires. The plate generally follows the contour of the bottom surface of the object. For the joystick, the base plate itself is a flat trapezoid with rounded corners.

A shallow rounded pocket will thus be drawn in the next few images to accommodate the plate. This task begins in Figure 3.11. First, the straight lines of a trapezoid are drawn that remain equidistant from the outside edges of the bottom. These will then be connected, next rounded, and then copied and repeated for the depth of the cavity.

In the first construction step, the line **i** is drawn parallel to the base, at a small distance **w** in the *y* direction from **a.** This line travels some distance past the outline

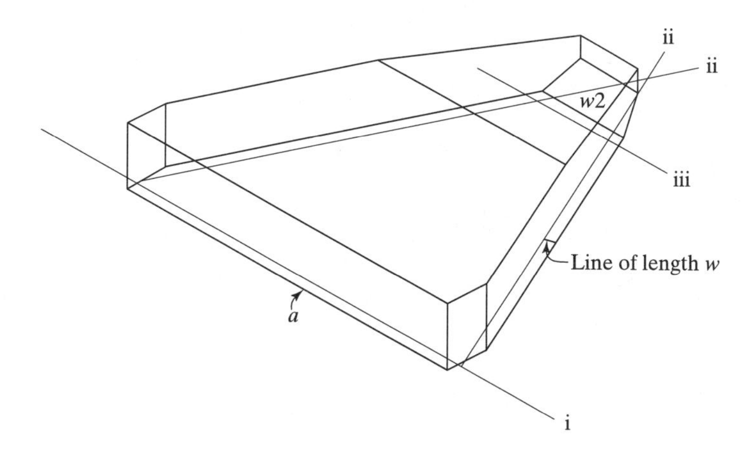

Figure 3.11 Initial formation of the cavity in the base exterior.

of the object on each side. Line **ii** begins by placing a tiny line of length **w** perpendicular to the other edge. The full extent of line **ii** is then constructed from the end point of this short construction line **w.** It goes past the front of the object, parallel to the angled bottom edge. The line is completed by extending it with the *extend* command such that it intersects line **i.** The extend command finds the intersection automatically. Line **iii** is constructed in a similar manner from the short construction line **w2** from the front face. Once one of the lines marked **"ii"** is constructed on the right side of the object, it is mirrored to the left, across the *y* axis, to form the image.

The next step is to trim lines **i, ii,** and **iii** so that only the trapezoid remains. The corners of the trapezoid are then rounded with the *fillet* command. The fillet command in AutoCAD can be quite confusing to a new user because the default fillet radius is always set to zero. The procedure is to enter the fillet command, hit the **"r"** key (for radius) when the dialog line appears in the command area, and then enter the desired fillet radius. When the fillet command is subsequently called, it will contain the new radius. The fillet command has been used on each corner of the trapezoid, selecting both intersecting lines at each corner. A consequence of using the fillet command is that a new line segment is created for the curved region of each fillet. The trapezoid has thus been transformed from four individual lines to eight segments: four lines plus four fillets.

At this point it is convenient to *join all eight segments comprising the trapezoidal pocket into a single line.* This is done with the *pedit* (polyedit) command. This procedure begins by typing pedit and hitting enter. When prompted, a segment of the rounded trapezoid is selected. AutoCAD says it is not a polyline and asks if the designer would like to make it one. To make it one, the designer can simply enter **y[es].** A set of polyline options is then presented, and the best choice is **"j"** for join. Each segment to be included in the polyline that caps the top of the handle is then selected, remembering there are eight segments in this case. Pressing "enter" joins all the segments into one entity. Joined polylines can be separated into their original segments by employing the *explode* command at any time.

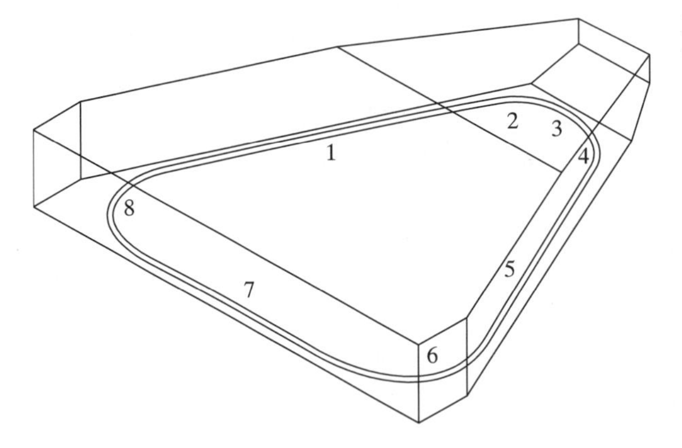

Figure 3.12 Completion of the cavity in the base exterior.

To create the depth of the cavity on the bottom, the *copy* command is used. The existing polyline is copied and repeated a short distance along the positive z axis. This gives two polylines, a small distance z apart. The distance z specifies the base plate cavity depth. The preceding procedures are used to generate Figure 3.12.

The outline of the base cavity could have been drawn with the *rectang* and/or *polygon* commands. These commands automatically make polylines; however, they are not preferred because individual segments might be edited out from the original polylines when the final smoothing command is applied. The only way to avoid this is to create the polyline from a set of line segments first. In other words, in most cases it is a simpler process to start by drawing line segments and then joining them into polylines.

There are additional cavities within the object that house the accelerometers and circuitry, as shown in Figure 3.13.

The process begins by drawing a series of lines in the shape of a large rectangle, a much smaller rectangle, and a trapezoid. These are shown in Figure 3.13 going in a "northeast" direction. First constructions are best done on the lower level with later extrusions in $+z$.

The corners of the large rectangle and trapezoid are then filleted (the radius needs to be reset to a smaller value). The common edge of the large rectangle and the very small rectangle is trimmed, leaving the open space between **a** and **c.** The common edge between the trapezoid and the smaller rectangle is also removed, leaving the open space between **b** and **d,** and the segments are joined as a polyline. The outline of the rectangles is then copied to a certain distance in the z direction, thus creating a pocket. Next, lines are drawn at the vertical intersecting edges—**a, b, c,** and **d**—to show depth. The trapezoid is also copied, but to a greater distance in the z direction. Progress thus far is illustrated in Figure 3.13, which shows the main block from earlier drawings, the shallow recess for the base plate, the medium depth pocket on the left, and the deeper trapezoidal pocket to the right. Wires can be run between the electronic components in each pocket through the small slot **abcd** in between the rectangular and trapezoidal pockets.

A few more details remain. The gripping handle is attached to the base with two fasteners. The fasteners come up through the base. They bolt into tapped holes

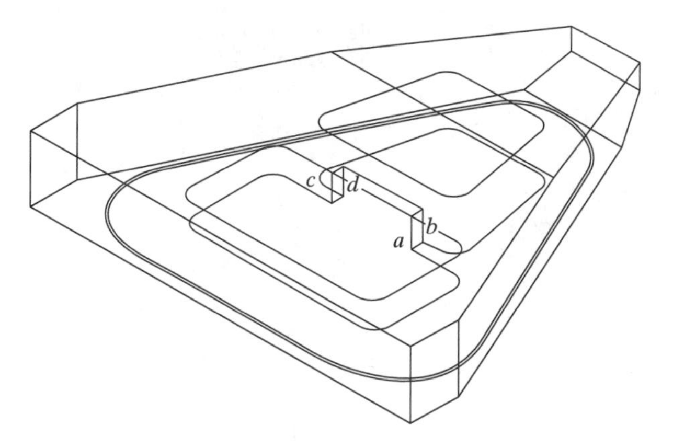

Figure 3.13 VR joystick base with some interior cavities.

in the bottom of the handle. Therefore, so as not to interfere with the electronics inside the inner cavities, it is necessary to provide large countersinks for the bolt heads.

The counter-bore for the large bolt head is represented by a large circle, and the hole for the screw thread part by a smaller concentric one. Thus, originally, two circles are drawn, one for the counter-bore that will accommodate the bolt head and one for the through-hole that will take the screw thread part.

Circles are created with the *circle* command. The user is prompted for the center and radius of the circle. The circles are then copied to the appropriate positions, namely, where a cylinder intersects a material surface. The dimension is known from the distance **ab** in the earlier figures. Thus in Figure 3.14, the upper smaller circles are coincident with the top surface, meaning they are through-holes for the screw threads.

Figure 3.14 also illustrates that circle **y** intersects one of the vertical surfaces of the rectangular pocket. Circle **y** was modified by trimming the entire right-side half. Such manual operations are a major disadvantage of wire frame modeling. Ambiguous drawings can result if they are not attended to, which may result in costly mistakes during manufacture.

Figure 3.14 is now completed to the extent required for this illustration. Screw holes for the base plate and a hole for the parallel port cable have been omitted here, but are eventually required.

To provide greater clarity, internal lines can be changed to hidden lines. With hidden lines, the object meets all the criteria required of typical multiview drawings.

Each of the six possible orthogonal viewpoints can be obtained with the *vpoint* command, but some hidden lines may have to be manually adjusted as the viewpoint changes. From the isometric viewpoint (Figure 3.14), however, the object is somewhat difficult to read. The cylinders (i.e., through-holes) removed from the base are not clearly characterized, and vertical edges at the fillets are not included. It is difficult to see depth and surfaces without concentrating.

Development of the wire frame model has made no reference to manufacturability or even physical feasibility. The drawing is simply a set of lines and curves located in three-dimensional space. Other AutoCAD commands could be applied at this

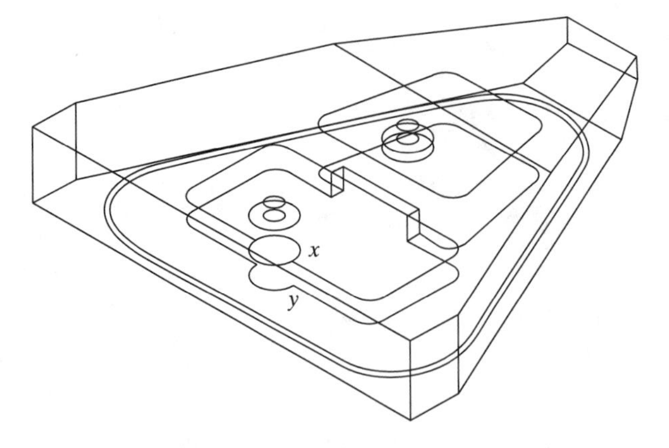

Figure 3.14 Complete 3-D drawing of the VR joystick base.

point to the wire frame model to render the surfaces. This rendering makes the object appear real to the human viewing the screen and is certainly an improved pictorial representation. However, since we are still working only with a wire frame, *this version of rendering does not make the computer understand the surfaces.*

Just to emphasize this last point: all the *understanding* so far is in the eye and the brain of the beholder—the human CAD designer. Actually, the human designer could have drawn an "Escher-like art image" that would be impossible to manufacture, and the *wire frame model* would have happily accepted it!

3.9 SOLID MODELING OVERVIEW

3.9.1 Introduction

In contrast to the previous wire frame methods, *solid modeling* creates objects that the computer "understands to be real solid bodies." There are several textbooks devoted to the formal aspects of these topics. For example, the details are comprehensively described in *Geometric and Solid Modeling* by Hoffmann (1989) and in *Computer Graphics: Principles and Practices* by Foley, van Dam, Feiner, and Hughes (1992). The notes below on Boolean set operations, b-rep, and CSG are arranged in a similar order as in these texts, and the permission to use diagrams in a modified form is gratefully acknowledged.

In the *wire frame* tutorial (Section 3.8), the person sitting at the CAD system's user interface mostly *clicks on points and connects them with lines* to create the joystick in Figure 3.14. By contrast, in the solid modeling tutorials, the person sitting at the CAD system's user interface *constructs* the joystick by adding, subtracting, or intersecting individual bodies (Section 3.10) or *destructs* the joystick by starting with a large block and subtracting smaller bodies (Section 3.11). Such construction or destruction operations combine objects to make new ones. This is like assembling or disassembling "Lego blocks" of different shapes and sizes. The operations are done with a modified version of *ordinary Boolean set operators,* called *regularized Boolean set operators.*

3.9.2 Regularized Boolean Set Operations

The goal is to carry out the constructions without creating superfluous or missing material. For example, careful examination of the second illustration of Figure 3.15 shows an extra *dangling* plane that is common to both solids after *ordinary Boolean intersection.* This is the undesirable dangling tab sticking up in the second illustration. The basic mathematics of ordinary Boolean set operators do not provide a perfect solid in all cases: they can create dangling lower dimensional objects.

To avoid such dangling features, it is necessary to use the regularized Boolean set operators for construction (Requicha, 1977). These are mathematically defined so that operations on solids always create closed solids with no dangling points, lines, or planes. The regularized operators are written with superscript stars* as:

- Regularized union operator = $\cup^*$
- Regularized intersection operator = $\cap^*$
- Regularized difference operator = $-^*$

Regularized operations are defined as follows:

$$(A\ \mathbf{op}^*\ B) = \text{closure (interior } (A\ \mathbf{op}\ B))$$

In the above equation, **op** is one of $\cup$, $\cap$, or $-$. For Figure 3.15 the **op*** is the intersection operator ($\cap^*$).

What does the expression mean in practice when computing $A \cap^* B$?

- In Step 1, Figure 3.15, the *ordinary* Boolean intersection, $A \cap B$, is computed to yield the volume of the object plus any dangling or lower dimensional faces, edges, or vertices.
- In Step 2, the set of points that is the *interior* space of $(A \cap B)$ = interior $(A) \cap$ interior (B) is found.
- In Step 3, the boundary points of $(A \cap \mathrm{B})$ are added (these will also be faces, edges, and vertices, but just those adjacent to any interior points of the intersection of A and B). The last cube in Figure 3.15 exhibits *closure.*
- Regularization eliminates any dangling lower dimensional objects that are not adjacent to any of the interior points of the new volume object, but keeps the set of points that *are* on the boundary.

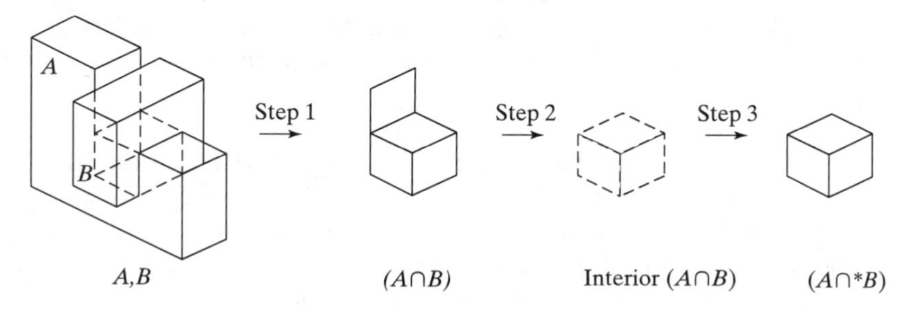

Figure 3.15 Intersection of two blocks A and B (from *Geometric and Solid Modeling: An Introduction* by Christoph M. Hoffman, © 1989 Morgan Kaufmann Publishers).

Figure 3.15 illustrates the problem with ordinary operators: the intersection of the two objects *A* and *B* contains the intersection of the interior and all the boundary of one object, with the interior and all the boundary of the other one. In Figure 3.15, object *A* is L-shaped, whereas object *B* is a smaller rectangular block. The *ordinary operator* is inclusive of all boundary intersections, therefore leaving the dangling face. By contrast the *regularized operator* contains (a) the intersection of the interiors, (b) the intersection of the interior of each of *A* and *B* with the boundary of the other, but (c) *only a subset of the intersection of their boundaries.*

Hoffman's example shows an external dangling face. For further clarification, consider Foley and associates' two-dimensional example in Figure 3.16, which shows, in cross section, an internal dangling edge **CD** that is removed when the regularization is done. In this figure, going from (a) to (b), the lighter object on the left is intersected with the darker, offset block on the right. The question is: Which parts of the intersection of their boundaries (between *A* and *D* in Figure 3.16c) are in the *regularized intersection* operation?

The criterion is that boundary-boundary intersections are included in the regularized Boolean intersection if and only if the interiors of both objects lie on the same side of this piece of the shared boundary. The reason is that since both of the objects have their interior regions on the same side of that part of the boundary (**AB** in the figure), the boundary must be included as well to maintain closure. In other words, when both objects have their interior regions on the same side of that part of the boundary, the interior of their intersection will include the same region with the same piece of boundary.

Next, those parts of one object's boundary that intersect with the other object's interior must be included. Thus, the small section **BC** was already part of the interior of the darker object. Now it is merged with the boundary of the lighter object and is included in the regularization as a part of the new object. By contrast, if the interiors of the original objects are on opposite sides of the shared boundary such as **CD,** that piece is excluded from the regularized operation. In this case, none of the interior points adjacent to the boundary are included in the intersection. Therefore, **CD** is not adjacent to any interior points of the resulting object and is not included after regularization.

Boundary points are defined as those points whose distance from the object and the object's complement is zero. However, boundary points need not necessarily be part of the object. A closed set contains all its boundary points, whereas an open set contains none. The union of a set with the set of its boundary points is known as the set's *closure.* The *boundary* of a closed set is the set of its boundary points. The *interior* consists of all the set's other points. The *regularization* of a set is defined as the closure of the set's interior points (Foley et al., 1992).

In summary, a *regular* set contains no boundary points that are not immediately adjacent to some interior point. Considering the closed cube of Step 3 in Figure 3.15, its top-left-back edge is adjacent to the interior and is included. But any of the points that were vertically above that edge and part of the dangling tab sticking up in Step 2 are not included since those points are not adjacent to any interior regions of the new object.

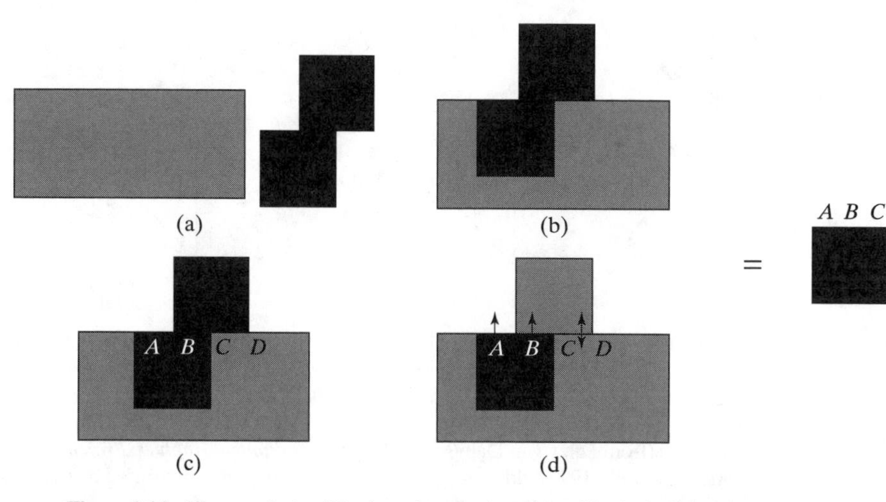

Figure 3.16 The *regularized* Boolean *intersection* of two blocks includes the following parts of the boundary: (1) section *AB*—because the interiors of both objects lie on the same side of it; (2) section *BC*—because it was already in the interior of one of the objects and interior-boundary intersection are included; but (3) not section *CD*—because the objects lie on opposite sides (from Foley/van Dam/Feiner/Hughes, *Computer Graphics: Principles and Practice*. © 1996, 1990 Addison-Wesley Publishing Company. Reprinted by permission of Addison Wesley Longman, Inc.).

3.9.3 The Boundary Representation Method

Several methods are available to represent solid objects: primitive instancing, sweeps, boundary representations, spatial partitioning, and constructive solid geometry (Foley et al., 1992). Step 3 in Figure 3.15 could represent a "cubic block of solid wood"; however, so far, no physical dimensions for the cube or its global position in space have been specified. Boundary representations, or *b-reps,* describe such an object in terms of its *surface boundaries.* The b-rep could be a list of the cube's faces, each represented by a list of vertex coordinates. The desirable properties needed to represent solids are described by Requicha (1980). Resolving *ambiguity* is one example in Requicha's list. Thus even for the simple cube it is important to list these vertices in such a way as to distinguish the outside and inside of the cube. To do this it has become customary to use the right-hand rule and list the vertices in a counter-clockwise (ccw) order as seen from the outside of the object.

Also, it is usual to only support solids whose boundaries are *2-manifolds.* The neighborhood of every point of a 2-manifold is *homeomorphic* (or topologically equivalent) to a two-dimensional disc. Figure 3.17a shows a point and its surrounding neighborhood on the surface of a triangular block. Whether the point is on a face, diagram (a), or on the edge, diagram (b), the neighborhood is still a disc. However, if there were an adjoined block as shown in Figure 3.17c, the neighborhood at the joint could be interpreted as two discs rather than one. Figure 3.17c is therefore not a 2-manifold.

There are several specific ways to store a b-rep. As mentioned above, the simplest possibility is to list all of the faces with their vertices. One drawback is that it is

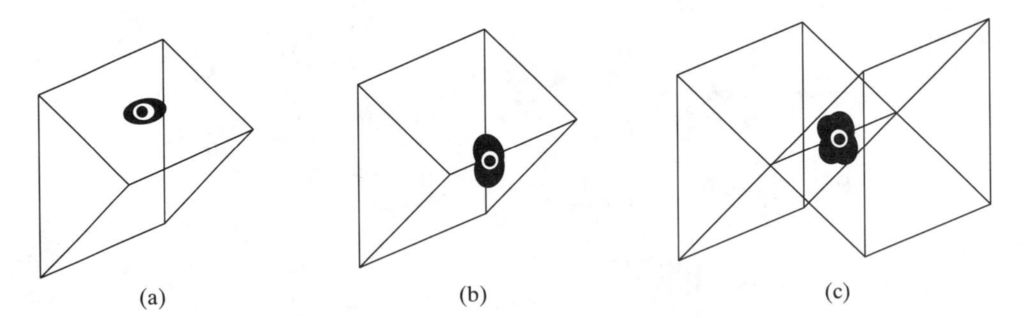

Figure 3.17 B-rep systems usually support only 2-manifolds. Every point on the surface is surrounded by a disc as shown in (a) and (b). The joined edge in (c) creates ambiguity for a topological disc. Object (c) is therefore not 2-manifold because the neighborhood of the joined edge in (c) is topologically equivalent to two discs (from Foley/van Dam/Feiner/Hughes, *Computer Graphics: Principles and Practice*. © 1996, 1990 Addison-Wesley Publishing Company. Reprinted by permission of Addison Wesley Longman, Inc.).

difficult to derive *adjacency* relationships from such a representation. For example, if it is desirable to know all the faces that are incident to a particular vertex, then it would be necessary to search through every single face in the desired object. Thus, other b-rep methods have been introduced to reduce the cost of such adjacency computation.

Baumgart's (1972, 1975) *winged-edge* data structure is one example of a b-rep aimed at compact representation and minimized computation costs. The winged-edge structure is shown in Figure 3.18. It is used to show that *edge e* is used by both faces *A* and *B* in the pyramid of Figure 3.19. A *right-hand rule* (ccw) is always used in CAD to keep track of the vertices, edges, and faces. For face *A* this gives *edge a* as the preceding edge and *edge d* as the succeeding edge. The right-hand rule for face *B* gives *edge c* as the preceding edge and *edge b* as the succeeding one. Table 3.1 gives other data for the pyramid. This is how the data are stored in the computer for compactness and accessibility. It also captures the geometric relationships in a brief manner.

3.9.4 Constructive Solid Geometry (CSG)

The general techniques for solid modeling in CAD/CAM were developed during the 1970s, responding to an obvious need in industry to move beyond the ambiguities of wire frame methods.

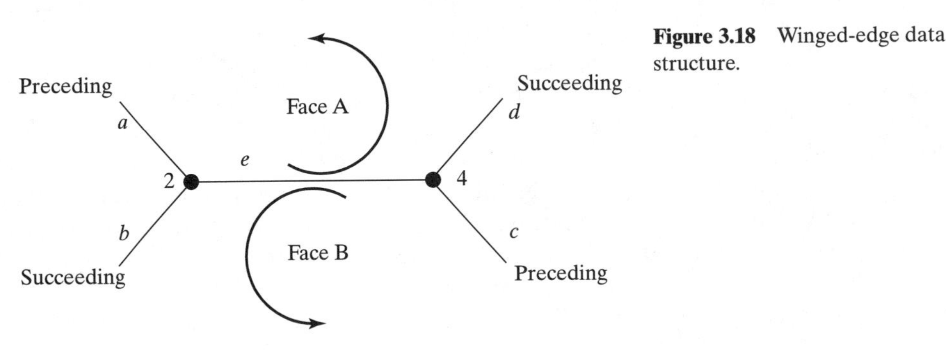

Figure 3.18 Winged-edge data structure.

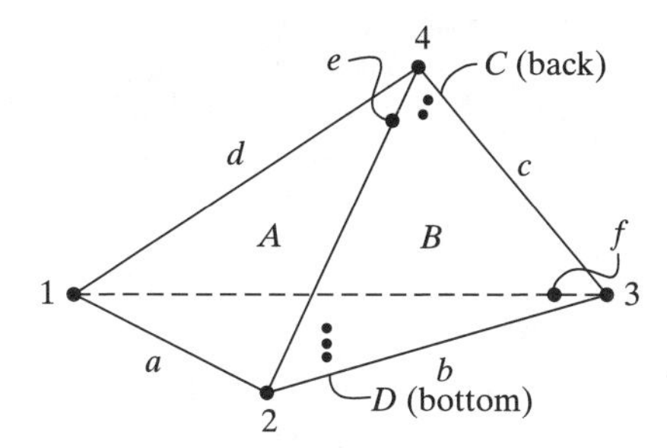

Figure 3.19 Pyramid example (from *Geometric and Solid Modeling: An Introduction* by Christoph M. Hoffman, © 1989. Morgan Kaufmann Publishers.).

TABLE 3.1 Defining the Pyramid with a Winged-Edge Data Structure

Vertices and edge			Faces		Edges for left face		Edges for right face	
Vertex	Vertex	Edge	Left	Right	Pred	Succ	Pred	Succ
1	2	a	A	D	d	e	b	f
2	3	b	B	D	e	c	f	a
3	1	f	C	D	c	d	a	b
3	4	c	B	C	b	e	d	f
1	4	d	C	A	f	c	e	a
2	4	e	A	B	a	d	c	b

Shah and Mantyla (1995) describe the two "camps" that evolved with the following quote. "Ian Braid and his colleagues at the University of Cambridge worked on *boundary representations,* models consisting of facets that were subsets of planar, quadric, or toroidal surfaces (Braid, 1979). Voelcker and Requicha at the University of Rochester introduced *CSG models,* consisting of a finite number of Boolean set operations applied to half-spaces defined by algebraic inequalities (Requicha, 1977; Requicha and Voelcker, 1977)." Both methods resulted in commercial developments by the early 1980s.

In CSG, blocks can be added together, subtracted from each other, and intersected with each other to create more complex shapes. An example of an L-shaped bracket is considered to illustrate the CSG method (Figure 3.20).

Two blocks and the hole are used to build up the more complicated solid. The two legs of the bracket are formed by a *union.* The hole is taken out of the solid leg by a *difference.* Figure 3.21 shows the CSG tree for the bracket. The shapes that made up the bracket are shown as the leaves of the tree. The nodes give information on which regularized Boolean operation should be carried out. For the bracket, the two blocks and the hole are at the leaves.

The second block is unioned with the first after being translated by (1) in the x direction. The hole is subtracted from the unioned pair after being translated by (5) in the x direction and (2) in the y direction.

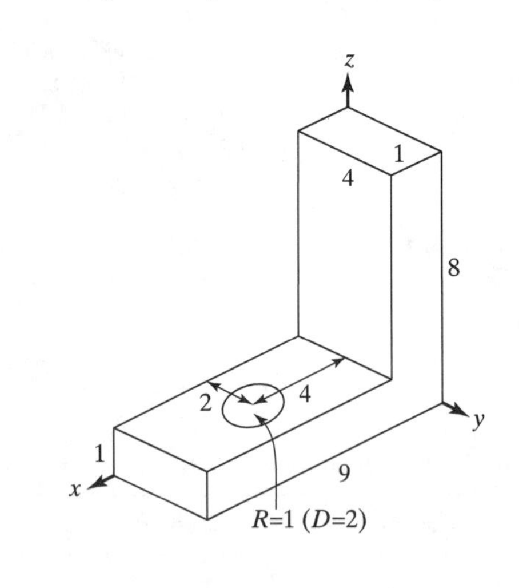

Figure 3.20 L-shaped bracket (from *Geometric and Solid Modeling: An Introduction* by Christoph M. Hoffman, © 1989. Morgan Kaufmann Publishers.).

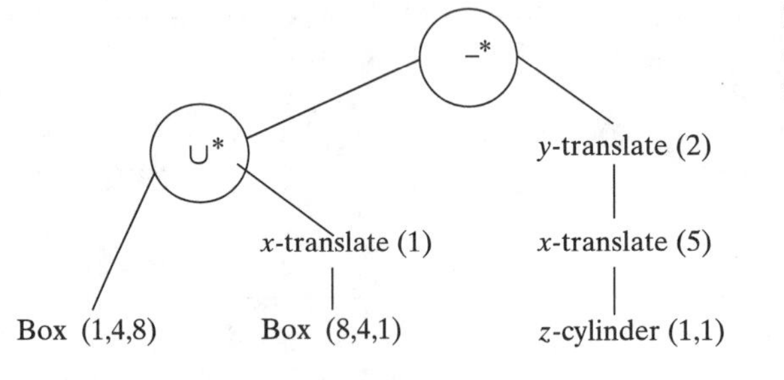

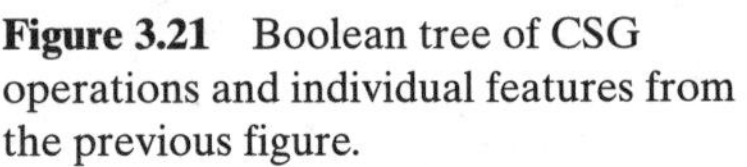

Figure 3.21 Boolean tree of CSG operations and individual features from the previous figure.

3.9.5 Feature-Based Design

An extension of CSG is to create special cases of building blocks that *in fact correspond to manufactured features.* The designer in this case calls upon a more restricted menu of features, such as holes and pockets, which correspond to the physical actions of a "downstream" manufacturing process. A menu that includes holes and pockets will correspond to machining.[2] This idea sets the stage for the destructive solid geometry (DSG)—Section 3.11—further on in this chapter. First, the text revisits the joystick example using solid modeling.

3.10 SECOND TUTORIAL: SOLID MODELING USING CONSTRUCTIVE SOLID GEOMETRY (CSG)

The joystick base will now be constructed using solid modeling techniques for comparison with the wire frame tutorial. The process begins by developing three-dimensional blocks of material to which other three-dimensional objects will be subtracted or added.

[2]Feature-based design is not just machining oriented. Sheet metal forming, forging, casting, and so on all exhibit process specific, standard shapes that can easily be made with a standard die set. If the designer knows which process is going to be used, it makes sense to design with this die set and corresponding shapes in mind. Of course, this is also the key philosophy of the MOSIS service for the design and rapid delivery of integrated circuits (MOSIS, 2000).

From a CAD/CAM viewpoint, the distinct advantage of CSG is that models are developed intuitively and give insights into manufacturability provided the designer is sympathetic to the manufacturing operations. Process planning, fixturing, and orientations also become much clearer. This is because the Boolean nature of CSG requires that features be added and subtracted in a logical manner. If done well, the ordering of the features can anticipate manufacturing. For communications, CSG models can be *rendered:* a process that takes the internal representation and creates a shaded, attractive picture on the screen.

Here is an important subtlety before the tutorial: *in terms of graphic representation, solid models are capable of being depicted as either wire frames or rendered solids.*

At the time of writing, the more expensive CAD tools running on high-end workstations show fully shaded solid blocks during every stage of the CAD procedures. On the other hand, in many other CAD packages, a solid might well be created with CSG methods but *temporarily displayed as a wire frame.* The previous sections indicate why this is in fact desirable: CSG programs incorporate complicated computer algorithms and therefore require considerable computer power. Thus while CSG procedures (even adding a simple hole) are being done on the partially completed CAD object, the computer will do the calculations faster if the temporary object is displayed as a wire frame. Once the CSG procedure is finished, the object can be rerendered.

For the purposes of illustrating CSG, Figures 3.22 onward are provided. Figures 3.22 and 3.23 show the base and the individual objects that are to be subtracted from it. Figure 3.22 is a wire frame representation of the solid models, while Figure 3.23 is a solid representation of the same model.

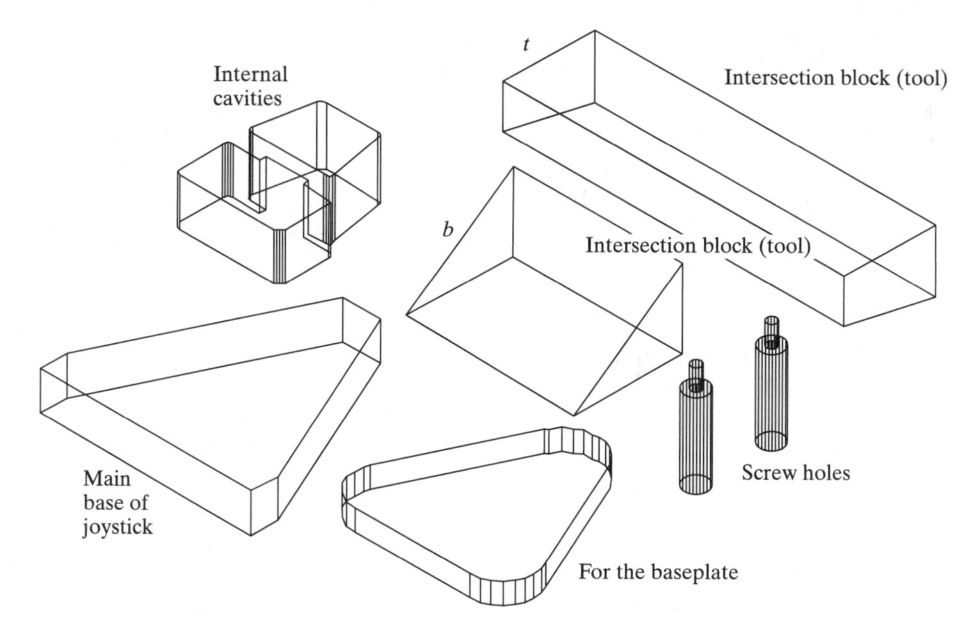

Figure 3.22 Wire frame drawing of VR joystick primitives.

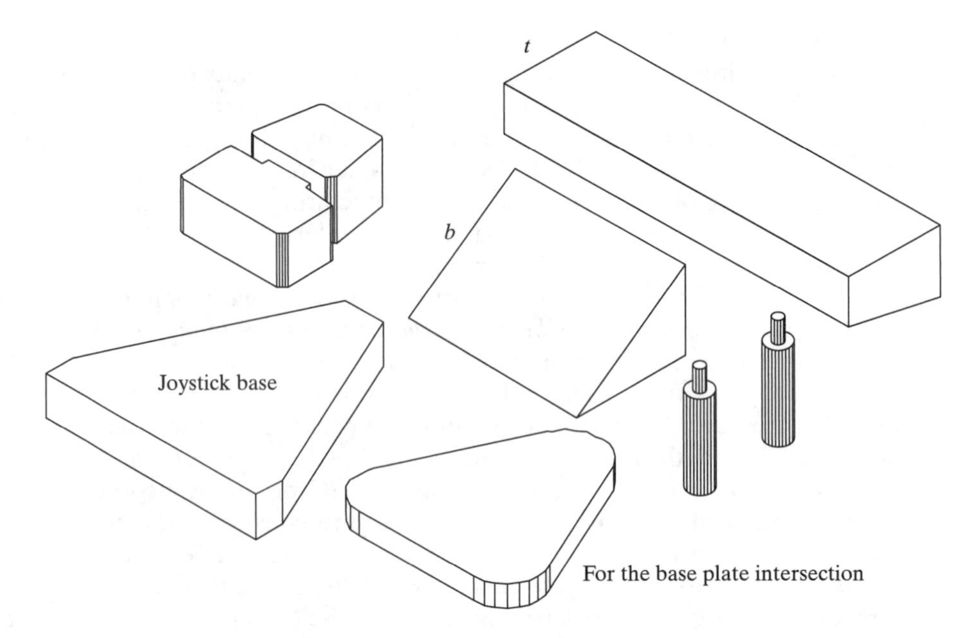

Figure 3.23 Solid model drawing of VR joystick parts with hidden lines removed.

These figures also include the *cutting blocks* or *intersection blocks* for the slopes on the front edges of the joystick. These are labeled **b** and **t** in the upper right of the figure. These are special geometric blocks that are created as separate operational tools by the user. They are used to cut or intersect with a virtual starting stock and create the sloping angles on the joystick.

The creation of the base on the left of the figures begins by tracing the outline of the block. Trace segments are then joined together as a polyline. The **extrude** command is then executed, and the block is extruded to the appropriate height. It is important to note that for the extrude command to work, the polyline has to be closed. At the bottom center of Figure 3.22, the outline of the base plate cavity is constructed in an identical manner with that of the wire frame. The outline of the plate is made into a polyline and extruded. Because the base plate cavity is constructed with Boolean subtraction, the height of the extrusion is arbitrary as long as its thickness is greater than the thickness of any intended recesses.

Since the design is for the bottom of the object, it makes sense to also construct the internal cavities for the sensors and printed circuit board at this time. The outline of the cavities is constructed in exactly the same way as in the wire frame. As before, the common edges are removed from the rectangular cavities. Like the base plate, the outlines are made into polylines and extruded to an arbitrary height. Two polylines, one for the rectangular part of the cavity and one for the trapezoidal part, are created and extruded.

The extrusion of the cavities remains sensitive to the differences in height between the rectangular and trapezoidal cavities. For simplicity, these two objects are then combined with the **"union"** command.

The next features to construct are the bores and counter-bores. Since the bore and the counter-bore are coaxial, it makes sense to construct them at the same time. Both holes are essentially cylinders and can be generated with the **cylinder** command. The command asks for the location of the center of the base, the radius of the cylinder, and then the height. The bore–counter-bore combination is created by defining the two cylinders and then unioning them with the **union** command used earlier.

To this point, all of the objects have been built from the world (or global) coordinate system. For example, this allows polylines to be built up on the specific *x-y* plane shown earlier in Figure 3.8. However, the *x-y* plane does not always have to be defined by the world coordinate system (WCS) shown in that figure. It can be redefined by a user coordinate system (UCS). A UCS is a user-defined or local coordinate system.

The sloping sections of the front of the object are good candidates for construction with intersection blocks along the world *x* axis. However, it is desirable to perform the construction from an *x-y* plane that is at right angles to the original WCS, *x-y* plane shown in Figure 3.8. It is therefore desirable to change the coordinate system from the WCS to a UCS. The command to do this is simply **ucs.** After entering **ucs,** the user is prompted to describe the change of coordinate system in Figure 3.8. This new reference plane corresponds to the small triangular face labeled **b** and the quadrilateral labeled **t** at the top (back) of Figure 3.22. In this particular case, the coordinate system is rotated 90 degrees about the original *x* axis and then 90 degrees about the new or modified *y* axis (because of the *x* rotation by 90 degrees) in Figure 3.8. It is emphasized that the order is important: when prompted to determine the axis of rotation, *x* is chosen first. Following the *x* rotation, the **ucs** command is chosen again to rotate the system 90 degrees about *y*.

Once the UCS has been adjusted, the *cutting* or *intersection* blocks labeled **b** and **t** can be extruded. As mentioned, these will be used as intersection blocks to create the angled faces on the front portion of the base. One polyline, **t,** is developed for the top slope of the base and another, **b,** for the bottom slope of the base. The top piece has been made from a quadrilateral, while the bottom has been made from a triangle. Once the polylines **b** and **t** are constructed, they can be extruded. A characteristic of the CSG subtraction process is that the height of the extrusion does not always have to be exact. It only has to be of sufficient height to cross the width at the largest part of the intersection between the two objects. Once the two intersection blocks have been extruded, they can be used to intersect the main block. These intersections give the main block its more "artsy," slanted faces at the front.

In Figures 3.24 and 3.25 the individual objects are shown in the appropriate locations for solid subtraction. The former is the wire frame rendition, while the latter is the solid. Figure 3.26 is the final part. Rendering can then be initiated with the **render** command. Be sure to compare Figure 3.26 with Figure 3.14.

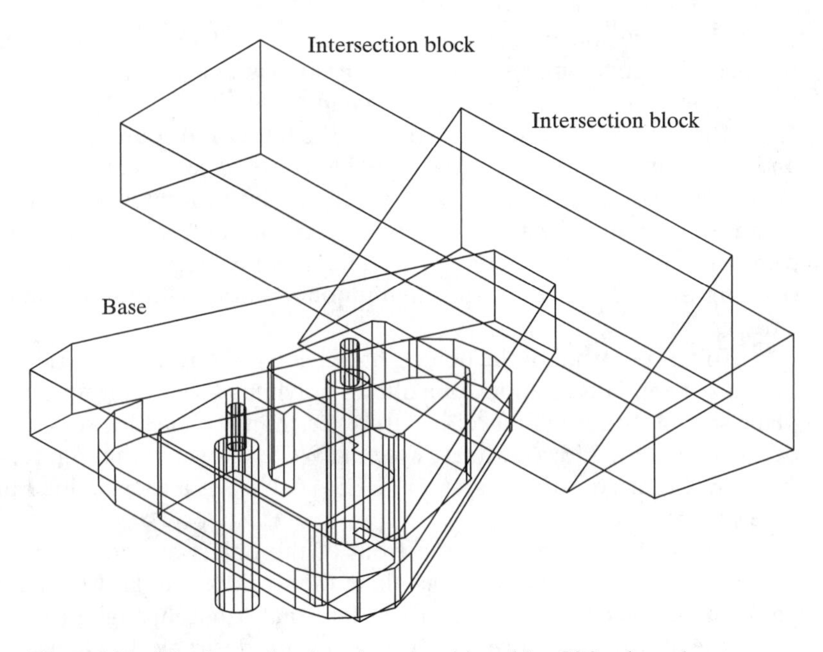

Figure 3.24 Wire frame drawing of parts positioned for CSG subtraction operations.

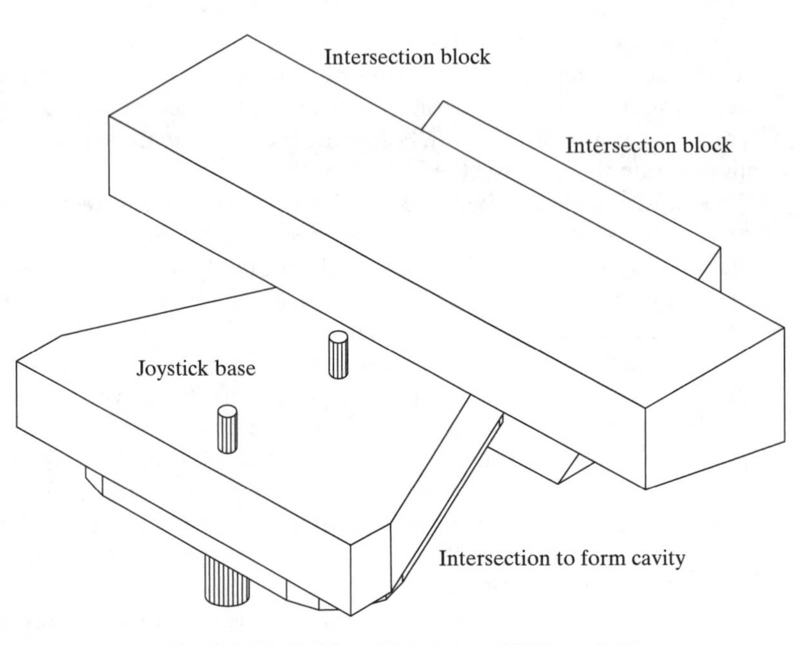

Figure 3.25 Solid model drawing of CSG method.

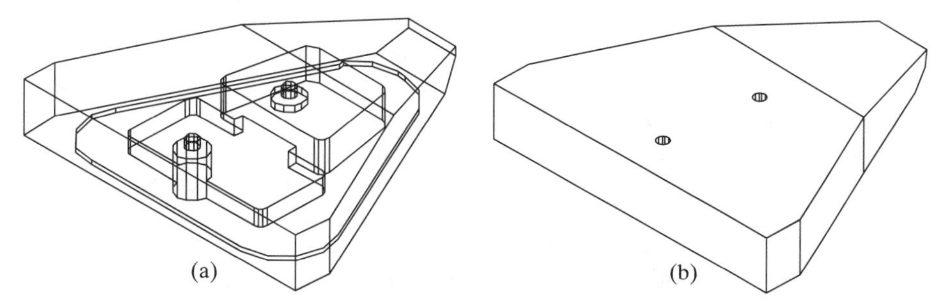

Figure 3.26 (a) Wire frame image compared with (b) solid model image. (Note: both are CSG.)

3.11 THIRD TUTORIAL: SOLID MODELING USING DESTRUCTIVE SOLID GEOMETRY (DSG)

3.11.1 DSG as a Feature-Based Design Environment

Destructive solid geometry (DSG) is a special case of conventional constructive solid geometry (CSG). In DSG only the regularized Boolean operations of *difference* are allowed (Requicha, 1980; Cutkosky and Tenenbaum, 1990; Shah and Mantyla, 1995; Woo, 1992; 1993; Regli et al., 1995). The key advantage of this procedure is that the design phase anticipates the fabrication-by-milling phase; that is, the differencing procedures from *graphical stock* during DSG mirror the cutting tool actions on *metal stock* during milling. Clearly, DSG is a way of carrying out feature-based design for the particular case of machining features. Other manufacturing processes would call for a feature-based design environment using, say, extrusion features or casting features.

In DSG, the differencing is done with a predefined group of features that the designer can select from a pull-down menu and then customize by adding parameters. Typical DSG features include the following eight shapes (see Wright and Bourne, 1988, 222): a through-hole, a blind hole, a pocket, a shoulder, a chamfer, a plane, a channel that goes completely through a block, and a slot or closed channel.

3.11.2 A Worked Example

Following the precedent set by the previous examples, DSG will now be demonstrated. This time the piece that caps the top of the handle will be built. This top piece is chosen because it is simpler and, in comparison with the base, easier to describe with DSG.

The object will be created with an "artsy-geometric" look rather than splined surfaces (Bartels et al., 1987; Riesenfeld, 1993; Puttre, 1992). Of the eight DSG primitives mentioned, the shapes that will be removed are mostly in the category of *chamfers.* In addition, one *pocket* will be created.

The first step is to establish and create the bounding block from which material will be removed by DSG (and later by machining). The SolidWorks CAD package will now be used rather than AutoCAD. As emphasized previously, this is merely to show some variation in this chapter, and any of the commercial CAD environments could have been selected for this tutorial.

The bounding block (or stock) is created by sketching its outline on one of the three coordinate planes and then extruding this outline through space to create a solid object. In SolidWorks, each of the three coordinate planes can be selected in the left window. When one of the coordinate planes is selected, it is highlighted in the right window and looks much like a playing card floating in space.

After selecting the desired coordinate plane, the *sketch* icon on the vertical toolbar on the right is clicked. This puts the user in sketch mode. A sketching toolbar on the right now becomes visible. A rectangle is sketched using the *rectangle* icon on this sketching toolbar. Dimensions and constraints are added so that the position and dimensions of the rectangle are fully defined. Note that in SolidWorks, the lines on the sketch turn black when they are fully constrained (meaning that there is no ambiguity as to their position and size). Once the sketch is fully defined, the *extrude boss/base* icon at the top of the left toolbar is clicked. The user specifies the distance that the sketched contour (a rectangle in this case) will be extruded out of the sketching plane. The contour is then extruded into a solid part, as shown in Figure 3.27.

Now that the bounding block has been created, material must be removed using DSG operations to create the final part. Most of the features to be removed in

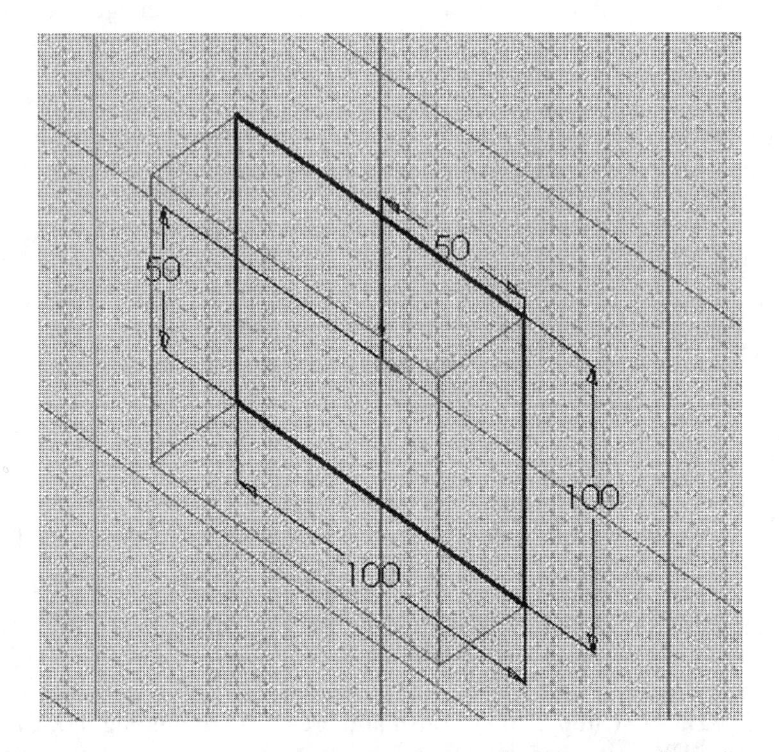

Figure 3.27 DSG method for constructing the joystick top: creation of bounding block. Thanks are due to Shad Roundy for his contributions.

the design of this part are *chamfer* features. In SolidWorks, chamfers can be created in two ways, contrasted in Figures 3.28 and 3.29 and described below:

- The first and most obvious (and the one that would usually be used) is to select the edge on the part that will be cut away and then click the *chamfer* icon on the left toolbar. The user is asked to specify two parameters to define the chamfer. For the one shown in Figure 3.28, the angle of the chamfer and the distance that it cuts into the top face were specified.
- Second, the same feature could be cut away using the *extrude cut* operation as well. In order to create the chamfer in this way, the side face of the bounding block is selected. The *sketch* icon on the right toolbar is then clicked. This allows the user to sketch on the face of the part. A triangle is sketched as shown in Figure 3.29. The triangle is then selected, and the *extrude cut* icon (second down on the left toolbar) is clicked. The user specifies "through all" as the cut depth, and the chamfer is created. Note that this is a more difficult way to create the chamfer and would not be used in most situations.

The chamfer operation is then repeated seven times to create the eight chamfers shown in Figure 3.30. This is an exploded view showing the geometry removed by each of the chamfers. The numbers shown represent the order in which the chamfer operations were performed.

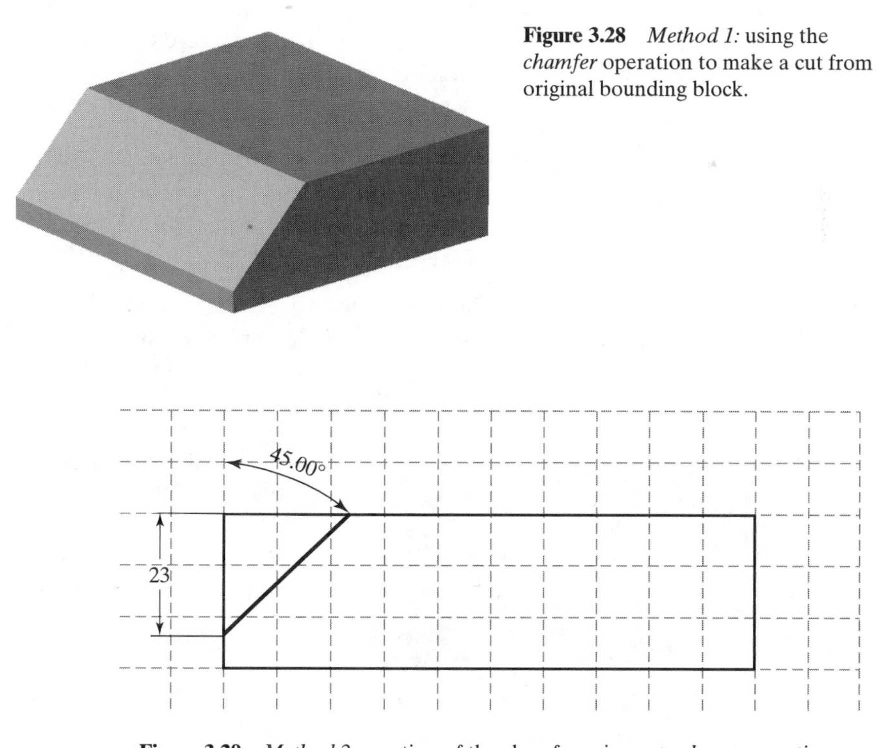

Figure 3.28 *Method 1:* using the *chamfer* operation to make a cut from original bounding block.

Figure 3.29 *Method 2:* creation of the chamfer using *extrude cut* operation.

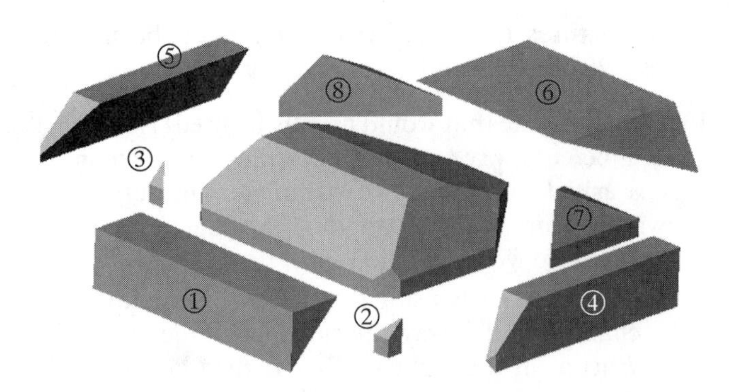

Figure 3.30 Exploded view of all chamfers.

The next, and final, DSG feature to remove is the recess for the top button. This recess is a *pocket,* in the language of DSG-related CAD/CAM. Unlike CSG, a pocket cannot be constructed by removing a prism of any height greater than a minimal height from the substrate. The depth of the DSG pocket has to be exact.

To create the pocket in SolidWorks, the contour of the pocket is sketched on the face into which the pocket will be cut. As explained earlier, to sketch on a face, the face is selected and then the *sketch* icon on the right toolbar is clicked. The sketch tools on the right toolbar are used to draw the desired contour. Once again, dimensions have to be added to fully define the position and size of the sketched contour. Once the contour is finished, the *extrude cut* icon on the left toolbar can be clicked. The desired depth of the pocket is then specified. The completed part with the pocket is shown in Figure 3.31. Note that the fillets on the corners can be added to the sketch before or after the cut operation. In either case, it is accomplished by selecting the corner to fillet and then clicking the *fillet* icon on the left toolbar. As in the base example, some details are still to be added, including the method of attachment by screws.

3.11.3 Summary of the DSG Approach

In concluding this subsection, it should be emphasized that the DSG approach asks the designer to work in a different way than CSG. The restriction to *difference* operators is somewhat surprising if the designer is more accustomed to general solid-

Figure 3.31 Solid model of the completed joystick top.

modeling techniques. However, it has been found that after some practice, designers gradually adjust to the new method and then feel no limitations. The burden of getting used to the new approach is outweighed by the main advantages of DSG: the design is usually easier to manufacture (Kim et al., 1999).

3.12 MANAGEMENT OF TECHNOLOGY

3.12.1 History of Design Environments

Which CAD system should be used by a particular firm? Even within a firm, which CAD system should be used by an individual project group or a solo engineer to create product drawings and specifications? How do MOT issues such as time-to-market, human resource allocations in a company, and long-term growth influence the decision on which CAD system to use?

A short summary of the history of different types of CAD systems can set the stage for an answer to these questions (Table 3.2).

3.12.2 Systems with "Low Overhead"

For student work, small design projects, and most modeling needs, there are a growing number of inexpensive CAD systems that create wire frame and rendered solid models without parameterization. These systems are the quickest to learn and function on today's personal computers with a fast processor and plenty of RAM. A

TABLE 3.2 General History of CAD with Approximate Dates (Jami Shah is acknowledged for his contribution to this table)

Date	Typical environment
Late 1950s to early 1960s (e.g., Sutherland, 1963)	Drafting on a computer screen rather than drafting with a T-square and paper. "Click on points and connect with lines."
1960s (e.g., Roberts, 1963)	Three-dimensional drafting with hidden-line removal: establish viewpoint and remove lines at "lower" two-dimensions.
1970s	Solid modeling techniques arrive in research work, followed by some commercial systems: *Boundary representations (Braid, 1979) *Constructive solid geometry or CSG (Requicha and Voelcker, 1977)
Mid-1970s to early 1980s (e.g., Grayer, 1976; Woo, 1982)	Feature-based systems relating CAD objects to manufacturable features Sculptured surfaces introduced
Mid to late 1980s (e.g., Pratt and Wilson, 1987)	Design with features in an explicit way
Late 1980s and 1990s	Commercial feature-based modeling and parametric design include: *Pro-Engineer by Parametric Technologies *IDEAS by SDRC *Unigraphics *CATIA

start-up company needing a modest CAD environment would also be wise to invest in these products, which include but are not limited to:

- AutoCAD commercially available from <**www.autodesk.com**>
- SolidWorks commercially available from <**www.solidworks.com**>
- IronCAD commercially available from <**www.ironcad.com**>

3.12.3 Systems with "High Overhead"

The next group of products have been built to do high-end solid modeling with real-time rendering, *and* they maintain a parametric model of the emerging design. This means that objects are initially created generically without specific dimensions. When objects are specifically instantiated, dimensions are added and everything scales up or down to suit. The user is able to define constraints between different parts of an object and then scale them.

For long-term company growth over several product variants this has enormous appeal. However, there is a major drawback. There is a huge learning time for such systems. Also, since they are updated every 18 months or so, further retraining on new "revs" is likely.

These are powerful design tools for a large automobile company or a national laboratory. In these environments, many similar components in a family are being designed. Their use in a bearing-manufacturing company like Timken Inc. is perhaps the easiest to visualize. Bore sizes, races, cover plates, and the like, can be created once and then "scaled up or down." For future revisions of a component, any existing parametric designs that might reside in a software library can quickly be reinstantiated to create a new object in the same family.

These larger systems also have direct links to supplementary packages that will do DFM/A and finite-element analysis. Most of them also include a CORBA-based open architecture that allows linking to other software applications (e.g., SDRC, 1996).

- ProEngineer commercially available from <**www.ptc.com**>
- IDEAS commercially available from <**www.sdrc.com**>
- Unigraphics commercially available from <**www.ugsolutions.com**>
- CATIA commercially available from <**www.catia.com**>

Translations between these different commercial CAD systems were once done with initial graphics exchange system (IGES) and can now be done with product definition exchange system (PDES/STEP). PDES/STEP is evolving into a useful worldwide standard (see ISO, 1989, 1993).

Other products such as Spatial Technology's ACIS (ACIS, 1993) play an intermediate role compared with the aforementioned applications. They have specialized in the "market niche" of creating an open de facto standard for solid representations. This is finding adoptions in other systems, including AutoCAD. The openness of ACIS is popular with the research community. From a management of technology viewpoint this direction toward open CAD systems is important.

3.12.4 Current Trends in CAD

The CAD field is developing very quickly indeed. At the time of this writing, "student editions" of PTC's Pro-Engineer and SDRC's IDEAS are becoming available for only $100. Thus, even these more sophisticated systems are becoming more readily available to the average user and are able to run on modest computer systems in the $1,500 to $3,000 price range for a well-configured environment. This still does not mean an end user should "jump right in" and use them. The big issues—discussed above—are the "learning curve" and the "library creation for parametric systems." These trade-offs are captured in Figure 3.32. On the other hand, used in a nonparametric way, these higher end packages can create excellent feature-based models. The governing factor seems to "boil down" to how much long-term interest a person or group has in using CAD tools. Here are three scenarios:

- For a start-up company, where a CAD system might be used only once to generate an idea and then an FDM prototype, the cheaper nonparametric approach is recommended.
- Also in small, newer companies, today's evidence is that the turnover among young engineers is high. It might not be worth investing the training time needed for the full parametric systems when quite satisfactory designs can be done with the cheaper systems like AutoCAD, SolidWorks, and so on, which have a short learning curve.
- But for large, stable companies, if several product revisions will be designed spanning several months or years, then the time invested in learning the parametric approach in ProEngineer, SDRC, and the like, will be worthwhile.

3.12.5 Future Trends in CAD: Multidisciplinary Concurrent Design/Engineering and Global Manufacturing

For a variety of cultural reasons, today's industrial growth is more and more dependent on situations where *large businesses are distributed.* Often these large business organizations are split up but then orchestrated over several continents, perhaps to take advantage of excellent design teams in one country and low-cost, efficient manufacturing teams in another. These trends place even more emphasis on concurrent engineering (or simultaneous design) and design for manufacturability and assembly (DFM/A). The goals are to coordinate all members of a design and manufacturing team at each stage of product development, manufacturing, sales, and service (see Urban et al., 1999).

To further complicate such trends, engineering products are more complex. Concurrent engineering is difficult enough when the product is nearly all mechanical (such as a gear box) or nearly all electronic (such as a television). But as automobiles, aircraft, robots, and computers become a highly complex mix of integrated circuits, power supplies, controllers, and mechanical actuators, concurrent engineering becomes even more challenging. It clearly demands the orchestration of

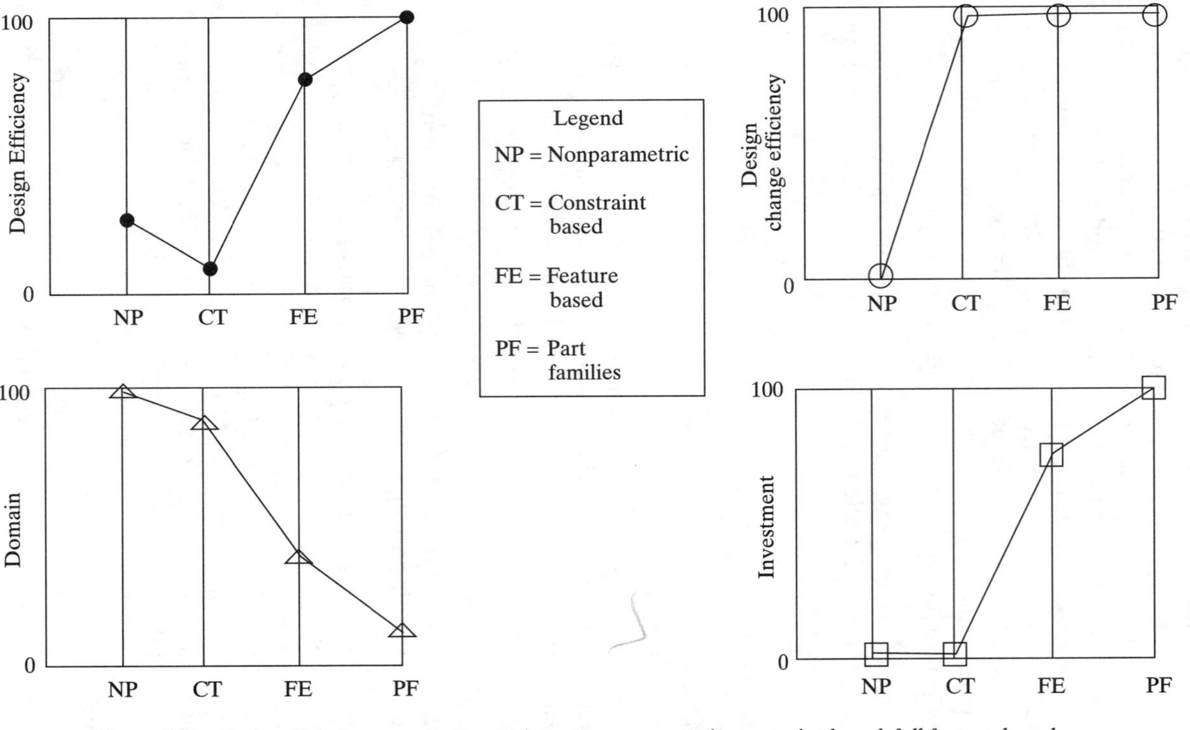

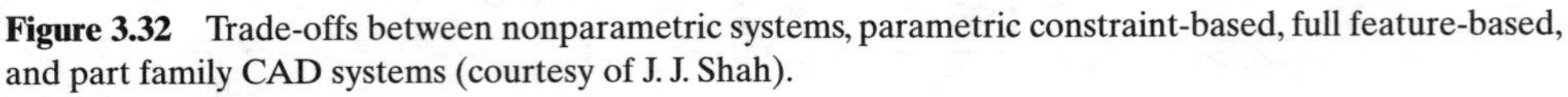

Figure 3.32 Trade-offs between nonparametric systems, parametric constraint-based, full feature-based, and part family CAD systems (courtesy of J. J. Shah).

multidisciplinary design teams. These trends will create the need for environments that allow, for example:

- The integration of electrical-CAD tools with mechanical-CAD tools. Chapter 6 describes a domain unified computer aided design environment (DUCADE) that facilitates multidisciplinary concurrent design for consumer electronic products.
- The creation of intelligent *agents* for Internet-based design. An example might be an agent for plastic injection-mold design (Urabe and Wright, 1997). Internet-based design environments allow the original part designer to import information on specific "downstream" processes—in this case, how to fabricate negative mold halves. Information could also include data on shrinkage factors, recommended draft angles for the mold, and snap fit geometries (Brock, 2000).

3.13 GLOSSARY

3.13.1 Boundary Edge Representation

Boundary representations, or *b-reps,* describe an object in terms of its surface boundaries: vertices, edges, and faces.

3.13.2 Creative Design

The formative, early phases of the design process, where market identification, concepts, and general form are studied.

3.13.3 Constructive Solid Geometry (CSG)

The addition, subtraction, or intersection of simpler blocklike primitives to create more complex shapes.

3.13.4 Destructive Solid Geometry (DSG)

A special case of CSG where the designer begins with "graphical stock" and changes its shape with only the subtraction or intersection commands, in order to suit later operations on a "downstream" machine tool.

3.13.5 Detail Design

The later phases of design in which specific shapes, dimensions, and tolerances are specified on a CAD system.

3.13.6 Design for Assembly, Manufacturability, and the Environment

The collection of terms used to encourage designers to adjust their design activities for ease of assembly (DFA), manufacturing (DFM), or environmentally conscious (E) issues. Often *DFX* is used to summarize all these "design for" activities.

3.13.7 Feature-Based Design

The use of specific primitive shapes in design that suit a particular "downstream" manufacturing process.

3.13.8 Ink-Jet Printing in 3-D

Rapid prototyping by rolling down a layer of powder and hardening it in selected regions with a binder phase that is printed onto the powder layer.

3.13.9 Injection Molding

Viscous polymer is extruded into a hollow mold (or die) to create a product.

3.13.10 Investment Casting

The word *investment* is used when time and money are invested in a ceramic shell that is subsequently broken apart and destroyed. The original positive master that is used to create the negative investment shell can be made by several processes. Lost wax and ceramic mold are the two most common.

3.13.11 Machining

General manufacturing by cutting on a lathe or mill; chip formation from a solid block rather than forging, forming, or joining.

3.13.12 Parametric Design

CAD techniques that represent general relationships (e.g., height-to-width), not necessarily specific dimensions.

3.13.13 Prototyping (Prototype)

"The original thing in relation to any copy, imitation, representation, later specimen or improved form" (taken from Webster's Dictionary).

3.13.14 Plastic Injection Molding

As "injection molding," described above. Note:Zinc die casting also involves "injection" into dies or molds.

3.13.15 Personal Digital Assistant (PDA)

Current, fashionable term for several handheld computing devices possibly with e-mail link, cell phone, and modest display.

3.13.16 Rapid Prototyping (RP)

A new genre of prototyping, usually associated with the SFF family of fabrication methods. Emphasis is on speed-to-first-model rather than fidelity to the CAD description.

3.13.17 Solid Freeform Fabrication (SFF)

A family of processes in which a CAD file of an object is tessellated, sliced, and sent to a machine that can quickly build up a prototype layer by layer.

3.13.18 Solid Modeling ("Solids")

CAD representations that correspond to real-world physical objects with edges, vertices, and faces. A CAD operation on a solid model will be consistent with a physical action or deformation that could be performed in the physical world. Wire frame CAD modeling does not guarantee this condition.

3.13.19 Tessellation

Representing the outside surfaces of an object by many small triangles, like a mesh thrown over and drawn around the object. This leads to an ".STL" file of the vertices and the surface normals of the triangles.

3.13.20 Wire Frame Modeling

CAD representations that correspond to abstract lines and points. An object can be drawn and even rendered, but the computer does not store an object that is "understood" in a physical sense.

3.14 REFERENCES

ACIS Geometric Modeler. 1993. Version 1.5, *Technical Overview.* Boulder, CO. Spatial Technology, Inc.

Baumgart, B. G. 1972. *Winged edge polyhedron representation.* Technical Report STAN-CS-320, Computer Science Department, Stanford University.

Baumgart, B. G. 1975. *A polyhedron representation for computer vision.* NCC 75: 589–596.

Berners-Lee, T. 1989. Information management: A proposal. CERN internal proposal.

Boothroyd, G., and P. Dewhurst. 1999. *DFMA software.* On CD from the company, or contact <**www.dfma.com**>.

Braid, I. C. 1979. *Notes on a geometric modeler.* CAD Group Document, 101, Computer Laboratory, University of Cambridge.

Brock, J. M. 2000. Snap-fit geometries for injection molding. Master's thesis, University of California, Berkeley.

Compton, W. D. 1997. *Engineering management,* "Creating and managing world class operations." Upper Saddle River, N.J.: Prentice-Hall.

Cutkosky, M. R., and J. M. Tenenbaum. 1990. A methodology and computational framework for concurrent product and process design. *Mechanism and Machine Theory* 25, no. 3: 365–381.

Finin, T., D. McKay, R. Fritzson, and R. McEntire. 1994. KQML: An information and knowledge exchange protocol. In *Knowledge building and knowledge sharing.* Edited by Kazuhiro Fuchi and Toshio Yokoi. Amsterdam, Washington D.C., and Tokyo: Ohmsha and IOS Press.

Foley, J. D., A. van Dam, S. K. Feiner, and J. F. Hughes. 1992. *Computer graphics: Principles and practice,* 2nd ed., Reading, Mass.: Addison Wesley.

Frost, R., and M. Cutkosky. 1996. An agent-based approach to making rapid prototyping processes manifest to designers. Paper presented at the ASME Symposium on Virtual Design and Manufacturing.

Grayer, A. R. 1976. A computer link between design and manufacture, Ph.D. diss., University of Cambridge.

Greenfeld, I., F.B. Hansen, and P. K. Wright. 1989. Self-sustaining, open-system machine tools. In *Proceedings of the 17th North American Manufacturing Research Institution* 17: 281–292.

Hauser, J. R., and D. Clausing. 1988. The house of quality. *Harvard Business Review* (May–June): 63–73.

Hazelrigg, G. 1996. *Systems engineering: An approach to information-based design.* Upper Saddle River, N.J.: Prentice-Hall.

Hoffmann, C. M. 1989. *Geometric and solid modeling.* San Mateo, CA: Morgan Kaufmann.

ISO. 1989. External representation of product definition data (STEP). ISO DP 10303-0.

ISO. 1993. Product data representation and exchange—Part I: Overview and fundamental principles. ISO DIS 10303-1, TC184/SC4/WG4 N193. Also see the following papers on PDES/STEP:

Wilson, P. 1989. PDES STEP forward. *IEEE Computer Graphics and Application* 79–80.
Eastman, C. 1994. Out of STEP? *Computer-Aided Design* 26, no. 5.

Kamath, R. R., and J.K. Liker. 1994. A second look at Japanese product development. *Harvard Business Review,* reprint number 94605.

Kim, J. H., F. C. Wang, C. Sequin, and P.K. Wright. 1999. Design for machining over Internet. *Design Engineering Technical Conference (DETC) on Computer Integrated Engineering,* Paper Number DETC'99/CIE-9082, Las Vegas.

Mead, C., and L. Conway. 1980. The CalTech intermediate form for LSI layout description. In *Introduction to VLSI Systems*, 115–127. Addison Wesley.

MOSIS. 2000. *University of Southern California's Information Sciences Institute—The MOSIS VLSI Fabrication Service,* **http://www.isi.edu/mosis/.**

Pratt, M. J., and P. R. Wilson. 1987. *Conceptual design of a feature-oriented solid modeler.* Draft Document 3B, General Electric Corporate R&D.

Puttre, M. 1992. Sculpting parts from stored patterns. *Mechanical Engineering,* 66–70.

Regli, W. C., S. K. Gupta, and D. S. Nau. 1995. Extracting alternative machining features: An algorithmic approach. *Research in Engineering Design* 7: 173–192.

Requicha, A. A. G. 1977. *Mathematical models of rigid solids.* Technical memo 28. Production Automation Project. New York: University of Rochester.

Requicha, A. A. G. 1980. Representations for rigid solids: Theory, methods, and systems. *ACM Computing Surveys*, 437–464.

Requicha, A. A. G., and H. B. Voelcker. 1977. *Constructive solid geometry.* Technical memo 25. Production Automation Project. New York: University of Rochester.

Richards, B., and R. Brodersen. 1995. InfoPad: The design of a portable multimedia terminal. In *Proceedings of the Mobile Multimedia Conference-2,* Bristol, England.

Riesenfeld, R. 1993. Modeling with NURBS curves and surfaces. In *Fundamental Developments of Computer-Aided Geometric Modeling*, 77–97. San Diego, CA: Academic Press.

Roberts, L. G. 1963. *Machine perception three-dimensional solids.* Technical report no. 315. Lincoln Laboratory, MIT.

SDRC. 1996. *The Open-IDEAS Programming Course Manual IMS 5282-5.* Milford, OH: Structural Dynamics Research Corporation.

Séquin, C. S. 1997. Virtual prototyping of Scherk-Collins saddle rings. *Leonardo* 30, no. 2: 89–96.

Shah, J. J., and M. Mantyla. 1995. *Parametric and feature based CAD/CAM.* Wiley. NY. (Also see Shah, J. J., M. Mantyla, and D. S. Nau. 1994. *Advances in feature based manufacturing.* New York: Elsevier.)

Sidall, J. N. 1970. *Analytical decision-making in engineering design.* Upper Saddle River, N.J.: Prentice-Hall.

Smith, C., and P. K. Wright. 1996. CyberCut: A World Wide Web based design to fabrication tool. *Journal of Manufacturing Systems* 15, no. 6: 432–442.

Stori, J. A., and P. K. Wright. 1996. A knowledge based system for machining operation planning in feature based, open architecture manufacturing. In *Proceedings (on Compact Disc) of the 1996 Design for Manufacturing Conference*, University of California, Irvine.

Suh, N. P. 1990. *The principles of design.* New York and Oxford: Oxford University Press.

Sungertekin, U. A., and H. B. Voelcker. 1986. Graphic simulation and automatic verification of machining programs. In *Proceedings of the IEEE Conference on Robotics and Automation.*

Sutherland, I. E. 1963. Sketchpad: A man-machine graphical communication system. In *Proceedings of Spring Joint Computer Conference,* 23.

Urabe, K., and P. K. Wright. 1997. Parting planes and parting directions in a CAD/CAM system for plastic injection molding. Paper presented at the ASME Design for Manufacturing Symposium, the Design Engineering Technical Conferences. Sacramento, CA.

Urban, S. D., K. Ayyaswamy, L. Fu, J. J. Shah, and J. Liang. 1999. Integrated product data environment: Data sharing across diverse engineering applications. *International Journal of Computer Integrated Manufacturing* 12, no. 6: 525–540.

Woo, T. 1992. Rapid prototyping in CAD. *Computer Aided Design* 24: 403–404.

Wright, P. K., and D. A. Bourne. 1988. *Manufacturing intelligence.* Reading, MA: Addison Wesley.

Wright, P. K., and D. A. Dornfeld. 1996. Agent based manufacturing systems. In *Transactions of the 24th North American Manufacturing and Research Institution,* 241–246.

3.15 BIBLIOGRAPHY

Bartels, R. H., J. C. Beatty, and B. Barsky. 1987. *An introduction to splines for use in computer graphics and geometric modeling.* San Mateo, CA: M. Kaufmann Publishers.

Hyman, B. 1998. *Fundamentals of engineering design.* Upper Saddle River, N.J.: Prentice-Hall.

Proceedings of the Institute of Mechanical Engineers. 1993. *Effective technologies for engineering success—Making CAD/CAM pay.* No. 1993-12.

Regli, W. C., and D. M. Gaines. 1997. A repository for design, process planning and assembly. *Computer Aided Design* 29, no. 12: 895–905.

Sequin, C. H., and Y. Kalay. 1998. A suite of prototype CAD tools to support early phases of architectural design. *Automation in Construction* 7: 449–464.

3.16 URLS OF INTEREST: COMMERCIAL CAD/CAM SYSTEMS AND DESIGN ADVISERS

1. Parametric Technology Corp, Pro/ENGINEER, **http://www.ptc.com**
2. Autodesk, AutoCAD, **http://www.autodesk.com**
3. SolidWorks, **http://www.solidworks.com**
4. Spatial Technologies, ACIS, **http://www.spatial.com**
5. 3D/EYE Inc, TriSpectives, **http://www.eye.com**
6. SDRC, I-DEAS, **http://www.sdrc.com**
7. EDS, Unigraphics, **http://www.edsug.com**
8. MSC, ARIES, **http://www.macsch.com**
9. DesignSuite by Inpart, Saratoga, California, **http://www.inpart.com**
10. Cambridge process selector, **http://www.granta.co.uk/products.html**

3.17 CASE STUDY

The goal of this case study is to reinforce the four levels of design described in Section 3.2. Specific ideas for a novel snow shovel are shown indented below the main design level. SDRC is the design tool being used in the example. Parametric design is highlighted. Reiterating a point made in the introduction, note that this chapter has attempted to move through a transition of design tools from simple wire frame to solid modeling, to solid modeling with rendering, and now to parametric design. Section 3.2 summarized four main phases of the design process. These are repeated below and used to guide the reader into the detailed steps using SDRC's IDEAS system.

1. *Art related and high-level:* "Design in any of its forms should be functional, based on a wedding of art and engineering" (W. A. Gropius, founder of the Bauhaus movement).

   The snow shovel will be designed in this case study as an attractive, colorful, lightweight, foldable device that mountaineers will buy at their local "outdoors shop."

2. *Engineering related and high-level:* "Design is the process of creating a product (hardware, software, or a system) that has not existed heretofore" (Suh, 1990).

   A collapsible snow shovel is designed in the next few pages with the purpose of improving the weight, cost, and usefulness over existing shovels. Emphasis is placed on the shovel head as the component with the most potential for improvement. The shovels that were found in the marketplace were separated into two primary design classifications. The first was the plastic shovel, which was lightweight and cheap but was not hard or stiff enough to be useful in ice or dense snow conditions. The second was the aluminum shovel, which was useful in all conditions but was significantly heavier and more expensive than a plastic shovel.

3. *Engineering related and at the analytical level:* "Design is a decision making process" (Hazelrigg, 1996).

   The new shovel incorporates the best of both shovel designs by combining a cheap, lightweight shovel scoop made out of polycarbonate with a hard, tough molded-in cutting blade made from aluminum 6061. Emphasis is placed on stiffening the shovel head through geometric features to allow a reduction in the shovel wall thickness (and thus a weight and cost reduction). This is accomplished by simulating load conditions using the ANSYS finite element analysis software and iterating the design to improve it.

4. *Detailed design:* "Design is to make original plans, sketches, patterns, etc." (Webster's Dictionary).

   The first step in the design of the shovel is the creation of a wire frame drawing to be extruded into the initial solid of the model. The most complex view of the part is generally selected for this wire frame, and in this case, the side view of the shovel is selected for the wire frame drawing. The final shovel design wire frame is shown in Figure 3.33.

   Notice the dimensions on the wire frame in Figures 3.33 and 3.34. Unlike conventional drafting packages where the dimensions are added to document a specific line length, parametric design controls the size of the part with these dimensions. They are therefore called *constraints* rather than dimensions in parametric design. Figure 3.34 shows the side view of the wire frame sketch of the shovel head after the angle constraint has been modified from 32 degrees to 45 degrees. Notice how this simple change dramatically alters the shape of the shovel. A standard drafting package would require the shovel head to be redrawn and then redimensioned to make this change. *The ability to rapidly change such design parameters is one of the key strengths of parametric design.* Also notice that in addition to the standard constraints of length, there are constraints for angular, radial, perpendicular, tangent, and coincident objects in Figures 3.33 and 3.34.

   Figure 3.35 shows the solid object from an isometric front view that is created when the wire frame shown in Figure 3.33 is *extruded* a distance of 225 millimeters (9 inches) and draft angles are added for strength and manufacturability. Figure 3.36 shows the same view after *fillets* have been added to the shovel head.

   The next step in the shovel design is to add cutouts to the bottom of the shovel head, which will become the stiffening ribs when the part is turned into a shell. Figure 3.37 shows a view perpendicular to the back edge of the shovel head

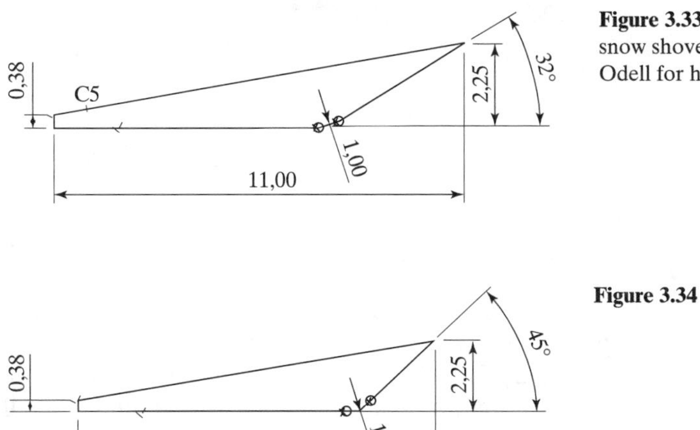

Figure 3.33 Wire frame model of snow shovel (Thanks are due to Dan Odell for his contributions).

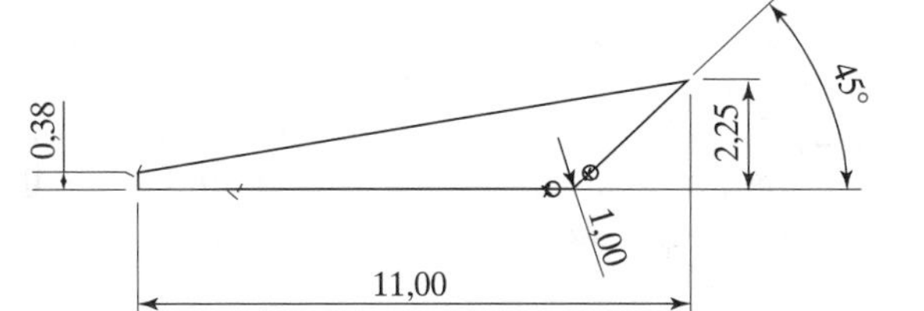

Figure 3.34

Figure 3.35

Figure 3.36

Figure 3.37

with the wire frame for the rib cutouts sketched on it. The wire frame cutouts extend past the actual part that they are intended to cut to ensure that they cut through the entire part.

Figure 3.38 shows an isometric view of the bottom of the shovel after the wire frame sketch has been extruded at an angle as a cutout to form the inverse of the rib profile. In a similar way, Figure 3.39 shows an isometric view of the top of the shovel after it has undergone a *shell* operation and a filleting operation. The shell operation takes a solid object and offsets its outer surfaces to generate a part of uniform wall

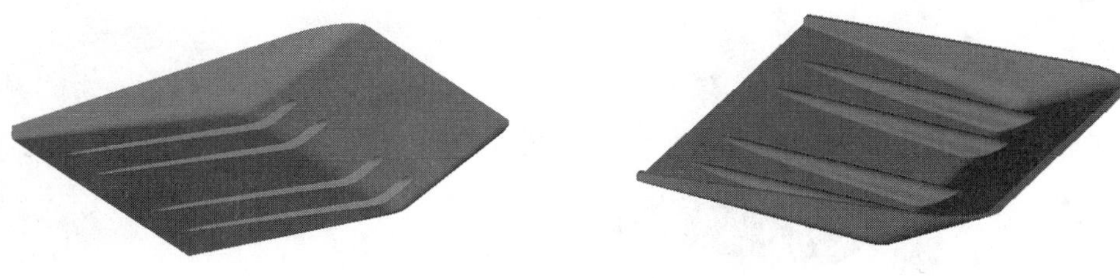

Figure 3.38

Figure 3.39

thickness. An option in the shell command allows specified surfaces to be removed from the shell operation, and in this case the top and front surfaces have been removed to create the scoop shape of the shovel. The shell command is very useful in design for the injection molding process where a uniform wall thickness is desirable to achieve uniform cooling and shrinkage. It can also be useful for processes such as thermoforming and sheet metal forming where a sheet of uniform thickness is used as the raw material.

The ribs were added to increase the moment of inertia of the shovel in the direction of the expected snow load and thereby to stiffen the shovel head. Stiffening the shovel through geometric changes allows a reduction in overall wall thickness, which relates to a reduction in cost and weight. Notice how the use of the shell command greatly simplifies the design of these ribs, which would have been quite difficult to model without use of this command.

The next step in the design of the shovel head is to add the interface between the shovel head and the shaft. The wire frame sketch of this interface is shown in Figure 3.40. To make this sketch and locate it properly, a reference plane has been added to the shovel. This plane is placed to be perpendicular to the rear face of the shovel so that wire frames that will be sketched and extruded

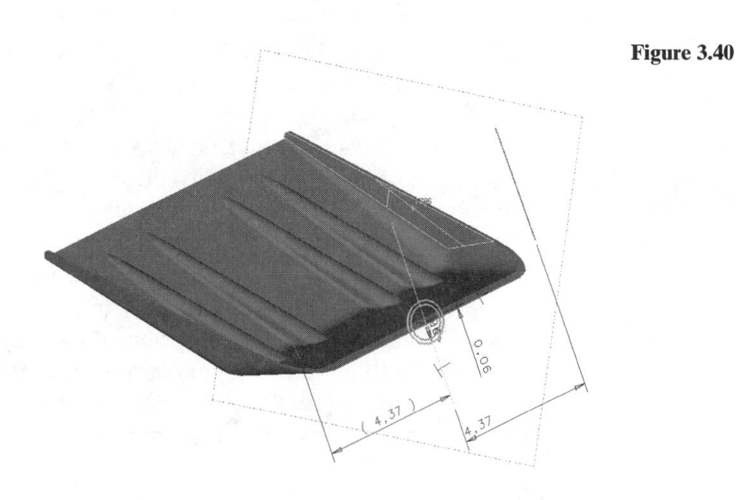

Figure 3.40

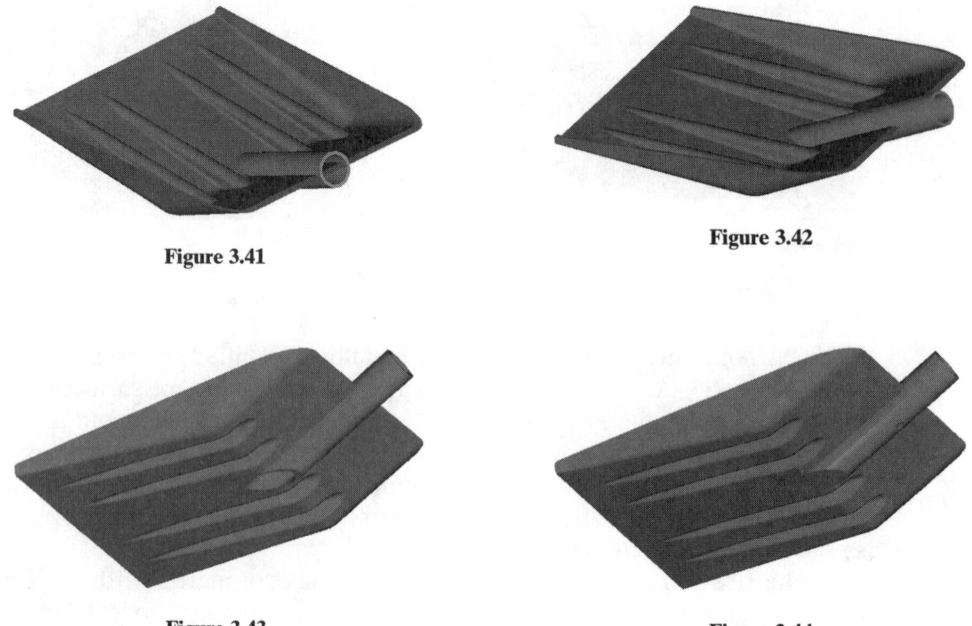

Figure 3.41

Figure 3.42

Figure 3.43

Figure 3.44

on it will be parallel to the rear of the shovel. Onto this plane, the outermost lines of the shovel head have been *focused* to give a reference for centering the interface. The interface is sketched as a tubelike structure to create a slip-fit with the shaft.

Figure 3.41 shows the shovel head after the interface section is extruded. This interface is extruded to a distance so that its full length extends past the shovel head. The interface is then cut off at an angle to match the bottom of the shovel. It is also lengthened, and the remaining wall of the shovel head inside the tube is removed. The result of these operations is shown in Figures 3.42 and 3.43.

Next, as shown in Figure 3.44, the resulting hole on the bottom of the shovel is sealed, and a hole is added for the fastener that attaches the shovel head to the shaft. This step completes the design of the shovel scoop itself. The next step is to add the molded-in aluminum cutting blade. Figure 3.45 shows a bottom view of the shovel with a wire frame sketch of the plastic section that will encase the cutting blade. Once this section is extruded, the wire frame for the blade itself is generated as in the bottom front isometric view in Figure 3.46. Notice that in this case the blade is drawn as an integral component of the shovel head. Later, a second model of the blade will have to be generated that includes the section that is encased in the plastic. This section will require several slots so that during injection molding, the plastic will flow through them and mechanically entrap the blade.

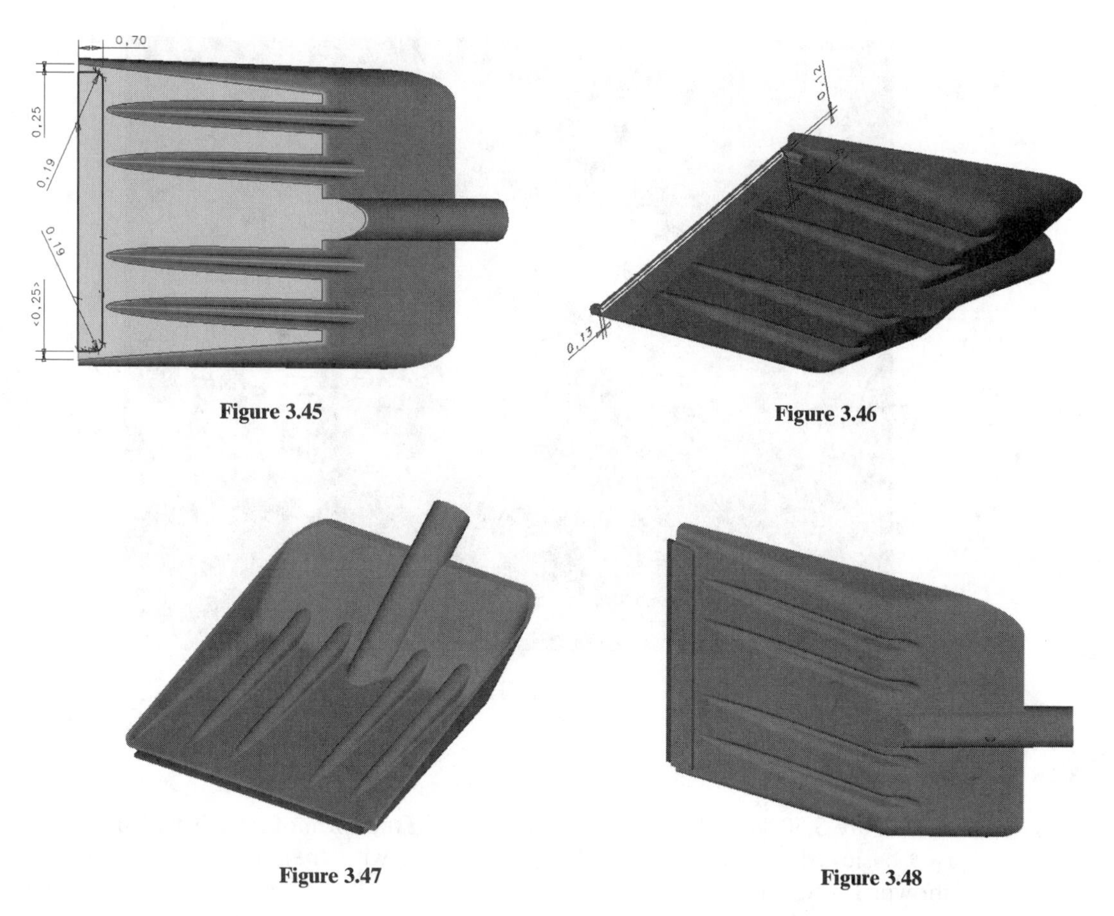

Figure 3.45

Figure 3.46

Figure 3.47

Figure 3.48

Once the blade is extruded, the remaining fillets are added to the part and the design is complete. The top and bottom isometric views of the completed shovel are shown in Figures 3.47 and 3.48.

It is important to note that the strength of parametric design is the ability to rapidly modify steps in the design to improve the final design. The steps that are documented for this case study are for the final shovel design, but many intermediate designs were modified to obtain the final one. If something in this design were found to be inadequate, any of these steps in the design could be reached and modified by accessing the "history tree" of the part.

For this design, the step of greatest interest was the design of the stiffening ribs. To help optimize these ribs, the solid model of the shovel head was imported into the ANSYS finite element analysis software and the loading was simulated under various conditions. Figure 3.49 shows this simulation for stress under a buckling load. This load produced the worst results for the shovel but is not expected to be encountered often, and the addition of the metal blade (which is not modeled in this simulation) will help to relieve some of this stress.

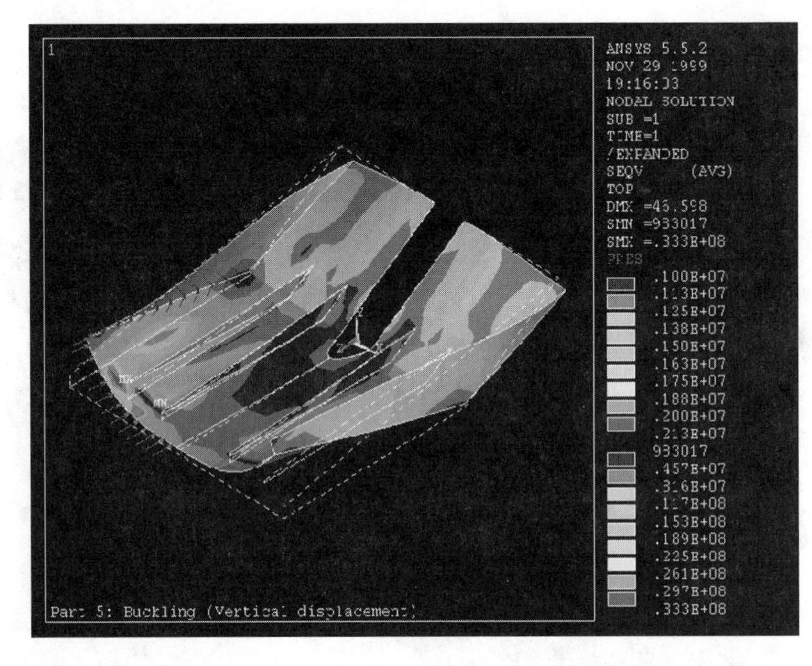

Figure 3.49

3.18 QUESTION FOR REVIEW

1. For Figure 3.50, give the CSG representation in the form of a CSG tree using the two basic primitives of a cylinder and a block with their local coordinates as shown. The origin of the world coordinates should be taken at the base of the cylindrical hole as shown. Any translations or rotations required must be clearly shown in the CSG tree.
   a. Does the point $P(6,0,3)$ lie in the object? Clearly show all necessary calculations using the CSG tree to make conclusions.
   b. After having marked all the vertices, determine how many edges and faces the object had.
   c. CSG representation is not unique. Prove this statement by creating another CSG tree for the above object. Use different-sized shapes. (Cylinders may be the same size.)

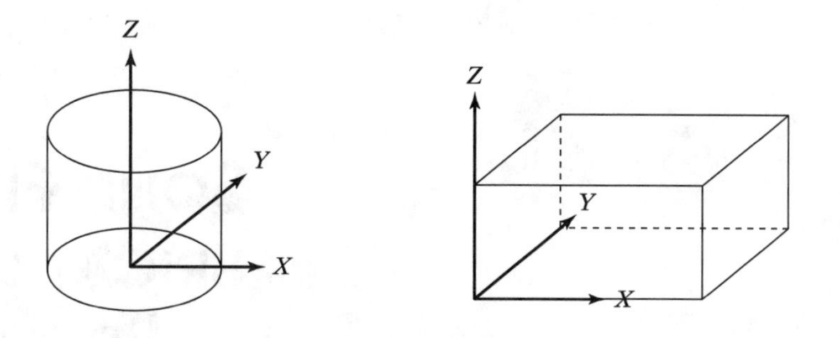

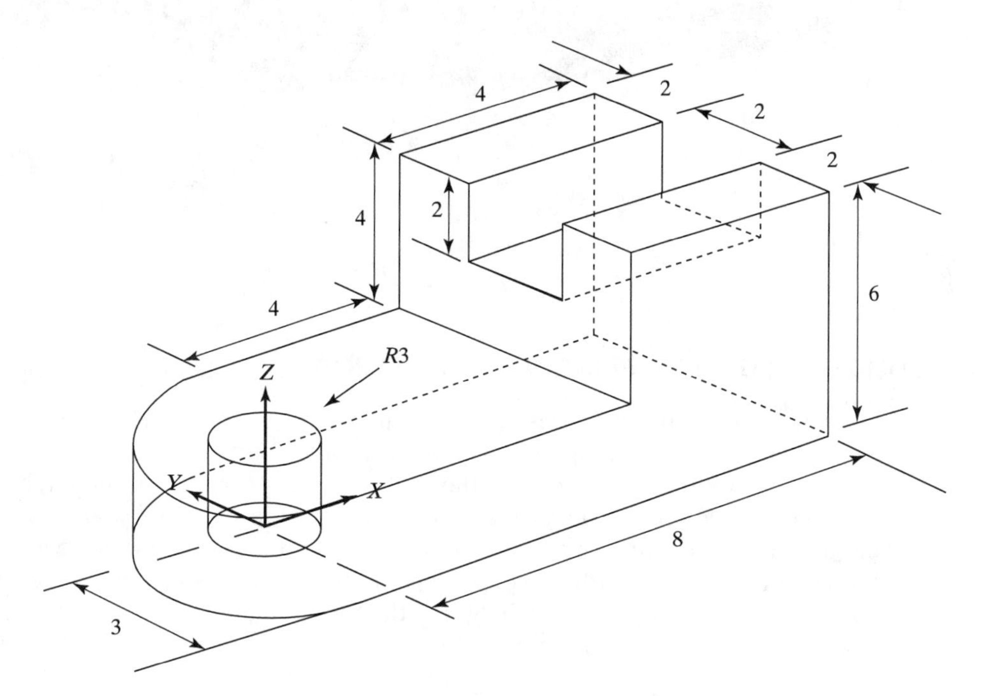

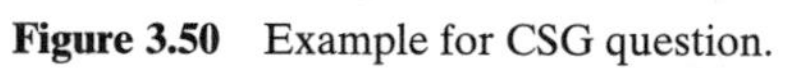
Figure 3.50 Example for CSG question.

CHAPTER 4

SOLID FREEFORM FABRICATION (SFF) AND RAPID PROTOTYPING

4.1 SOLID FREEFORM FABRICATION (SFF) METHODS

Several manufacturing processes are available to make the important transition from computer aided design (CAD) to a *prototype* part. Several new technologies began to make their appearance after 1987. In that year, stereolithography (SLA) was first introduced by 3D Systems Inc., and over the next five years several rival methods also appeared. This created the family of processes known as solid freeform fabrication (SFF). As with most new technologies at the beginning of the "market adoption S-shaped curves" in Chapter 2, the SFF domain is accompanied by a relative amount of advertising "hype." SFF processes are sometimes described as:

- Parts on demand
- From art to part
- Desktop manufacturing
- Rapid prototyping

At the time of this writing, stereolithography (SLA), selective laser sintering (SLS), fused deposition modeling (FDM), and layered object modeling (LOM) are being used on a *day-to-day basis* by commercial prototyping companies. The three-dimensional (3-D) printing process in cornstarch, plastic, and ceramics is also being used commercially. The methods lower on the list show promise but do not seem to be in great use by third party prototyping houses to make their daily income. Casting is a special case. It is still used to make one-of-a-kind prototypes. Furthermore, for batch runs in the 10 to 500 category it is a very cost effective method to use once an original mold has been made by a process such as stereolithography. Machining is also used to make one-of-a-kind or several prototypes.

4.1.1 Summary of SFF and Rapid Prototyping Processes

In daily commercial use:

- Stereolithography (SLA)
- Selective laser sintering (SLS)
- Laminated object modeling (LOM)
- Fused deposition modeling (FDM)

More at the research and development (R&D) stage:

- 3-D printing in cornstarch, plastic, or ceramic
- 3-D printing with plastics followed by planarization using machining
- Solid ground curing (similar to SLA)
- Shape deposition modeling (a combination of addition and subtraction)

Non-SFF (traditional):

- Machining
- Casting

Comparisons done in the early 1990s by the Chrysler Corporation revealed that the SLA process was ahead of its rival nontraditional prototyping methods in terms of cost and accuracy (these studies excluded an evaluation of machining and casting). Following the technical descriptions in this chapter, additional figures and tables are thus included to compare these costs and accuracies. Over the last decade, SLA has further emerged as the most used SFF process, especially for the generation of the master patterns for casting and injection molding. At the time of this writing, SLS, FDM, and LOM have the most visibility after SLA.

4.1.2 The History of SFF Methods

During the late 1970s, Mead and Conway (1980) created the groundwork for the fast prototyping of very large scale integrated (VLSI) circuits. Designers were encouraged to think in terms of five two-dimensional (2-D) patterns. These patterns defined three stacked interconnection layers on a metal-oxide-semiconductor (MOS) wafer and their mutual connections through holes. The patterns described the actual geometry of the connection runs and via holes that one would see when looking down onto the circuit chip, regardless of the exact process and number of masking steps that were used to implement the chip (see MOSIS, 2000).

Inspired by this success, beginning in the 1970s, several companies tried to create layered manufacturing for mechanical parts. Also by the mid-1980s, several U.S. government studies analyzed the possibilities of a "mechanical MOSIS" (Manufacturing Studies Board, 1990; Bouldin, 1994; NSF Workshop I, 1994, and II, 1995).

The prospects for a mechanical MOSIS were thus frequently linked to the fabrication processes in the lists mentioned (Ashley, 1991, 1998; Heller, 1991; Kruth, 1991; Woo, 1992, 1993; Au and Wright, 1993; Kochan, 1993; Kai, 1994; UCLA, 1994; Weiss and Prinz, 1995; Cohen et al., 1995; Dutta, 1995; Jacobs, 1992, 1996; Beaman et al., 1997; Kumar et al., 1998; Sachs et al., 2000).

The introduction of the first commercial SFF technology—stereolithography—was accompanied by the advent of the *ST*ereo*L*ithography (.STL) representation of a CAD object. ".STL" is a modified CAD format that suits a subsequent slicing operation and the "downstream" laser-scanning paths on a physical SLA, FDM, or SLS machine.

Is a soccer ball round? The answer depends on how carefully the ball is measured. Nominally, it is a perfect sphere. However, on closer inspection, the leather is sewn together from about 20 little hexagonal patches and a few pentagonal patches to create the curvature. In reality it is an approximation to a sphere.

Likewise, the ".STL" format approximates the boundary surfaces of a CAD model by breaking it down into interconnected small triangles—a process called *tessellation.* Each triangle is represented by the *x/y/z* coordinates of each of its three vertices, enumerated by the right-hand rule—that is, counterclockwise (ccw) order as viewed from the outside of the body. The vector normal to the surface of each triangle is also specified. This tessellated surface is stored as an "*.STL* file." This file, perhaps containing up to 200,000 triangles, is sent over the Internet to a prototyping shop.

As shown in Figure 4.1, this tessellated CAD model is then sliced like a stack of playing cards. For 3D Systems' machines this is known as the *SLI* or sliced file. Other rapid prototyping machines use the slicing technique but have their own file creation details and names. Each slice for the imaginary soccer ball will thus be a circle. However, because of the tessellation procedure it will not be a perfect circle. The slicing action cuts through the triangles on the boundary. Thus, each circular slice (or disc) will actually be a multisided polygon running inside the "bounding circle." The number of sides on this inner polygon is of course related to how finely divided the original tessellation was made.

Inside the SLA machine, the laser first creates the outer boundary of each slice and then "weaves" across each slice in a hatching pattern to create the layer. The number of slices and the style of the weaving pattern are chosen by each rapid prototyping shop. Especially for SLA and SLS a certain amount of trial and error, or craftspersonship, begins to play a role at this stage. This is reviewed in more detail over the next few pages.

".STL" is now the standard exchange format for SFF processes. However, it is inadequate for many reasons. First, the files are large due to the tessellation method. Second, there are redundancies in the ".STL" format. One example of redundancy is as follows: the triangles are represented by the "counterclockwise rule" so that it is clear in which direction the outer-surface normal acts. However, it has also become customary to specify the surface vector as well. Inconsistency can be introduced as a result of this redundancy, and no rules exist for resolving it.

McMains (1996) describes how ".STL" does not capture topology or connectivity, making it difficult to fix some of the common errors found in files—such as cracks, penetrating or extraneous faces, and inconsistent surface normals—without resorting to guessing the designer's original intent. More general digital interchange formats have also been used with SFF. These include ACIS (1993) and IGES (Heller, 1991). However, as described in NSF (1995), problems arise with these formats, too.

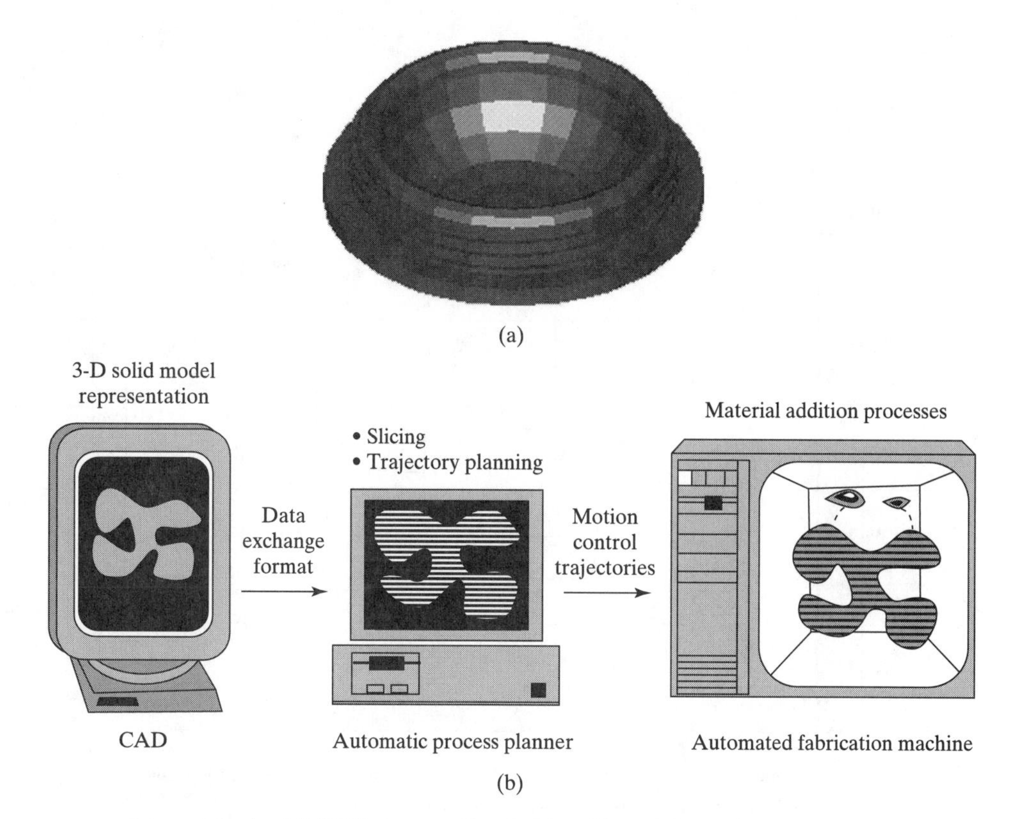

Figure 4.1 An ".STL" file is a tessellated object. The top figure shows a contact lens holder represented by many surface patches. The ".STL" file is then sliced. Laser motions then harden the part (courtesy of Lee Weiss).

One aspect of ongoing research is thus to improve this representation language (McMains et al., 1998).

4.2 STEREOLITHOGRAPHY: A GENERAL OVERVIEW

4.2.1 Background

Stereolithography (SLA) was launched commercially by 3D Systems Inc. in 1987 (see Jacobs, 1992, 1996). The process is shown in Figure 4.2.

The commercial launch followed from the studies of several independent programs on the curing of photopolymers. Some of these are mentioned in Table 4.1.

Also of historical interest is that the photocurable liquid was first developed for the printing industry and for furniture lacquers or sealants. In the latter case, to avoid carcinogenic solvents, an ultraviolet (UV) curing process was developed for furniture sealants.

One can imagine how SLA grew out of these developments: the SLA inventors must have seen how layers of the photocurable liquids could be built up on a chair

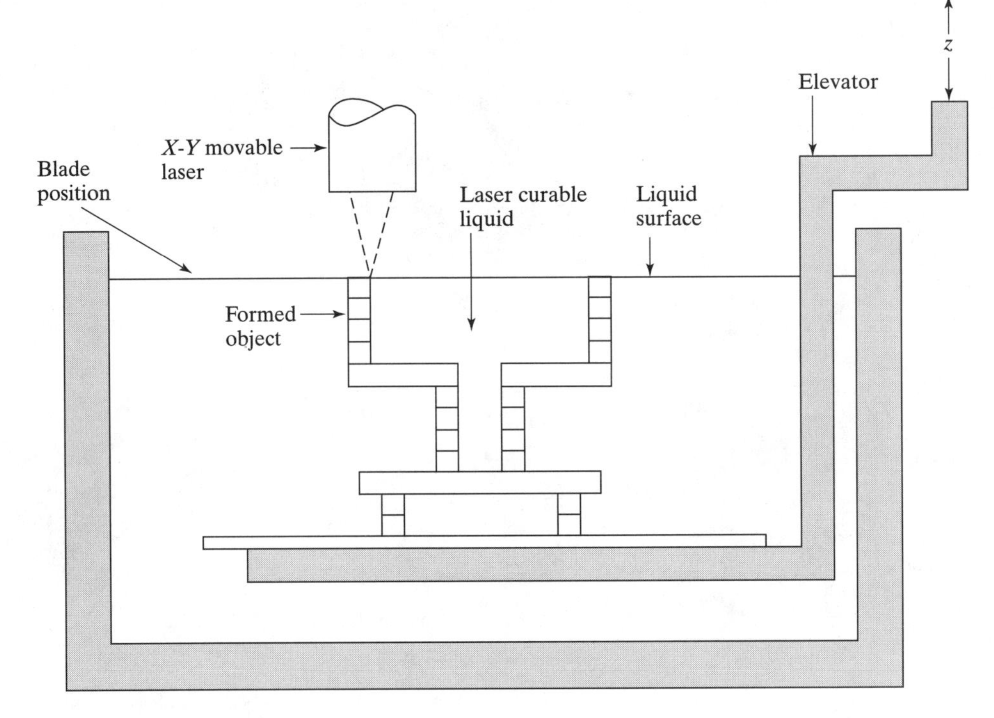

Figure 4.2 Stereolithography (SLA), based on the commercially published brochures of 3D Systems Inc. The helium-cadmium laser in the SLA-250 cures and fuses successive layers of resin. These descend on top of each other on the elevator until the whole part is formed, at which point the object is lifted from the vat.

TABLE 4.1 History of SLA

Date	Person(s)	Company and location	Activity
1970s	A. Herbert	3M, Minneapolis	R&D
1970s	H. Kodama	Nagoya Prefecture Research, Japan	R&D
1970s	C. Hull	Ultra Violet Products, California	R&D
1986	C. Hull and R. Freed	3D Systems Inc., formed from Ultra Violet Products, California	Patent secured
November 1987	3D Systems	3D Systems	Demonstration of the SLA-1 (which became the SLA-250) at the Autofact show in Detroit

leg, forming a solid shape. Then, the use of a helium-cadmium laser created more energy and more focused solidification patterns than a simple UV arc lamp. Finally, the rapidly decreasing costs of microprocessors during the early 1980s paved the way for the tessellation routines during CAD and the control of the lasers in the actual SLA machine.

One can also imagine the excitement these first inventors felt, as they saw the first layer of SLA material solidifying on the surface of a vat of resin: viewed at an angle it resembles the first layers of ice solidifying on a pond in early winter.

In production, once this first layer is cured, the elevator type stage lowers by 50 to 200 microns (0.002 to 0.008 inch) depending on the desired accuracy, and further layers are cured and connected by self-fusing to the previous ones. At the end of the process, the elevator rises and the component is lifted out and cured in its entirety. Postcuring is needed, probably overnight, before the prototype is ready for use. Hand sanding may be required to mitigate the *stair-stepping* effect described later.

Note that the object in Figure 4.2 has overhanging areas about halfway down its height dimension. During the actual process these need to be supported by slender sacrificial columns. Without these, the horizontal part of the component sags. Additional hand finishing is needed to snap out these slender sacrificial columns and sand any small stubs away from the surface.

4.2.2 Stereolithography Details: The ".STL" File Format

Introduced by 3D Systems Inc. in 1987, the ".STL" file format has become the de facto standard, even though other "direct slice" methods have been tried. The ".STL" method tessellates the CAD model with triangles just like the hexagons and pentagons on the surface of a soccer ball.

The ".STL" file is (a) a header, (b) the number of triangles, and (c) a list of the triangle description by vertices and the normal vector to the triangle. Table 4.2 shows the layout. The size of the ".STL" file is (50 × number of triangles) + 84. Thus a 10,000-triangle object needs 500,084 bytes.

TABLE 4.2 The ".STL" File Format

Entity	Described by
The header	80 bytes
The number of triangles	Unsigned long integer (4 bytes)
For each tessellation triangle (50 bytes of information)	See below
Normal vector I	Floating point integer (4 bytes)
Normal vector J	Floating point integer (4 bytes)
Normal vector K	Floating point integer (4 bytes)
First vertex X	Floating point integer (4 bytes)
First vertex Y	Floating point integer (4 bytes)
First vertex Z	Floating point integer (4 bytes)
Second vertex X	Floating point integer (4 bytes)
Second vertex Y	Floating point integer (4 bytes)
Second vertex Z	Floating point integer (4 bytes)
Third vertex X	Floating point integer (4 bytes)
Third vertex Y	Floating point integer (4 bytes)
Third vertex Z	Floating point integer (4 bytes)
Attribute	Unsigned integer (2 bytes)

Two rules govern the triangle descriptions (Figures 4.3 and 4.4).

1. The right-hand counterclockwise rule, or "*ccw* rule," is a corkscrew acting outward on the soccer ball, to order the vertices and the normal vector.
2. The vertex-to-vertex rule, which insists that the vertices on an adjacent triangle link to the neighbor and that no vertices meet a neighboring edge.

4.2.3 Stereolithography Details: C-Slice Processing

When the ".STL" file arrives at the rapid prototyping bureau, the slicing begins as follows:

- Sort the ".STL" triangles into "*z* values" (this establishes the layers).
- Find the boundary segments (gives contiguous internal and external pocket/shape contours).
- Create boundary polylines.
- Apply edge compensations (based on operator's knowledge of laser physics).
- Compare with adjacent layers to minimize stair-stepping on chamfered sides.
- Smooth boundaries.

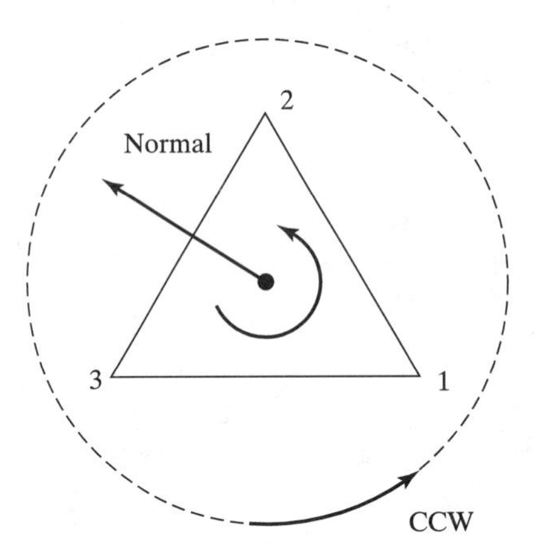

Figure 4.3 Rules for tessellation.

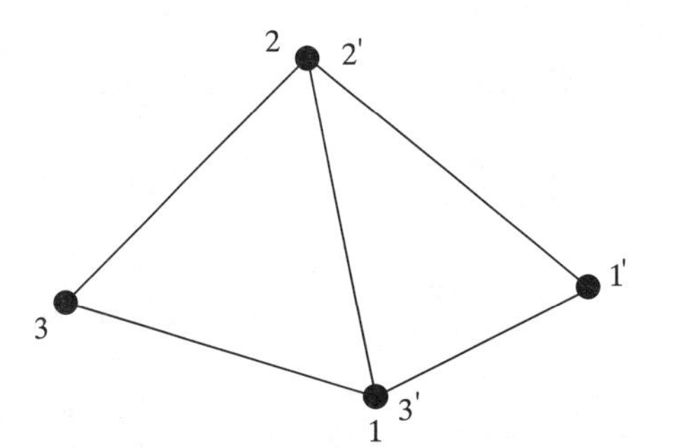

Figure 4.4 Vertex-to-vertex rule means that a vertex cannot join to a random point somewhere on an edge. Each vertex has to meet another vertex on the neighboring triangle.

- Output boundary data.
- Treat next cross section.

4.2.4 Stereolithography Details: The Resin

The photocurable liquid was developed for printing and for furniture lacquer/sealant. To avoid liquid that had carcinogenic solvents, the UV curing process was developed. Lasers provide more direct energy and allowed the invention of SLA once computers were powerful enough to create a tessellation. SLA is a low-energy curing process compared with SLS (using a CO_2 laser).

Photopolymerization is defined as linking small molecules (monomers) into larger molecules (polymers) comprised of many monomer units. Vinyl monomers have a carbon-carbon (C=C) double bond attached to complex groups donated by "R." In the original resin, the monomer groups are only weakly connected to their neighbors by van der Waals bonds. As the laser acts on the bonds, the C=C bonds break. The broken monomer groups connect to each other, forming long chains (see Table 4.3).

TABLE 4.3 Polymerization

Weak van der Waals bonds between the adjacent chains	Strong covalent bonds along chains
$H_2C = CH$ $H_2C = CH$ (with R attached to CH)	$-CH_2-CH-CH_2-CH-$ (with R attached to the first CH)

The bonding between such chains then creates three key effects:

- The liquid gels into a solid.
- The density increases.
- The shear strength increases.

Although the original vinyl monomers are already cross-linked, they get much more strength from the formation of the covalent bonds in the long chains.

4.2.5 Stereolithography Details: The SLA Manufacturing Process

To create any individual layer, the laser traces out the boundaries of a layer first. This is called bordering; imagine a large elastic band or loop lying on the surface. Second, a hatching or weaving pattern crosses the entire area. Third, the hatched areas are filled in, causing the final gelling and solidification (Figure 4.5).

After each layer is formed, the laser scanning moves to the next layer. However, some careful process planning is needed to create the accuracy of only a few thousandths of an inch. Details that control accuracy are presented after Figure 4.5.

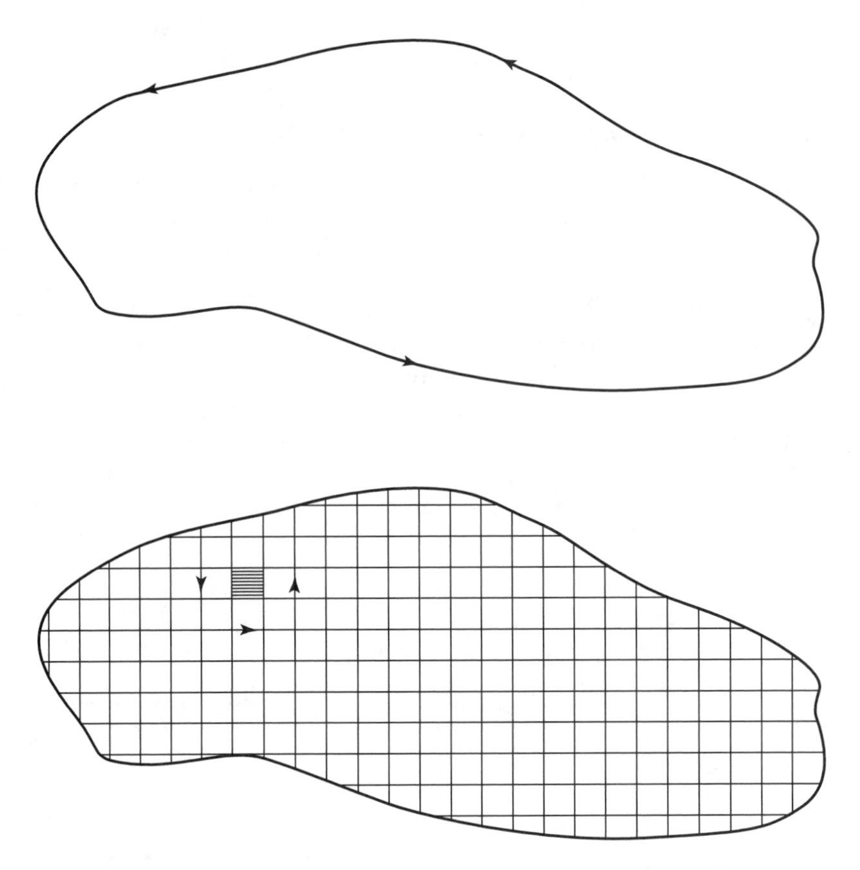

Figure 4.5 Establishing the border, then hatching and filling (filling is shown on just one square).

Note: These steps are for an SLA-500 machine at the time of writing. The details for the SLA-250 are slightly different, and, in addition, new refinements are constantly taking place.

4.2.5.1 Step 1. Preparation of the Script

The part building needs instructions on the desired accuracy. Typically, a layer thickness of 100 microns (0.004 inch) is the average build layer. However, it may range from 50 to 200 microns (0.002 to 0.008 inch) depending on the desired accuracy. Also the Zephyr blade sweeping times, and the "*z*-wait" times, need to be programmed in. These are described later.

4.2.5.2 Step 2. Leveling and Laser Calibration

SLA resins undergo 5% to 7% total volume shrinkage, and of this amount, 50% to 70%, occurs in the vat during polymerization (Jacobs, 1992). Since the liquid level is always shrinking down, a sensor must be installed to follow the level. If the vat is not at the desired height for the beginning of the run, a plunger mechanism adjusts the fluid level by fluid displacement. Also, it is crucial to adjust the laser position with reflective "eyes" at the corners of the machine's stage. In fact, this check of laser position occurs just before each layer is done.

4.2.5.3 Step 3. Making the Initial Supports

The first few runs with the laser are not for the part itself but for small supports that the actual part will rest upon. The supports can be viewed as small feet, rather like those on a heavy sofa or piano: they are needed on the bottom of the part to lift the lowest layer off the floor of the elevator platform. In particular, the supports are needed:

- So that the Zephyr blade will not hit the platform
- To compensate for platform distortion
- So that it is easier to remove the finished part
- Internal supports are also needed for any "overhanging" structures

When making the supports, after the first laser cured layer is formed, the stage needs to be pulled down about 12 mm (0.5 inch) for the SLA-500 (Jacobs, 1992). This "deep dip" allows the viscous, honeylike fluid to more easily flow over the surface of the first layer of the supports. The elevator then rises up to be positioned 100 microns (0.004 inch) below the surface. It is usual to wait about 5 seconds and then do the laser curing again. This creates the second layer—but still, this is concerned with the supports, not the part itself. This procedure repeats until the supporting stubs are large enough. The operator usually makes these decisions.

4.2.5.4 Step 4. Creating the Actual Part

The procedure to make the actual part (not the supports) is somewhat different. Once the supports are finalized, the first bottom surface of the part is generated by the "bordering + hatching + filling" described earlier.

The elevator descends by 100 microns (0.004 inch) and then waits typically for 45 seconds. This time is programmed in by the operator. It is a recommendation from the SLA fluid supplier as the time needed for the full curing to occur of a part layer. Note that although the laser has begun the polymerization process, it still takes up to 45 seconds for the full effect of polymerization to occur and to harden the layer enough to build subsequent layers on top of it. After the 45-second wait, the first layer is hardened enough for the Zephyr blade to sweep over the surface and precisely set the 100 micron (0.004 inch) layer of liquid for the second polymerization.

4.2.5.5 Step 5. Sweeping Using the Zephyr Blade

At first glance, the Zephyr blade looks like a "hard squeegee" used to clean a car window. In fact, it has a long, hollow cavity between two adjacent blades, and this cavity is under the influence of a slight vacuum pump. This draws SLA liquid into the bottom of the blade. Thus, as the blade sweeps over the surface, it is "charged" with liquid and more easily and uniformly deposits the next liquid layer onto the first. At the same time the sweeping blade distributes the SLA liquid evenly. Note that the honeylike SLA fluid is very viscous, and it needs the distribution of the vacuumized Zephyr blade to get an even surface.

As the Zephyr blade traverses the whole vat, it removes excess resin in some areas, and yet because it is "charged" with resin, it distributes and fills any areas that lack resin. The sweep takes about 5 seconds (Jacobs, 1992) unless a hollowlike part is being made where the viscous fluid inside the hollow takes longer to follow the blade. The sweep gives a uniform thin layer, but given the viscosity of the fluid, there is a tendency for resin to adhere to the blade, followed by separation and a "bulge" just downstream from the part's leading edge.

4.2.5.6 Step 6. "Z-Wait" of about 15 Seconds

Even after all the adjustments and sweeping, a "crease" exists around the edge of the part at the solid-liquid interface. The "z-wait" allows a relaxation of this effect to a flatter, smoother resin surface.

4.2.5.7 Step 7. Extra Skin Filling

At the very end of the process, more intense hatching may be desirable on the top surface of the part. Very closely spaced line vectors cause more intense solidification structures on the up-facing surfaces. It is likely that similar patterns would have been done on the down-facing outer skin in Step 4.

4.2.5.8 Step 8. Final Steps

The final steps include:

- Draining excess resin from any inner or depressed cavities
- Cleaning and rinsing with solvents
- Snapping out bridgeworks
- Hand sanding and polishing
- Postcuring in a broad spectrum UV light source

4.2.6 Stereolithography Details: Laser-Based Manufacturing and Prototyping

During stereolithography, selective laser sintering, or any laser-based process, many details of the "*laser energy delivered*" to the resin (or powder for SLS) control solidification and the accuracy that can be achieved. First consider penetration depth. Note that the bottom of each SLA layer has to adhere to the previous layer, and so the topic of main interest is the "energy at depth z" of the laser. Lasers give much more energy (i.e., are able to cause more "polymerization by irradiance") than regular arc lamps. But as they travel down through the resin or powder they do nevertheless decay exponentially by the Beer-Lambert exponential law of absorption:

$$H_{(x,\,y,\,z)} = H_{(x,\,y,\,0)} \exp\left(-\frac{z}{D_p}\right) \tag{4.1}$$

A critical exposure $H_{(c)}$ is needed to "gel" the resin. D_p is a resin constant defined by the depth of a particular resin that results in a reduction of irradiance level to $1/e$ ($= 1/2.718$) of the H_o level on the surface (Figure 4.6). That is, at a depth of $z = D_p$ the

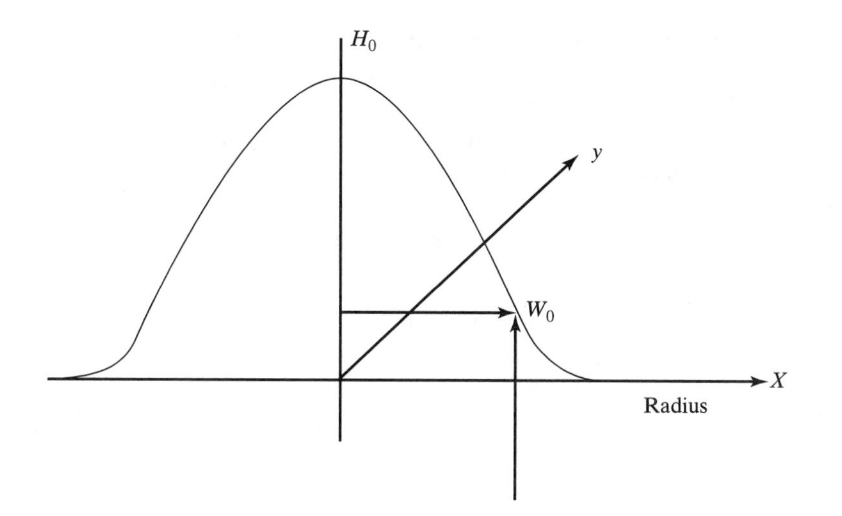

Figure 4.6 Gaussian decay of laser across the surface.

irradiance is ~ 37% of H_o. For the SLA-250, typical values are given by Jacobs (1992) as follows, and it is of interest to relate laser behavior to resin solidification:

$$\text{Nominal laser power} = (P_L) = 15 \text{ milliwatts}$$

$$\text{Central spot size} = (2W_0) = 0.25 \text{ millimeters}$$

For the whole spot the Gaussian irradiance curve controls the basic physics, and like any point source of light it decays from the center.

Across the surface, as opposed to down through the surface (Equation 4.1), the laser decays as follows:

$$H_{(x,\,y,\,0)} = H_{(r,\,0)} = H_0 \exp\left(-\frac{(2)r^2}{W_0^2}\right) \tag{4.2}$$

where W_0 is the $\frac{1}{e^2}$Gaussian half width (Figure 4.6).

Thus, at $r = W_0$,

$$H = H_0 e^{-2} = 0.135 H_0$$

It can also be shown that

$$H_{\text{average}} = \frac{P_L}{\pi W_0^2} \tag{4.3}$$

$$= 30.56 \text{ watts per cm}^2 \tag{4.4}$$

If the scan speed is 200 mm per second, the scanning laser's exposure time on a given area is

$$t_e = \frac{2W_0}{V} = 1.25 \text{ milliseconds} \tag{4.5}$$

The laser exposure's average energy density is then

$$E_{(\text{average})} = H_{\text{average}} \times t_e = 38.2 \text{ mJ/cm}^2 \tag{4.6}$$

The analysis now proceeds to calculate the polymerization ability from the laser's photon flux. It is necessary to use Planck's equation to find the photon energy first:

$$E_{(\text{photons})} = \frac{hc}{\lambda} = \frac{(6.62 \times 10^{-34} \text{J} \cdot \text{s}) \times (3 \times 10^{10} \text{cm/s})}{3.25 \times 10^{-5} \text{ cm}} \tag{4.7}$$

$$E_{(\text{photons})} = 6.1 \times 10^{-19} \text{ Joules per photon} \tag{4.8}$$

λ is the laser wavelength

c is the speed of light

h is Planck's constant

Denoting that N_{ph} = number of photons per square centimeter hitting the resin surface:

$$N_{\text{ph}} = \frac{E_{\text{average}}}{E_{\text{photons}}} = 6.3 \times 10^{16} \text{ photons per cm}^2 \tag{4.9}$$

This flux of photons penetrates into the resin and acts on the polymer chains to cause polymerization. Even if the photochemical efficiency is only 50%, the {C=C} bonds will polymerize to {C—C—C}.

4.2.7 Selective Laser Sintering (SLS)

Another very popular method is selective laser sintering (SLS), commercialized by the DTM Corporation. In many respects SLS is similar to SLA except that the laser is used to sinter and fuse powder rather than photocure a polymeric liquid.

The first step is to prepare the ".STL/SLI" files as described earlier. Inside the SLS machine, a thin layer of fusible powder is laid down and heated to just below its melting point by infrared heating panels at the side of the chamber. Then a laser sinters and fuses the desired pattern of the first slice of the object in the powder. Next, this first fused slice descends, the roller spreads out another layer of powder, and the process repeats (Figure 4.7).

In comparison with SLA, this process can rely on the supporting strength of the unfused powder around the partially fused object. Therefore, support columns for any overhanging parts of the component are not needed. This allows the creation of rather delicate, lacelike objects. Nevertheless hand finishing is still needed to improve the inevitable stair-stepping. Also, SLS parts have a rough, grainy appearance from the sintering process, and it is often preferable to hand smooth the surfaces. Another difficulty is maintaining the temperature of the powder at a few degrees below melting. This is done with the infrared panels, but maintaining an even temperature over a large mass of powder requires long periods of stabilization before sintering by the laser can be started.

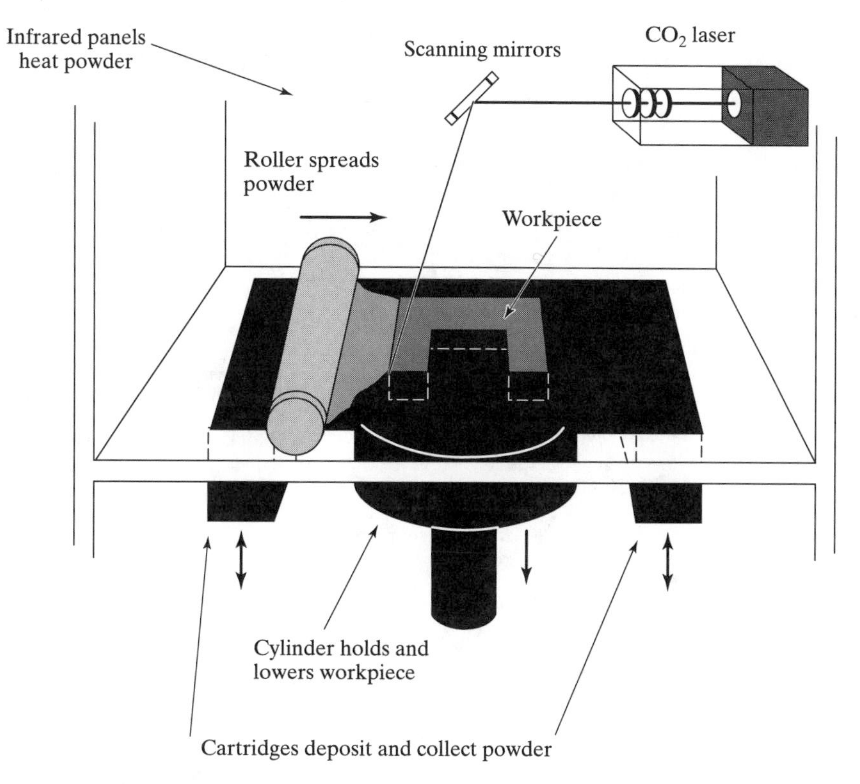

Figure 4.7 Selective laser sintering (SLS), based on commercially published brochures from the DTM Corporation.

4.2.8 Laminated Object Modeling (LOM)

Laminated object modeling (LOM) was developed by Helisys Inc., and like SLA and SLS, it was first offered commercially in the period from 1987 to 1990. In LOM, the laser is used to cut the top slice of a stack of paper that is progressively glued together. After each profile has been cut by the laser (shown at the bottom right of Figure 4.8), the roll of paper is advanced, a new layer is glued onto the stack, and the process is repeated. After fabrication, some trimming, hand finishing, and curing are needed. For larger components, especially in the automobile industry, LOM is often preferred over the SLA or SLS processes.

4.2.9 Fused Deposition Modeling (FDM)

Fused deposition modeling (FDM) was developed by Stratasys Inc. and is executed on machines called the FDM 1650, 2000, or 8000 series. Figure 4.9 shows that the material is supplied as a filament from spool. The overall geometry and system are reminiscent of icing a cake. The filament melts as it flows through a heated delivery head and emerges as a thin ribbon through an exit nozzle. The nozzle is guided around by CNC code, and the viscous ribbon of polymer is gradually built up from a

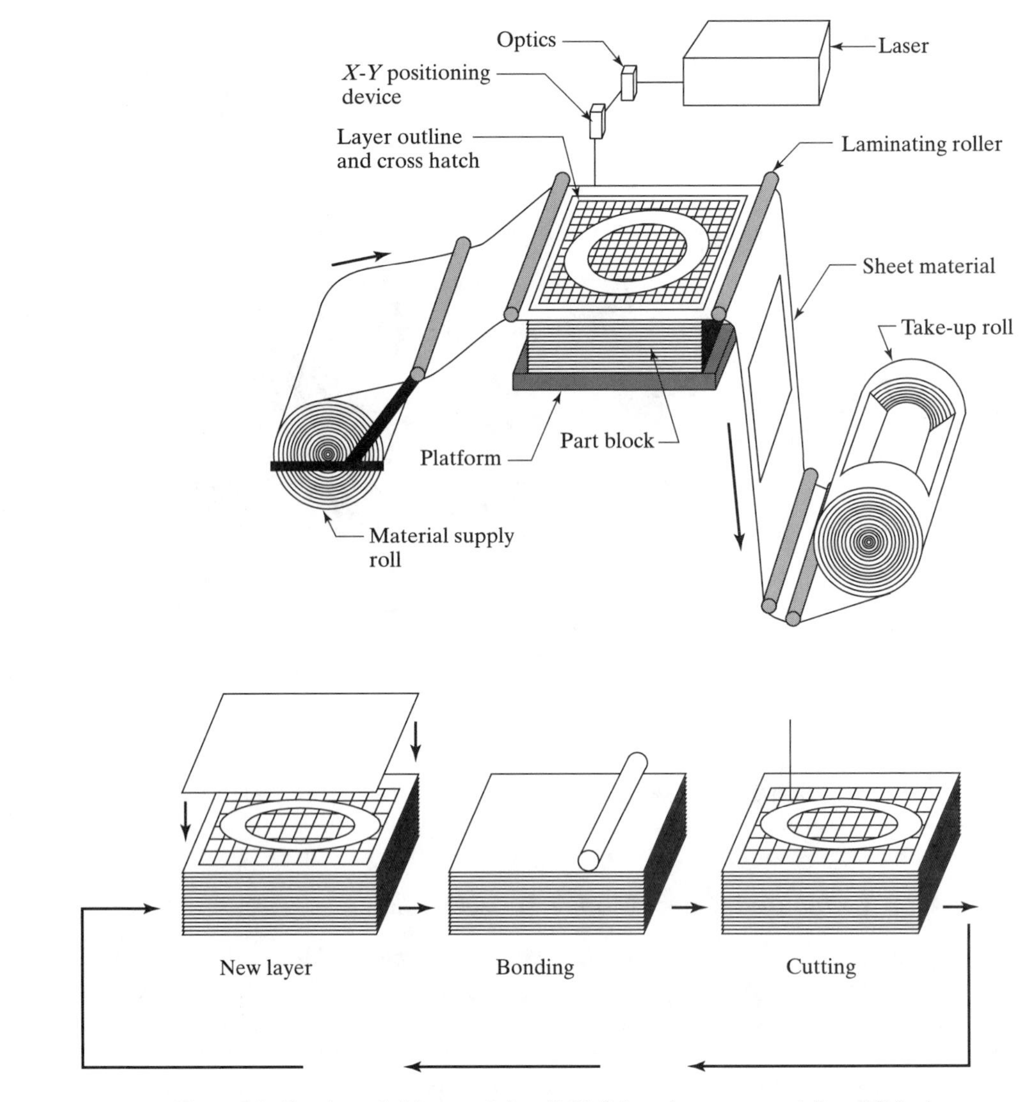

Figure 4.8 Laminated object modeling (LOM), based on commercially published brochures of Helisys Inc.

fixtureless base plate. In terms of motion control, FDM is more similar to CNC machining than SLA or SLS. For simple parts, there is no need for fixturing, and material can be built up layer by layer. The creation of more complex parts with inner cavities, unusual sculptured surfaces, and overhanging features does require a support base, but the supporting material can be broken away by hand, thus requiring minimal finishing work. Thus, despite the similarities with the CNC machine from the point of view of control, the resulting parts that can be made are more in the SFF family. A similar deposition machine, the Model-Maker 3-D plotter, has been devel-

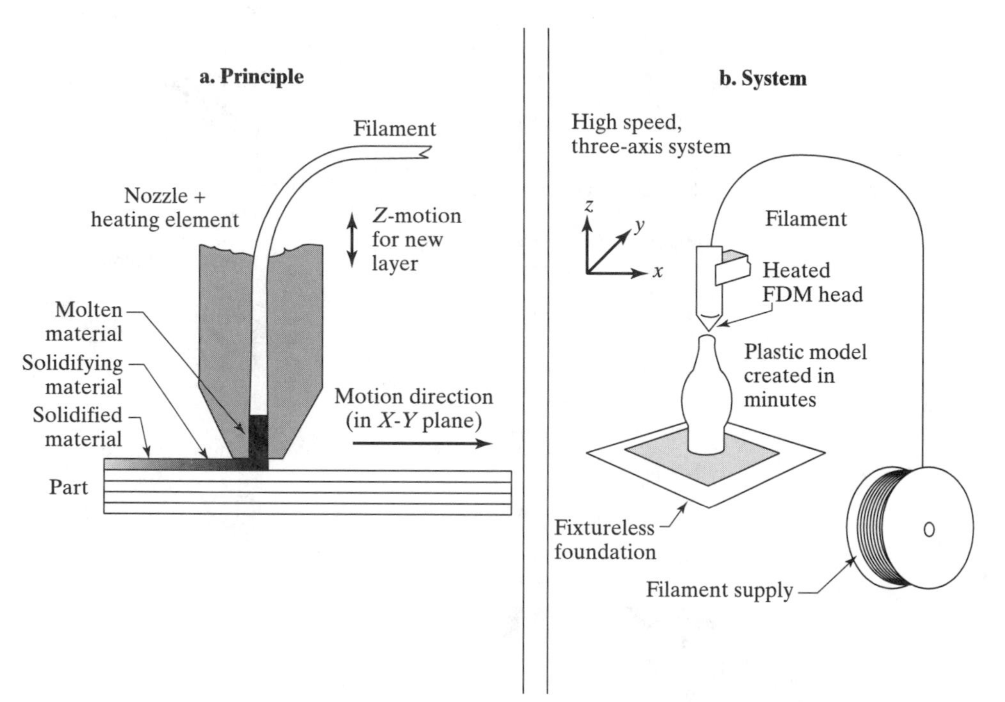

Figure 4.9 Fused deposition modeling (FDM), based on published brochures of Stratasys Inc.

oped by Sanders Inc. In the Sanders machine, nozzles are used to deposit viscous polymers. It is difficult to control the flow of the viscous polymer and obtain an evenly distributed layer, just as it would be to control the flow of toothpaste onto a flat surface to obtain an even thickness. Each formed surface layer is thus machined (or planarized) with a milling cutter prior to the application of the next layer.

4.2.10 3-D Plotting and Printing Processes

Several types of 3-D printing processes have been developed in recent years and are constantly being updated at the time of this writing. Some of these are aimed at the educational CAD/CAM market where students are invited to obtain quick models of an emerging design. At the same time, such machines might be useful in an industrial design studio, where artists might want to generate and regenerate a quick succession of prototypes for the "look and feel" of an emerging design. Examples include:

- 3-D printing of cornstarch, followed by layer-by-layer binder hardening, is the basic principle behind the Z-Corporation machine. The first step in Figure 4.10a is to spread a thin layer of powder of the desired material across the top of the bed. The next step hardens the desired geometry into this layer of powder. The hardening is not done with a laser (like SLS) but with a binder phase. Fine

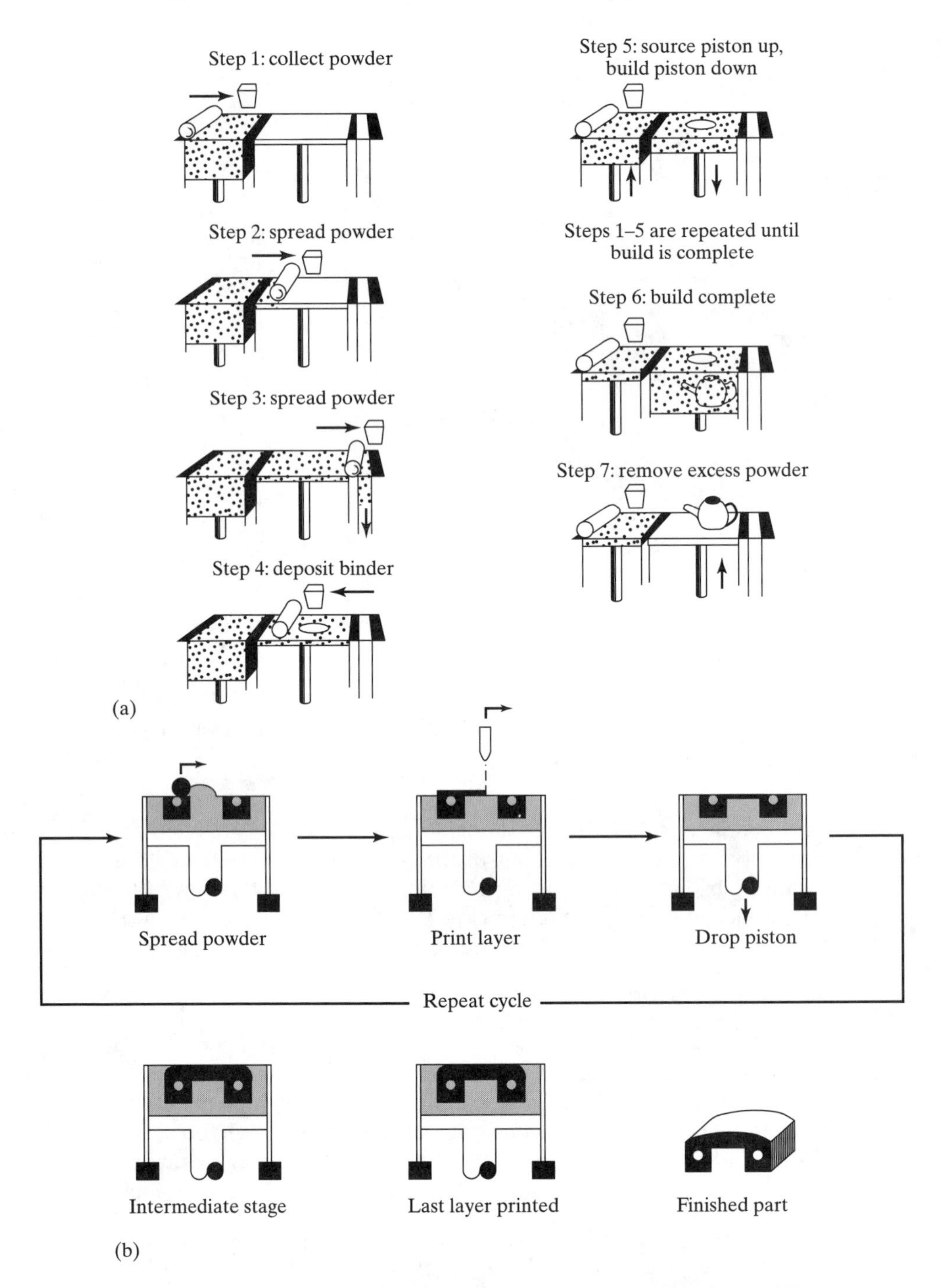

Figure 4.10 (a) 3-D printing (based on commercially published brochures of Z-Corporation Inc.) and (b) the method developed by Sachs and colleagues (2000).

droplets of the binder stream are printed down through a continuous-jet nozzle carried by the print head. Since material is built up layer by layer in an *x/y* plane, the process resembles the motions of the ink-jet printing heads on a conventional word processing printer.

- A more accurate 3-D printing process, developed by Sachs and colleagues at MIT, was the forerunner to this technology (Figure 4.10b). This process is being used to build the ceramic molds for metal castings and powder-metal tooling for injection molding dies. Commercial applications of this process are growing (Smith, 2000; Sachs et al., 1992, 2000).

4.2.11 Solid Ground Curing (SGC)

Solid ground curing was introduced by Cubital Inc. A schematic diagram of the process is shown in Figure 4.11. The quickest way to understand the sketch is to begin with the operation at the "cross roads," where the mask is being used to photocure the uppermost layer of liquid of the block. Solid ground curing uses exactly the same physical process as SLA to photocure the polymer liquid. The key difference is that SLA does it by using a laser *point source,* whereas SGC does it by exposing a *complete plane* at once through the mask. Cubital's machine is one integrated unit. However, two interlaced processes occur simultaneously. On the left of the schematic the following steps are shown: CAD files are sliced, a mask plate is prepared for a single layer, and the finished mask rotates into the exposure area. On the right of the schematic, a thin layer of photopolymer is spread over the surface of a block. This moves into the exposure area to be hardened. Postprocessing steps on the right include wiping off the residual photopolymer, using supporting wax to fill in any

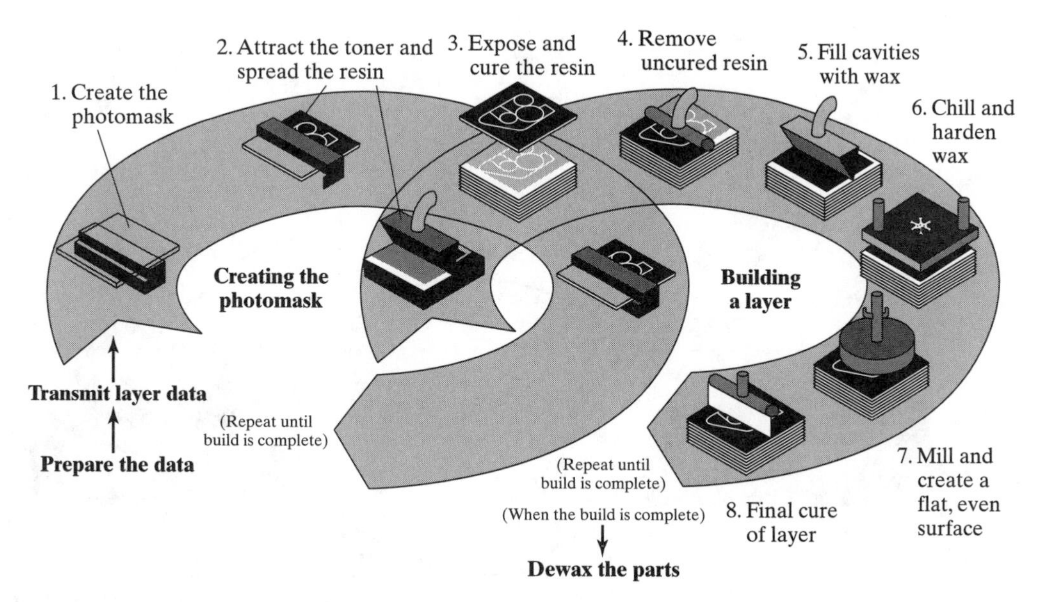

Figure 4.11 Solid ground curing process, based on commercially published brochures of Cubital America, Inc.

hollow areas, cooling, and planarization before returning to the first position again for more spreading of the photocurable liquid. Meanwhile, as shown on the left, the first pattern on the mask plate is erased and the next pattern is applied.

4.2.12 Shape Deposition Manufacturing (SDM)

In some of the two processes described earlier, for example, 3-D printing by Sanders and SGC by Cubital, a combination of material additive and material removal takes place. Shape deposition manufacturing (SDM) also exploits this paradigm (Weiss et al., 1990; Weiss and Prinz, 1995; Weiss et al., 1997; Weiss and Prinz, 1998). The goals for SDM are to combine the advantages of SFF (i.e., easy to plan, does not require special fixturing, arbitrarily complex shapes, and heterogeneous structures) with the advantages of machining (i.e., high accuracy, good surface finish, and wide-scale availability of existing CNC machines and infrastructure).

In SDM, a CAD model is again sliced into 3-D layered structures. Layered segments are deposited as near-net shapes and then machined to net shapes before additional material is deposited. The sequence for depositing and shaping the primary and support materials is dependent upon the local geometry (Figure 4.12). The idea is to decompose shapes into layered segments such that undercut features need not be machined but are formed by previously shaped segments. SDM can use alternative deposition sources from welding to extrusion. Producing smooth surface transitions between layers, however, remains a challenge, due in part to the layer-by-layer accumulation of residual stresses.

SDM can therefore combine complex surfaces and high accuracy. In the future it also promises to fill a niche for creating "wearable computer" products with multiple materials and even with embedded electronics (Smailagic and Siewiorek, 1993).

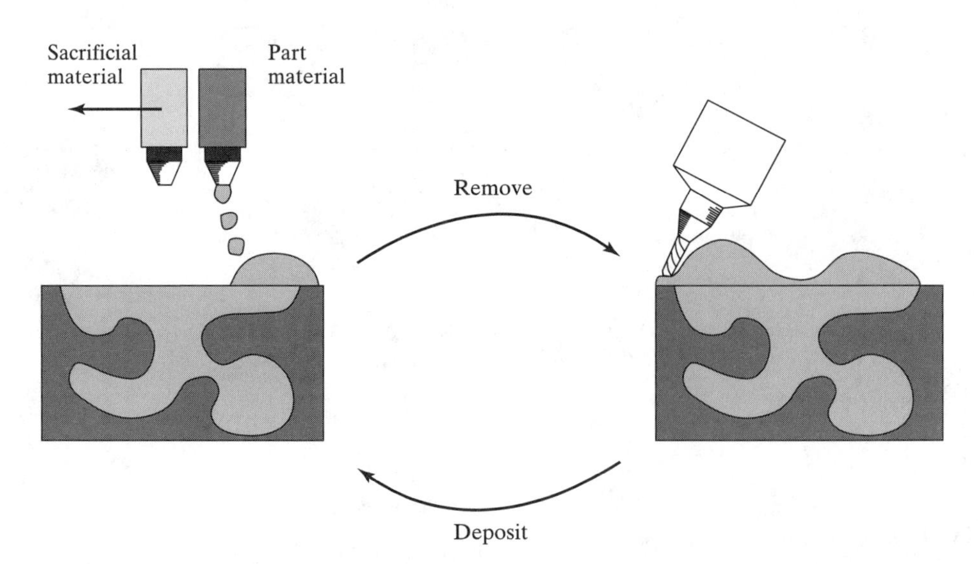

Figure 4.12 Shape deposition manufacturing (SDM) (courtesy of Lee Weiss).

4.3 COMPARISONS BETWEEN PROTOTYPING PROCESSES

4.3.1 Materials That Can Be Formed with the Various Processes

The SLA process uses photocured polymers that do not exhibit great strength or toughness. Nevertheless, in the SFF family, the SLA process is the most accurate and has emerged as the industry standard for creating a *master pattern* that might then be used as the basis for a casting or injection mold.

On the other hand, if a *single prototype* needs to be tested to destruction, or carried around for a while, it really has to be made from metal or a structural plastic such as ABS. If only one or two prototypes are needed, the FDM process is an ideal choice. FDM can extrude ABS polymers and create prototypes that are between 50% and 80% of full ABS strength. For full strength plastic or metal prototypes, the standard machining process is the preferred choice, despite the more limited range of geometric shapes that can be made by machining. CNC machining is also the most likely prototyping process for the *small batch manufacturing* of 2 to 10 components. If CNC machining is out of the question because of geometric complexity, SLS metal-powder parts might be the best choice. Beyond batch sizes of 10, it is worth considering the use of small batch casting methods. This decision will be influenced by desired accuracy, machining being better than casting. Some developments in shape deposition manufacturing and 3-D printing are leading to direct mold making (e.g., Weiss et al., 1990; Sachs et al., 2000).

4.3.2 Accuracy

Accuracy is perhaps the next key feature that distinguishes the various prototyping processes. The list that follows gives some very general values for a variety of SFF processes and other more traditional processes that can be used to make one or two components.[1]

- Hot, open die forging: +/− 1,250 microns (0.05 inch)
- Laminated object modeling: +/− 250 microns (0.010 inch)
- Investment (lost-wax) casting: +/− 75 microns (0.003 inch)
- Selective laser sintering: +/− 75 to 125 microns (+/− 0.003 to 0.005 inch)—depends on part geometry
- Stereolithography: +/− 25 to 125 microns (+/− 0.001 to 0.005 inch)—depends on part geometry
- Plastic injection molding from a machined mold (prototyping version): +/− 50 to 100 microns (+/− 0.002 to 0.004 inch)
- Rough machining: +/− 50 microns (0.002 inch)
- Finish machining: +/− 12.5 microns (0.0005 inch)

[1]The first entry corresponds to the age-old village blacksmith's prototyping shop. See Wright and associates (1982) for the CNC controlled version.

- Electrodischarge machining: +/− 2.5 microns (0.0001 inch)
- Lapping and polishing: +/− 0.25 microns (0.00001 inch)

When comparing the everyday prototyping methods, the most accurate remains machining, with easily achieved accuracies of +/− 25 microns (0.001 inch) and even half this with a good craftsperson. The next most accurate is prototyping by plastic molding from a machined mold, with an accuracy of +/− 50 microns (0.002 inch).

After that the SLA and SLS processes are listed. For a typical component, selective laser sintering and stereolithography average out at +/− 50 to 125 microns (0.002 to 0.005 inch). This is different from the accuracies of 25 microns (0.001 inch) quoted by the suppliers of SLA equipment, and confrontational e-mails will probably be a result of the obvious differences used in this text. However, these advertised accuracies of 25 microns (0.001) are for simple linear objects. Some users have to make complicated computer casings and medical monitors where open shell structures "warp and shrink all over the place," to quote one user. In some cases this warping and shrinking worsens the SLA accuracy to as much as +/− 375 microns (0.015 inch).

For SFF, the other accuracy consideration is stair-stepping. Mentally picture the soccer ball again, but this time with perfectly smooth surfaces. Now approximate the soccer ball by representing it as a stack of thin slices. The largest diameter slice is at the equator; the smallest slice is at the poles. Figure 4.13 from Jacobs (1996) shows that the approximation to the soccer ball becomes worse as the bounding curve comes up around the object toward the poles. In addition, the loss in accuracy/fidelity is related to layer thickness. Since SLA processes are improving all the time, layers down to 25 to 50 microns (0.001 to 0.002 inch) are now possible, therefore giving better and better accuracies. As with all manufacturing processes, the process then does take longer and more cost is involved.

Investment (lost-wax) casting is listed at +/− 75 microns (0.003 inch). Thus if the casting process is used to make a short-run prototyping mold and then the part is injected in plastic, it would seem to offer about +/− 125 microns (0.005 inch), but some hand finishing and some cosmetic work on the mold will give as good a plastic part as the cast mold.

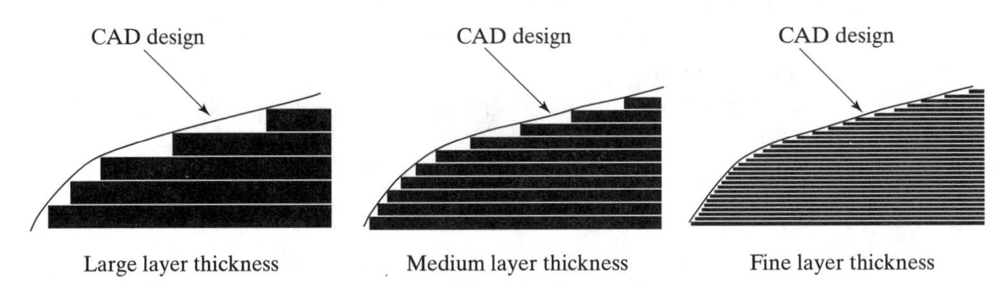

Figure 4.13 The stair-stepping approximation in SFF processes.

4.3.3 Lead Time of Prototypes

With an in-house dedicated FDM machine, a part can be produced within a 24-hour period. For an ongoing design activity where a design team needs a series of prototypes—for the look and fit of subcomponents and subassemblies—the FDM process is ideal.

An in-house integrated CAD/CAM system for machining can generate a simple part in a "morning's work," whereas more complex parts will take two to three days. An in-house stereolithography machine will also create the same parts in two or three days, measuring the time from receiving the ".STL" file to a fully cured product. The curing time, incidentally, is an added time factor, often overlooked when rival companies develop their advertising literature and compare their particular process with others.

If an in-house machine is not available, it should be realized that the SLA service bureaus are swamped with business in today's economy. Unless a special customer relationship exists, turnaround time of one to three weeks is more probable. Given the need for some negotiation with a client, and the need to check incoming computer files, the actual turnaround time may be longer still. Nevertheless, the rapid prototyping shops are selling "service and speed" rather than "fidelity."

For small batches (10 to 500) of injection molded plastic parts, customers can expect a three- to six-week turnaround time. The steps might be (a) an SDRC/IDEAS or Pro-Engineer CAD file is received from the Internet, (b) files are checked, (c) an SLA master is made, (d) an aluminum mold is cast, and (e) the finished batch of 10 to 100 is injection-molded in ABS plastic.

4.3.4 Batch Size

Chapter 2 describes the influence of batch size. For just one component, SFF processes—such as stereolithography, fused deposition modeling, and selective laser sintering—or machining is the obvious choice. Small-batch casting in metal, or small-batch injection molding in plastic, is used for batch runs between 50 and 500.

4.3.5 Cost

In general, cost increases with fidelity and accuracy needed, for all the rapid prototyping processes. Figure 4.14 shows why this is so. Specifically, if the designer desires more accuracy, the ".STL" files will need to be of finer resolution, the slicing will also be thinner, the laser will make more scanning paths, and the time and hence costs will increase. Also, all prototyping processes (SFF or machining) require some hand finishing, sanding, and deburring. Obviously, costs increase if the designer prefers a smoother surface finish. In all prototyping processes there is also a relationship between complexity, surface finish, accuracy, and cost. For SFF, Figure 4.14 shows that overhanging features require explicit support especially for SLA. For the arrangement on the right of Figure 4.14, the support columns have to be broken off by hand after manufacture. This usually leaves small stubs on the surface, which must then be sanded away.

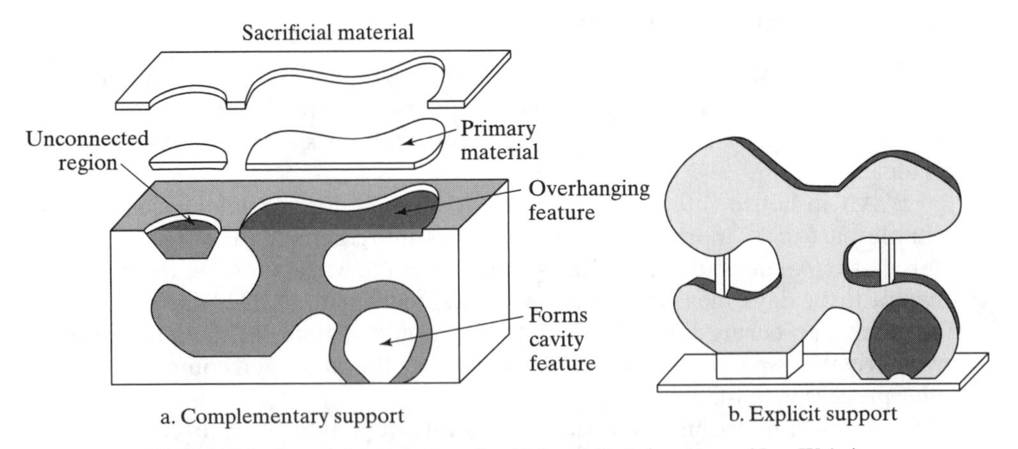

Figure 4.14 Supporting structures for SLS and SLA (courtesy of Lee Weiss).

TABLE 4.4 Rapid Prototyping Machine Cost—Also See Section 4.3.6 for Installation and the Like (as of March 2000)

Rapid prototyping machine (RP process)	Machine cost
SLA-250 (SLA)	$210,000
SLA-350 (SLA)	400,000
SLA-500 (SLA)	500,000
FDM 2000 (FDM)	120,000
LOM-2030H (LOM)	275,500
SGC 4600 (SGC)	275,000
SGC 5600 (SGC)	450,000
Sinterstation 2500 (SLS)	200,000
Sinterstation 2500^{plus} (faster than the above) (SLS)	310,000

4.3.6 Ancillary Costs

The costs shown in Table 4.4 are the base cost of the machine. It should be emphasized that there are also additional miscellaneous costs of a warranty, installations, and so on. For example, the Helisys 2030H LOM machine has a base price (as of March 2000) of $275,500, which actually includes a first-year service warranty. Installation is estimated at $3,000; training at $2,000. Additional options include a chamber heating module at $4,499 and an initial supply package at $5,995. Thus the total for the complete package is $292,494. This example is not meant to endorse or criticize the LOM machine; rather it shows the real cost of doing business. All the machines in the table have such setup costs, which add 10% to 20% onto the base price. Some processes such as SLS also require a supplementary room for powder preparation and venting. Further data on cost comparisons (Table 4.4), materials (Table 4.5), part size (Table 4.6), and total part cost (Table 4.7) now follow. Figure 4.15 compares accuracy.

4.3.7 Commercial Comparisons of Cost and Capability

TABLE 4.5 Modeling Material Comparison

Rapid prototyping	Liquid photocurable polymers	Sintered metal powder and waxes	Sheet materials	Polymer spool	Viscous solidifying polymers
Stereolithography	X				
Selective laser sintering		X			
Laminated object modeling			X		
Fused deposition modeling				X	
Solid ground curing	X				
3-D printing followed by machining					X

TABLE 4.6 Maximum Part Size Comparison (as of March 2000)

Machine	Company	Part size capability (in.)
SLA-250	3D Systems, Inc.	$10 \times 10 \times 10$
SLA-350	3D Systems, Inc.	$13.8 \times 13.8 \times 15.7$
SLA-500	3D Systems, Inc.	$20 \times 20 \times 23$
FDM 2000	Stratasys, Inc.	$10 \times 10 \times 10$
LOM-2030H	Helisys, Inc.	$32 \times 22 \times 20$
SGC 5600	Cubital America, Inc.	$20 \times 14 \times 20$
Sinterstation 2000	DTM Corp.	12×15
Sinterstation 2500	DTM Corp.	$15 \times 13 \times 16.7$

TABLE 4.7 Rapid Prototyping Process, Speed and Cost Comparison—Chrysler Benchmarking Test Reported In "Rapid Prototyping Report," Vol. 1, No. 6, June 1992. Note That This Comparison Was Done with 1992 Machines Such as the 3D Modeler by Stratasys. Many Machines Such as the Sinterstation 2500^{plus} Have Become Much Faster Since Then.

Rapid prototyping process	Machine	Total process time (hr:min)	Total part cost
Stereolithography	SLA-250	7:25	$133.94
Stereolithography	SLA-500	7:03	187.95
Fused deposition modeling	3-D Modeler	12:39	344.94
Laminated object modeling	LOM-1015	11:02	109.40
Solid ground curing	Solider 5600	11:21	88.70*
Selective laser sintering	Sinterstation 2000	4:55	199.23

*Assumes 35 parts built simultaneously.

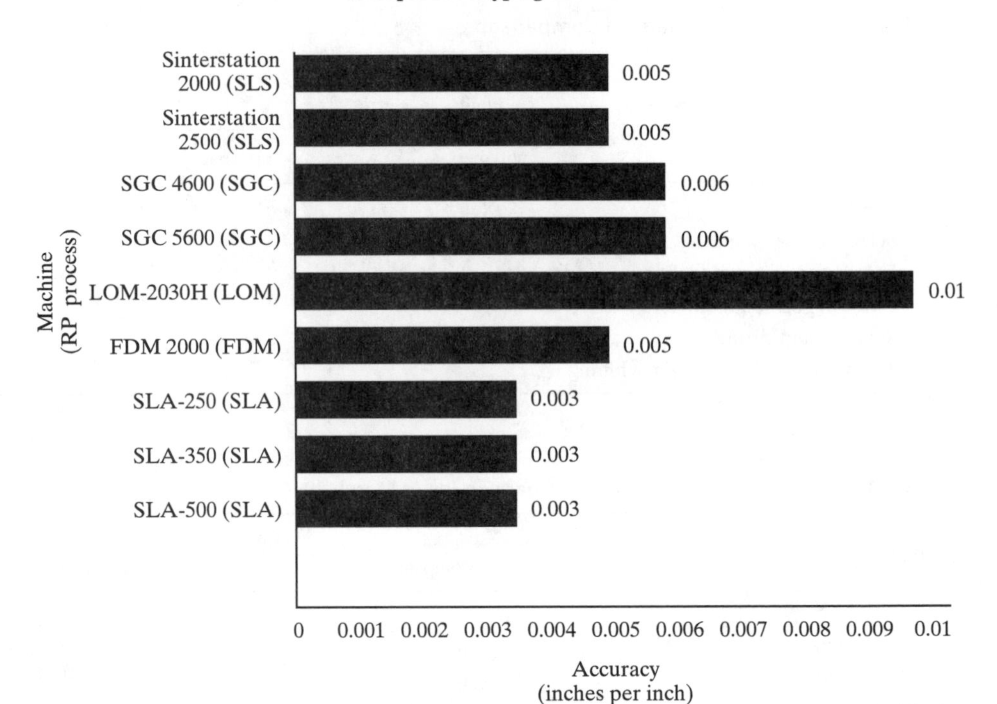

Figure 4.15 Comparison of accuracy (as of March 2000).

4.4 CASTING METHODS FOR RAPID PROTOTYPING

4.4.1 Introduction

The classic manufacturing texts by DeGarmo and associates (1997), Kalpakjian (1997), Schey (1999), and Groover (1999) are remarkably comprehensive in their coverage of the casting process. The several methods of casting include:

- Lost-wax investment casting
- Ceramic-mold investment casting
- Shell molding
- Conventional sand molding
- Die casting

Rather than duplicate the material found in other books, this section focuses on casting as it is done by rapid prototyping companies. Batch sizes from 50 to 500 are typical. The key market strategy is that casting is cheap and fast. However, it may not be the choice for the final product because of its tolerances. Depending on the type of casting chosen, the tolerances vary from +/− 75 microns (0.003 inch) for lost-wax processes to +/− 375 microns (0.015 inch) for standard sand castings (also see Chapter 2).

4.4.2 Lost-Wax Investment Casting

As mentioned in Chapter 1, the fundamentals of casting were invented by Korean and Egyptian artists many centuries ago. The following steps are known as the *lost-wax investment casting* process (Figure 4.16): (a–c) a master pattern of an engineering or art object is first carved from wax; (d–f) it is surrounded by a ceramic slurry that soon sets into solid around the wax; (g) the wax is melted out through a hole in the bottom, leaving a hollow cavity; (h) this hole is plugged, and liquid metal is poured into the open cavity from the top; (i) after a while, the metal solidifies and the ceramic shell can be broken away to get the part; (j) some cleaning, deburring, and polishing are needed before the object is finished.

The process was greatly improved and made more accurate during World War II for aeroengine components. Today it is used for products such as jet engine turbine blades and golf club heads. On the top line of Figure 4.16, wax patterns are formed from injection molds, assembled on treelike forms, and then treated with the slurry.

Alternate layers of fine refractory slip (zircon flour at 250 sieve or mesh size) are applied, followed by a thicker stucco layer (sillimanite at 30 sieve or mesh size). The coated components are dipped in fluidized beds that contain isopropyl silicate and liquid acid hardener. Drying takes place in ammonia gas. The next step is to eliminate the wax in a steam autoclave at 150 °C, fire the mold for 2 hours at 950 °C, then pour in the liquid steel or aluminum.

In summary, the modern lost-wax method has one of the best tolerances in the casting family because the original wax patterns are made in nicely machined molds. Today, tolerances of +/− 75 microns (0.003 inch) are readily obtainable. Also the as-cast surface is relatively smooth and usable for the same reason. Other advantages include:

- No parting lines if the wax original is hand finished.
- Waxes with surface texture can give direct features such as the dimples on a golf club.
- Automation of the slurry dipping is possible using robots, thereby reducing costs.
- Products such as turbine blades can be unidirectionally solidified, giving good mechanical properties in the growing direction.

4.4.3 Ceramic-Mold Investment Casting Procedures

The snag about the previous method is that the wax pattern is destroyed. The *ceramic-mold investment casting* technique therefore employs reusable submaster patterns in place of the expendable wax patterns. This version of investment casting ideally involves *five* steps to make it efficient and to retain, as much as possible, the fine care and expense that go into creating the original master positive in Step 1. The steps are as follows:

- Step 1. Positive: make an original *master pattern* with stereolithography or machining.

Mold to make pattern

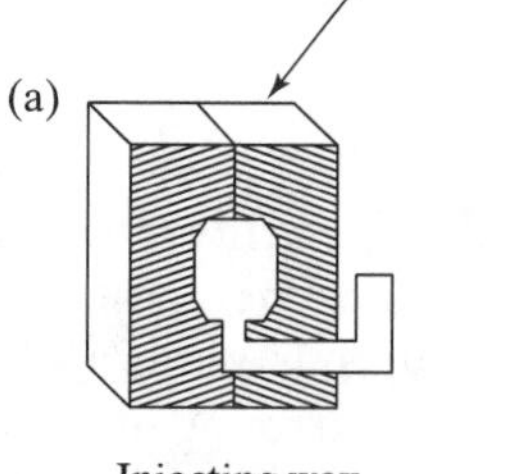

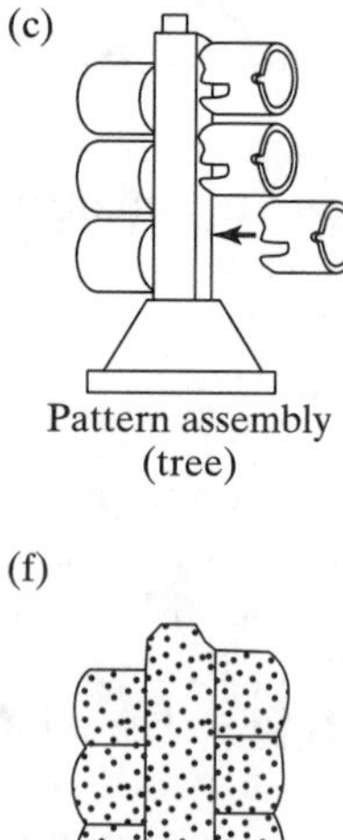

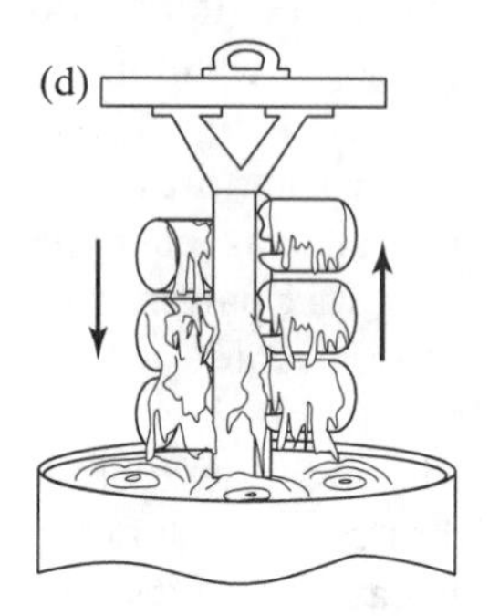

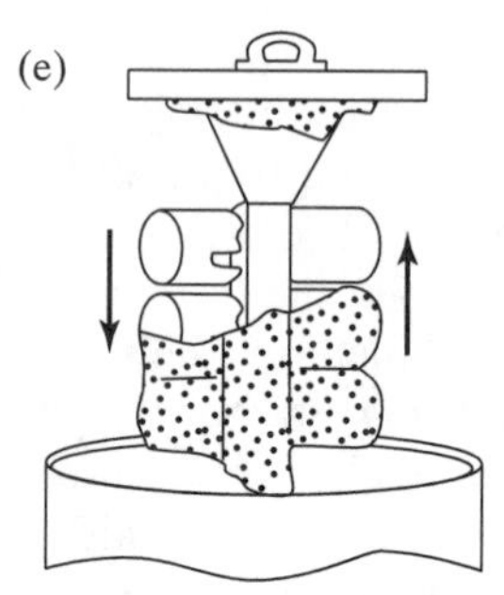

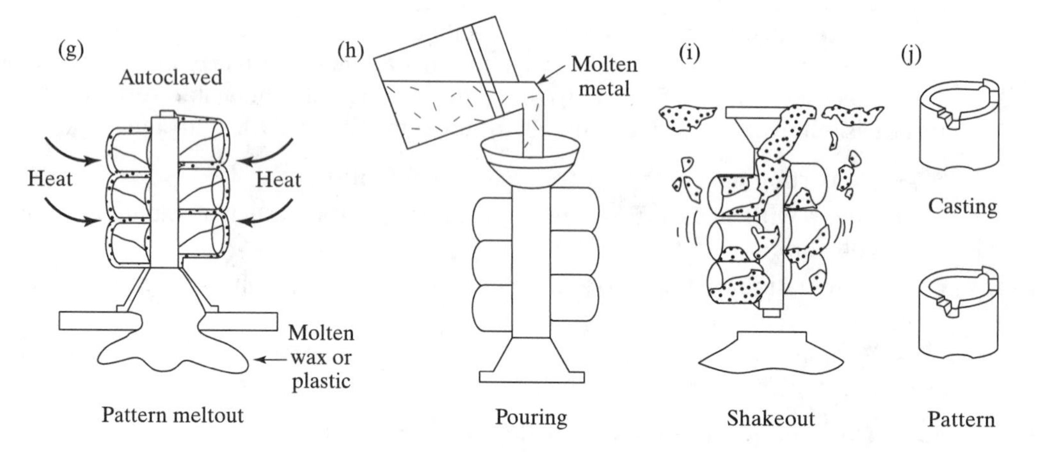

Figure 4.16 The lost-wax investment casting process. Upper diagrams (a) through (c) lead to the tree of wax master patterns. Middle diagrams show the slurry and stucco being applied. Lower diagram shows the casting (adapted from literature of the Steel Founders' Society of America).

- Step 2. Negative: create a shell around the master with highly stable resin. A negative space is created around the original positive master pattern. This shell can be pulled apart to give a parting line.
- Step 3. Positive: create reusable *submaster rubbery molds* from the shells.
- Step 4. Negative: create the destroyable slurry/ceramic molds.
- Step 5. Positive: pour metal into the ceramic molds, which are then broken apart to get the components, which must then be degated and deburred.

SLA can be used to make the original master pattern, or a CNC machine can be used to mill the master from brass, bronze, or steel. Of course, the process can start at Step 3, but this might damage the original master, especially if it is SLA. Also, to get high productivity in the factory, it is preferable to have many molds at Step 3, all of which can be made from the stable resin negative in Step 2.

Prototyping companies like to use the hard resin to fabricate the negative in Step 2, because the resin has good dimensional stability. Note that it is typical to have two resin molds, one for each side of the casting, separable by a parting line.

Once the hard resin shells have set, they can be filled with a slurry gel that solidifies to a hard "rubbery positive" for Step 3. This intermediate submaster mold can be stripped away from the resin shells while it is still "rubbery." The material is ideal for the rather rough handling environments of a foundry, and the rubbery properties mean that no draft angles are needed for stripping these submasters off the resin shells.

The Step 4 negative mold is made from a graded aluminosilicate with a liquid binder (ethyl silicate) and isopropyl alcohol. This is poured around the submasters from Step 3. Once the slurry has set, the two ceramic halves are joined to create the inner cavity, the slurry is fired at 950 °C to give it strength, and the casting process, say with molten aluminum, can begin.

After solidification, the component is broken out of the ceramic, cleaned up, and deburred. The parting line can cause problems, but in general, good accuracy is obtained: +/− 125 to 375 microns (+/−0.005 to 0.015 inch).

4.4.4 Shell Molding

An alternative form of high-accuracy casting is *shell molding.* Metal pattern plates are first heated to 200°C to 240 °C. A thin wall of sand, 5 to 15 millimeters (0.25 to 0.75 inch) thick, is then sprayed over the plates. The sand is resin-coated to ensure adhesion to the metal plate. Phenolic resins, with hexamethylene-tetramine additives, are combined with the silica to ensure rigid thermosetting of the sprayed sand. The next steps are to cure, strip, and dry the sand molds, which are comparably very accurate for casting. Once the excess sand is removed and casting is finished, accuracies can be as low as +/− 75 microns (0.003 inch).

4.4.5 Conventional Sand Molding

The cruder, cheaper version of casting starting with wooden or plaster patterns is called *sand casting.* A sand impression is made around the pattern with gates and risers for the poured metal. This gives tolerances of +/− 375 microns (0.015 inch). Newer developments include:

1. A high-pressure jolt-and-squeeze method: Here mechanical plungers push the sand against the mold at a jolt of 400 psi. This gives a tighter fit of the sand against the pattern and hence better tolerances after casting.
2. Carbon dioxide block molding: Here the interfacing between the sand and the pattern is made up of a special material about 12 millimeters (0.5 inch) thick. It is a refractory mix of zircon or very fine silica, bonded with 6% sodium silicate, which is then hardened by the passage of carbon dioxide.

4.4.6 Die Casting

Die casting is predominantly done by the high-pressure injection of hot zinc into a permanent steel die. Today, the die or mold for this type of casting is almost certain to be milled on a three- or five-axis machine tool.

Die costs are relatively high, but smooth components are produced with accuracies in the range of +/− 75 microns (0.003 inch). However, these high costs for the permanent molds mean that die casting does not really fit into the rapid prototyping family. It is mostly used for large-batch runs of small parts for automobiles or consumer products. Since low melting point materials such as zinc alloys are used in the process, component strengths are relatively modest.

Today, the injection molding of plastics (Chapter 8) is often preferred over zinc die casting.

4.5 MACHINING METHODS FOR RAPID PROTOTYPING

4.5.1 Overview

Chapter 7 deals with the generalized machining operation including the mechanics of the process. This chapter focuses on advances in CAD/CAM software that allow CNC machining to be more of a "turnkey rapid prototyping" process. One goal is to fully automate the links between CAD and fabrication. Another goal is to minimize the intensely hands-on craft operations (e.g., process planning and fixturing) that demand the services of a skilled machinist.

CyberCut™ is an Internet-based experimental fabrication test bed for CNC machining. The service allows client designers on the Internet to create mechanical components and submit appropriate files to a remote server for process planning and fabrication on an open-architecture CNC machine tool. Rapid tool-path planning, novel fixturing devices, and sensor-based precision machining techniques allow the original designer to quickly obtain a high-strength, good-tolerance component (Smith and Wright, 1996).

4.5.2 WebCAD: Design for Machining "on the Internet" on the Client Side

A key idea is to use a "process aware" CAD tool during the design of the part. This prototype system is called *WebCAD* (Kim et al., 1999). Sun Microsystems' Java™—a portable, object-oriented, robust programming language similar to C++—is being used as a framework for serving mini-applications. The GUI is a 2.5D feature-based[2] design system that uses the destructive solid geometry (DSG) idea introduced in the last chapter (Cutkosky and Tenenbaum, 1990; Sarma and Wright, 1996). Recall that the user starts out with a prismatic stock and removes primitives or "chunks" of material. By contrast, conventional constructive solid geometry (CSG) means building up a part incrementally from "nothingness." In the "destructive" paradigm, instead of allowing arbitrary removal, the user is also constrained to removing certain shapes of material, referred to as features. These features take the form of pockets, blind holes, and through-holes.

WebCAD also contains an expert system capturing rules for machinability. At the top of Figure 4.17, the designer is shown being guided by these rules. For example, a "forbidden zone" is imposed around a through-hole feature to prevent it from being designed too close to an edge. In the event that the designer violates a rule, a "pop-up" window advises on an appropriate remedy by moving the hole further into the block—typically by its radius dimension. WebCAD also uses a WYSIWYG ("what you see is what you get") environment, with explicit cutting tool selection and visible corner radii on pockets. At the time of this writing, further improvements also include freeform surface editing and selection of different cutting tool sets depending on final fabrication location (Kim, 2000).

The rationale for imposing destructive features upon the designer is that each of these features can readily be mapped to a standard CNC milling process. The scheme thus resembles the interaction between a word processor and a printer regarding the "printability" of the document. It is easy to criticize that the restriction to DSG limits the set of parts that can be designed. However, the key advantage of this design environment is that the design-to-manufacture process is more deterministic than conventional methods, which rely on unconstrained design and on looser links between design, planning, and fabrication. Experience shows that designers are somewhat concerned at first that they are constrained; however, the opportunity to be provided with the correct part very quickly proves to be attractive.

4.5.3 Planning on the Server Side

When the client's design is finished, the resulting geometry can be sent over the Internet to a process planner residing on a remote server. An automated software pipeline takes the geometry and determines in which order the features should be cut, the exact tool paths to traverse, cutting feeds, and spindle speeds for a machine tool.

Macroplanning orders the individual features and creates the specific machining setups in fixtures. CyberCut's current macroplanner is a feature recognition module that can reliably extract the volumetric features from 2.5D parts. The output of the module is not just a machining feature set but a rich data structure that also gives important connectivity information that relates one feature to another. A recent advance in the macroplanner is its ability to recognize and process features containing freeform surfaces (Sundararajan and Wright, 2000).

[2]A 2.5D *feature* is a machinable feature with arbitrary outside peripheral contour and a uniform sweeping depth into the block being machined.

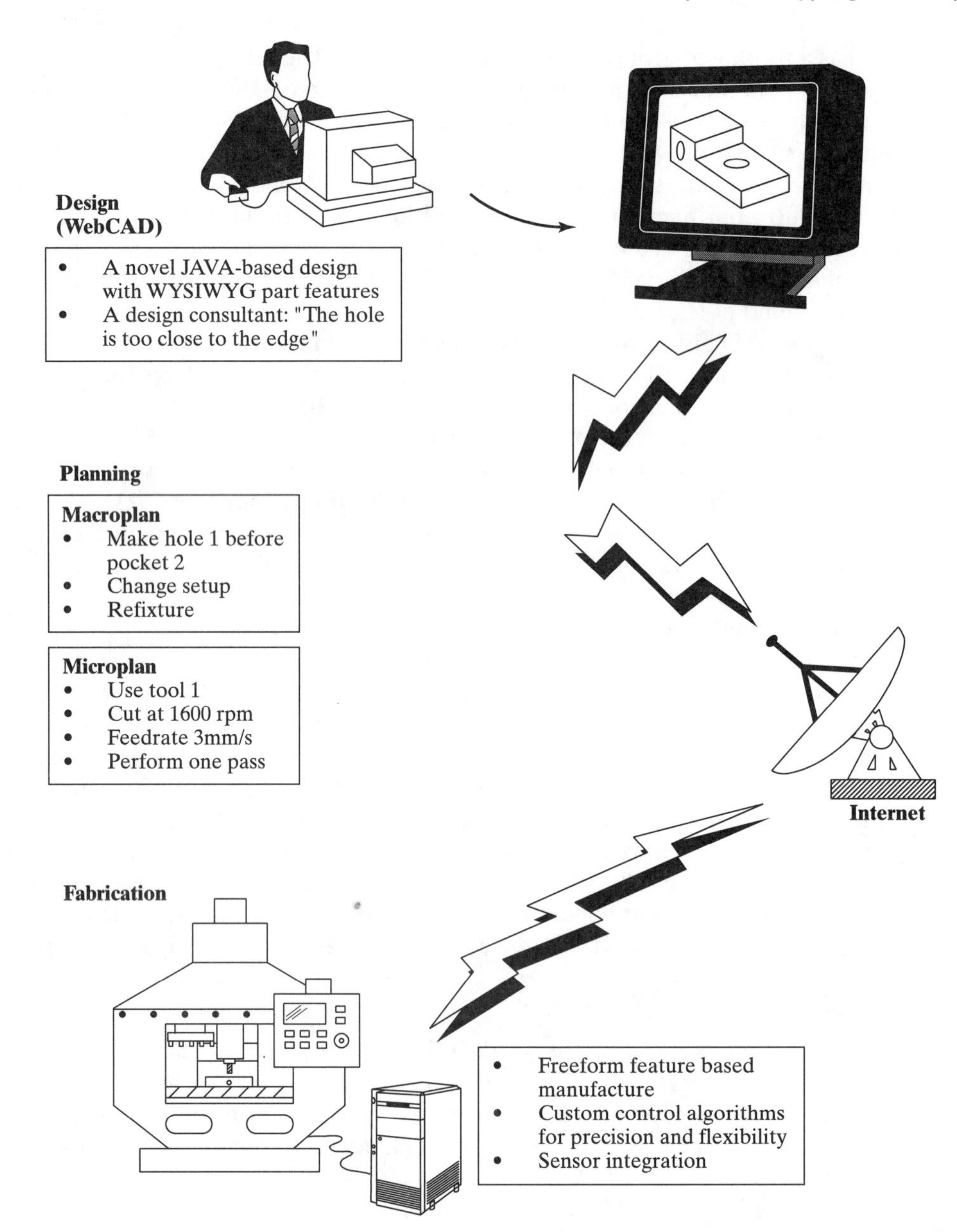

Figure 4.17 The CyberCut project: integrated design, planning, and fabrication.

Microplanning and tool-path planning decompose the DSG volumes into specific tool motions. Colloquially speaking, this is the step that is like lawn mowing: each volume has to be carved out with a specific tool diameter, and the overlap between each strip has to be considered in relation to part tolerance and surface roughness. The corners of pockets (just like lawns) might require special methods.

Freeform surfaces must be divided into flat and steep regions. The flat regions are machined with a projecting spiral tool-path pattern, and the steep regions are machined with a slicing tool-path pattern. This yields good tool-path uniformity and only moderate computational complexity. The decomposition aims at minimizing machining time within the constraints of the specified surface roughness, tolerance, and machine tool safety. The time of individual operations can also be estimated, which can be sent back to the designer, providing an early estimation of the machining costs.

4.5.4 Fabrication by Milling on the Server Side

Finally, a stream of NC commands performs the machining on an open-architecture milling machine. (By contrast, if it had been determined along the way that the client would have been better served by SFF technology, CyberCut can connect to a fused deposition modeling [FDM] machine.) The particular milling machine being used is an open-architecture machine that can execute advanced tool-path trajectories. One example is a machine path interpolator that can traverse complicated freeform paths represented by NURBS. This ability brings a richer surface generation capability to the ostensibly traditional machining process. By doing so, it continues to compete with the SFF methods from the point of view of geometric complexity (Hillaire et al., 1998). More details of the open-architecture machine tool itself are reserved for Chapter 7 on machining.

4.6 MANAGEMENT OF TECHNOLOGY

4.6.1 Summary

Solid freeform fabrication (SFF) techniques and conventional rapid prototyping techniques such as machining and casting are key technologies for improving product realization cycles and reducing time-to-market. The availability of Internet-based software tools (Berners-Lee, 1989; Java, 1995) has accelerated the links between CAD and prototype creation. The Internet has also allowed access to manufacturing sites in many different countries (Smith and Wright, 1996; DeMeter et al., 1995; Mitsuishi et al., 1992; Frost and Cutkosky, 1996; Finin et al., 1994). In summary:

- SLA emerges as the most commercially accepted of the newer SFF prototyping methods.
- SLS emerges as a very useful, commercially accepted alternative to SLA for overhanging structures needing support and for stronger materials that can be sintered rather than photocured.
- FDM is an excellent choice for an in-house machine that can be used by an industrial design team for an iterative series of prototypes.
- LOM is excellent for larger components.
- 3-D printing and planarization using the inexpensive Sanders and Z-Corporation machines are gaining commercial acceptance at the time of this writing. Very inexpensive 3-D digitizers coupled with miniature milling

machines are also entering the market (see URL for Roland Digital Group at the end of the chapter).

- Machining and casting remain central to the rapid prototyping field, especially for high-strength prototypes and longer batch runs of several prototypes.

4.6.2 Future Trends

The accuracy of processes such as stereolithography and selective laser sintering is improving as time goes by. These processes are being used more and more in the creation of *the original, first master* for casting and for plastic injection molding. As consumer products such as stereos, cellular phones, personal digital assistants (PDAs), and handheld computers (Richards and Brodersen, 1995) begin to look more aerodynamic, there is a need to create molds that have unusual curves and reentrant shapes; these are easy to create in SLA or SLS, especially in comparison with machining.

It has nonetheless been emphasized that SFF's accuracy is poor in comparison with machining. Overhanging structures may be hard to support during fabrication, and there are problems with component warping during curing. While simple shapes might have accuracies of +/− 25 to 75 microns (0.001 to 0.003 inch), the range for complex shapes might be as high as +/− 125 to 375 microns (0.005 to 0.015 inch).

While the strength of SFF parts is today less than machined parts, new trends are closing the gap. The FDM parts made by the Stratasys machine can be formed in near full strength ABS and similar polymers. Cheung and Ogale (1998), for example, have increased the strength of photopolymers by fiber reinforcement. Also, research at Sandia Laboratories on a process called laser engineered net shaping (LENS) is permitting direct fabrication of high-strength metal molds. This and similar projects are modified versions of DTM Inc.'s SLS process.

At the same time, CAD/CAM techniques for machining are advancing rapidly. For example, the CyberCut freeform design tools linked to open-architecture milling machines will continue to expand machining's capability (Greenfeld et al., 1989; Schofield et al., 1998; Hillaire et al., 1998). There is a subtle point to be made here: much of the increased activity in the SFF prototyping methods was originally prompted by the poor communication between CAD and CNC machine tools. During the late 1980s, stereolithography's competitive edge over machining came from the fact that the CAD model could be instantly "sliced" and then turned into laser scanning paths for rapid part production. With the CyberCut methodology and open-architecture control, the conventional machining process can be equally competitive from an art-to-part speed standpoint, and it continues to give the high-accuracy and product integrity qualities that it always gave.

The evidence is thus clear that the capabilities in both the machining and the SFF fields are constantly improving. It has also been noted that several of the methods such as 3-D printing with planarization, SGC, and SDM combine deposition with machining "to get the best of both worlds." Perhaps in a similar way, the 3D Systems' QuickCast method uses "the best of SLA combined with the best of investment casting." In QuickCast, *disposable SLA patterns* are fabricated with distinctly hollow internal structures. When the ceramic shells for casting are created around these hollow SLA patterns, the latter collapse inward, leaving the casting mold intact and ready for use. This process, described by Jacobs (1996, 183–252), is gaining rapid acceptance commercially.

The issues mentioned are predominantly technical. As this chapter draws to a close, it is important to recall an earlier point from Chapter 2 that "prototypes structure the design process" (see Kamath and Liker, 1994). Physical prototypes focus the efforts of a distributed design team, especially if subcontracting is a big part of the process.

Perhaps the most important conclusion is this: *each manufacturing process will play a vital role at different points in the product development cycle.* SFF techniques will be more evident at the front end, machining will be more evident partway through to create highly accurate molds, and plastic injection molding will be most evident in the final high-volume production method for the consumer's product. Once again it must be emphasized that "manufacturing in the large" is an integration of many software tools, physical processes, and market strategies.

In summary, rapid prototyping dramatically accelerates time-to-market.

- Psychologically, it focuses the attention of the members of the design team in a "learning organization" (see Chapter 2).
- Physically, it reduces the time necessary to make a full production die from hardened steel and to launch into mass production.

4.7 GLOSSARY

4.7.1 Die Casting

Low-pressure casting, often of liquid zinc, into a machined mold.

4.7.2 Electrodischarge Machining (EDM)

The use of an electrode to melt and vaporize the surface of a hard metal. Usually restricted to low rates of metal removal of very hard metals.

4.7.3 G-Codes

The standard low-end machine tool command set that gives motion, for example, G1 = linear feed.

4.7.4 Injection Molding

Viscous polymer is extruded into a hollow mold (or die) to create a product.

4.7.5 Ink-Jet Printing in 3-D

Rapid prototyping by rolling down a layer of powder and hardening it in selected regions with a binder phase that is printed onto the powder layer.

4.7.6 Investment Casting

The word *investment* is used when time and money are invested in a ceramic shell that is subsequently broken apart and destroyed. The original positive master that is used to create the negative investment shell can be made by several processes. Lost-wax and ceramic mold are the two most common.

4.7.7 Laminated Object Modeling (LOM)

Rapid prototyping by laser cutting the top layer of a stack of paper, each layer of which is glued down.

4.7.8 Lapping and Polishing

Final finishing and smoothing of a surface that has already been machined. The surface-lapping operation is usually done on flat lapping plates loaded with diamond paste down to 0.25 microns in diameter.

4.7.9 Machining

General manufacturing by cutting on a lathe or mill; chip formation from a solid block rather than forging, forming, or joining.

4.7.10 M-Codes

The standard low-end command set for machine tool operations that are not related to x, y, or z motion of the axes—for example, M6 = call tool into spindle.

4.7.11 Near-Net Shape

Forming, forging, or sintering operations that produce an object "nearly" to its final shape so that only minor finish machining is needed.

4.7.12 Plastic Injection Molding

As "injection molding," described earlier. Note: Zinc die casting also involves "injection" into dies or molds.

4.7.13 Prototyping (Prototype)

"The original thing in relation to any copy, imitation, representation, later specimen or improved form" (taken from Webster's dictionary).

4.7.14 Rapid Prototyping (RP)

A new genre of prototyping, usually associated with the SFF family of fabrication methods. Emphasis is on speed-to-first-model rather than fidelity to the CAD description.

4.7.15 Selective Laser Sintering (SLS)

Rapid prototyping by laser sintering of polymer or ceramic powders. The laser moves as a point source across the surface of the powder, first sintering the bottom slice of the desired object. A roller spreads more powder and a second layer is sintered, also fusing to the one below.

4.7.16 Shape Deposition Manufacturing (SDM)

Rapid prototyping with alternative deposition runs followed by machining runs across an object to build up complex prototypes.

4.7.17 Solid Freeform Fabrication (SFF)

A family of processes in which a CAD file of an object is tessellated, sliced, and sent to a machine that can quickly build up a prototype layer by layer.

4.7.18 Solid Ground Curing (SGC)

Rapid prototyping, also by laser curing of photocurable polymers, but done layer by layer through a photomask rather than by laser point sources.

4.7.19 Stereolithography (SLA)

Rapid prototyping by laser curing a photocurable liquid. The laser moves as a point source across the surface of the liquid, first curing the bottom slice of the object. This slice moves down on an elevator by 50 to 375 microns (0.002 to 0.015 inch), depending on desired accuracy. The next layer is then photocured, also fusing to the one below.

4.7.20 Tessellation

Representing the outside surfaces of an object by many small triangles, like a mesh thrown over and drawn around the object. This leads to an ".STL" file of the vertices and the surface normals of the triangles.

4.7.21 3-D Printing/Plotting

Rapid prototyping by printing/plotting thin jets of polymer onto a fixtureless baseplate followed by simple machining/planarization.

4.8 REFERENCES

ACIS Geometric Modeler. 1993. *Technical Overview.* Spatial Technology, Inc., Version 1.5.

Ashley, S. 1991. "Rapid prototyping systems." *Mechanical Engineering* 34–43.

Ashley, S. 1998. "RP industry's growing pains." *Mechanical Engineering* 64–67.

Au, S., and P. K. Wright. 1993. A comparative study of rapid prototyping technology. In *Intelligent concurrent design: Fundamentals, methodology, modeling, and practice.* ASME Winter Annual Meeting, New Orleans, Louisiana. 73–82.

Beaman, J. J., J. W. Barlow, D. L. Bourell, R. H. Crawford, H. L. Marcus, and K. P. McAlea. 1997. *Solid freeform fabrication. A new direction in manufacturing: With research and applications in thermal laser processing.* Norwell, MA: Kluwer Academic Publishers.

Berners-Lee, T. 1989. Information management: A proposal. CERN Internal Proposal.

Bouldin, D., ed. 1994. Report of the 1993 workshop on rapid prototyping of microelectronic systems for universities. National Science Foundation Workshop.

Cheung, T. S., and A. A. Ogale. 1998. Processing of multilayer fiber reinforced composites by 3D photolithography. In *Proceedings of the 1998 NSF Grantees Design and Manufacturing Conference, Monterrey, Mexico,* 557-559. Arlington, VA: National Science Foundation.

Cohen, E., S. Drake, L. Gursoz, and R. Riesenfeld. 1995. Modeling issues in solid freeform fabrication. NSF Solid Freeform Fabrication Workshop II, Design Methodologies for Solid Freeform Fabrication, June 5–6, Pittsburgh, PA.

Cutkosky, M. R., and J. M.Tenenbaum. 1990. A methodology and computational framework for concurrent product and process design. *Mechanism and Machine Theory* 25 (3):365–381.

DeGarmo, E. P., J. T. Black, and R. A. Kohser. 1997. *Materials and processes in manufacturing,* 8th ed. New York: Prentice-Hall.

DeMeter, E. C., Q. Sayeed, R. E. DeVor, and S. G. Kapoor. 1995. An Internet model for technology integration and access part 2: Application to process modeling and fixture design. MTAMRI Report 1995. University of Illinois.

Dutta, D. 1995. Layered manufacturing in Project Maxwell. NSF Solid Freeform Fabrication Workshop II, Design Methodologies for Solid Freeform Fabrication, June 5–6, Pittsburgh, PA.

Finin, T., D. McKay, R. Fritzson, and R. McEntire. 1994. KQML: An information and knowledge exchange protocol. In *Knowledge building and knowledge sharing,* edited by Kazuhiro Fuchi and Toshio Yokoi. Tokyo, Japan: Ohmsha and IOS Press.

Frost, R., and M. Cutkosky. 1996. An agent-based approach to making rapid prototyping processes manifest to designers. ASME Symposium on Virtual Design and Manufacturing.

Greenfeld, I., F. B. Hansen, and P. K. Wright. 1989. Self-sustaining, open-system machine tools. In *Proceedings of the 17th North American Manufacturing Research Institution* 17:281–292.

Groover, M. P. 1999. *Fundamentals of modern manufacturing.* Upper Saddle River, NJ: Prentice-Hall.

Heller, T. B. 1991. Rapid modeling—What is the goal? In *The Second International Conference on Rapid Prototyping,* 242-244. Dayton, OH: Rapid Prototype Development Laboratory (RPDL), 242–244.

Hillaire, R., L. Marchetti, and P. K. Wright. 1998. Geometry for precision manufacturing on an open architecture machine tool (MOSAIC-PC). In *Proceedings of the ASME International Mechanical Engineering Congress and Exposition,* 8: 605-610. Anaheim, CA: MED.

Jacobs, P. F. 1992. *Rapid prototyping and manufacturing: Fundamentals of stereolithography.* Dearborn, MI: Society of Manufacturing Engineers.

Jacobs, P. F. 1996. *Stereolithography and other rapid prototyping and manufacturing technologies.* Dearborn, MI: Society of Manufacturing Engineers.

Java, 1995, is a trademark of Sun Microsystems, Incorporated. Documentation can be found at **http:// www.javasoft.com**. (Also refer to J. Gosling and H. McGilton. "The Java Language Environment: A White Paper," Technical Report, Sun Microsystems, 1995.)

Kai, C. C. 1994. Three-dimensional rapid prototyping technologies and key development areas. *Computing & Control Engineering Journal* 5 (4):200–206.

Kalpakjian, S. 1997. *Manufacturing processes for engineering materials,* 3d ed. Menlo Park, CA: Addison Wesley Longman.

Kamath, R. R., and J. K. Liker. 1994. A second look at Japanese product development. *Harvard Business Review.* (Reprint Number 94605).

Kim, J. H., F. C. Wang, C. Sequin, and P. K. Wright. 1999. Design for machining over Internet. Paper presented at the Design Engineering Technical Conference (DETC) on Computer Integrated Engineering, Las Vegas, NV. Paper Number DETC'99/CIE-9082.

Kim, J. H. 2000. WebCAD 2000: Distributed CAD tool for machining. Master of Science Thesis, Department of Computer Science, University of California, Berkeley.

Kochan, D. 1993. *Solid freeform manufacturing: Advanced rapid prototyping.* New York and Amsterdam: Elsevier.

Kruth, J. P. 1991. Manufacturing by rapid prototyping techniques. *Annals of the CIRP* 40 (2):603–614.

Kumar, V., P. Kulkarni, and D. Dutta. 1998. Adaptive slicing of heterogeneous solid models for layered manufacturing. University of Michigan Technical Report, UM-MEAM-98-02.

Manufacturing Studies Board (National Research Council). 1990. Rapid prototyping facilities in the U.S. manufacturing research community. Edited by T. C. Mahoney.

McMains, S. 1996. Rapid prototyping of solid three dimensional parts. Master of Science Thesis, Computer Science, University of California, Berkeley.

McMains, S., C.S. Sequin, and J. Smith. 1998. SIF: A solid interchange format for rapid prototyping. Paper presented at the 31st CIRP International Seminar on Manufacturing Systems. University of California, Berkeley.

Mead, C., and L. Conway. 1980. The Caltech intermediate form for LSI layout description. In *Introduction to VLSI Systems,* 115–127. Addison Wesley.

Mitsuishi, M., S. Warisawa, Y. Hatamura, T. Nagao, and B. Kramer. 1992. A user friendly manufacturing system for hyper-environments. In *Proceedings of the 1992 IEEE International Conference on Robotics and Automation,* 25–31.

MOSIS. 2000. University of Southern California's Information Sciences Institute—The MOSIS VLSI Fabrication Service. **http://www.isi.edu/mosis/**.

NSF. 1994. Solid Freeform Fabrication Workshop I. New Paradigms for Manufacturing, Arlington VA.

NSF. 1995. Solid Freeform Fabrication Workshop II. Design Methodologies for Solid Freeform Fabrication, June 5–6, Pittsburgh, PA.

Richards, B., and R. Brodersen. 1995. InfoPad: The design of a portable multimedia terminal. In *Proceedings of the Mobile Multimedia Conference, 2.* Bristol, England.

Sachs, E., M. Cima, J. Bredt, A. Curodeau, T. Fan, and D. Brancazio. 1992. CAD-casting: Direct fabrication of ceramic shells and cores by three dimensional printing. *Manufacturing Review* 5 (2):117–126.

Sachs, E., N. Patrikalakis, D. Boning, M. Cima, T. Jackson, and R. Resnick. 2000. The distributed design and fabrication of metal parts and tooling by three dimensional printing. In *Proceedings of the 2000 NSF Grantees Design and Manufacturing Conference.* Arlington, VA: University of British Columbia and National Science Foundation.

Sarma, S., and P. K. Wright. 1996. Algorithms for the minimization of setups and tool changes in "simply fixturable" components in milling. *Journal of Manufacturing Systems* 15, (2):95–112. (Also see S. Sarma, S. Gandhi, and P. K. Wright. 1995. Reference free part encapsulation: A universal fixturing technology for rapid prototyping by machining. In *Concurrent Product and Process Engineering.* 1:339–351. Anaheim, CA: MED.).

Schey, J. A. 1999. *Introduction to manufacturing processes.* New York: McGraw-Hill.

Schofield, S., F. C. Wang, and P. K. Wright. 1998. Open architecture controllers for machine tools part 1: Design principles, part 2: A real-time, Quintic spline interpolator. *Journal of Manufacturing Science and Engineering* 120:417–432.

Smailagic, A., and D. P. Siewiorek. 1993. A case-study in embedded system design: The VuMan 2 Wearable Computer. *IEEE Design and Test of Computers,* 56–67.

Smith, C., and P. K. Wright. 1996. CyberCut: A World Wide Web based design to fabrication tool. *Journal of Manufacturing Systems* 15 (6):432–442.

Smith, D. 2000. 3DP prototyping at Motorola. Personal communication.

Sundararajan, V., and P. K. Wright. 2000. Identification of multiple feature representation by volume decomposition for 2.5 D components. *Transactions of the ASME. Journal of Manufacturing Science and Engineering* 122 (1): 280–290.

University of California, Los Angeles (UCLA). 1994. University Extension, Department of Engineering, Information Systems, and Technical Management, Short Course Program, Rapid Prototyping: Technologies and Applications.

Weiss, L. E., E. L. Gursoz, F. B. Prinz, P. S. Fussell, S. Mahalingam, and E. P. Patrick. 1990. A rapid tool manufacturing system based on stereolithography and thermal spraying. *Manufacturing Review* 3 (1):40–48.

Weiss, L. E., and F. B. Prinz. 1995. Shape deposition processing. NSF Workshop II. Design Methodologies for Solid Freeform Fabrication, June 5–6, Pittsburgh, PA.

Weiss, L. E., R. Merz, F. B. Prinz, G. Neplotnik, P. Padmanabhan, L. Schultz, and K. Ramaswami. 1997. Shape deposition manufacturing of heterogeneous structures. *SME Journal of Manufacturing Systems* 16:239–248.

Weiss, L. E., and F. B. Prinz. 1998. Novel applications and implementations of shape deposition manufacturing. Paper presented at the Naval Research Conference.

Woo, T. 1992. Rapid prototyping in CAD. *Computer Aided Design* 24:403–404.

Woo, T. 1993. *Rapid automated prototyping: An introduction.* Industrial Press.

Wright, P. K., D. A. Bourne, J. A. E. Isasi, G. C. Schatz, and J. G. Colyer. 1982. A flexible manufacturing cell for swaging. *Mechanical Engineering* 104 (10):76–83.

4.9 BIBLIOGRAPHY

Benett, G., ed. 1996. *Developments in rapid prototyping and tooling.* Mechanical Design Publication Ltd. London: Bury-Saint Edmunds.

Koenig, D. T. 1987. *Manufacturing engineering: Principles for optimization.* Washington, New York, and London: Hemisphere Publishing Corporation.

4.10 URLS OF INTEREST

4.10.1 Rapid Prototyping

1. **http://www.biba.uni-bremen.de**
2. **http://www.metalcast.com**
3. **http://www.motorola.com**
4. **http://cybercut.berkeley.edu**
5. **http://www.cs.hut.fi/~ado/rp/rp.html**
6. **http://www.cubital.com/cubital/**
7. **http://www.helisys.com**
8. **http://www.stratasys.com**
9. **http://www.3dsystems.com/**
10. **http://www.dtm-corp.com**
11. **http://www.rolanddga.com** (see products Modela and Picza)
12. **http://www.cs.berkeley.edu/~sequin/CAFFE/cyberbuild.html**

4.11 INTERACTIVE FURTHER WORK

4.11.1 Internet-Based CAD/CAM

The CyberCut WebCAD design environment can be found at the following URL: **http://cybercut.berkeley.edu**. Jaeho Kim and Ashish Mohole have been the creators of WebCAD, **<ashish@kingkong.me.berkeley.edu>**, and we welcome your comments and suggestions.

4.11.2 The Assignment

Enter this design environment to design a part. Also answer the basic question that follows but read the next section first.

4.11.3 Getting Started

It is possible to get this to run at home on a PC comparable with:

- Dell Optiplex Gxi with 28,800 bps modem connection
- Microsoft's Internet Explorer 5.0

There have been difficulties in getting WebCAD to run on Solaris versions of Netscape.

First-time users should click on the "Gallery" on the main menu to get some idea of the kind of parts that can be made or type in the location **http://Cybercut.berkeley.edu/html/gallery.htm**. Next, click on "WebCAD" in the "Design" section of the front page. From here it is possible to click on the "WebCAD 2000 Quick Start Manual" to view a tutorial or click on "Run WebCAD 2000" to begin using the CAD tool.

The "WebCAD 2000 Quick Start Manual" can be printed out for easy reference.

- At a university the download will take a minute or so.
- But for a design from home it takes a long time to get the byte code. It could take at least 5 minutes to download all the code.
- *After clicking on "Run WebCAD 2000" go and make a cup of tea and watch ESPN or MTV for 5 minutes.*
- A "Welcome to WebCAD 2000" window and a "Tool bar" window should appear.
- Click on "Make new part" and proceed with selecting stock as indicated. The program forces the selection of stock before any design can begin.

4.11.4 Questions to Add to the Assignment

At a minimum, try to design a simple paperweight out of aluminum that is a block with some initials on one side of it. Print out the design and attach it to the questions (recall from the MAS assignment about how to print from a browser!).

1. Where was the work done?
2. What kind of machine?

3. Which Internet browser?
4. How long did the full download take?
5. Did the "WebCAD 2000 Quick Start Manual" get the design on track quickly?
6. Did the gallery page help?
7. How long did it take to feel comfortable with the graphic user interface GUI?
8. How long did the design take?
9. What part type (or name) was designed?
10. Provide suggestions for the software development in the future.

CHAPTER

5

Semiconductor Manufacturing

5.1 INTRODUCTION

As already described, the material in this book is organized as "a journey along the product development path with emphasis on the fabrication techniques." Chapters 2 through 4 focused on the needs of the customer, conceptual design, detailed design, and rapid prototyping. The emphasis was on mechanical computer aided design/manufacturing (CAD/CAM) rather than electronics. However today, nearly every consumer product has an integrated circuit (IC) in it. And the more "high-tech" a product is, obviously the more sophisticated the inner electronics become. So it makes logical sense to study the inner workings of today's consumer products as discussed in the next chapters in the book. Colloquially speaking, integrated circuits and their associated electronics are the "brains" that will fit into the outer casings or "bodies" that were just prototyped in Chapter 4, by a method such as fused deposition modeling (FDM), or mass-produced in Chapter 8, by injection molding.

5.2 SEMICONDUCTORS

For the automobile and aerospace industries of today, it is not too much of an exaggeration to state that "an automobile is a computer on wheels" or that "an airplane is a flying computer." Even very traditional manufacturing industries such as steelmaking can survive only by using computer control and sensors to obtain exacting quality assurance. In today's "information age revolution" all manufacturing industries have the integrated circuit and the microprocessor at their core. In the same way, in the industrial revolution 200 years ago, all manufacturing industries had the steam engine at their core.

The next two chapters examine the technology and management issues in two critical areas of manufacturing. Chapter 5 focuses on the fabrication of *semiconductors.* These are the basic building blocks of integrated circuits (also called ICs or microchips or just chips). Semiconductor manufacturing has a worldwide compounded annual growth rate of 18%. Today, semiconductors are a $150 billion a year industry (see Appendix 1 of this chapter, Section 5.18). This rises to as much as $200 billion if the semiconductor *equipment* manufacturers are included. The projections take semiconductors to a $1 trillion industry over the next decade (see the Semiconductor Industry Association, 1997). Semiconductors and ICs are the "brains" of every computing device manufactured today, from supercomputers to personal computers to kitchen appliances. Chapter 6 will examine *computer manufacturing* including the assembly of ICs, printed circuit boards, and other key components into a device. It is worth glancing ahead to Chapter 6 to see how everything fits together.

5.3 MARKET ADOPTION

Tremendous leaps in computer science and electrical engineering technology over the past few decades have made possible the wide range of computing devices that are available today. These go vastly beyond the original transistor invented in 1947, the first IC in 1958, and the first microprocessor introduced commercially in 1971. For example, comparisons between the original microprocessor, the Intel 4004, and one of today's processors show the dramatic progress (Figure 5.1a has 2,300 transistors on a 3 × 4 mm chip; Figure 5.1b has 1.3 million on an 11 × 15 mm chip).

Increasingly, though, the main force driving the computer industry is not new basic knowledge but the market. Computer users are more sophisticated and demanding than ever. They want higher performance, more functionality, smaller dimensions, and interconnectivity with other products—all at supercompetitive prices.

These two chapters on semiconductor and computer manufacturing are therefore placed in this sequence in the book for another reason besides the journey along the product development life cycle. Already, many aspects of semiconductor and personal computer manufacturing have placed these technologies well along the market adoption S-shaped curves in Chapter 2. Gone are the days when the semiconductor industry or a computer manufacturer could rely solely on innovative designs and disregard the integration with manufacturing and efficient production processes. Those were the days, say around 1980, when consumers marveled at the user-friendly icons of Apple's machines in comparison with the nonintuitive DOS commands. Such friendly graphical interfaces created a brand loyalty for Apple that won and kept customers well into the late 1980s. But today, semiconductors and computers are headed toward the market maturity of autos and steel, which are dealt with in later chapters. To compete today, semiconductor manufacturers, computer makers, and their suppliers must therefore make better products at lower cost. Semiconductor fabs must now focus on high-yield manufacturing systems. Computer makers need to design customer-focused products, "outsource" to agile assembly factories, and orchestrate rapid supply chains. At the time of this writing around the year 2000, it is clear that

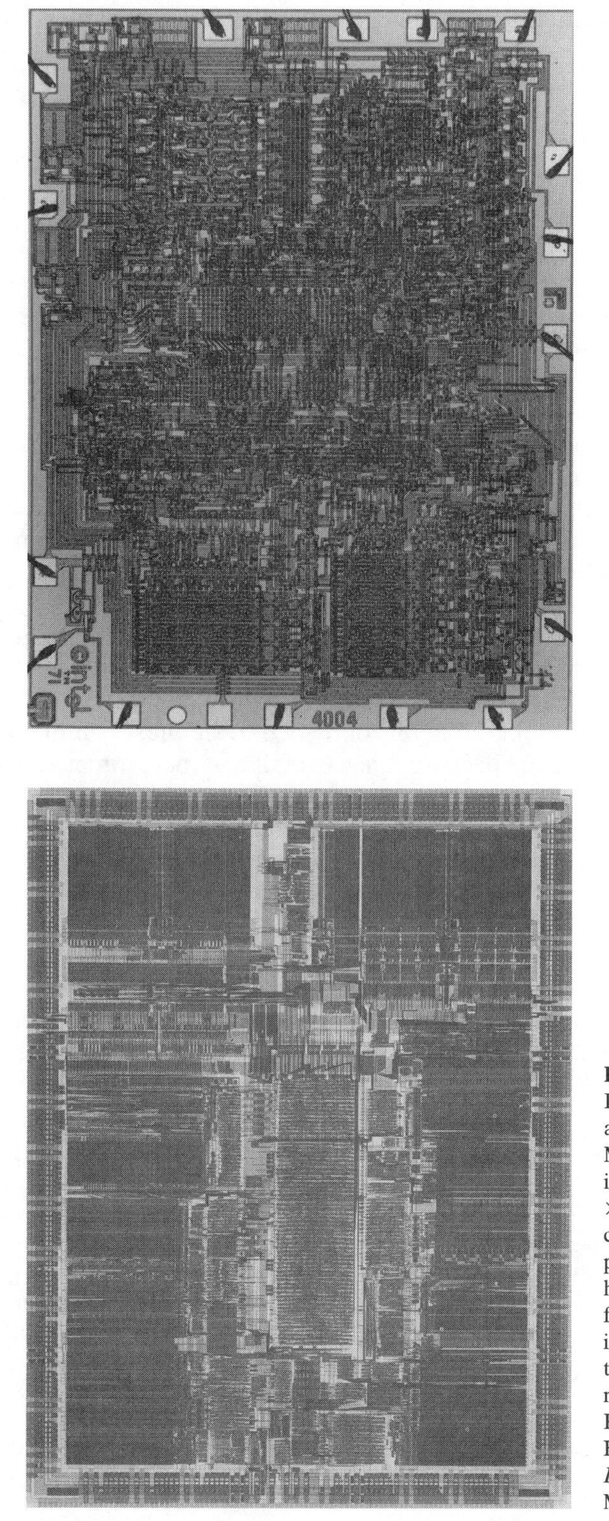

Figure 5.1 (a) The first microprocessor, Intel's 4004, includes 2,300 transistors on a 3 × 4 mm chip (courtesy Intel). (b) The MIPS Technology Inc. R4000 processor includes 1.3 million transistors on an 11 × 15 mm chip. The right-hand side contains the data path for the integer portion of the microprocessor. The left-hand side contains the data path for the floating point processor. The central area is the controller. The two large blocks on the top contain the fast, short-term cache memory (courtesy MIPS Technology Inc. Reprinted from D. A. Patterson and J. L. Hennessy's *Computer Organization and Design: The Hardware/Software Interface*, Morgan Kaufmann Publishers, 1994, 24).

some companies—such as Intel, Compaq, Dell, and Gateway—have understood and exploited this new landscape,whereas other companies—such as Apple—have had more mixed success in recent years.

5.4 THE MICROELECTRONICS REVOLUTION

The key to building faster, cheaper, smaller, and more powerful computers is to miniaturize electronic circuit components. Smaller devices have superior performance characteristics: more components in a small area increase the circuit's energy efficiency and processing speed. The usual way to measure miniaturization is by the length, L_G, of the polysilicon gate bridging the source and drain region of a transistor. This dimension is shown in later figures.

A key component of an integrated circuit (IC) is the transistor. Transistors are the largest member of a family of solid-state devices called "semiconductors." They are built from a special class of materials with electrical properties somewhere between those of conductors and those in insulators. Pure semiconductor material exhibits high resistance, which can be lowered by adding small amounts of impurities called "dopants."

When fabricating an integrated circuit, the transistors, resistors, and capacitors, as well as their interconnections, are fabricated together—integrated—in a continuous substrate of semiconductor material. Active circuit elements are formed by *doping* selected regions of the material. Silicon is by far the most commonly used semiconductor substrate material because it has overall cost, performance, and processing advantages.

With each new IC generation, device geometries have become smaller and ICs have become more powerful. In 1965, Gordon E. Moore, then with Fairchild Corporation but later an Intel cofounder, observed an important trend that was later elevated to a "law" in the popular electronics press. He predicted that the number of transistors that could be integrated on a single die would grow exponentially with time, roughly doubling every 18 to 24 months. Moore correctly anticipated today's ICs, which can hold several millions of transistors on a chip, providing far more functionality (Figure 5.2). The reciprocal view is the dramatically lower cost for the same functionality. A log plot of "dollars per function" over time measured in years shows a linear decrease. In simplest form this means that any chip with a given functionality will be about half its original cost in 18 to 24 months.

Producing miniaturized devices requires precise and sophisticated design and microfabrication. Computer aided design tools have significantly improved the precision and level of complexity achievable in circuit layout planning. Automated process technologies, advanced clean room systems, and testing equipment have helped bring chip fabrication to submicron levels. The explosion in IC applications is also producing a boom in advanced manufacturing equipment. It includes advanced lithography equipment, specialized ion-beam machines, chemical-

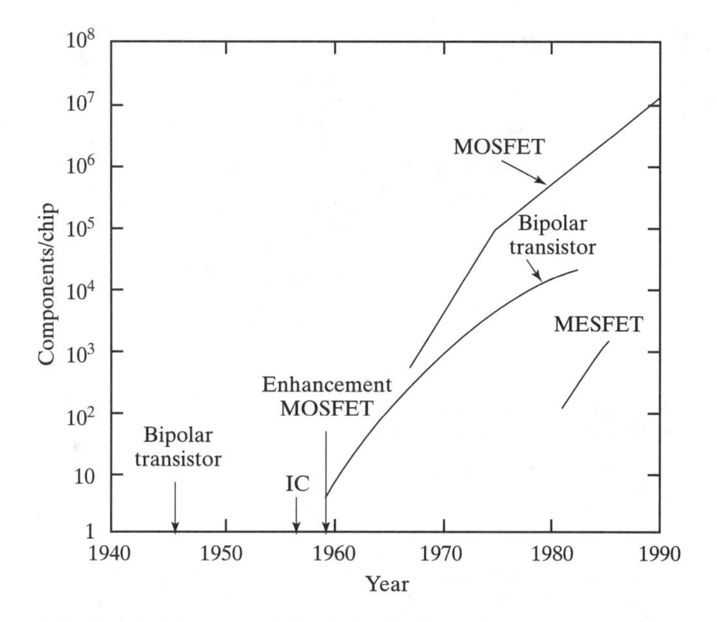

Figure 5.2 Trends in integrated circuit density (from *Digital Integrated Circuits* by Rabaey, © 1996. Reprinted by permission of Prentice-Hall, Inc., Upper Saddle River, NJ).

mechanical polishing equipment to achieve ultraflat surfaces, lasers, and high-vacuum systems.

The semiconductor industry is currently focused on producing larger wafers and smaller process geometries. Larger wafers reduce raw material costs and increase chip processing outputs. Current state-of-the-art semiconductor manufacturing systems produce 200 millimeter (8 inch) wafers with 0.25 to 0.35 micron line geometries. At the time of this writing, the major manufacturers are developing and beginning to use 0.13 to 0.18 micron processes. This will further accelerate the trend shown in Figure 5.2.

During the time this book goes to press and gets published, some of the first 300 millimeter (12 inch) wafers will be in production. By the year 2010, the Semiconductor Industry Association (1997) predicts the production of 0.03 micron line widths on 450 millimeter (18 inch) wafers. Quite simply, these larger wafers mean more chips per batch, which means lower processing costs per chip. Actually, this is not entirely new news. Henry Ford applied analogous principles to automobile manufacturing. For example, specialized tooling and more efficient transfer lines in Dearborn, Michigan, meant more cars per hour and lower processing costs per batch of cars. The comparison with wafer production is not exactly aligned, but in both cases, the simple economics are about spreading the fixed costs of the factory, the people, and the manufacturing equipment over a greater number of individual products (see Figure 5.3).

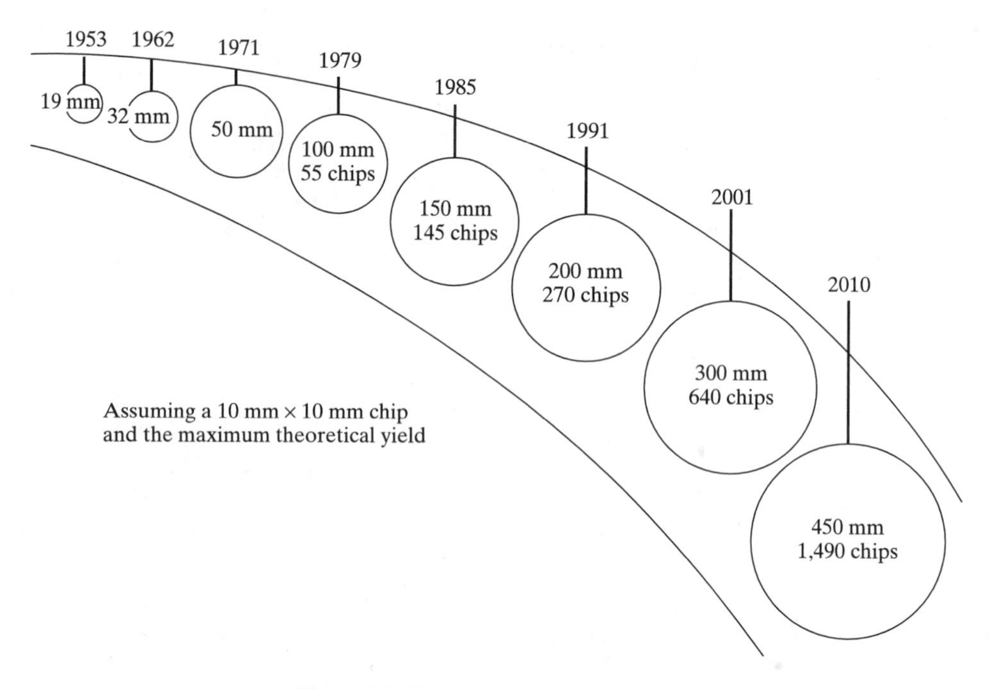

Figure 5.3 Trends in silicon wafer diameter.

5.5 TRANSISTORS

5.5.1 Historical Background

The earliest electronic computers used bulky vacuum tubes resembling short neon lights to create the rapid on/off electric switching that is necessary to perform binary computations and logical functions. In the 1940s, it took thousands of vacuum tubes to create the famous computers that occupied several rooms. Not surprisingly, this was a rather costly and tedious way to go about building a calculating machine.

In 1947, vacuum tube computing was rendered obsolete, almost overnight, by the *transistor.* Three Bell Labs scientists—William Shockley, Walter Brattain, and John Bardeen—are credited with a series of inventions that introduced, refined, and then commercially launched the transistor.[1] Their invention was smaller, faster, and cheaper; handled more complex operations; and generated less heat than its predecessor. The transistor could amplify electric signals moving through circuits embedded in a solid piece of semiconductor material. Transistors were thus called "solid-state" devices because electric current flows through a solid semiconductor rather than through a vacuum tube.

[1]The importance of the vision of M. Kelly, Bell Labs' research director at the time, is also usually stressed. He understood that vacuum tubes were holding back the electronics industry and fostered an innovative research atmosphere to find an alternative. A popular anecdote is that if one of today's cell phones were made from vacuum tubes, the device would be as big as the Washington Monument in the District of Columbia.

Transistor technology started the microelectronics revolution by making high-performance inexpensive electronics possible. Transistors showed up in a burgeoning array of electronic products—from rockets to portable radios—throughout the 1950s. Also, with fewer power, heat, and size constraints, computer designers could build faster, more reliable computers that occupied much less space. But properly connecting hundreds of transistors with thousands of other electric circuit components was an enormous design, manufacturing, and performance problem.

The problems of interconnecting the discrete devices in computers were overcome with the invention of the integrated circuit in 1958 by Jack Kilby at Texas Instruments. This enabled the fabrication of circuit components and their interconnections on a single chip.

Integrated circuits are classified into analog and digital. Analog integrated circuits include a large family of circuits used in power electronics, instrumentation, telecommunications, and optics. Digital integrated circuits are usually classified into two types, memory and logic chips:

- *Memory chips* consist of memory cells and associated circuits for address selection and amplification. Process technologies are extremely well developed for 16 and 64 megabyte dynamic random access memories (DRAMs). DRAMs are inexpensive commodity products differentiated by speed, power consumption, configuration, and package type. From an integrated design and fabrication viewpoint, specialty DRAMs and video RAMs are the more emerging technologies of interest.
- *Logic chips* contain the circuits needed to perform arithmetic, logic, and control functions central to the microprocessor. Application specific integrated circuits (ASICs) are tailored to a customer's particular requirement, as opposed to one of the standard "Intel-inside" microprocessors.

Rapid advances in IC design and process technologies meant that chips could be made at commercially viable scales by the early 1960s. Improvements in miniaturization technology permitted ever-increasing numbers of components to fit on smaller and smaller chips (Table 5.1).

By 1971, a single integrated circuit (IC) was built that combined logic functions, arithmetic functions, memory registers, and the ability to send and receive data. This device was called the *microprocessor.* It was used in many applications and spurred

TABLE 5.1 Trends in IC Integration Levels

Integration scale	Abbreviation	Devices per chip
Small scale integration	SSI	2 to 50
Medium scale integration	MSI	50 to 5,000
Large scale integration	LSI	5,000 to 100,000
Very large scale integration	VLSI	100,000 to 1,000,000
Ultra large scale integration	ULSI	>1,000,000
Future capabilities	?	>1,000,000,000

the factory-floor robotics revolution of the late 1970s (see Figure 1.2). For the robotics industry, the *microcontroller* was a cheap and reasonably powerful specialized control system built around the microprocessor. Of course, the microprocessor also made possible the development of the *microcomputer*—or the personal computer (PC).

5.5.2 Semiconductors: *p*-Type and *n*-Type

A semiconductor is a crystalline material (usually silicon) with electrical properties lying between conductors, such as aluminum and copper, and insulators, such as rubber and glass. Silicon crystallizes in a diamond-shaped lattice, with each atom surrounded by four other atoms in a tetrahedron. The atoms share valence electrons, which give each atom a complete valence shell. In its pure state, a semiconductor material exhibits relatively high electrical resistance. Adding controlled amounts of certain chemical impurities (*dopants*) to the crystal structure of the semiconductor lowers its resistivity and allows current to flow through the material. The atomic structure of the dopant determines whether the resulting material will be "*n*-type" or "*p*-type."

- *n*-type silicon is typically created by doping silicon with phosphorus, which has five electrons in its outer shell. In comparison with the four-electron silicon, this creates additional free electrons in the material, which readily move in response to a voltage. Since most of the conduction is carried by negatively charged electrons, the material is called *n*-type.
- *p*-type silicon is typically created by doping silicon with boron. Boron has only three electrons in its outer shell. Since all the silicon atoms were nicely balanced with four[2] electrons in their outer shell, the presence of the boron intruder creates additional vacancies, or "holes," in the material. These holes can be thought of as positive charges. Surrounding electrons can move in and fill this hole and, in doing so, leave behind another hole. The holes thus seem to move in a direction opposite to the electron flow. Since most of the conduction occurs by way of the positively charged holes, the material is called *p*-type.

Modifying the concentration of dopants controls the resulting change in semiconductor conductivity. The process of doping semiconductor materials to selectively increase their conductivity is fundamental to the manufacture of advanced semiconductor devices because it makes possible the fabrication of basic circuit substructures.

Silicon is the material of choice for microelectronics devices because of its numerous advantages. As one of the most abundant elements on the planet, silicon is cheap and readily available. It can be subjected to higher temperatures than germanium, the next most popular semiconductor resource. Silicon also has critical processing advantages. It easily oxidizes to form silicon dioxide, an excellent insulator among circuit components. Silicon dioxide is also extremely useful during the fabrication process because it is an effective barrier layer during multiple doping opera-

[2]Strictly, each Si atom shares its four electrons with its neighbors, creating eight in the outer shells.

tions. Gallium arsenide rather than silicon is increasingly used in optoelectronic and high-frequency communication devices.

5.5.3 The Transistor

The region where p-type and n-type semiconductors meet forms a crucial structure known as a *pn* junction (Figure 5.4). A *pn* junction is basic to the operation of most electronic devices. For example, a diode is a *pn* junction that allows the flow of current in one direction and blocks it in the opposite direction. A bipolar junction transistor (BJT) is made by sandwiching three different semiconductor slices into one solid block, such that the center slice is of one type and the two outer slices are of the opposite type. In effect, this creates two *pn* junctions. Depending on how the junctions are combined, the transistor is either "*npn*" or "*pnp*" (see Figure 5.5). In an *npn* transistor, electrons can flow from the emitter (n), across the base (p), to the collector (n). More significantly, applying a voltage to the base vigorously rips electrons from the emitter and sends them rocketing across the base into the collector—in effect, amplifying the input current to the base. The stronger the voltage on the base, the stronger the resulting flow of current through the transistor. This amplification is more utilized in analog devices such as an electric guitar. For the ICs in computers, the primary function is the ultrafast switching ability for logic.

Figure 5.5 shows a simple sandwichlike *npn* arrangement. By contrast, Figure 5.6 shows the horizontal layout of the field effect transistor (FET). The terminology of the *npn* transistor—emitter, base, and collector—is now changed to *source, gate,* and *drain* for the FET. To activate the transistor, voltage is applied to the polysilicon control gate (center of Figure 5.6). Electrons flow out of the source region (marked n^+) through the *channel* (part of the p-type substrate) and into the drain (also marked n^+). The amount of flow is precisely controlled by the voltage applied to the gate. For the n-type device (NMOS) a positive voltage is applied to the polysilicon gate. The gate and the p-type substrate form the plates of a capacitor with the gate oxide (SiO_2) as the dielectric of the capacitor. The reader is referred to a text such as Rabaey's (1996) for the relationship between the applied gate voltage and the current flow between the source and the drain.

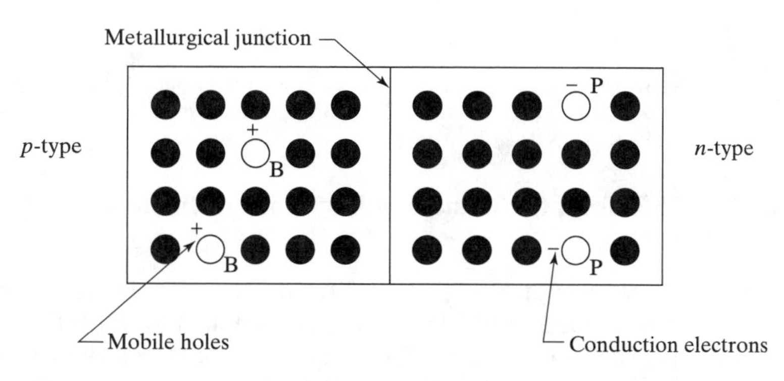

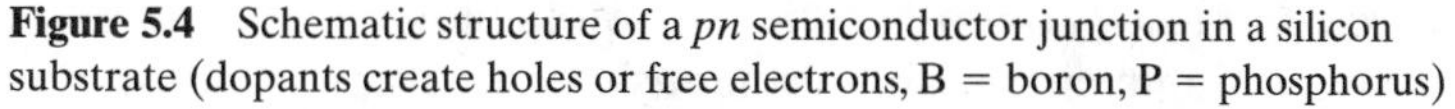

Figure 5.4 Schematic structure of a *pn* semiconductor junction in a silicon substrate (dopants create holes or free electrons, B = boron, P = phosphorus)

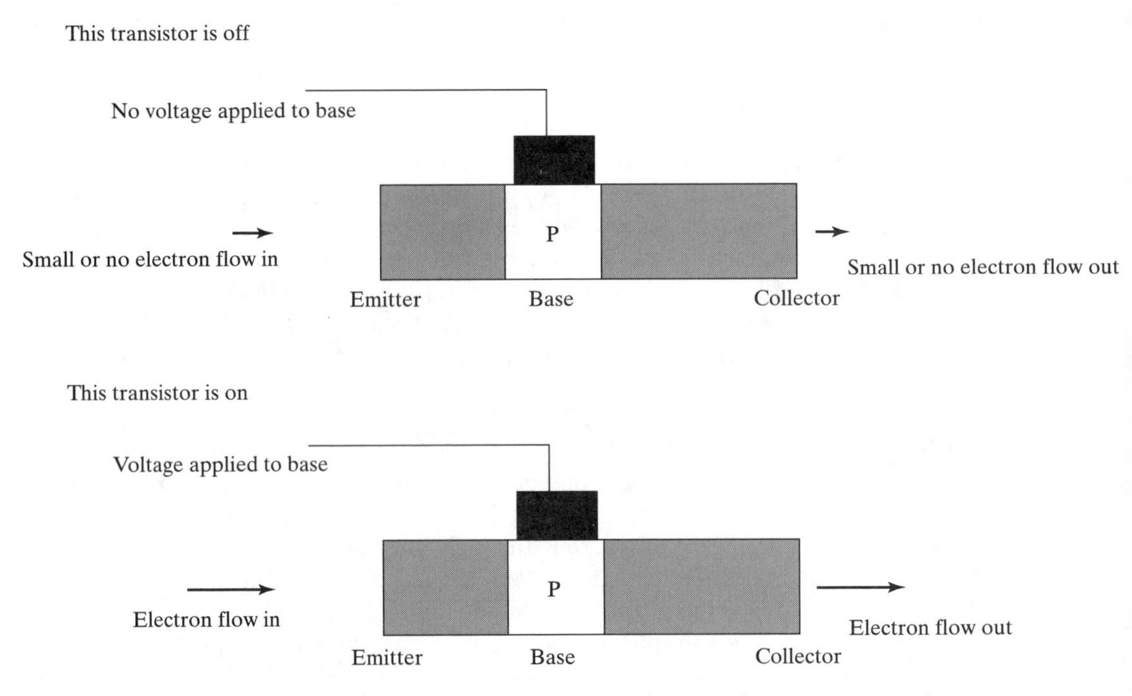

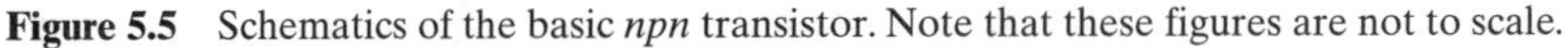

Figure 5.5 Schematics of the basic *npn* transistor. Note that these figures are not to scale.

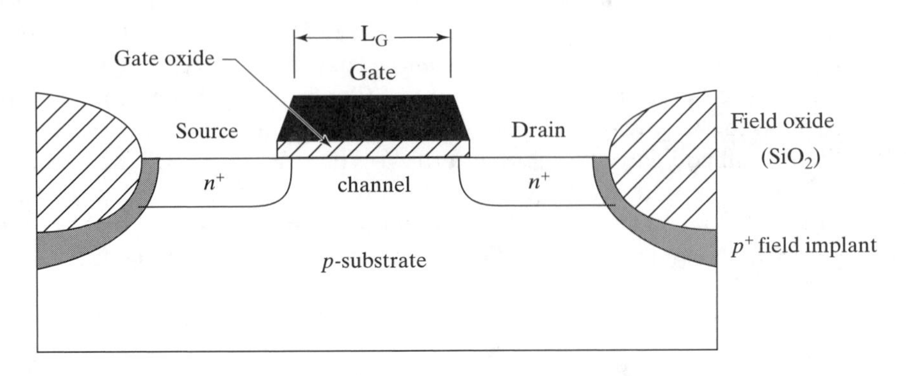

Figure 5.6 Basic structure of an *n*-type NMOS IC (from *Digital Integrated Circuits* by Rabaey, © 1996. Reprinted by permission of Prentice-Hall, Inc., Upper Saddle River, NJ).

5.5.4 MOSFETs as the Basic Building Block of the Integrated Circuit

Metal-oxide semiconductors (MOSFETs) are one type of field effect transistor. They are the essential building blocks of integrated circuits. MOSFETs can be made either *p*- or *n*-type. The *n*-type devices (NMOS) are faster than *p*-type devices (PMOS). In practice, the most common type is complementary MOS-type (CMOS) circuits. In this case, a single circuit simultaneously controls pairs of *n*-type and *p*-type transistors. CMOS circuits are the most popular because of their integration-density and power-consumption efficiencies (see review on p. 4 of Rabaey, 1996). The precise, high-speed switching of MOSFET devices allows transistors to carry out the rapid binary data processing that lies at the foundation of modern computing.

5.5.5 The NMOS Transistor

Key terminology includes:

- *Substrates,* which are p-type for NMOS
- *Active* transistor areas, which are $n+$ in NMOS
- *Polysilicon* layers for the gate electrode
- *Select* regions, or *field implant* regions, which are p^+ in NMOS
- *Field oxide* regions of silicon dioxide (SiO_2)
- *Interconnect* layers, usually of aluminum
- *Contact* layers for interconnections between different layers
- *Wells,* which are n-type within a p-type substrate for CMOS transistors

The basic structure of an IC depends on the specific transistor technology used. In MOS-based chips, source and drain regions are formed by selectively "doping" portions of a p-type or n-type substrate surface to the opposite type of material. The NMOS device is made up of n^+ source/drain, these n^+ areas arise from the selective *doping* of desired regions in a p-type substrate. The conductive gate is made with a thin film of polycrystalline silicon (usually referred to as polysilicon). Comparatively thick layers of silicon dioxide, called the *field oxide,* and highly doped *field implants* (select regions of p^+ in NMOS) insulate neighboring n^+ areas. Aluminum layers provide the interconnections among circuits. Copper will increasingly be used for this purpose.

5.5.6 The CMOS Circuit

The complementary MOS process is preferred over basic NMOS because it leads to the creation of more circuits on a chip. This is shown in Figure 5.7. The process starts with a p-substrate, which will eventually be doped in certain areas for n^+type transistors (on the left). A mask is used early on in the process to define many additional *n-wells*—shown on the right—which will then contain p^+ transistors.

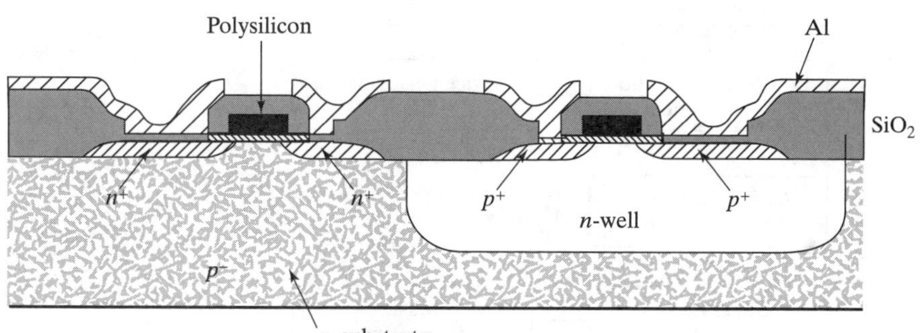

Figure 5.7 The CMOS transistor (from *Digital Integrated Circuits* by Rabaey, © 1996. Reprinted by permission of Prentice-Hall, Inc., Upper Saddle River, NJ).

5.6 DESIGN

The design of integrated circuits—say, for the embedded systems in cell phones, PDAs, and cameras—is outside the scope of this book. However, typical design levels for such devices are shown in Figures 5.8 and 5.9. In Figure 5.8, a hierarchy is shown that breaks down a simple IC's description into several levels of abstraction. These include:

- The defined global function of the device.
- Subfunctions, which must coordinate with the global goal. Therefore iterative high-level simulations are needed. These iterations are indicated by the feedback loop at the top of the diagram.
- The assembly of these subfunctions into cells or functional blocks.
- The creation of specific transistor and circuit layouts that deliver the performance of the desired functional blocks while still being manufacturable in a standard "fab."

Figure 5.9 is similar but for a more complex device such as a wireless networked computer or a wireless PDA. Such a device needs three main divisions (shown in three columns) of the design abstraction for (a) analog data processing, (b) digital data processing, and (c) protocols and control (see **http://bwrc.eecs.berkeley.edu**). Some common development tools from Figure 5.9 are listed in Table 5.2. For one of these complex devices, with more than a million transistors, today's IC designers target the gate level netlist description in the fifth row of the

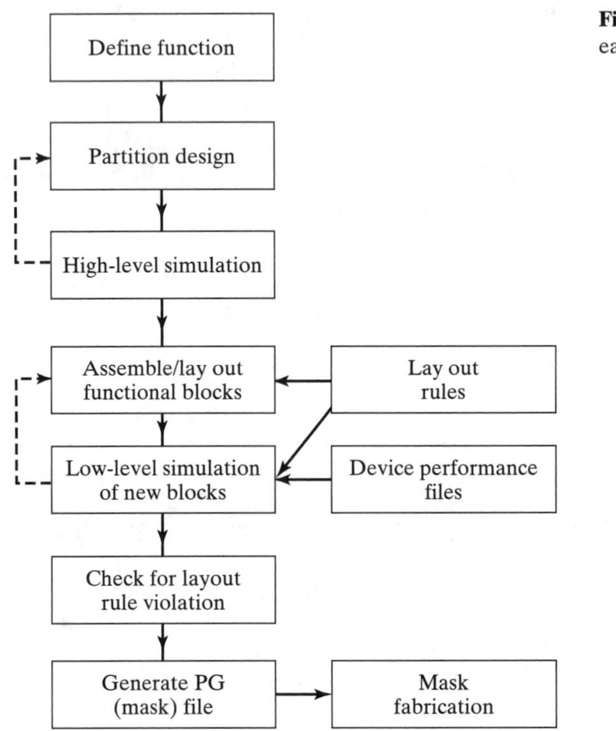

Figure 5.8 Design flow, typical of the early 1990s, for a simple device.

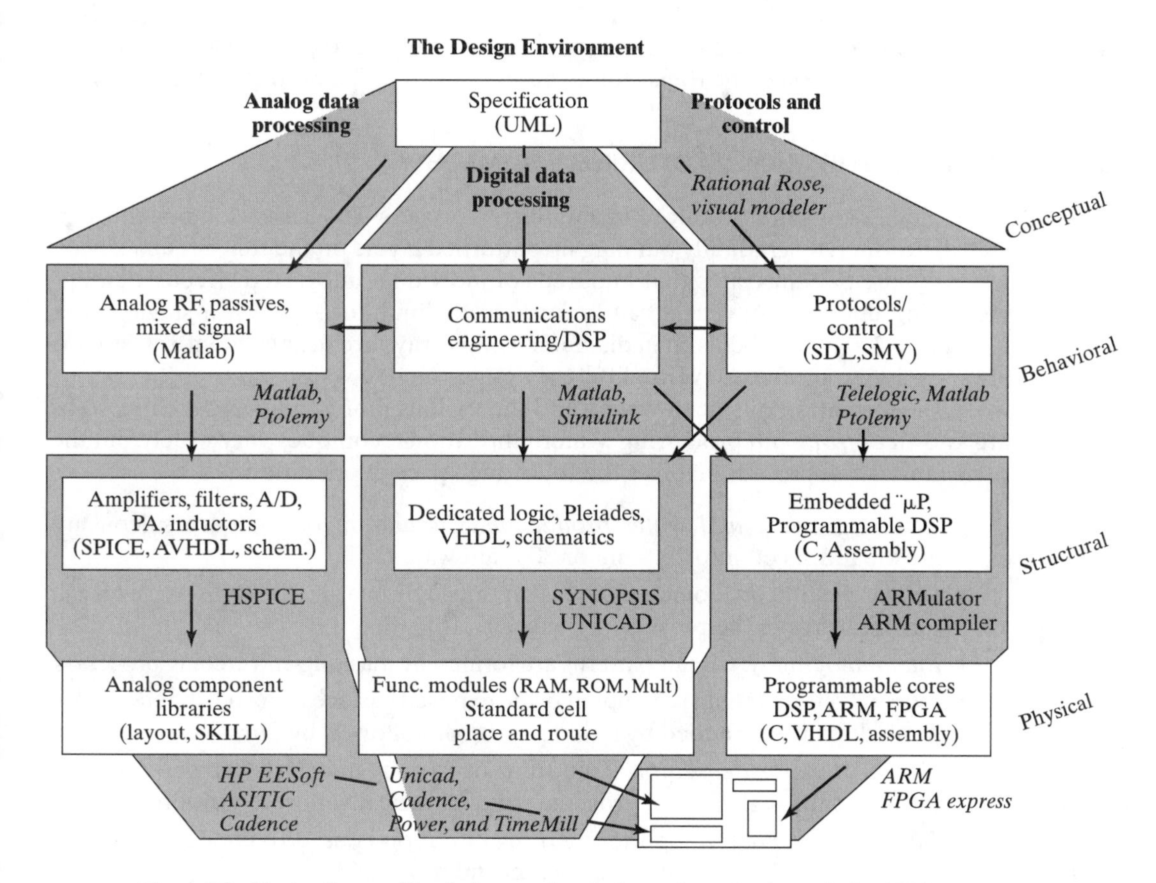

Figure 5.9 Design flow and levels for complex devices—for example, a wireless PDA (courtesy R. Brodersen, 2000).

TABLE 5.2 Design Flow, Typical for Today's More Complex Devices Shown in Figure 5.9. The Table Was Prepared with the Help of Rhett Davis.

Level of design	Company (tool in parentheses)
Functional specifications	Mathworks (Matlab, Simulink); Cadence (SPW,VCC)
Register transfer level (RTL) coding and behavioral simulation	Synopsys (VSS); Cadence (Verilog-XL); Mentor Graphics (VHDL simulator)
Logic synthesis	Synopsys (Design Complier; Module Compiler; Behavior Compiler)
Test insertion and automated test pattern generation	Synopsys
Gate level netlist simulation	Synopsys (VSS); Cadence (Verilog-XL); Mentor Graphics (VHDL)
Floor planning	Cadence (Design Planner, Pillar); Avant! (Apollo)
Placement and route	Cadence (Silicon Ensemble, IC Craftsman)

table. By contrast, Figure 5.8 is more typical of the early 1990s design flow for simpler devices, where the specific transistor layout is created at the bottom of the diagram.

5.7 SEMICONDUCTOR MANUFACTURING I: SUMMARY

Integrated circuits are produced in a multistep process. The active components of a chip are formed by gradually building up patterned layers on the thin circular silicon wafer. Figure 5.3 illustrates that hundreds of individual integrated circuits, perhaps measuring under 1 cm^2 each, can be produced simultaneously on a single wafer measuring 200 mm or 300 mm in diameter. Chip arrays are usually identical, but it is also possible to produce several different designs on one wafer.

The formation of silicon wafers and the various lithography and etching steps are known as *front-end* processing. Wafer fabrication techniques vary widely, but the basic fabrication process involves the following series of operations:

- *Crystal growing and wafer production.* Circular ingots of pure silicon are grown and sliced into 200-mm or 300-mm wafers.
- *Oxidation.* Silicon dioxide (SiO_2) is produced by heating the wafer to very high temperatures in the presence of oxygen.
- *Photolithography.* Circuit patterns are formed by masking and etching processes.
- *Doping.* After etching is completed, the exposed surfaces may be doped. The n^+ or p^+ dopants are added by ion implantation followed by diffusion processes.
- *Chemical vapor deposition.* Thin films of various materials are deposited on the wafer through several processes (e.g., chemical vapor deposition [CVD]).
- *Interconnect creation.* Sputtering or evaporation is used to create conducting circuits between individual transistors and devices.
- *Testing and packaging.* Individual ICs are tested for quality and placed in protective packages that can later be connected on a printed circuit board.

The fabrication process is always carried out in a clean room environment, a confined area in which dust, temperature, and humidity are precisely controlled to varying degrees. Different classes of clean rooms define the maximum number of particles per cubic foot in the space (Table 5.3). Reducing dust and other variables is essential to avoid contaminating the chip's circuitry and lowering chip yields.

TABLE 5.3 Trends in Class Ratings for Clean Rooms

"Class" of clean room	Number of 0.5 micron particles per cubic foot	Number of 0.5 micron particles per cubic meter
Class 10,000	10,000	350,000
Class 1,000	1,000	35,000
Class 100	100	3,500
Class 10	10	350
Class 1	1	35

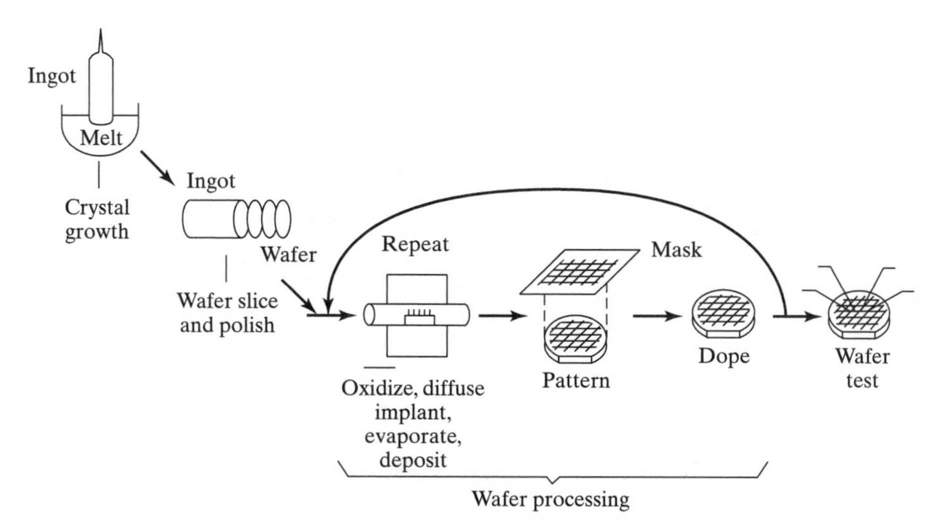

Figure 5.10 IC manufacturing process flow.

Figure 5.10 shows the process flow and general fabrication sequence for the manufacturing. Once again, note that the lithography, etching, and layering steps in the central parts of the diagram are repeated many times to build up the several layers indicated in the schematics of Figures 5.6 and 5.7.

Once all active components and circuits are fully formed, the wafer is protected with a layer of insulating dielectric film. Finally, a patterned passivation layer is added, with revealed bonding pads to which bonding wires can be attached. These bonding wires link to the IC package during the phase known as *back-end* processing. Eventually, the chips are assembled onto printed circuit boards (see Chapter 6).

5.8 SEMICONDUCTOR MANUFACTURING II: NMOS[3]

Figure 5.11 shows the multilayer structure of an NMOS circuit. It is useful to summarize the step-by-step manufacturing flow before the details of each process such as lithography are considered in Section 5.10. For the new student in this area, it is important to emphasize that Steps 3 through 6 are not concerned with creating the transistor itself but with defining the peripheral areas that will contain the p^+ field implants and field oxide on the very outside of Figure 5.6. Importantly, and by contrast, the inner channels that will eventually be the {n^+pn^+ source-gate-drain}

[3]*Important:* By far the most common manufacturing method today is CMOS. In CMOS, both NMOS and PMOS transistors are produced in tandem. After the first oxidation and nitriding steps, the *n*-wells are formed so that at high magnification, the wafer will have a checkerboard appearance. Manufacturing then involves "hopping back and forth" between the *p*-substrate and the *n*-well shown in Figure 5.7. To a new student in manufacturing, this alternation between the *n*-type and *p*-type transistors can get complicated. Thus, for simplicity, the text stays with the "single-type" NMOS.

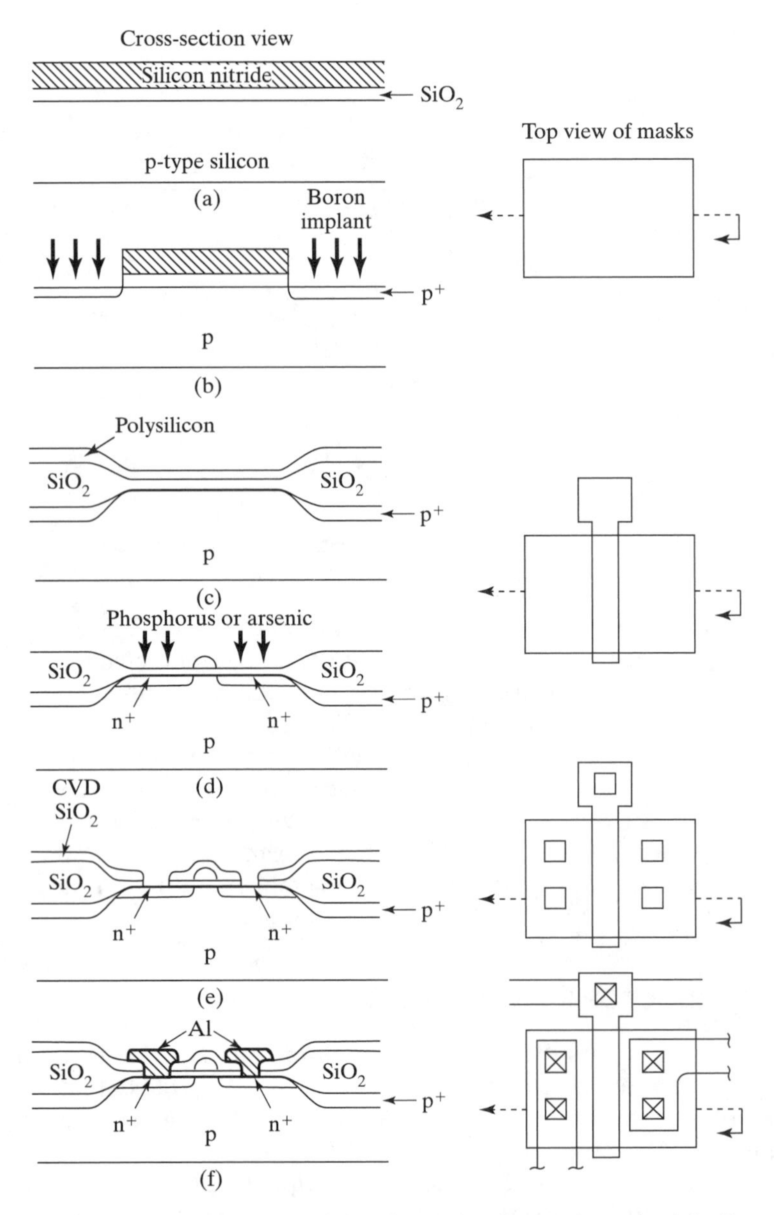

Figure 5.11 NMOS: wafer on the left and corresponding masks on the right (from *Introduction to Microelectric Fabrication* by Jaeger, © 1988. Reprinted by permission of Prentice-Hall, Inc., Upper Saddle River, NJ).

transistor areas remain covered and protected by the inert, hardened photoresist. Work does not begin on these areas until Step 8.

Step 1: A standard wafer of *p*-type silicon with specific resistance is scrupulously cleaned and then heated in an oxygen rich furnace to create silicon dioxide

(SiO_2). This creates a thin layer—sometimes called the pad oxide—on top of the substrate material.

Step 2: A layer of silicon nitride is added using chemical vapor deposition (CVD).

Step 3: Ultraviolet-light-sensitive photoresist is applied to the wafer. A mask pattern is shown on the top right of Figure 5.11a for the outer p^+ field implants and field oxide areas. It is shown as a simple rectangle of four thin lines. The lines are transparent, but every other region of the mask, including the center of the rectangle, is opaque. The UV light only passes through the lines of the mask, and it deliberately damages the photoresist into this pattern. The damaged photoresist is then sloughed away in chemical solutions. This leaves a rectangular pattern of four connected trenches with naked silicon nitride at the bottom. In Steps 4 and 5, the silicon nitride and silicon dioxide at the bottom of these trenches are etched away.

Step 4: The wafer is dry etched with a plasma process to create the vertical-wall trenches. Speaking colloquially, the NMOS wafer under a microscope will now look like Manhattan: tall buildings still protected by oxide/nitride, standing next to avenues that have been etched away down to the *p*-type substrate. Following the etching process, these avenues are then doped with boron to form *select* p^+ type regions (Figure 5.11b).

Step 5: Boron, a *p*-type dopant, is ion-implanted into these naked avenues. The dopant is then driven in deeper by a supplementary diffusion. This combination of ion-implantation and diffusion creates the p^+ field implant regions shown on the left and right of Figures 5.6 and 5.11b. Why is this necessary? It means that the transistor—in the center part of the figure—is "boxed in." Electrons will be constrained to flow from the source to the drain rather than getting loose and being attracted off to some other circuit. The p^+ regions are also called the *channel-stop implants.*

Step 6: Thermal field oxidation then covers these p^+ regions with the thick SiO_2 layers shown at the left and right of Figure 5.11c.

Step 7: Let's pause for a moment! The p^+ areas covered in the thick layer of SiO_2 are rather like "sidewalls" that establish the boundaries of the transistor. Work can now begin on the central area where the source, gate, and drain will be built up.

Step 8: Recall from Step 3 that the central area was left *protected* by the photoresist and still has the oxide/nitride sandwich on it from Steps 1 and 2. This oxide/nitride sandwich now has to be etched away so that the gate can be constructed.

Step 9: A very thin layer of silicon oxide (SiO_2) is grown on the whole wafer as the foundation of the gate. This thinner layer of oxide is shown in the middle of Figure 5.6, just under the polysilicon gate region. Recall from Section 5.5.3 that this layer of SiO_2 acts as the dielectric between the gate and the *p*-type substrate. Controlled growth of this layer to precise dimensions is critical to the performance characteristics of the device.

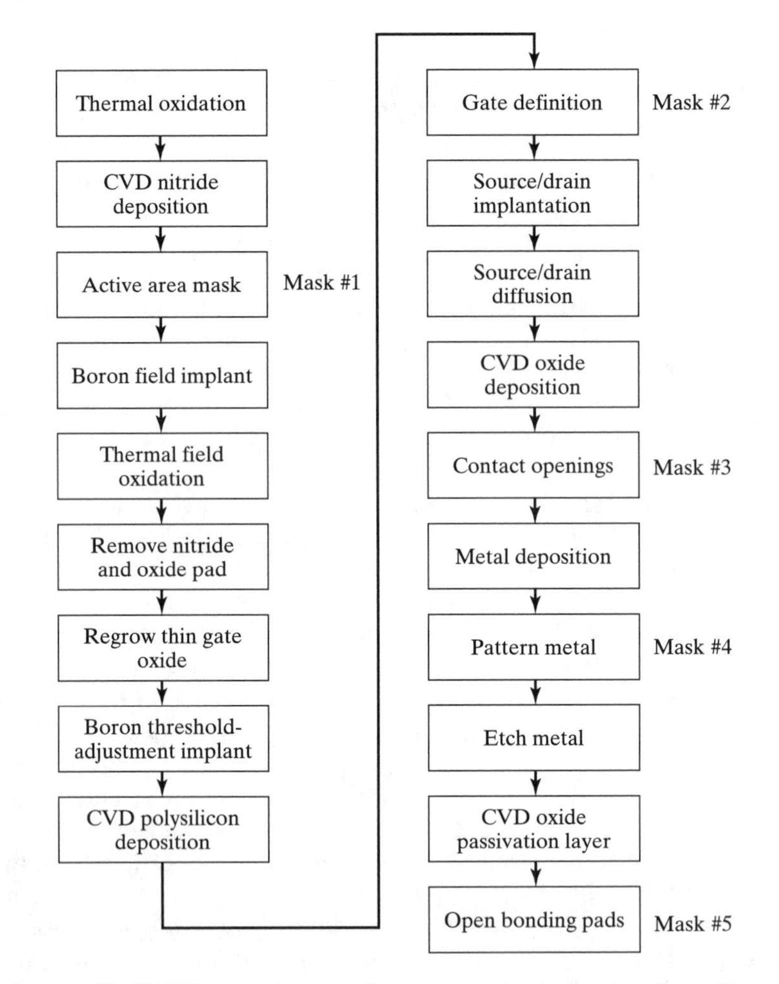

Figure 5.12 NMOS processing steps that correspond to the previous figure (from *Introduction to Microelectric Fabrication* by Jaeger, © 1988. Reprinted by permission of Prentice-Hall, Inc., Upper Saddle River, NJ).

Step 10: Further doping is done with boron, which can penetrate through this thin oxide layer, to adjust the necessary threshold voltage of the gate region.

Step 11: Polysilicon is deposited using CVD to create the contact for the gate itself. Figure 5.11c represents the wafer at this point in the proceedings. Also, the process has reached the lower left column of the box diagrams in Figure 5.12.

Step 12: The second photolithography mask is then used to create the precise and delicate features in the central transistor region. This is shown as the two tracks in the center of the mask in Figure 5.11d. The region that will end up as the polysilicon gate is covered by an opaque region of the mask during photolithography. However, the thin channels that correspond to the source and drain are exposed, and in those regions the photoresist is selectively damaged and removed. Etching then removes the polysilicon from the top surface of these "naked" regions. Figure 5.11d shows the wafer at this stage: the polysilicon is etched away except for the small central gate region.

Step 13: This next step is an important one: phosphorus or arsenic, n-type materials, are ion-implanted into these "naked" regions of the originally p-type wafer/substrate. A diffusion process is then used to drive in the phosphorus to create the highly doped n^+ regions. The transistor is now defined. The two n^+ channels in Figure 5.10d are the source and the drain. They are straddled by the polysilicon gate.

Step 14: Let's pause again. The important work is now finished purely from the point of view of building the transistor itself. The remainder of the steps that follow are more concerned with creating the interconnections and protecting the surface.

Step 15: A further CVD oxidation step covers the wafer shown at the top left of Figure 5.11e. Then, the third mask is used to open up the contact areas to the source, gate, and drain. This third mask is shown on the right of Figure 5.11e. The contact openings for the n^+ regions are the four small squares on each side of the gate lines.

Step 16: The surface of the wafer is then metallized with aluminum. Sputtering or evaporation is used for this step.

Step 17: A fourth mask shown on the bottom right of Figure 5.11 is used to create the pattern of interconnections to the other transistor circuits. The shaded areas of Figure 5.11e show the cross-sectional views of the aluminum contacts.

Step 18: The remainder of the metal not needed for the interconnections is etched away.

Step 19: A passivation layer of CVD oxide is deposited on the wafer.

Step 20: Small windows for bonding pads are opened up at the periphery of the IC. Thin wires are connected to these bonding pads. In later operations, during the back-end processing (Section 5.11), these are connected to the lead frame of the package. Final cleaning and passivation create the final IC ready for such back-end packaging.

5.9 LAYOUT RULES

As in the mechanical world of Chapters 3 and 4, the design of the transistor layout is constrained by the physics of manufacturing. Design rules are required to account for variations in mask alignment, depth of focus problems in lithography, etching, and lateral diffusion. The main constraint is the minimum mask dimension that can, with fidelity, be transferred to a wafer. Design rules specify the minimum horizontal *intralayer* spacings between features on the same layer. Other design rules govern the *interlayer* transistor layouts between layers. The dimensions of the contacts, vias, and wells are also governed by rules.

Distances in the figures that follow are measured in a *scalable design length,* λ, following a general procedure[4] proposed by Mead and Conway (1980). In fact, these scalable design lengths do not scale in an entirely linear manner, and they err on the conservative side. As a result, today's industries tend to use the micron rules, which directly prescribe preferred dimensions. Nevertheless, for an introductory text such as this, the Mead and Conway scalable ideas are presented because they are generic.

5.9.1 Intralayer Design Rules

In the first Figure 5.13, the active transistor areas are 3λ in dimension and 3λ apart. The polysilicon gate region is 2λ. The metal regions are typically 3λ. The contacts through to the active transistors and the metal-to-metal via holes are 2λ. CMOS wells are larger at 10λ.

5.9.2 Interlayer Design Rules

Figure 5.14a shows a polysilicon layer (*po*) overlapping and defining the channel on an n^+ diffusion region. It means that the minimum length of the active transistor will be 2λ—as defined by the minimum width of polysilicon in the previous Figure 5.13. The minimum width will be 3λ—as defined by the minimum width of diffusion to create an active region. The dimension from the active region to the well boundary is 5λ. Note that such design rules specify minimum distances from a "feature to an

[4]In the current MOSIS system the minimum line width is set to 2λ. For example, for a 1.2 micron process (i.e., a process with a minimum line width of 1.2 microns), $\lambda = 0.6$ micron.

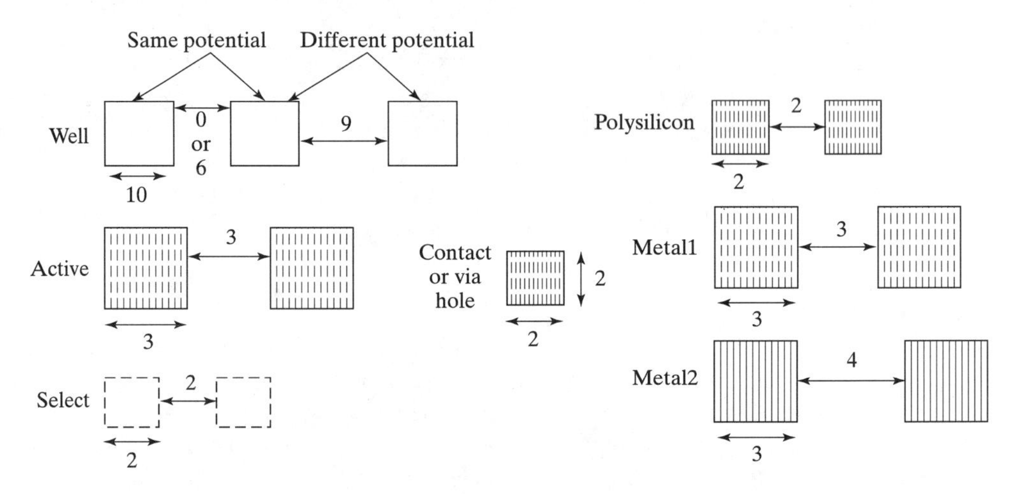

Figure 5.13 Minimum dimensions and spacings of intralayer design rules (from *Digital Integrated Circuits* by Rabaey, © 1996. Reprinted by permission of Prentice-Hall, Inc., Upper Saddle River, NJ).

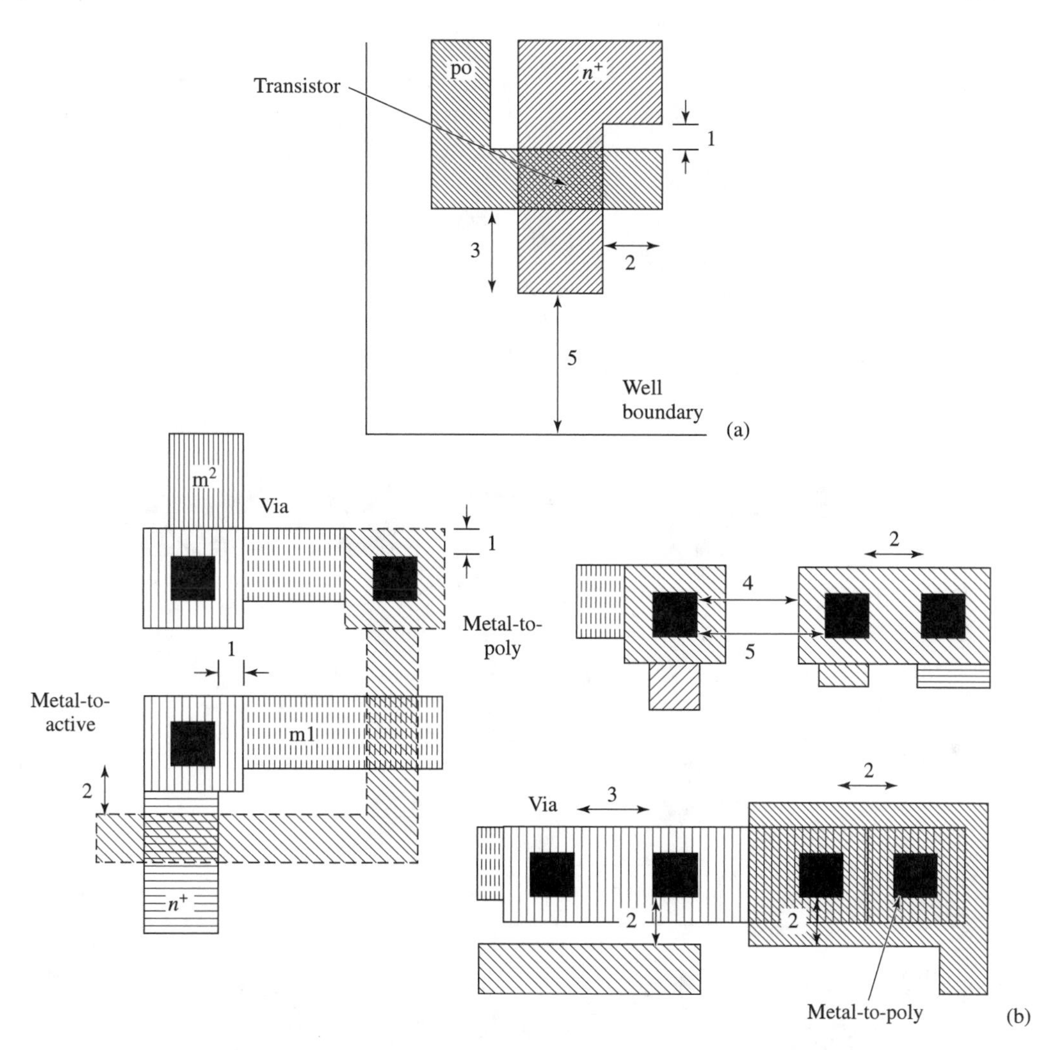

Figure 5.14 Interlayer rules (from *Digital Integrated Circuits* by Rabaey, © 1996. Reprinted by permission of Prentice-Hall, Inc., Upper Saddle River, NJ).

edge." This is the same concept used in Figure 4.17, where in the mechanical CAD/CAM system, the minimum distance from a "feature to an edge" is specified.

In Figure 5.14b, the design rules for the vertical openings that allow aluminum to penetrate down to the active transistor are specified, along with the dimensions of the vertical vias that interconnect different metal layers. For example, the label "metal-to-active" on the left of the diagram is next to a darkened 2×2 square that represents a vertical opening that allows the connection between the m1 aluminum and the n^+ active transistor site. Figure 5.15 shows a combined view of a simple device. It includes (a) layout, (b) cross section, and (c) circuit diagram.

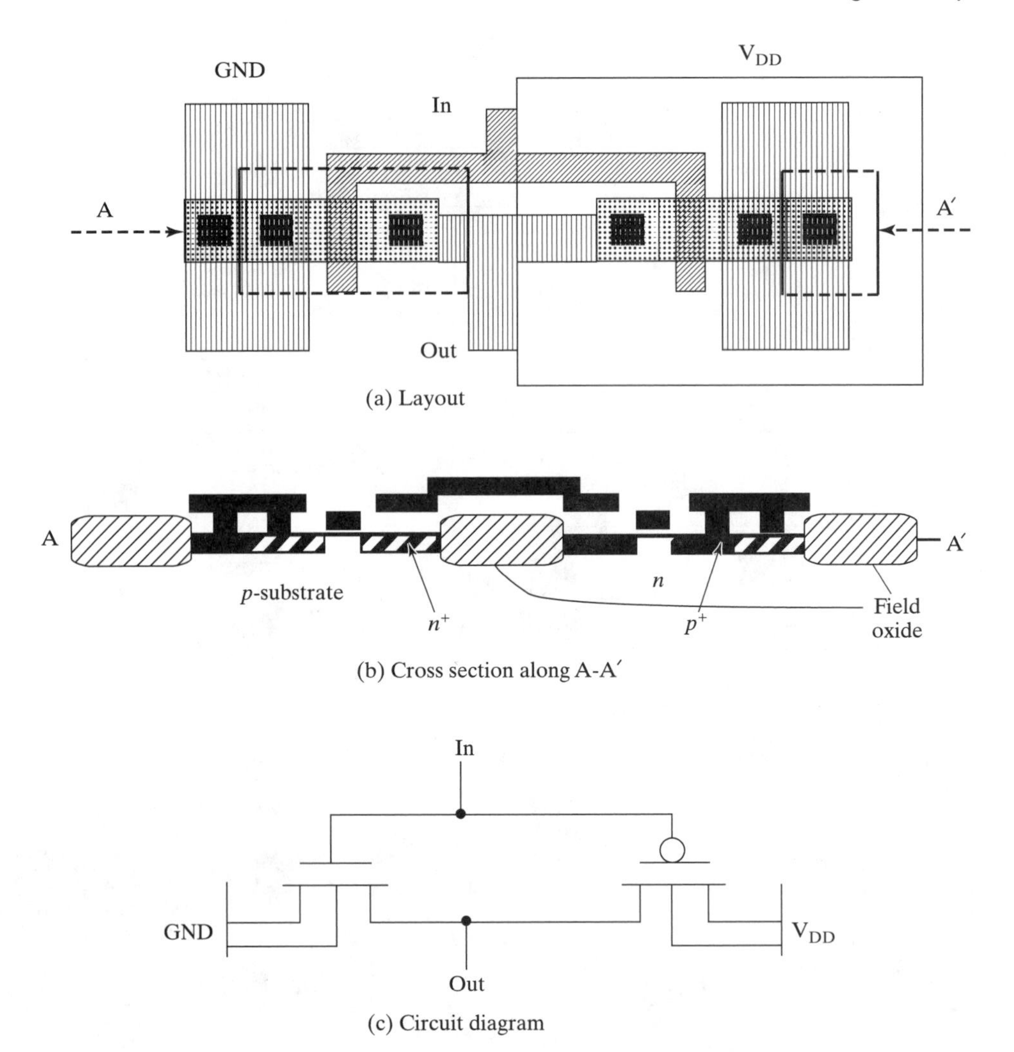

Figure 5.15 Example layout including the vertical process cross section and corresponding circuit diagram (from *Digital Integrated Circuits* by Rabaey, © 1996. Reprinted by permission of Prentice-Hall, Inc., Upper Saddle River, NJ).

5.10 MORE DETAILS ON FRONT-END PROCESSING

A variety of excellent texts present comprehensive details of the front-end processes: Mueller and Kamins (1986), Jaeger (1988), Pierret (1996), and Campbell (1996). Some key aspects are summarized in the next few pages.

5.10.1 Silicon Ingots and Wafer Preparation

The process begins by heating relatively low grade silicon, or ferrosilicon, with carbon in an electric furnace. A series of reduction steps creates impure silicon. Conversion to liquid silicon chloride allows purification by distillation. Heating $SiCl_4$ in a hydrogen atmosphere creates ultrapure polycrystalline silicon.

A single crystal silicon ingot is next produced by the Czochralski method of unidirectional casting. A solid silicon *seed finger* is dipped into the vat of molten, ultrapure silicon and then slowly withdrawn (Figure 5.16). When growing the single crystal, the solidification direction is usually aligned to the <111> or the <100> direction. This causes the molten silicon to cool as a single crystal around the finger and be drawn out into a long cylinder of the required diameter.

This polycrystalline silicon ingot can be ground to a uniform smoothness and polished. It is then sliced with a diamond saw into circular wafers about 0.5 mm thick and 200 or 300 mm in diameter. The wafer surfaces are also ground and polished. Wafers are chemically cleaned to remove all traces of particles, bacteria, and other impurities. The procedure involves dipping a rack of wafers into successive boiling chemical and deionized water baths. This is an extremely toxic and hazardous process requiring extensive safety and environmental protection measures.

5.10.2 Thermal Oxidation Procedures

As described already, the starting sequence for a typical MOS process is to grow a thin padding of silicon dioxide onto the wafer. Wafers are heated in furnaces and exposed to purified oxygen under carefully controlled conditions as shown in Figure 5.17a.

In later processing steps for the thicker field oxide, the wafers can be heated in a water vapor atmosphere. The water vapor method shown in Figure 5.17b builds layers more quickly than dry oxygen heating.

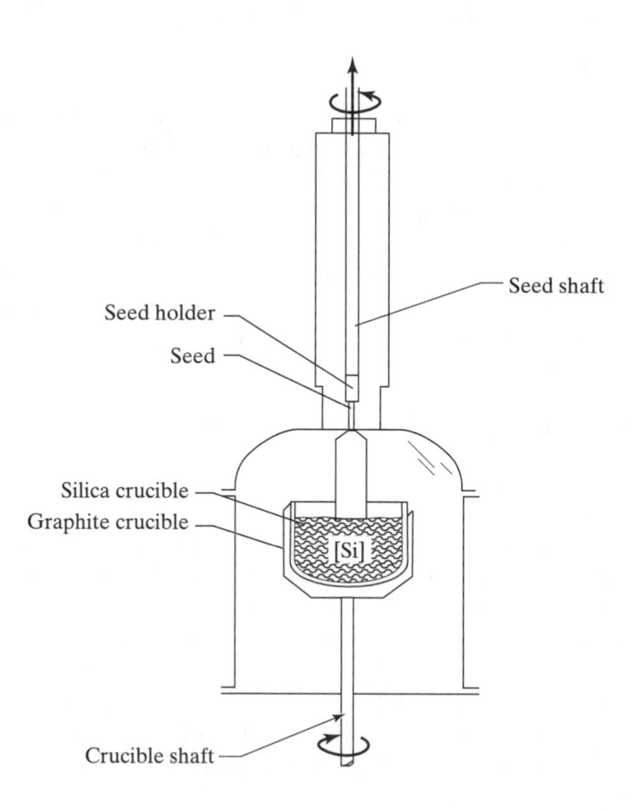

Figure 5.16 Simplified Czochralski method for unidirectional solidification.

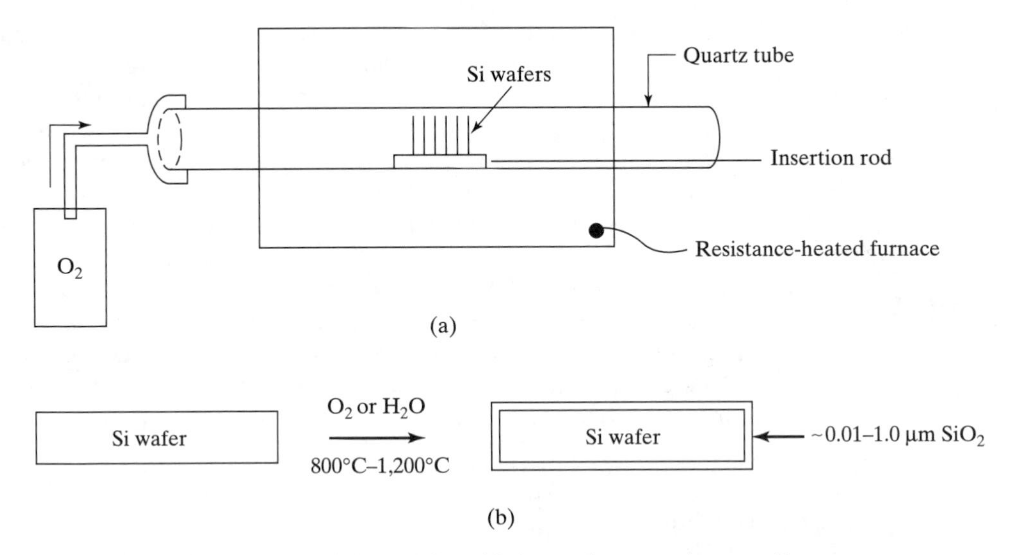

Figure 5.17 Simplified views of the oxidation equipment and process (from *Semiconductor Device Fundamentals* by Pierret, © 1996. Reprinted by permission of Prentice-Hall, Inc., Upper Saddle River, NJ).

However, dry oxygen is again preferred for growing the gate oxide's SiO_2 because it gives better Si to SiO_2 interface properties. Rapid thermal oxidation (RTO) allows short time oxidation at suitably high temperatures (Campbell, 1996).

5.10.3 Creating Photomasks

The CAD files containing the desired circuit patterns are transferred to a set of photographic plates or *photomasks*. To do this, the CAD files are first fed into a *pattern generator*—a computer controlled exposure machine. The generator uses flash exposure to transfer the IC pattern onto a light-sensitive plate known as the mask. This step is similar to photographic developing. The generator flashes onto the plate a large series of rectangles that correspond to the circuit diagram. The plate is covered in an emulsion/photoresist material, which deliberately breaks down under the exposure. Then, once the exposed resist is sloughed off, the plate is transparent just in those areas that correspond to the circuit.

5.10.4 Photolithography: Projecting the Mask Pattern onto the Wafer

Many steps follow to transfer the pattern in each photomask to the wafer. The wafer surface is coated with light-sensitive photoresist material. Typically, photoresist liquid is poured onto the center of the round wafer, which is spun at 1,000 to 5,000 rpm in order to produce a uniform, thin adhesion. The thickness of the film can be controlled by altering liquid viscosity and spinning speed. The photoresist is dried in a warm nitrogen or plain air oven.

Photolithography is shown in Figures 5.18 through 5.20. In the early days of IC manufacture, *contact* and *proximity* printing were used (Wolf and Tauber, 1986). In such methods the photomask was in contact with, or very close to, the wafer.

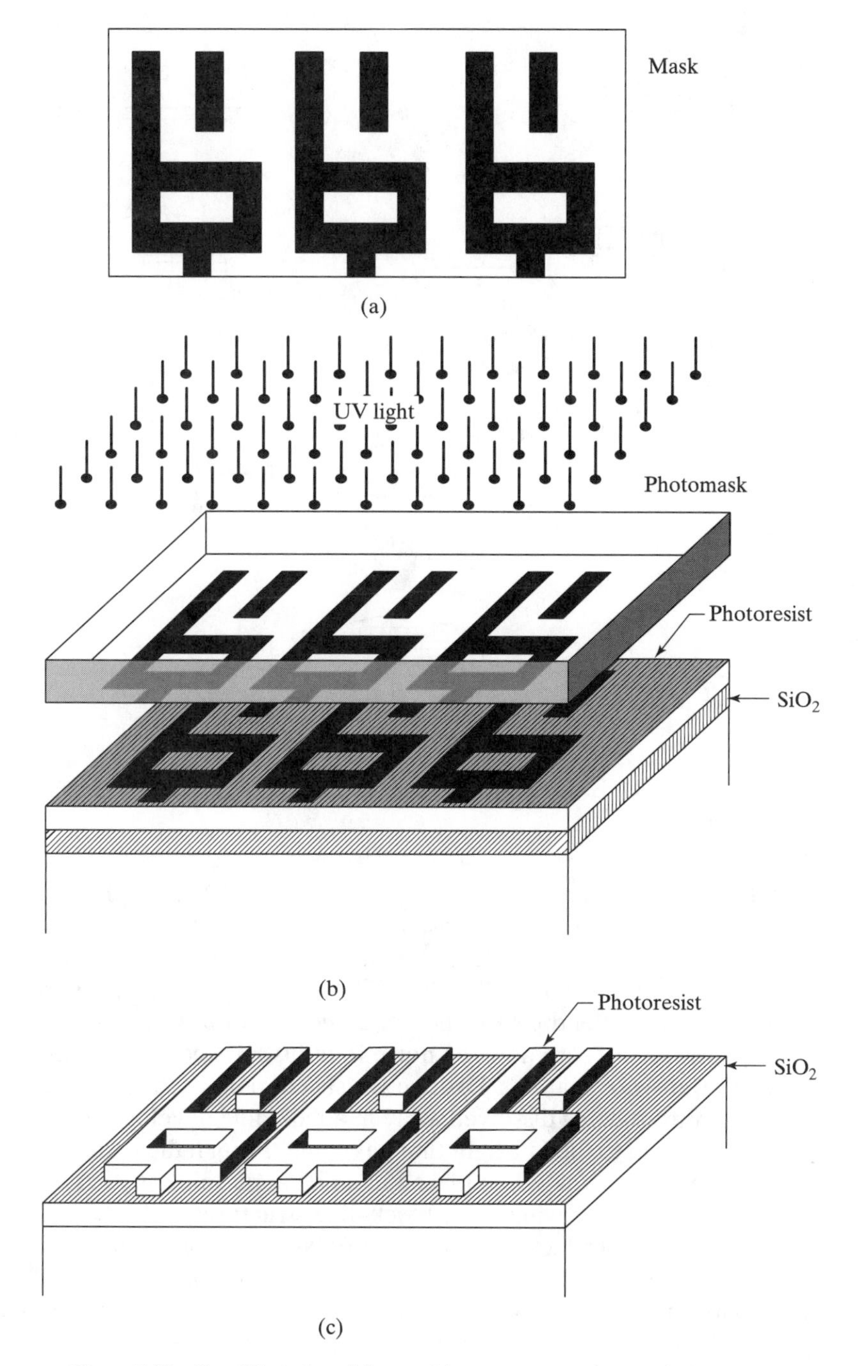

Figure 5.18 Simplified photolithographic process steps for proximity printing.

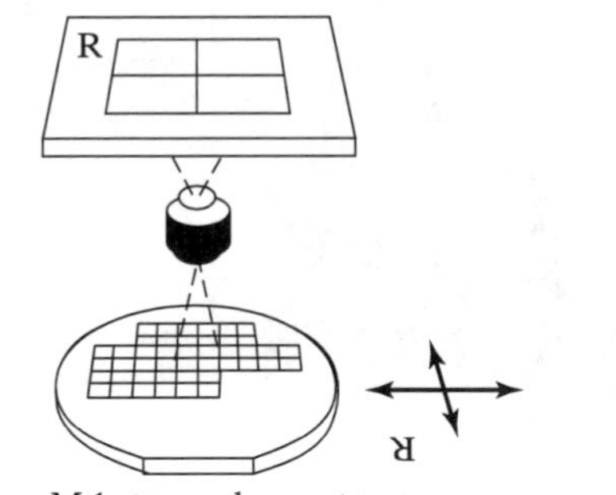

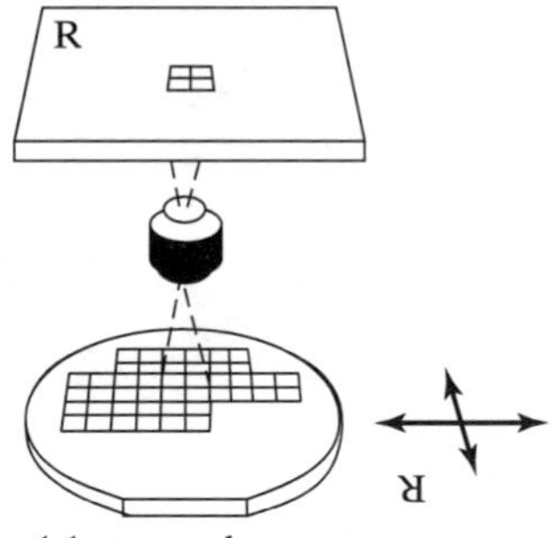

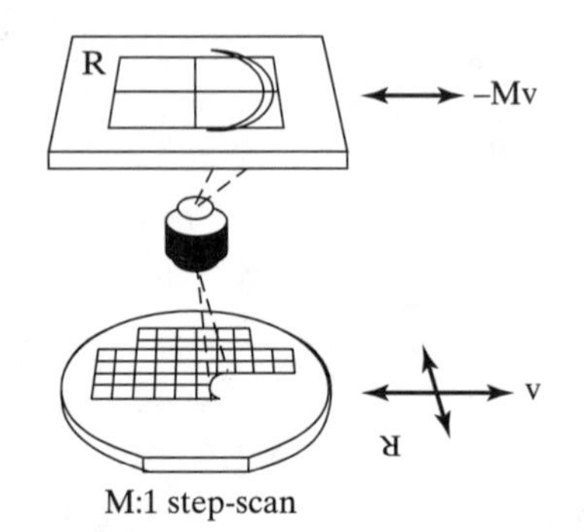

Figure 5.19 Schematics of the different stepper configurations in the lithography process.

Figure 5.18 is more in keeping with the *proximity-type* photomask. This shows (a) a mask at the top, (b) the mask positioned above the photoresist on the wafer, and (c) the preferentially damaged photoresist on top of the silicon dioxide layer.

During photolithography, ultraviolet (UV) light exposes the photoresist in a prescribed way, depending on whether a *negative* or *positive* resist material is being used. Positive resists are now the norm in industry because they give better control for small transistor features. Positive resists contain a sensitizer that normally prevents them from being dissolved away in an alkaline developer solution bath. But if they are exposed to the UV light that has come through the patterns in the mask, the sensitizer breaks down. When placed in an alkaline solution, these regions are preferentially removed, leaving a city block–like structure.

In early IC production, it was also possible to expose many dice at once. One simultaneous exposure was done with a mask that contained many repeats of the same pattern. However, as IC features became smaller, it was found difficult to achieve the registration from one photomask to the next. Also, wafers can get thermally distorted during the intermediate CVD or doping/diffusion steps.

Of course, precise alignment of the wafer and the mask is absolutely essential for each subsequent layer to match up with the previous one. For that reason, today, the process is usually fully automated, and one die at a time, or even one area of each

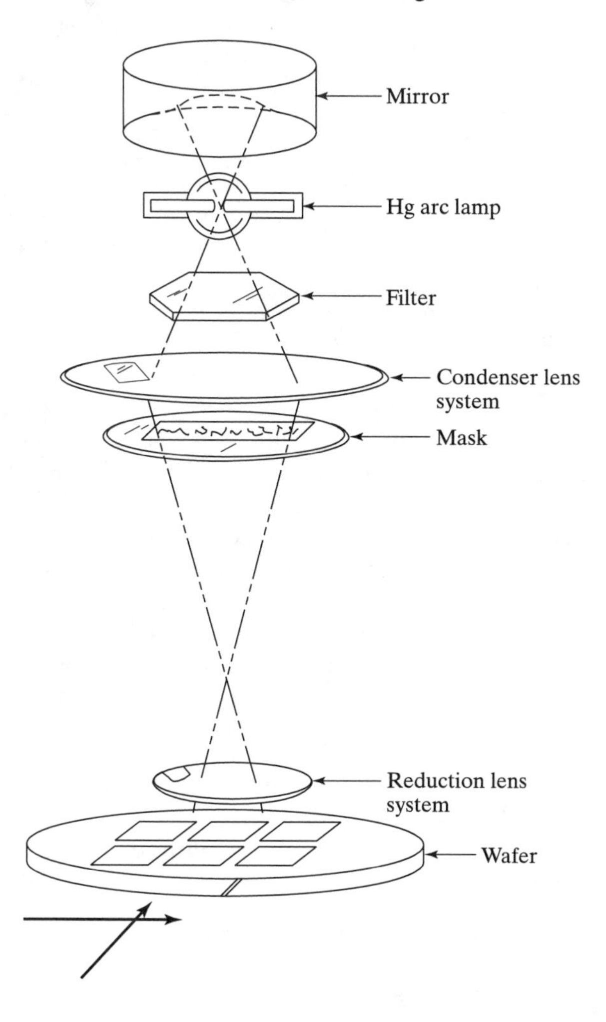

Figure 5.20 Schematic of the reduction step and repeat technique for projection printing.

die on the wafer, is exposed. This is the method called *projection printing* using a lens system that is mounted well away from the wafer surface. The photomask is inserted in a *step-and-repeat* camera, or the *optical wafer stepper.* This transfers the pattern by beaming a light through the lens system to the photoresist. The images are demagnified through the lens down onto the wafers (Figures 5.19 and 5.20).

5.10.5 Etching: Creating the Transistor Channels

Once the pattern is transferred, the light-exposed areas of the photoresist are removed by developing, and the remaining photoresist is baked to harden and consequently protect the desired pattern underneath, as indicated in Figure 5.11b. Wet chemical solutions or a dry plasma gas can be used to selectively etch away those regions of the nitride/oxide barrier layers that are left unprotected. By contrast the areas of unexposed photoresist protect the underlying areas of the pattern that are to be temporarily kept in place.

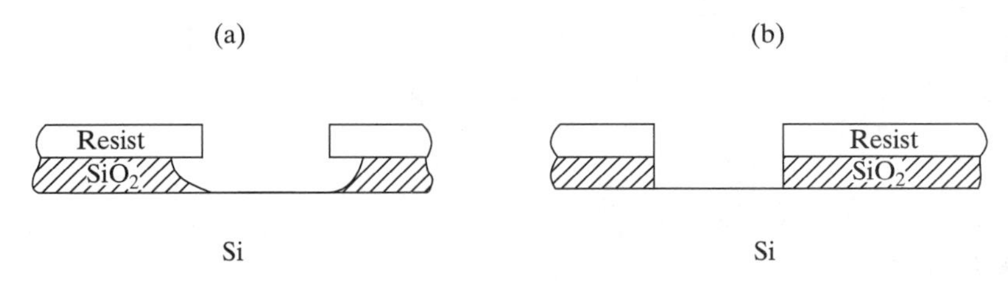

Figure 5.21 Wet etching (left) creates undercutting as opposed to dry, reactive ion etching (right) (from *Introduction to Microelectric Fabrication* by Jaeger, © 1988. Reprinted by permission of Prentice-Hall, Inc., Upper Saddle River, NJ).

Wet etching with a solution containing hydrofluoric acid (HF) creates the valleys or windows in nitride/oxide layers. Temperature, time, and solution strength are monitored carefully so that the nitride/oxide on the substrate is etched quickly, and yet the photoresist on the other surfaces is not damaged.

Wet etching causes undercutting into the walls underneath the photoresist, as shown on the left of Figure 5.21. Thus, although wet etching might still be done in small prototyping laboratories, *dry etching* is preferred today for commercial operations because it does not create the undercutting. It can be done with a variety of plasma beams. For example, reactive ion etching (RIE) simultaneously attacks the surface with chemical and physical effects. The plasma is excited in a radio-frequency electric field, and a stream of reactive ions hits the surface to achieve the following for a silicon surface:

- Gases containing fluorine or chlorine interact chemically with a silicon compound and weaken the inherent structure.
- The ions in the plasma have enough energy to knock out the exposed, weakened atoms, thereby eroding the surface.

Note that dry etching may also be used in the later stages to etch patterns into the aluminum interconnect layers.

An examination of Figure 5.22 shows the result of some of the processing steps:

- The photoresist is protecting a layer of silicide on top of a layer of polysilicon on top of the silicon wafer (dark gray).
- The protected areas are 0.5 micron wide.
- The unprotected areas are 1.5 microns wide.
- The dry etching prevents undercutting, but there is still some undesirable tapering of the vertical walls.
- The silicide (TiS_2) is 0.18 micron thick, and the polysilicon is 0.26 micron thick.

5.10.6 Doping: Selectively Isolating the Active Transistor and Select Areas

Doping can be accomplished by bombarding the silicon with dopant atoms from a particle accelerator (ion implantation) followed by further controlled drive-in diffusion.

Ion implantation uses a high-voltage accelerator to induce dopant atoms into the wafer surface. Ion implantation is easier to handle than basic diffusion, can be

Figure 5.22 Result of using photoresist to protect the polysilicon covered channels 0.5 micron wide (from *Device Electronics for Integrated Circuits*, Richard S. Muller and Theodore I. Kamins, Copyright © 1986. Reprinted by permission of John Wiley & Sons, Inc.).

more precisely controlled, and allows a wider range of barrier layer materials. Figure 5.23 shows that the desired dopant atoms are ionized (bottom right) and then accelerated by an electric field (center of figure) to energies that typically range from 25 to 200 keV. When this beam hits the exposed surfaces, the dopant atoms penetrate the first 1 to 2 microns of the surface layer. This high-energy bombardment in fact also damages the crystallographic lattice of the silicon. The structure is therefore annealed, and this also has the effect of locating the dopants at the substitutional rather than interstitial sites to create n^+ or p^+ regions.

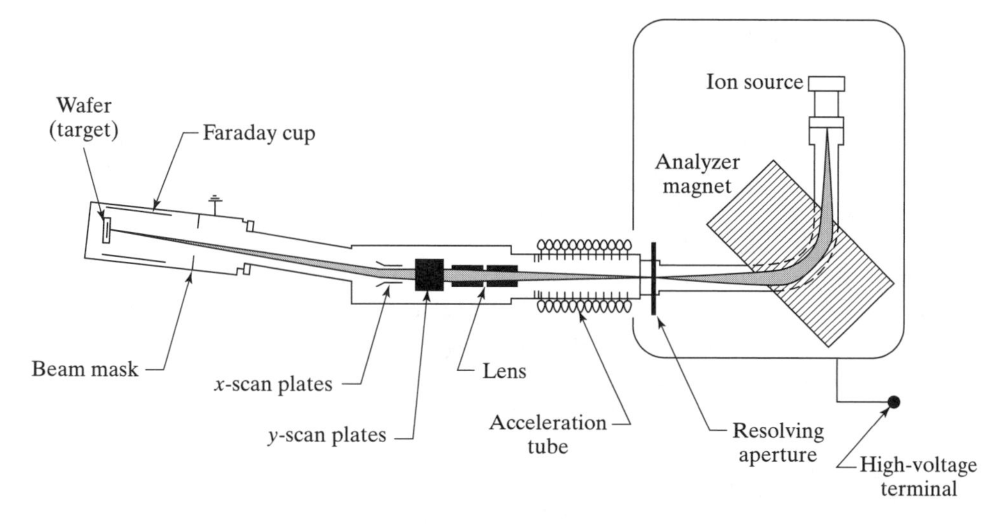

Figure 5.23 Ion implantation device (adapted from Runyan and Bean, 1990).

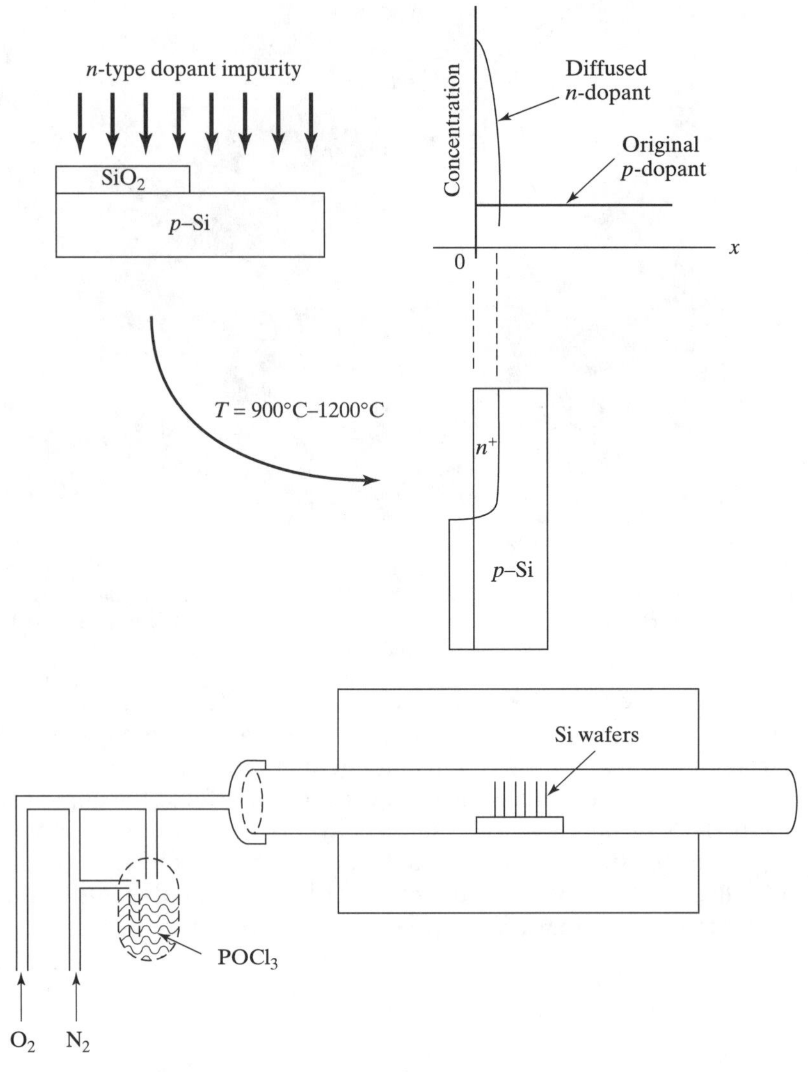

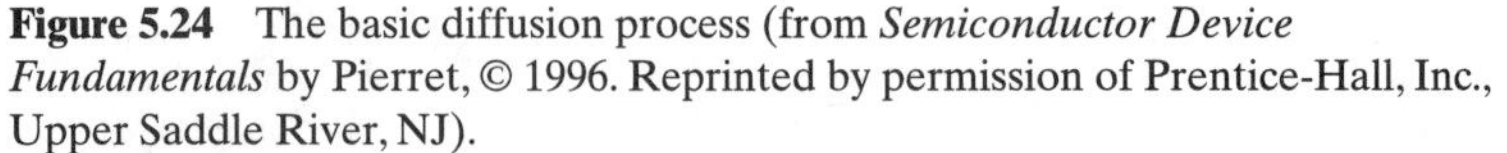

Figure 5.24 The basic diffusion process (from *Semiconductor Device Fundamentals* by Pierret, © 1996. Reprinted by permission of Prentice-Hall, Inc., Upper Saddle River, NJ).

The ion implantation method allows good control of the dopant concentration: dopant concentrations are measured in the number of atoms per square centimeter of surface. High purity of dopants is possible because a mass spectrometer (labeled analyzer magnet in the figure) near the dopant source acts as a sorting agent, allowing only the desired dopant species to reach the wafer target.

Also it should be mentioned that ion implantation can penetrate through an existing layer such as the thin oxide shown in Figures 5.6 and 5.11. Thus additional doping can be done after the high-temperature cycles that form the SiO_2. Muller and Kamins (1986) and Campbell (1996) provide graphs of depth versus ion energy.

Drive-in diffusion follows ion implantation to create a deeper penetration. Figure 5.24 schematically shows the diffusion at around 1,000°C of an n-type dopant into an existing p-type doped substrate. The concentration of dopant is governed by Fick's law of diffusion (see, for example, Muller and Kamins, 1986).

5.10.7 Chemical Vapor Deposition (CVD): Creating Layers for Barriers and Circuits

Thin layers and other materials can be sequentially deposited on the wafer with chemical vapor deposition (CVD). Out of interest, also note that CVD is used extensively to create hard abrasion-resistant coatings on cutting tools (see Chapter 7). This involves thermal reactions or breakdowns of gas compounds to coat the substrate. This is a popular method for depositing barrier layers. Polysilicon, silicon dioxide, and silicon nitride are routinely deposited using CVD. In Figures 5.25 and 5.26 two CVD processes are shown. The high-capacity, low-pressure (LPCVD) process is

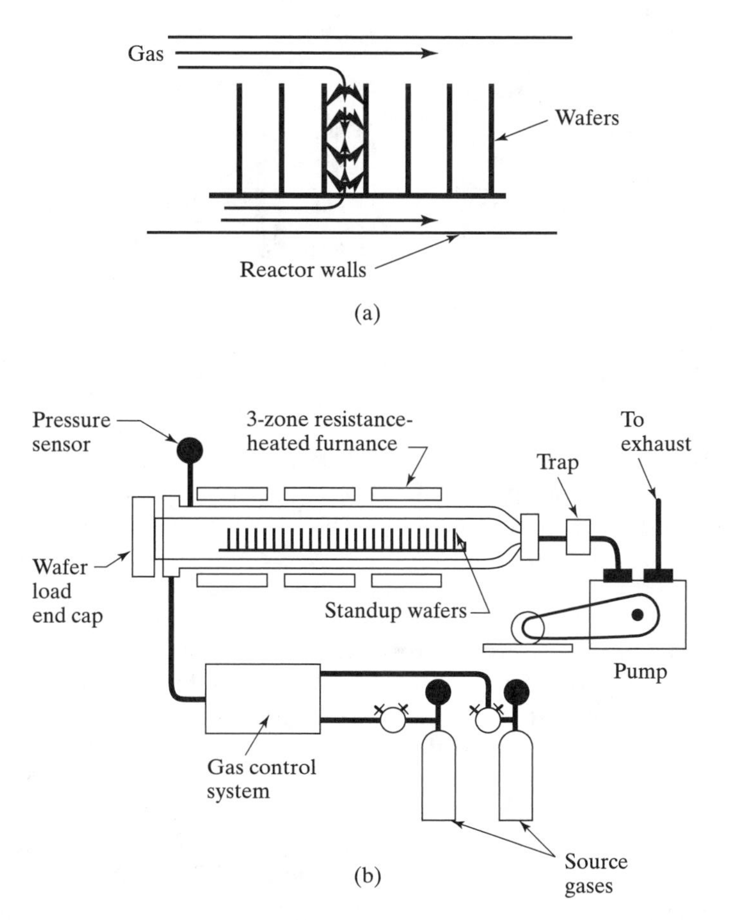

Figure 5.25 Low-pressure CVD (from *Device Electronics for Integrated Circuits,* Richard S. Muller and Theodore I. Kamins, Copyright © 1986. Reprinted by permission of John Wiley & Sons, Inc.).

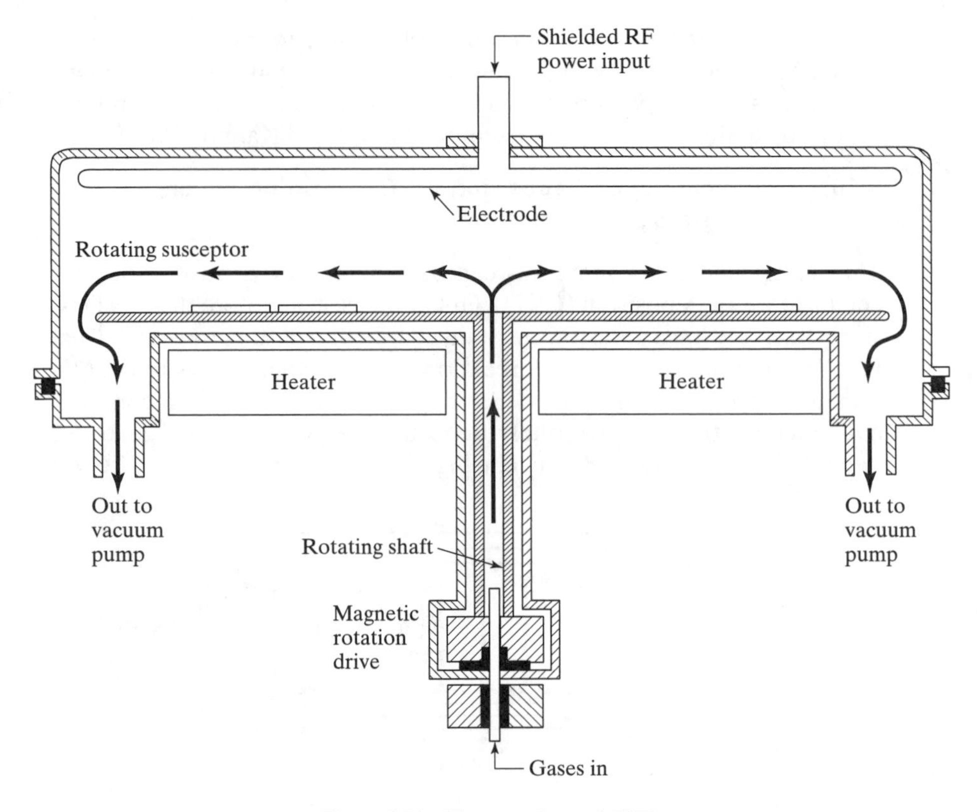

Figure 5.26 Plasma enhanced CVD.

routinely used to deposit SiO_2 and Si_3N_4 and polysilicon. The diagram shows the gases flowing around the vertical wafers where surface reaction and diffusion take place. The plasma enhanced (PECVD) process is chosen because it occurs at lower temperatures and is well suited to certain operations. For example, the final deposition of the passivation layer on top of the aluminum interconnections must occur below 500°C so that the aluminum does not melt.

A metallurgical cross section through the films created by the CVD methods reveals amorphous or polycrystalline transitional layers that build from the pure substrate. By contrast, *epitaxy* differs from these previous CVD methods because it is an extension of the underlying crystallography of the substrate. Epitaxy is most commonly used to grow a thin layer of single crystal silicon onto the silicon wafer. Vapor phase silicon tetrachloride ($SiCl_4$) or silane (SiH_4) is used to form additional silicon on top of a preexisting structure. It is especially useful when needing to grow a lightly doped layer of silicon on the top of a heavily doped substrate, particularly in bipolar transistors. It is also useful in CMOS techniques to grow lightly doped wells on top of existing heavily doped substrates. Vapor phase epitaxy techniques are described in detail in Campbell (1996, see Chapter 14).

5.10.8 Interconnections and Contacts

To produce a functioning integrated circuit, the millions of transistors and devices fabricated through the repeated photolithography-etching-doping-deposition cycles must finally be interconnected. Interconnections are made with metals that adhere well to substrate materials. Aluminum or aluminum-silicon-copper alloys are generally used. Copper will increasingly be used to achieve smaller submicron circuit geometries (Singer, 1997; Braun, 1999). Between two and six layers (shown in Figure 5.27) of metal are deposited over the entire surface of the wafer, with each layer insulated by a dielectric layer. Metal penetrates to the active transistor regions to form the interconnections to, say, the n^+ region shown on the left of Figure 5.27. The second and other layers create circuits between different transistors and devices. Different layers of metal are connected to each other with the vertical channels called *vias,* also indicated in the right side of the figure.

Sputtering deposits thin films onto the wafer surface in vacuum conditions. A *source* of the desired deposition material is bombarded with ions, typically ions of argon, Ar^+. This knocks out atoms from the source, which then sputter onto the wafer and create the thin film. The general setup for a conductive material such as aluminum is shown in Figure 5.28. At the top of the chamber, the target is a cathode. The wafers are mounted on the system's anode in the lower part of the figure.

Evaporation processes can alternatively be used to deposit a thin surface film on the wafer for the aluminum interconnections. As shown in Figure 5.29, an aluminum source is heated and vaporized inside a vacuum chamber. The wafers are placed like a target opposite this vaporizing source, and with the reduced pressure, aluminum vapor travels through the chamber to be deposited on the wafer. Careful control of temperature, atmosphere, and placement is obviously needed to create layers of even thickness. Several methods are available for heating and vaporizing the source.

- In *filament evaporation,* short samples of aluminum wire are heated in a tungsten boat or are hung from the loops of a resistance-heated tungsten filament. Resistance heating vaporizes the aluminum source.
- In *flash evaporation* a spool of the aluminum wire is constantly fed into the vacuum chamber. A heated ceramic bar vaporizes the incoming wire.
- In *electron-beam evaporation,* a fixed source is heated and vaporized with a 15 keV "e-beam." The filament and flash heating methods are subject to the purity of the source.

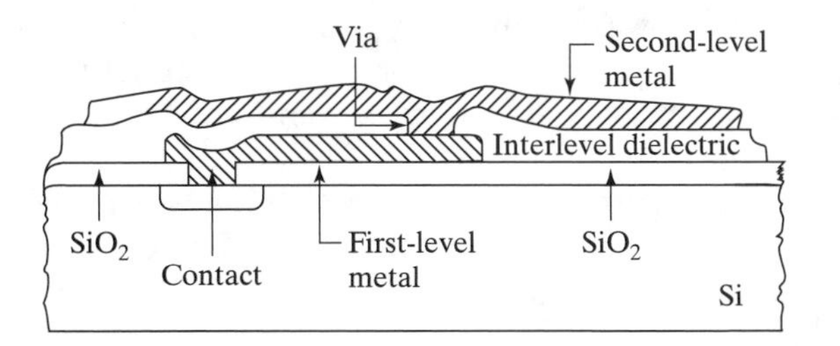

Figure 5.27 Basic two-level metallization (from *Manufacturing Processes for Engineering Materials* by Kalpakjian, © 1997. Reprinted by permission of Prentice-Hall, Inc., Upper Saddle River, NJ).

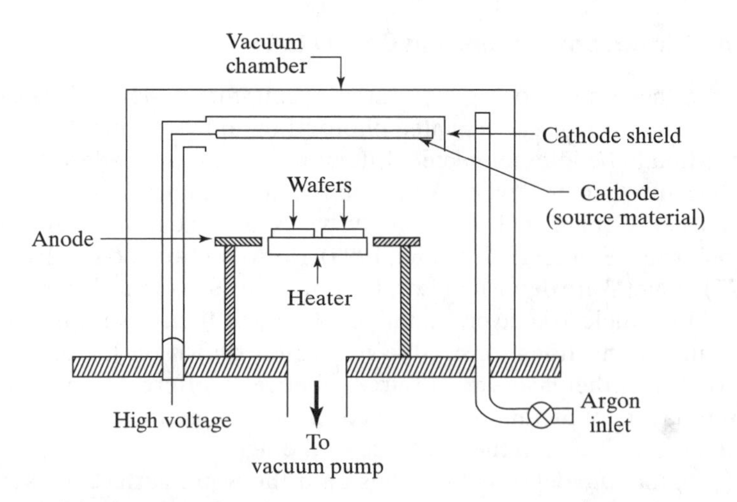

Figure 5.28 Sputtering for the interconnect layers (from *Introduction to Microelectric Fabrication* by Jaeger, © 1988. Reprinted by permission of Prentice-Hall, Inc., Upper Saddle River, NJ).

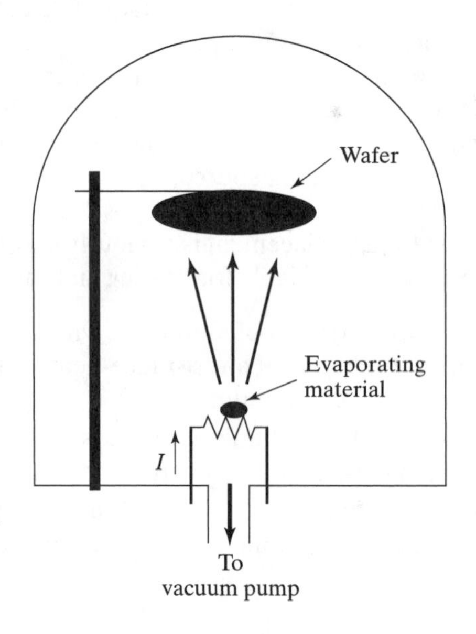

Figure 5.29 Hot filament evaporation (from *Semiconductor Device Fundamentals* by Pierret, © 1996. Reprinted by permission of Prentice-Hall, Inc., Upper Saddle River, NJ).

The e-beam method can cause wafer damaging x-rays. In general practice all these evaporation techniques are less favored in today's commercial fabs. Sputtering is used for its superior topological coverage and moderate pressure requirements.

After the last layer of metal is patterned, a final passivation layer is deposited in order to protect the IC from contamination and damage. Small openings are then etched through the layer to expose square aluminum bonding pads, from which wires will be attached to the package.

5.11 BACK-END PROCESSING METHODS

5.11.1 Summary

Wafers are electronically tested for functionality and separated into individual dice. Each die is set into a chosen package, wire-bonded to the outer perimeter of the package, and finally tested ready for assembly onto a printed circuit board (PCB). This segment of semiconductor production is called *back-end* processing. Figure 5.30 provides an overview of these back-end steps for one of the most common package types, the dual-in-line package. The single IC is shown on the front right side. It is set onto the base with epoxy or a metal alloy. The wire bonds (shown darker) run from bonding pads on the IC to the lead frame of the package. The lead frame connections go through to the J-leads or gull-wings that subsequently are attached to the PCB. The outer cover (labeled molding compound) completes the package.

5.11.2 Testing and Separation

IC designers include special test dice on the wafer that are subjected to all the same oxidation, etching, layering, and doping processes as the desired IC. These special test dice are monitored as much as possible after each of the processing steps described earlier. At the very end of wafer production these test dice are put through an additional series of computer-controlled tests in which fine, needlelike probes contact the aluminum bonding pads of the test dice. If this first check shows that the processing parameters were within proper limits, then each die is tested for functionality. Dice that need to be rejected are marked with an ink spot.

After preliminary testing is completed, each die is separated from the wafer, usually by a diamond saw. In this process the wafer is held down on a sticky sheet of Mylar and the diamond saw is used either to saw between the dice completely

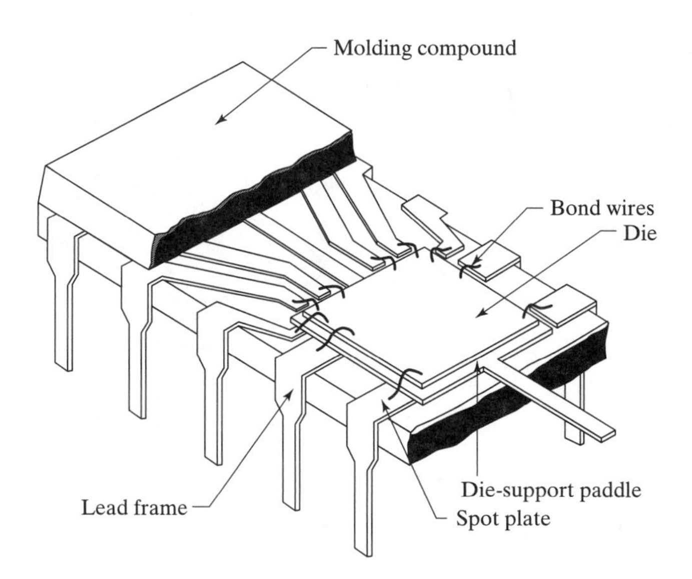

Figure 5.30 The dual-in-line package (DIP) (from *Manufacturing Engineering and Technology* 3/e by Kalpakjian, © 1995. Reprinted by permission of Prentice-Hall, Inc., Upper Saddle River, NJ).

through the wafer or to scribe the wafer and provide continuous notches. In the latter approach, the wafer can be turned upside down on a soft pad. A lightly pressurized roller passes across the back of the wafer, and controlled cracks separate the dice. This method is related to the <100> wafer growing direction. In this orientation, natural cleavage planes run normal to the through thickness direction and to the die separation lines on the wafer surface. Once all the dice are separated, any inked chips are discarded, while the remaining chips are inspected visually, under a microscope, for defects. The *die yield* from basic wafer production, wafer testing, die separation, and retesting is considered in the next main section.

5.11.3 Attachment, Wire Bonding, and Packaging

The good dice are then seated into a desired die package. The bottom of the die is secured with a metal-filled epoxy, or with a 96% gold–4% silicon eutectic alloy that melts and then solidifies in the range 390°C to 420°C to secure the die to the surface.

Wire bonding makes the electrical contacts between the top of the die and the surrounding *lead frame* of the package. Figure 5.31 shows the delicate wires running from the bonding pads (typically 100 to 125 microns in size) to the frame of the protective package. Of the methods available to attach the thin wires to the bonding pads, thermosonic bonding has emerged as the most efficient method of attachment. In *thermosonic* wire bonding, delicate, 25 micron, gold or aluminum wires are pressure-welded to the pads with a blunt indenter. The bond is made secure by simultaneously heating the substrate to 150°C and ultrasonically vibrating the joint. Solid-state welding thus occurs from a combination of pressure, vibration, and warm-plastic deformation of the soft gold or aluminum. Thermosonic bonding machines are easily automated for high-speed production.

5.11.4 Dual-in-Line Packages (DIPs)

The package and packaging material chosen for a chip depend on the IC's size, number of external leads, power and heat dissipation requirements, and intended operating environment. Dual-in-line packages (DIPs) are common packaging styles. They are

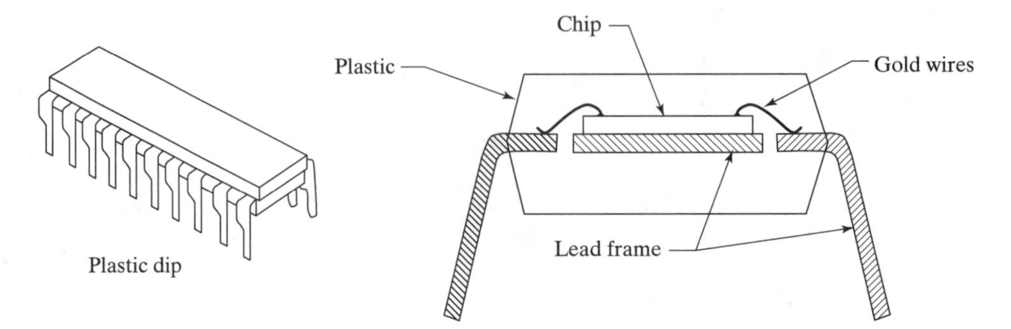

Figure 5.31 The DIP packaging method (from *Manufacturing Engineering and Technology* 3/e by Kalpakjian, © 1995. Reprinted by permission of Prentice-Hall, Inc., Upper Saddle River, NJ).

inexpensive, easy to handle, and made from a variety of materials to suit the application including epoxy, plastic, metal, or ceramic. Also the DIP continues to be a workhorse for prototype circuit design. The usual form factor is a plastic rectangle with the I/O leads placed at approximately 0.1 inch spacings along the perimeter edges (Figure 5.31).

5.11.5 Quad Flat Packages (QFP)

Quad flat packages (QFP) in either plastic or ceramic are today the most often seen commercial packages for gate arrays, standard logic cells, and microprocessors. Such flat packs are especially favored for computer systems with several stacked printed circuit boards (PCBs), which demand low-profile chips to reduce the vertical packing space. Figure 5.32 shows the standard layout. The upper part of the figure shows that wire bonds will connect the bonding pads to the external leads at the periphery of the ceramic (or plastic) package. The lower figures show the periphery layouts including the *gull wing* in the center diagrams or the *J-lead* at bottom left. Despite the popularity of the QFP, close inspection of these diagrams points out the next technology trend. If the spacing of the leads gets too small, an individual lead might get bent during handing, or, in later processing, *solder shorts* might form on the PCB between adjacent legs. Further developments to address this issue are reserved for Chapter 6.

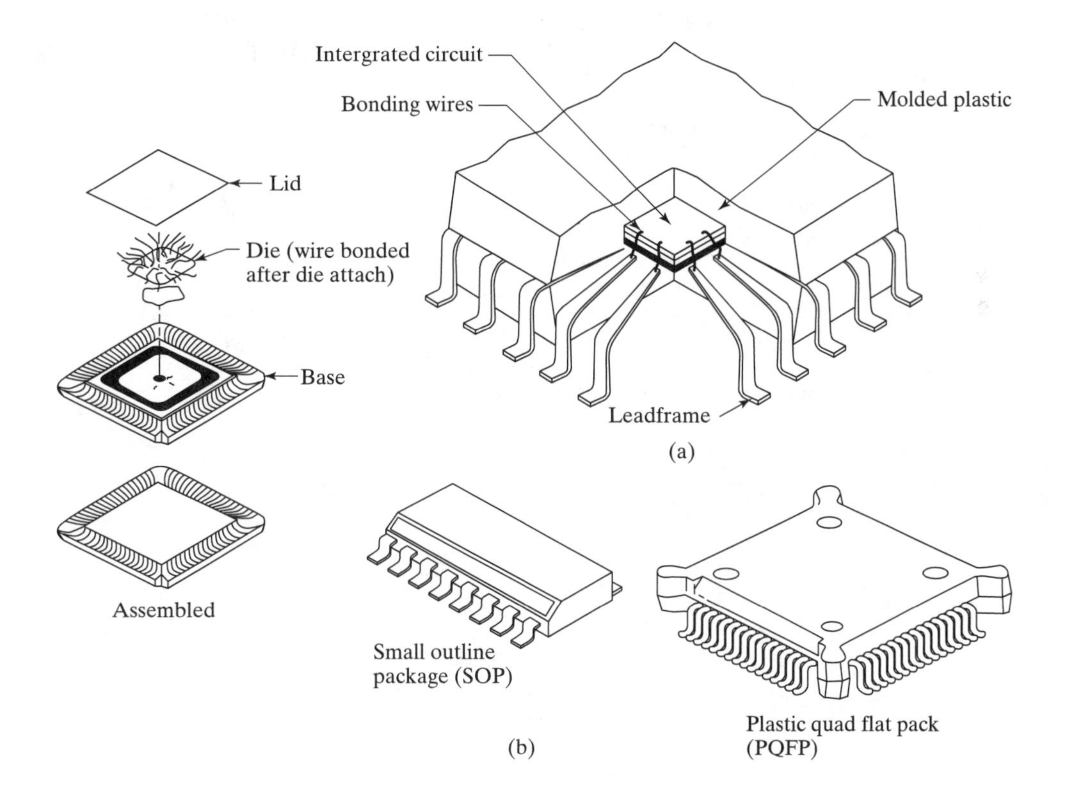

Figure 5.32 Quad flat packaging (QFP) (adapted from Kalpakjian, 1995).

5.12 COST OF CHIP MAKING[5]

5.12.1 Overview

Manufacturing involves many processing steps, and each step adds to the cost of the wafer. Therefore, although the cost of a raw unprocessed wafer is only $15 for a 200-mm wafer, the final processed wafer often costs several thousand dollars after about 100 processing steps. The wafer costs depend on the number of masks used, the complexity of the circuits, and the clean room requirements of the process. The cost increases with the number of layers in a nonlinear fashion, since each additional masking layer introduces more defects and decreased yield. The cost of the wafer also increases with smaller feature sizes due to stringent requirements on lithography and process control. However, the cost per chip might then be lower due to the larger number of chips that can be "squeezed onto the real estate."

Table 5.4 shows that lithography is the most expensive aspect of processing. Furthermore, to further reduce line width, lithography is the area where the greatest research effort is needed. Lithography processes and their associated costs will thus continue to be a main focus area in the management of technology.

5.12.2 Cost of a Single IC

The calculation of the cost of a single IC involves the three main costs in Equation 5.1, modified by the *final integrated yield*—that is, the number of good dice leaving the final testing area:

Cost of an individual die on a wafer + cost of testing + cost of packaging (5.1)

[5] *Important:* Throughout this section the data are based on mid-1990s costs. As time goes on, the costs will change. Also, yields will creep toward the ideal 100% level. At the same time, newer designs of chip will experience lower yields—perhaps nearer 50%—while the manufacturing start-up problems are resolved and debugged. The yields shown in the later examples are from Patterson and Hennessy (1996b). By today's standards these are extremely low, but they would still arise in pilot plants. Dataquest's annual Market Analysis of Semiconductor Supply and Pricing Worldwide, including its own "Cost Model," is one of the best sources for current data. Therefore a recent example for a 64-Mb DRAM in the year 2000 is included in Appendix 2 of this chapter (Section 5.19).

TABLE 5.4 Relative Costs of Production Processes

Manufacturing process step	Percentage of wafer processing cost per cm^2 (excludes packaging, test, and design costs)
Lithography	35%
Multilevel materials and etching	25
Furnaces/implants	15
Cleaning/stripping	20
Metrology	5

The following subsections are based on Patterson and Hennessy (1996b). The point to always keep in mind is that the "good dice" leaving each step of the IC fabrication process have to bear the processing costs of all the "bad dice" that were discovered and rejected along the way. Obviously, all efforts are made to detect these bad dice as soon as possible. Nevertheless, some time, effort, and cost will have gone into creating mistakes. For example:

- Perhaps a complete wafer has to be rejected. Possible causes include a poorly calibrated stepper, a faulty vacuum system, chemical impurities in a CVD system, or an atmosphere control problem. Detecting this larger scale problem is the function of the test dice on the wafer. These are tested as soon as possible after each processing step to avoid wasting time and resources on a wafer that might already be ruined.
- Or, in a more isolated manner, perhaps a dust problem has created several bad dice on an otherwise satisfactory wafer.
- Or, alternatively, during back-end processing, an otherwise good die has been misaligned and damaged.

At each step some time and cost will have gone into creating these bad dice. And so this cost has to be shouldered by the good dice. Thus the final costs of a single integrated circuit are obtained by dividing Equation 5.1 by the final die yield from the final test.

For each processing step, an *intermediate die yield* can be specified. It is usually stated as a percentage or a value between zero and one. So, in Equation 5.2, if 90% of the dice on the wafer are good dice, by multiplying the "dice per wafer" by 0.9 in the denominator, it can be seen that the cost of each die is higher than if the yield were perfect at 100% or 1.

5.12.3 The Cost of an Individual Die on a Wafer

The cost of an individual die on a perfect wafer involves three main items:

- How many dice fit on a single wafer
- What percentage of these actually work correctly—namely, the *process die yield*
- An allowance for a few test dice on the wafer—not included in the following equations for simplicity

$$\text{Cost of die} = \frac{\text{cost of wafer}}{\text{dice per wafer} \times \text{die yield}} \tag{5.2}$$

Step 1: calculate the "dice per wafer."

$$= \left[\frac{\pi \times \left(\frac{\text{wafer diameter}}{2}\right)^2}{\text{die area}}\right] - \left[\frac{\pi \times \text{wafer diameter}}{\text{die diagonal}}\right] \tag{5.3}$$

The second term allows for the dice around the edge of the wafer. Rings of dice at the outside lose the tip of their outside corners due to the "square peg in a round hole" problem. Strictly, this outside ring might not be exposed during lithography, thereby saving some time, but still some costs will go onto the wafer during processes such as CVD and diffusion.

The preceding equations are very dependent on wafer size, prompting the move to the 300 mm wafers in the new fabs.

The equation gives the following dice per wafer:

- 1 square centimeter die on a 150 mm or 6 inch wafer = 138 dice
- 1 square centimeter die on a 200 mm or 8 inch wafer = 269 dice
- 1 square centimeter die on a 300 mm or 12 inch wafer = 635 dice

Or for a larger IC:

- 2.25 square centimeter die on a 150 mm or 6 inch wafer = 56 dice
- 2.25 square centimeter die on a 200 mm or 8 inch wafer = 107 dice
- 2.25 square centimeter die on a 300 mm or 12 inch wafer = 269 dice

However, note that this calculation gives only the *maximum* number of dice produced if the fab could achieve 100% yield. The next question is: How many of these are good?

Step 2: calculate the "die yield."

$$= [\text{Wafer yield}]\left[1 + \frac{\text{defects per unit area} \times \text{die area}}{\alpha}\right]^{-\alpha} \tag{5.4}$$

where the wafer yield accounts for wafers that are so bad they need not be tested. Next, the value of α is an empirical factor corresponding to the number of masking levels and the complexity of the manufacturing process being used. Typically, in today's multilevel CMOS processes, $\alpha = 3$.

Factory measurements indicate that the defects per unit area lie somewhere between 0.6 and 1.2 depending on the maturity of the individual processes used. Although these data are empirical rather than analytical, the method assumes that (a) the defects are randomly distributed over the wafer and that (b) the yield is inversely proportional to the complexity of the fabrication process, as measured by the factor α obtained by collecting factory-floor data from CMOS manufacturing.

So, for example, using Patterson and Hennessy's (1996b) data, if:

- The wafer yield is 100% or 1 (for the sake of simplicity)
- The defects per unit area are 0.8 per square centimeter
- The die area is 1 square centimeter

$$\text{Die yield} = 1 \times (1 + [0.8 \times 1] / 3)^{-3} = 0.49$$

From these calculations, it can be concluded that the number of good 1 cm^2 dice on a 200 mm (8 inch) diameter wafer reduces from the maximum possible of 269 to a reduced figure of (269 $\times$ 0.49) = only 132.

Again using 1996 data from Patterson and Hennessy (1996b, see p. 63), manufacturing a 200 mm (8 inch) wafer in CMOS costs between $3,000 and $4,000 depending on the complexity and brand of the microprocessor. Therefore, using \$3,500 as the average wafer cost, the individual die cost for a 1 cm^2die, with 0.8 defects per square centimeter, on an 8-inch wafer = \$3,500 / (269 × 0.49) = \$26.55.

Before the chip is ready to be used in a computer, further costs of testing, packaging, retesting, and shipping must be invested. And, of course, these are just the *variable costs* of the manufacturing processes (see Equation 2.1). The *fixed costs* of research and development (R&D), capital expenditures, personnel, and marketing add considerably more.

Note that if the die size is increased to 2.25 square centimeters, the painful result for the 200 mm wafer is (107 × 0.24) = only 25 good ones. This reduced number makes the individual costs considerably higher at \$140—nearly five times higher. Die designers realize that they cannot easily influence the daily costs of running the factory and controlling the yield from individual CMOS operations. But they can influence the die area and strive to reduce it by considering the functions that are included on the die and the number of I/O pins.

5.12.4 Additional Costs of Testing the Die after Processing and Slicing

Producing the dice is one set of costs. However, the dice must be tested after the CMOS processing and subsequent slicing up procedures to ensure customer satisfaction. A few dice will be damaged just from testing. So, again, since the bad dice have to be tested before it is known they are bad, the good dice must bear this cost.

$$\text{Cost to test a die} = \frac{\text{cost of testing per hour} \times \text{average test time}}{\text{die yield after the test}} \tag{5.5}$$

In Patterson and Hennessy's 1996 examples, the quoted testing costs vary from \$50 to \$500 per hour depending on the type of test needed. Testing time also varies with die complexity, from 5 to 90 seconds. Expensive microprocessors with many pins need a longer test with more expensive equipment.

5.12.5 Cost of Packaging

The next set of costs involves the back-end packaging of the finished die. These costs are determined by the packaging material and its design, the number of pins, and the die size. The cost of the packaging material depends in large part on the desired heat dissipation rate from the operating IC when it is being used in the computer. For example, in 1996 data:

- A plastic quad flat pack (PQFP) that will dissipate less than 1 watt of heat from a 1 cm^2die with 208 pins will cost about \$2.
- Alternatively, a ceramic pin grid array (PGA) might have 300 to 600 pins for a larger 2 cm^2 die dissipating much more heat, and the costs will rise to as much as \$30 to \$70 per package.

Table 5.5 includes examples:

TABLE 5.5 Package and Test Costs (Courtesy of MIPS Technologies)

Package type	Pin count	Package cost ($)	Test time (sec)	Test cost per hour ($)
PQFP	<220	12	10	300
PQFP	<300	20	10	320
Ceramic PGA	<300	30	10	320
Ceramic PGA	<400	40	12	340
Ceramic PGA	<450	50	13	360
Ceramic PGA	<500	60	14	380
Ceramic PGA	>500	70	15	400

The data for die area and different packages can be seen in some recent products (Table 5.6):

TABLE 5.6 Microprocessors and Characteristics for Some Products with Wafer Cost

Microprocessor	Die area (mm^2)	Pins	Estimated wafer cost ($)	Package
MIPS 4600	77	208	3,200	PQFP
PowerPC 603	85	240	3,400	PQFP
HP 71 $\times$ 0	196	504	2,800	Ceramic PGA
Digital 21064A	166	431	4,000	Ceramic PGA
SuperSPARC/60	256	293	4,000	Ceramic PGA

Finally, there is the cost for assembly labor, bonding pads to pins, burn-in testing, and further failure analysis.

Therefore the total costs are computed in Table 5.7 for a 200 mm wafer, a wafer yield of 95%, and $\alpha = 3$:

TABLE 5.7 Total Costs for Some Microprocessors around the Year 1995

Type	Die yield	Dice per wafer	Good chips	Cost per good die ($)	Final cost including testing packing ($)
MIPS 4600	0.4787	357	171	18.71	32.45
PowerPC 603	0.4495	321	144	23.53	45.51
HP 71 $\times$ 0	0.2102	128	27	103.62	181.55
Digital 21064A	0.2535	154	39	101.95	157.08
SuperSPARC/60	0.1492	94	14	282.35	318.31

So for one example in detail, the MIPS 4600 has a die area of 77 mm^2.

- For a 95% wafer yield and alpha equal to 3, the die yield comes out to be 0.4787.
- The number of dice per wafer, assuming a 200 mm wafer, is 357. Thus, the number of good chips per wafer is 171.
- In Table 5.6, MIPS 4600 wafers cost $3,200 each.
- From each wafer, the price for a good chip is thus $18.71.
- The physical package for this chip costs $12.
- There are also labor costs: the average testing-time cost per good chip is $0.833 and the average packaging-time cost is $0.907. These testing and assembly-time costs add up to $1.74.

Altogether, the costs are (18.71 + 12.00 + 1.74) = $32.45. The costs for other processors are much higher. The Sun SPARC/60 is given as $318.31. Also these are *manufacturing costs not retail costs.*

In future years, costs will be lower—much lower! However, the basic idea will still hold that each failure makes the good dice cost more and the costs escalate with die size.

5.12.6 Conclusion: Relation to Integrated CAD/CAM

It is worth summarizing with some key conclusions from these calculations.

5.12.6.1 Design

- With $\alpha = 3$, the cost of the die is a function of the fourth power of the die area. Therefore, the circuit designer's final choice of die area is dramatically important to die cost.
- This die area depends on a variety of issues including the specific technology being used, the number of functions and hence transistors on the chip, and the number of pins on the border of the die.

5.12.6.2 Manufacturing

The manufacturing process itself dictates the wafer cost, the wafer yield, α, the defects per unit area, and the final integrated yield after packaging and testing. In the next section, the history of the semiconductor industry reinforces the fact that design *and* manufacturing are of equal importance in the "best practices" for the semiconductor industry.

5.13 MANAGEMENT OF TECHNOLOGY

5.13.1 Historical Trends in the Business

The semiconductor industry has gone through tremendous structural and technological change over the past three decades—since, say, the first 1K DRAM 1103 chip made by Intel in 1970. Once a small market dominated by a few companies in Boston,

Texas, and California, semiconductors became an intensely competitive global industry by the 1980s, with Japanese producers steadily usurping the market lead.

In the 1980s the U.S. semiconductor industry's competitive slide was caused in large part by persistent manufacturing weaknesses. The slide was initially blamed on unfair trade practices by Japan. But while their Japanese competitors focused intensely on process improvements that enabled them to boost chip yield and lower production costs, U.S. firms concentrated on improving chip miniaturization and functionality and largely neglected the efficiency of the production process. Lagging productivity and product quality sharply undercut the competitiveness of U.S. semiconductors.

By 1985, things looked especially grim for much of the U.S. industry. Excess fabrication capacity led to huge industry losses, and many semiconductor start-ups were forced out of the market. Routine production then moved out of the United States to Japan, Malaysia, South Korea, and Taiwan in order to take advantage of low labor costs but also because of excellent production methods. The loss of market share and cumulative production experience appeared to doom the U.S. semiconductor industry.

The competitive picture for the U.S. semiconductor industry is very different today. Macher and associates (1998) identify the following "corrective" issues:

- The improvements in quality assurance in all aspects of U.S. fabrication
- Many innovative fabrication methods in lithography, etching, and doping
- Important changes in the worldwide demand for semiconductors
- The fact that in the mid-1980s, the United States withdrew from some IC product lines—certain memory products were examples; these were products in which *design innovations* could not compensate against the superior capital investments that other countries had made in *manufacturing excellence* in their foundries

The change in U.S. quality is clearly shown in the period between May and November 1993 as measured by comparison with Japanese and Korean fabs (Figure 5.33). This graph is the integrated yield for 0.7 to 0.9 micron CMOS memory chips. Leachman and Hodges (1996) show similar trends for all chip designs, both logic and memory. The reader is referred to Leachman and associates' extensive report series at <**http://euler.berkeley.edu/csm**>.

Tremendous growth in new applications has also boosted demand for ICs. Memories and PC-oriented microcomponents still take up most of the market, indicating the computer industry remains the most important consumer of ICs.

In addition, ICs are also at the heart of a burgeoning array of new products including high-definition television (HDTV), interactive multimedia, integrated services digital networks (ISDNs), cellular and wireless communication systems, automotive electronics, and handheld computers. There are major new sources of mass demand for electronics, computers, and communications products in Asia, Latin America, India, Eastern Europe, and other regions. At the same time, many IC users are demanding products tailor-made to their specifications, creating an array of specialty and niche markets. All of these trends have resulted in a substantial surge in global production in nearly every type of semiconductor IC manufacturing.

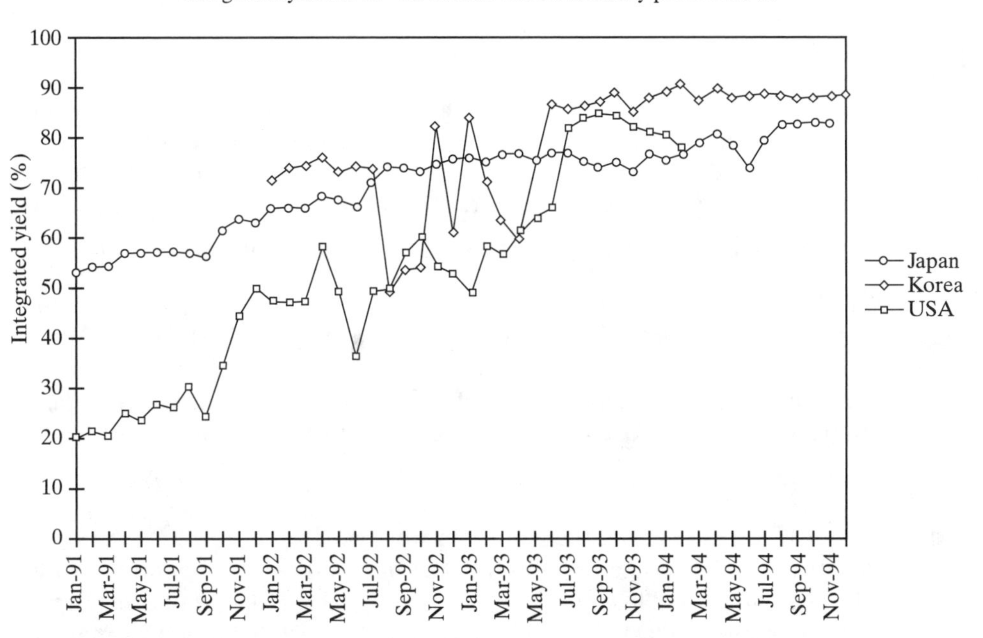

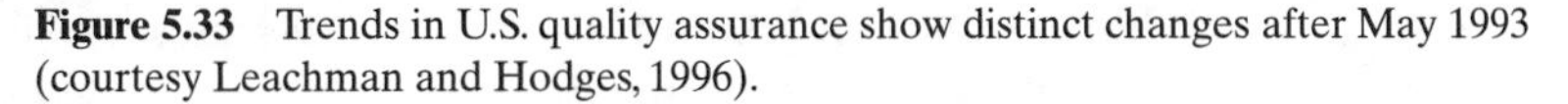

Figure 5.33 Trends in U.S. quality assurance show distinct changes after May 1993 (courtesy Leachman and Hodges, 1996).

In summary, U.S. producers have recovered and maintained significant market share. They have done so in part by reengineering from commodity chips to high-value-added products, particularly microprocessors. The 1980s market was still relatively small for these products, but now they are in great demand. However, manufacturing quality and efficiency still seem to be the key factors. This statement is true for both the automobile and the semiconductor industries as they continue to grow out of the doldrums of the mid-1980s.

5.13.2 The $2.5 Billion Fab

Staying ahead in the semiconductor market today is extremely expensive. Constant product innovation forces companies to invest more heavily in product design and planning. At the same time, constructing and equipping a new manufacturing facility costs twice as much as it did 10 years ago. For example, a high-volume fabrication plant for DRAMs has risen from about $400 million in 1990 to approximately $1 billion today.

Part of the cost is due to the fact that semiconductor making is a highly toxic process, heavily regulated by environmental and worker protection laws (see Sidhaye, 1999). Moving to submicron and large wafer process technologies will drive costs up even further. Over the next five years the scenario for a fabrication plant is:

- 0.13 to 0.18 micron features
- 300 mm (12 inch) wafers

- 25,000 wafers per month
- Projected cost of $2.5 billion

Moore has noted (see Leyden, 1997) that other observers in the semiconductor equipment manufacturing field have updated his "law" with respect to manufacturing. The predictions are that the construction and equipping costs will rise exponentially and be even more dramatic than in the past decade. Although anecdotal, this "new law" states that the cost of a semiconductor fabrication plant will roughly double every two years.

5.13.3 Trends and "Alliances" in Advanced Lithography

These investments are daunting even for the deep pockets of Intel, Lucent, and IBM. And the future, beyond these 300 mm fabs, is even more daunting. Therefore, consortia projects, or "alliances," between these larger companies are beginning to emerge. This is especially the case of advanced lithography, which, as can be seen in Table 5.4, already accounts for the largest fraction of front-end costs.

5.13.3.1 UV and Deep-UV Lithography

The 0.35 micron lines of the late 1990s were generated from UV sources with wavelengths of 365 nanometers. Today's 0.25 micron lines are generated with deep-UV (DUV) sources at 248 nanometers. Generally, the cited limit of commercial deep UV with high-purity glass lenses is a wavelength of 193 nanometers that can produce lines 0.13 micron wide, although recent trade reports indicate that 0.08 micron line widths might be feasible with alternating aperture phase shift masks (see Semiconductor International, 1998).

5.13.3.2 EUV Lithography

One alliance for future miniaturization is between Intel, Motorola, Advanced Micro Devices, three national laboratories, and several semiconductor equipment manufacturers. Their project is utilizing shorter wavelength, extreme ultraviolet (EUV) lithography rather than ordinary UV lithography. The goal is a 0.03 to 0.1 micron feature size.

In EUV, laser generated plasmas produce a source at wavelengths of 13 nanometers. Highly reflective molybdenum/silicon mirrors, rather than glass lenses, focus the 13 nanometer waves through the mask and demagnifiy them onto the wafer to create the features (Figure 5.34). For the beta version of the manufacturing equipment, the aim is to produce 300 mm wafers, 26×52 mm dice, 0.1 micron features, and 40 wafers per hour.

5.13.3.3 X-Ray Lithography

X-ray lithography uses 0.01 to 1 nanometer wavelength sources and has been successfully used to build devices in the 0.02 to 0.1 micron range. The process requires a synchrotron to accelerate the high-energy electrons for the source. The process is being developed at IBM and Sanders (see DeJule, 1999). While the technical feasibility has been well proved in dedicated locations, other observers argue that commercial fabs—accustomed to DUV lithography—will not rush to install and maintain a synchrotron (see Peterson, 1997).

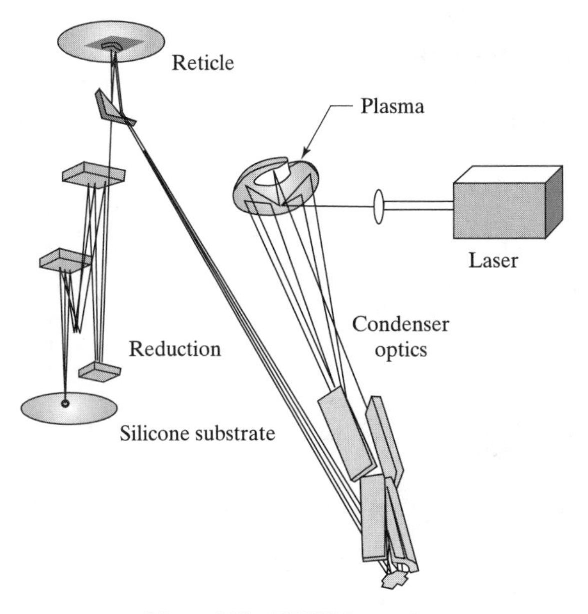

Figure 5.34 EUV lithography.

5.13.3.4 Scattering with Angular Limitation Projection Electron-Beam Lithography: SCALPEL

In this method, an electron beam is used to direct a high-energy, finely focused "pencil source" onto the substrate. Rather than use a photomask, the beam can be directly guided by the data in the CAD files. Lucent Technologies' Bell Labs has developed a version of the process that is called SCALPEL (scattering with angular limitation projection electron-beam lithography).

Table 5.8 summarizes the values discussed earlier, and Figure 5.35 shows the projected technologies needed to push EUV to greater limits. Advances in resist and mask technologies are also needed to achieve such "deep submicron" levels.

TABLE 5.8 Lithography Summary

Method	Wavelength (nanometers)	Feature size (nanometers and microns)
UV	365	350 (0.35 micron)
Deep UV	248	250 (0.25 micron)
Deep UV refined	193	130–180 (0.13–0.18 micron)
Extreme UV	10–20	30–100 (0.03–0.1 micron)
X-ray	0.01–1	20–100 (0.02–0.1 micron)
SCALPEL (electron beam)	—	80 (0.08 micron)

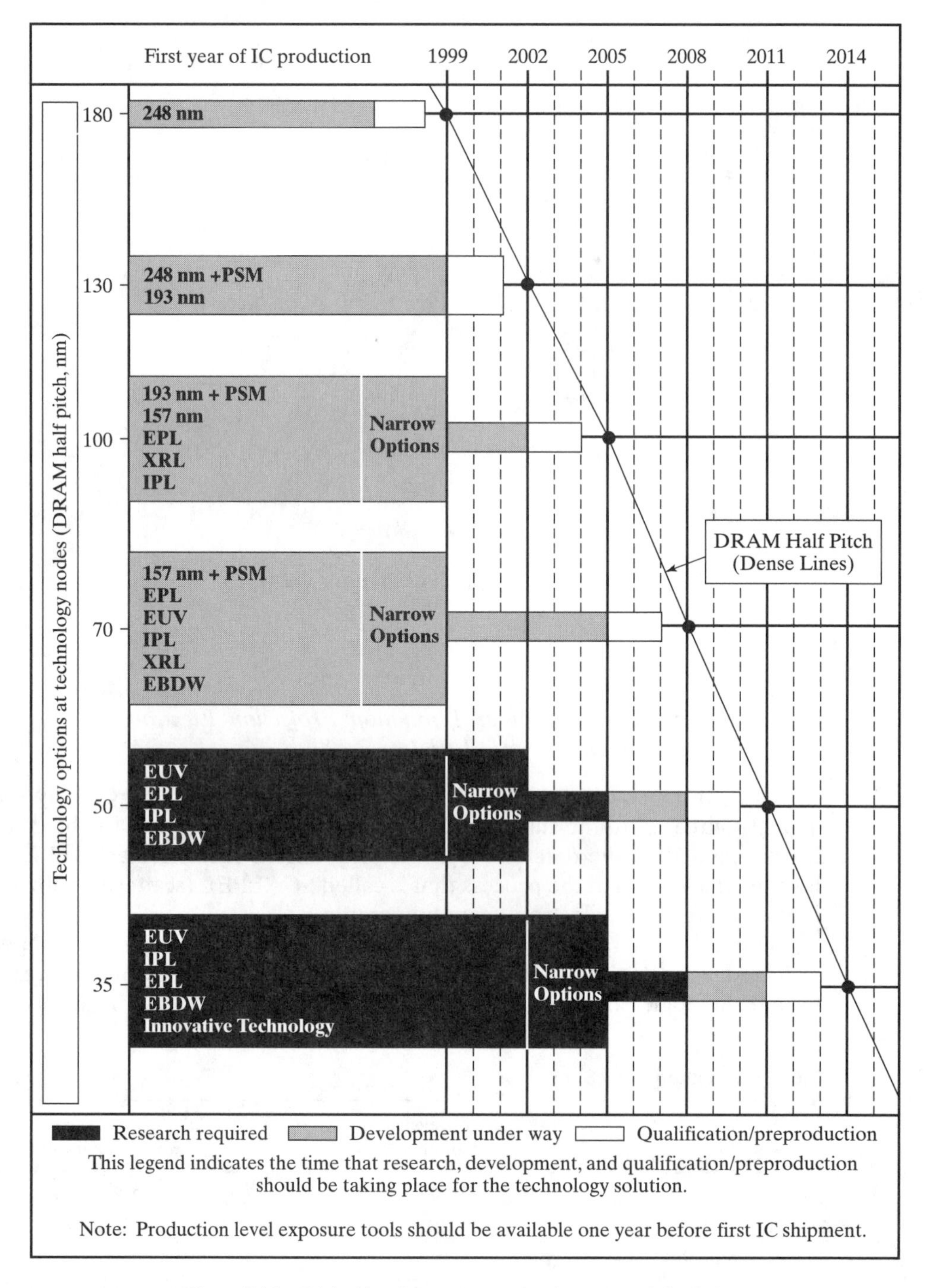

Figure 5.35 Critical level exposure technology potential solutions (courtesy of Semiconductor Industry Association, 1998).

5.13.4 Trends in Advanced Materials and Processing

In addition to lithography, advances are needed in materials science because there are natural limits to the "scalability" of the generic technology based on silicon. Some of these trends in materials are summarized in Figure 5.36.

Examples that are often discussed include the following (see Bohr, 1998; and Semiconductor Industry Association, 1997):

- The creation of new substrate techniques, especially silicon-on-insulator (SOI). The procedure replaces the bulk silicon substrate. Instead, a thin layer of silicon is created on an insulating surface. In the 1960s, sapphire was tried as the backing substrate for the layer of silicon. But later, the "bond and etch-back"

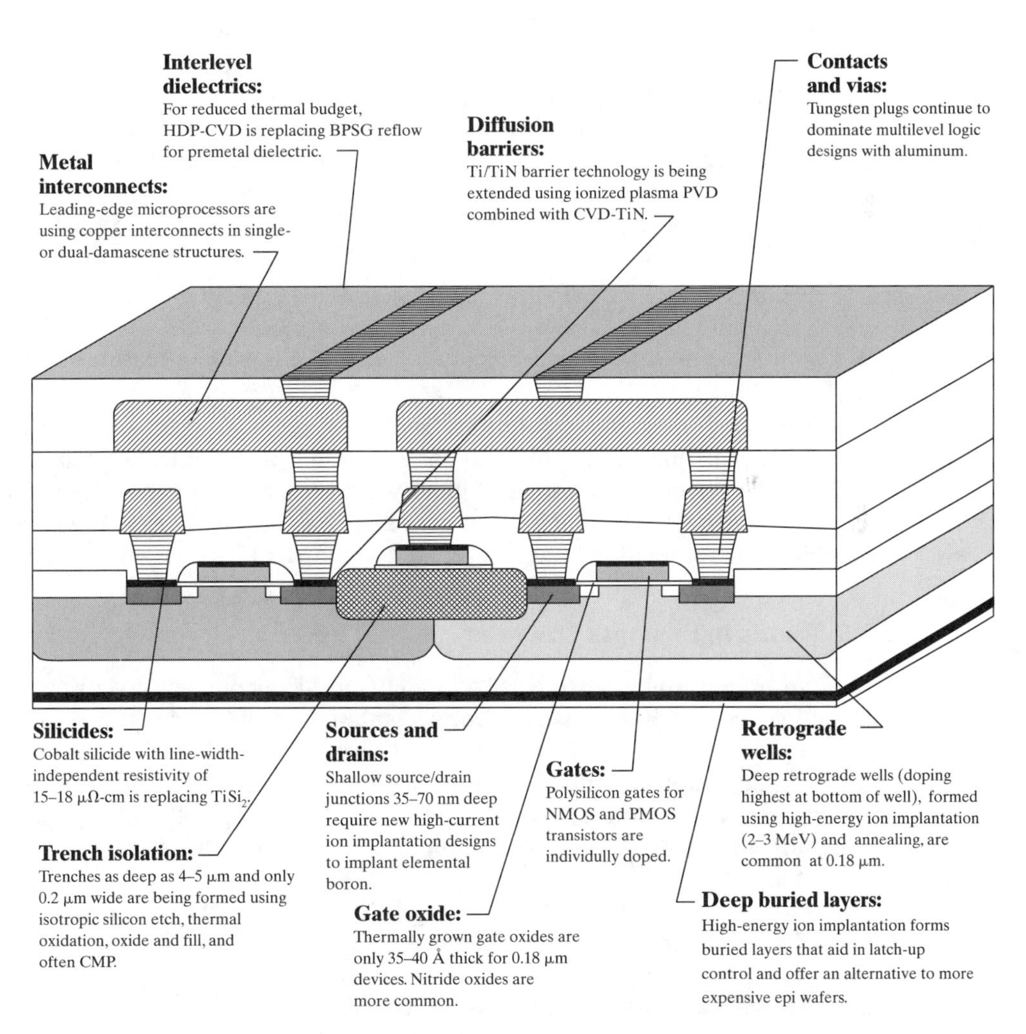

Figure 5.36 Potential new materials in 0.18 micron CMOS (adapted from literature of Semiconductor International).

method was developed. In this method two silicon wafers—one with a pre-grown oxide layer—are first bonded together. One of the layers of silicon is then gradually etched down, to leave thousands of angstroms of silicon on oxide. After appropriate doping, the silicon is then ready to have the transistors built on it. In comparison with standard CMOS, this provides low power and high speed for logic circuits (see Bohr, 1998; DeJule, 1999b).

- Replacing the silicon dioxide under the gate with other materials that have better dielectric properties. Specifically, for features smaller than today's 0.25 to 0.35 micron, the layer of silicon dioxide below the polysilicon gate is only 2 nanometers—that is, 4 to 5 atoms—thick. At this thickness of SiO_2, it is possible that electron tunneling can occur. Materials such as tantalum oxide are cited in the literature as an alternative.
- Replacing the polysilicon of the gate with materials that reduce the gate delay.
- Gradually replacing aluminum as the main interconnect material with other metals such as copper that have better conductivity (Braun, 1999). Copper has twice the conductivity of aluminum. If problems with contamination of the silicon substrate can be solved, copper may well be used for the next few years. Nevertheless, it is emphasized in the literature that if the feature sizes do indeed approach 0.05 micron toward the year 2012 (see Semiconductor Industry Association, 1997), then copper alloys will also fall short of the necessary gate speeds. Copper may "only buy us one or two generations" of product (Spencer, 1998).
- Developing multilayer resist technologies that allow taller features to be built on the substrate.
- Developing processing solutions such as chemical mechanical polishing (CMP). This has great potential for increasing the number of interconnect layers and hence circuit density in advanced ICs and microprocessors. If successive layers can be planarized between deposition steps, then it will be possible to increase interconnect layers to as many as 12, as shown in Figure 5.37.

5.13.5 Trends in Business Practices

In 1987, an organization called SEMATECH (the SEmiconductor MAnufacturing TECHnology consortium) was created. A combination of federal government and industry funding began SEMATECH, and although the federal funding ended in 1996, the organization continues to represent an important alliance for the semiconductor industry as a whole. It especially creates an effective collaboration between U.S. manufacturers of ICs and the equipment suppliers who support them (see Macher et al., 1998).

It is likely that the early success of SEMATECH augured well for a "culture of alliances." As the 1990s unfolded and it was evident that R&D costs were so astronomical, most companies realized they could not go it alone. Thus, as discussed earlier, the EUV lithography alliance has considerable public visibility. However, there are many forms of alliances emerging that include licensing agreements, fab/assembly/testing agreements, joint funding of new IC designs, and joint manufacturing process ventures.

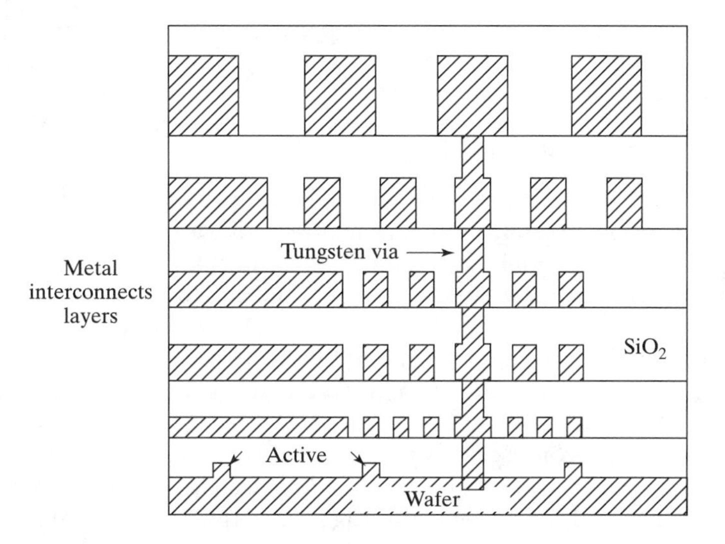

Figure 5.37 Increase in interconnect layers made possible by chemical mechanical polishing (adapted from Bohr, 1998).

Another example of an alliance is the International 300-mm Initiative (I300I), which includes chip makers from the United States, Europe, Korea, and Taiwan (see Ham et al., 1998). This alliance is focused specifically on the R&D needed to make the bigger 300 mm wafers. Difficulties include avoiding internal voids in the large diameter silicon ingots, avoiding warpage during slicing, maintaining flatness during polishing, and avoiding further warpage during the etching and baking cycles.

A related trend is the emergence of "fabless" chip-design studios—often called the IP model for the intellectual property thus created. These firms create specialized designs but then subcontract their fabrication work to external "foundries." There is currently much debate over the long-term competitiveness of such partnerships and the separation of product design and manufacturing. At first glance, it seems to be contrary to the concurrent engineering philosophy advocated throughout this book. However, it appears likely that semiconductor production in the United States will increasingly be carried out by (a) on the one hand, large capital-intensive firms such as Intel, and (b) on the other hand, a flexible, changing network of R&D alliances among smaller entrepreneurial firms. And in these alliances, integrated design and manufacturing will persist across several companies that cover the spectrum from "IP to fab." As summarized by Macher and associates (1998), the fabless design studio is more a North American model with roughly 500 such companies in the United States in 1998, while the state-of-the-art foundries that support them are located in Asia. The Taiwan Semiconductor Manufacturing Company (TSMC) is perhaps the most visible of these "foundries" (see <**www.tsmc.com**>).

A large, technologically aggressive, and financially strong semiconductor company such as Intel can compete by investing heavily in its own R&D, participating in alliances with other companies, adding new capacity, and taking advantage of its market-leader position. The prospects for new firms seeking to break into the industry are much narrower. Even so, several new semiconductor ventures have managed to carve out market niches by designing and marketing technically advanced products. For a small company to identify and exploit a niche, it generally requires significant design advantages, easy access to fab capacity, and strong marketing capabilities.

5.13.6 "Best Practices"

The lesson of the 1980s—that technological prowess alone is not enough—will continue to apply to semiconductor manufacturing at any point in the future. During the "dark days" of the mid-1980s, the U.S. semiconductor industry stayed afloat in large part because it had access to an unparalleled R&D infrastructure that produced a steady stream of technological breakthroughs and commercial applications. In a nutshell, new designs kept the United States from dying altogether in the 1980s, but it took dramatic improvements in manufacturing to re-create the sustained growth of the 1990s (Macher et al., 1998).

U.S. semiconductor firms still tend to be driven by an attitude in which each company wants to boast about new chip or transistor development ahead of the competition. New developments in design[6] can often be found in the business sections of daily newspapers and publications such as the *Economist.* Perhaps this strategy impresses Wall Street analysts and boosts stock value.

But long-run profitability still depends on a company's ability to maintain high product quality and productivity and minimize time-to-market for these new innovations. In this regard, a study funded by the Sloan Foundation has identified several semiconductor manufacturing best practices (Leachman and Hodges, 1996). Examples of best practices include:

- Using quality control with "6 sigma" quality assurance
- Integrating design, process planning, and production
- Accelerating time-to-market with commercial applications of high-quality design innovations
- Rapidly accommodating customer requests—that is, flexibility
- Using alliances to build stronger supplier relations
- Implementing inventory control and just-in-time systems to cut overhead
- Improving machine throughput and usage

Other industry observers (Spencer, 1998) note that the semiconductor industry's 18% compounded growth rate over the last three decades has been based on a combination of new technologies and best practices—for example,

- Reduced feature sizes
- Lithography improvements—moving from projection printing to steppers

[6]At the time of this writing, around the year 2000, the following are in the news: (a) silicon on insulator, (b) building dynamic RAM into separate trenches of the processor chip, (c) the use of silicon germanium in communication chips, and (d) vertical transistors with very small gate lengths.

- Improvements in resolution that occurred from wet to dry etching
- Improvements from ion beam implantation versus plain furnace diffusion
- The quality assurance push of the late 1980s

However, Spencer argues that these technologies and best practices have to some extent been "used up." One important area that does remain, in order to maintain the 18% growth rate, is the last of Leachman and Hodges's recommendations—*machine throughput and utilization* (Leachman et al., 1999). With a $2.5 billion fab, it is more important than ever to ensure that *integrated* design and manufacturing—not design alone—adds strategic value to a single company or several companies in an alliance.

5.14 GLOSSARY

5.14.1 Active

Regions of the MOSFET where the surface is doped to form the components of a transistor.

5.14.2 Application Specific Integrated Circuit (ASIC)

ASICs are designed to suit a customer's particular requirement, as opposed to a DRAM or microcontroller, which are general-purpose parts. Examples include programmable array logic devices, electrically programmable logic devices, field programmable gate arrays (FPGAs), and fully customized IC designs.

5.14.3 Back End

The manufacturing processes to test and package individual chips.

5.14.4 Bipolar

A fast semiconductor device that is relatively power and space hungry. Bipolar has been mostly replaced by CMOS technology except in certain high-performance applications.

5.14.5 Bonding Pads

The place where the bonding wires are attached, which eventually link to an IC package.

5.14.6 Channel

The region separating the source and drain of a field-effect transistor.

5.14.7 Central Processing Unit (CPU)

The main microprocessor of a computer system.

5.14.8 Chemical Vapor Deposition (CVD)

Process of using thermal reactions or breakdowns of gases to coat the substrate.

5.14.9 Complementary Metal-Oxide Semiconductor (CMOS)

A MOS technology in which both *p*-channel and *n*-channel components are fabricated on the same die to provide integrated circuits. It is the dominant semiconductor manufacturing process because of its high density and low power attributes relative to other MOS or bipolar processes.

5.14.10 Contact

Layers for providing the interconnects between different layers.

5.14.11 Depletion Layer

Region of the silicon that experiences the migration of holes due to applied voltages across the substrate. In the MOS transistor, a voltage applied at the gate initially creates a depletion layer. However, once the voltage is high enough, the layer is "inverted" and the transistor turns on.

5.14.12 Die

Individual chips on a wafer.

5.14.13 Diode

A *pn* junction device that controls the direction in which electron flow can occur, depending on the polarity of the voltage applied.

5.14.14 Doping

Diffusing the silicon substrate with donors or acceptors to lower the resistivity of the semiconductor.

5.14.15 Drain

One of the three regions that form a field-effect transistor. Carriers originate at the source, traverse the channel under the gate, and are collected at the drain to complete the current path. The flow between source and drain is controlled by the voltage applied to the gate.

5.14.16 Dual-in-Line Packaging (DIP)

An inexpensive industrial style of packaging chips (see Figure 5.30).

5.14.17 Dynamic Random Access Memory (DRAM)

The lowest cost per bit and the most popular type of semiconductor read/write memory chip, in which the presence or absence of a capacitive charge represents the state of a binary storage element (zero or one). It is "dynamic" because the electrical charge must be periodically refreshed in order to retain stored data.

5.14.18 Etching

Removing material to create a patterned layer.

5.14.19 Fab or Wafer Fabrication

A plant or factory where semiconductors are made (fabricated).

5.14.20 Field-Effect Transistor (FET)

A planar solid-state device in which current between source and drain terminals is controlled by voltage applied to a gate terminal.

5.14.21 Field Programmable Gate Array (FPGA)

A complex programmable logic device configured to customer specifications. These devices allow systems manufacturers to add value beyond the microprocessor.

5.14.22 Front End

Fabrication processes dealing with creating the component. It includes photolithography, doping, oxidation, and the like.

5.14.23 Gate

The control electrode in the FET. A voltage applied to the gate regulates the conducting properties of the semiconductor channel region located directly under the gate. In a MOSFET, it is separated from the semiconductor by a very thin oxide layer.

5.14.24 Gate Array

An ASIC customized to a specific application by interconnecting thousands of gates.

5.14.25 Integrated Circuit (IC)

An electronic circuit in which many active elements—such as transistors, resistors, capacitors, and diodes—are fabricated and connected together on a continuous substrate.

5.14.26 Interconnect

The metal wiring connecting the various transistors and components together. Aluminum is the chief metal of choice, although copper is being considered for some applications.

5.14.27 Microcontroller

A cheap and reasonably powerful specialized control system that uses a microprocessor. Embedded in a system, a microcontroller performs a specific function, such as advancing the paper feed in a printer.

5.14.28 Microprocessor

A single integrated circuit (IC) that combines logic functions, arithmetic functions, memory registers, and the ability to send and receive data.

5.14.29 MOSFET

Metal-oxide semiconducting field-effect transistor. A particular type of FET that has a polysilicon gate separated from the substrate by a gate oxide.

5.14.30 NMOS

NMOS is an *n*-type MOSFET transistor. NMOS transistors are faster than their PMOS counterparts.

5.14.31 *n*-Type Semiconductor

n-type behavior is induced by the addition of donor impurities, such as arsenic or phosphorus, to the crystal structure of silicon.

5.14.32 Photolithography

Circuit patterns formed by masking and etching processes.

5.14.33 Photomask

A mask that shields parts of the substrate, and exposes other parts, for processes such as dry etching or doping. Masks hold the key to the layout patterns and the resulting circuit.

5.14.34 Photoresist

Material used to react with incoming UV rays and undergo chemical reaction. Subsequent upon these reactions, the photoresist can be dissolved, leaving the substrate ready for etching.

5.14.35 PMOS

PMOS is a *p*-type MOSFET transistor. Though slower than their NMOS counterparts, they are needed to complement the NMOS in CMOS transistors.

5.14.36 Polysilicon

Short for polycrystalline silicon. Used for constructing the gate in MOS devices.

5.14.37 Printed Circuit Board (PCB)

Surface where individual dice are assembled and joined in a functioning product.

5.14.38 *p*-Type Semiconductor

The *p*-type behavior is induced by the addition of acceptor impurities, such as boron, to the crystal structure of silicon.

5.14.39 Quad Flat Package (QFP)

A type of IC package most often used for packaging microprocessors.

5.14.40 Select

Step in the front-end processing in which dopants form outer barriers for the transistor.

5.14.41 Semiconductor

A class of materials, such as silicon and germanium, whose electrical properties lie between those of conductors (such as copper and aluminum) and insulators (such as glass and rubber). The material exhibits relatively high resistance in a pure state and much lower resistance when it contains small amounts of certain impurities or *dopants.*

5.14.42 Semiconductor Device

The term is also used to denote electrical devices made from semiconductor materials.

5.14.43 Solid State

Refers to the electronic properties of crystalline materials, generally semiconductors—as opposed to vacuum tubes that function by flow of electrons through ionized gases.

5.14.44 Source

One of three regions that form a field-effect transistor. Electrons in an *n*-type MOSFET originate at the source and flow across the *p*-channel to the drain.

5.14.45 Sputtering

Processing method of depositing thin films onto the wafer by knocking out atoms from a source, or a target, which sputter onto the wafer.

5.14.46 Substrate

The underlying silicon where the various front-end processing techniques are performed. Substrates can be *p*-type or *n*-type depending on the types of dopants used.

5.14.47 Transistor

The basic circuit building block from which various integrated circuits are composed. The bipolar transistor is two *pn* junctions combined to form *npn* or *pnp* transistors.

5.14.48 Via

Vertical channel connecting different layers of metal.

5.14.49 Wafer

A thin slice, sawed from a cylindrical ingot of bulk semiconductor material (usually silicon) 4 to 12 inches in diameter. Arrays of ICs are fabricated in/on the wafers during the manufacturing process.

5.14.50 Well

Most typically, wells are used in CMOS. A large volume of *n*-type material on a *p*-type substrate allows localized PMOS transistors to be built (see Figure 5.7).

5.14.51 Wire Bonding

Refers to the attachment of thin wiring from the chips to the lead frame of the package.

5.14.52 Yield

The percentage or fraction (zero to one) of wafers, dice, or packaged units that conform to specifications. Yield metrics in semiconductor manufacturing include:

- Wafer yield—the fraction of the wafers that complete wafer processing
- Die yield—the fraction of the dice on a wafer that meet specifications
- Assembly yield—the fraction of units that are assembled correctly
- Final test yield—the fraction of packaged units that pass all device specifications

5.15 REFERENCES

Bohr, M. 1998. Silicon trends and limits for advanced microprocessors. *Communications of the ACM* 41 (3):80–87.

Braun, A. E. 1999. Aluminum persists as copper age dawns. *Semiconductor International,* August, 58–66.

Brodersen, R. W. 2000. <www.bwrc.eecs.berkeley.edu>.

Campbell, S. A. 1996. *The science and engineering of microelectronic fabrication.* Oxford and New York: Oxford University Press.

Colclasser, R. A. 1980. *Microelectronics: Processing & device design.* New York: Wiley.

DeJule, R. 1999a. Next generation lithography tools. *Semiconductor International,* March, 48–52.

DeJule, R. 1999b. SOI comes of age. *Semiconductor International,* November, 67–74.

Einspruch, N. G. 1985. *VLSI handbook.* Orlando, FL: Academic Press.

Ham, R. M., G. Linden, and M. M. Appleyard. 1998. The evolving role of semiconductor consortia in the U.S. and Japan. *California Management Review* 41 (1):137–163.

Jaeger, R. C. 1988. *Introduction to microelectronic fabrication.* Reading, MA: Addison Wesley.

Kalpakjian, K. M. 1995. Fabrication of microelectronic devices. In *Manufacturing Engineering and Technology,* edited by Serope Kalpakjian. Reading, MA: Addison Wesley.

Leachman, R. C., and D. A. Hodges. 1996. Benchmarking semiconductor manufacturing. *IEEE Transactions on Semiconductor Manufacturing* 9 (2):158–169.

Leachman, R. C., and D. A. Hodges. 1998. Benchmarking semiconductor manufacturing. Third Report, Engineering Systems Research Center Report No. CSM-31, University of California, Berkeley.

Leachman, R. C., and C. H. Leachman. 1999. Trends in worldwide semiconductor fabrication capacity. Engineering Systems Research Center Report No. CSM-48, University of California, Berkeley.

Leachman, R. C., J. Plummer and N. Sato-Misawa. 1999. Understanding fab economics. Engineering Systems Research Center Report No. CSM-47, University of California, Berkeley.

Leyden, P. 1997. Interview with Gordon Moore. *Wired Magazine,* May, 164–166.

Macher, T. J., D. C. Mowery, and D. A. Hodges. 1998. Reversal of fortune? The recovery of the U.S. semiconductor industry. *California Management Review* 41 (1):107–136.

Mahajan, S., and L. C. Kimerling. 1992. *Concise Encyclopedia of Semiconducting Materials and Related Technologies.* Oxford and New York: Pergamon Press.

Mead, C., and L. Conway. 1980. *Introduction to VLSI systems.* Reading, MA: Addison Wesley.

Muller R. S., and T. I. Kamins. 1986. *Device electronics for integrated circuits.* New York: Wiley and Sons.

Patterson, D. A., and J. L. Hennessy. 1996a. *Computer architecture: A quantitative approach.* San Francisco, CA: Morgan Kaufman Publishers.

Patterson, D. A., and J. L. Hennessy. 1996b. *Computer organization and design: The hardware/software interface.* San Francisco, CA: Morgan Kaufman Publishers.

Peterson, I. 1997. Fine lines for chips. *Science News* 152: 302–303.

Pierret, R. F. 1996. *Semiconductor device fundamentals.* Reading, MA: Addison Wesley.

Rabaey, J. M. 1996. *Digital integrated circuits.* Upper Saddle River, NJ: Prentice-Hall Electronics and VLSI Series.

Red Herring. 1995. Special issue on semiconductors, September.

Rosler, R. S., W. C. Benzing, and J. Baldo. 1976. Plasma enhanced CVD. *Solid State Technology* 19: 45–53.

Runyan, W. R., and K. E. Bean. 1990. *Semiconductor integrated circuit processing technology.* Reading, MA: Addison Wesley.

Semiconductor Industry Association. 1997. *The national technology roadmap for semiconductors: Technology needs.*

Semiconductor International. 1998. Sub-100 nm features with single layer 193 nm resist. April, 20.

Siddhaye, S. V. 1999. Design for the environment in electronics manufacturing: Product optimization for waste stream minimization. Ph.D. Dissertation, University of California, Berkeley.

Singer, P. 1997. Copper goes mainstream: Low k to follow. *Semiconductor International,* 67–70.

Spanos, C. 1999. *Processing and design of integrated circuits.* Course Reader for EECS 143 at University of California, Berkeley.

Spencer, W. J. 1998. *Regents' Lecture Series,* University of California, Berkeley (available on videotape at Haas School of Business).

Wolf, S., and R. N. Tauber. 1986. *Silicon processing for the VLSI era. Vol. 1, Process technology.* Sunset Beach, CA: Lattice Press.

Zuhlehner, W., and D. Huber. 1982. *Czochralski grown silicon crystals,* Vol. 8. New York: Springer Verlag.

5.16 BIBLIOGRAPHY

5.16.1 Technical

Angel, D. P. 1994. *Restructuring for innovation: The remaking of the U.S. semiconductor industry.* New York: Guilford Press.

Augarten, S. 1983. *State of the art: A photographic history of integrated circuits.* New Haven and New York: Ticknor and Fields Press.

Beadle, W. E., J. C. C. Tsai, and R. D. Plummer, eds. 1985. *Quick reference manual for silicon integrated circuit technology.* New York: Wiley.

Hodges, D., and H. Jackson. 1988. *Analysis and design of digital integrated circuits,* 2d ed. New York: McGraw-Hill.

U.S. Industrial Outlook. 1994. *Chapter 15: Electronic components, equipment, and superconductors.*

Van Sant, P. 1985. *Microchip fabrication: A practical guide to semiconductor processing.* San Jose, CA: Semiconductor Services.

Wolf, W. *Modern VLSI design: A systems approach.* Upper Saddle River, NJ: Prentice-Hall.

5.16.2 Social

Kaplan, D. A. 1999. *The silicon boys and their valley of dreams,* San Francisco: William Morrow.

Reid, T. R. 1984. *The chip: How two Americans invented the microchip and launched a revolution.* New York: Simon & Schuster.

5.16.3 Recommended Subscriptions

Embedded Systems Programming, Miller Freeman, 1601 West 23rd St., Lawrence, KS, 66046, <**www.embedded.com**>.

IEEE Transactions on Semiconductor Manufacturing, 3 Park Avenue, New York, NY, 10016, <**www.ieee.org**>.

Semiconductor International, 8773 S. Ridgeline Blvd., Highlands Ranch, CO., 80126, <**www.semiconductor.net**>.

5.17 URLS OF INTEREST

For design tools: <**http://bwrc.eecs.berkeley.edu.**>

5.18 APPENDIX 1: Worldwide Semiconductor Market Share

TABLE 5.9 Worldwide Semiconductor Market Share Estimates in the Mid-1990s

Top 10 companies	1995 revenue ($B)	1996 revenue ($B)	1995–96 growth (%)
Intel	13.17	16.94	29
NEC	11.31	10.58	−6
Motorola	8.73	8.44	−3
Hitachi	9.14	8.06	−12
Toshiba	10.08	7.98	−21
Texas Instruments	7.83	7.09	−9
Samsung	8.33	6.20	−26
Fujitsu	5.54	4.51	−19
Mitsubishi	5.27	4.20	−20
SGS-Thomson	3.39	4.20	24
Industry totals	151.27	140.69	−7

5.19 APPENDIX 2: Cost Model Variables in Year 2000—Example for a 64-Mb DRAM (Courtesy Dataquest)

TABLE 5.10 Wafer Data and Chip Yields (# = Test Seconds Per Die)

Wafer size (diameter in inches)	A	8
Capacity utilization (%)	B	100.00
Geometry (microns)	C	0.35
Processed wafer cost ($)	D	1650.00
Die area (square millimeters)	E	155,494.00
Active area factor	F	1.00
Number of masks (type of process)	G	18.00
Defect density per square inch/per mask	H	0.111
Gross die per wafer	=I=(0.75*pi(A/2)^2*10^6)/E	242.00
Processed wafer cost per gross die ($)	=J(D/I)	6.81
Test cost per hour ($)	K	120.0
Wafers tested per hour	=L=1/((#*I)/3600)	0.08
Wafer sort cost per gross die ($)	M=(K/L)/I	6.00
Cost per gross die at wafer sort ($)	=N=(J+M)	12.81
Wafer sort yield (%)	=O=2.718^((−H*G)*E)	73
Cost per sorted die ($)	=P=N*100/O	17.48

TABLE 5.11 Assembly Data

Material cost/sorted die + package cost ($)	Q	0.48
Number of package pins	R	44
Assembly yield (%)	S	90
Cost per assembled die ($)	=T=(P+Q)/S*100	19.95

TABLE 5.12 Final Test Data

Test time per die (sec)	U	60.00
Cost per hour of testing ($)	V	120.00
Test cost per die ($)	=W=U*V/3600	2.00
Final test yield (%)	X	90
Cost per final tested unit ($)	=Y=(T+W)/X*100	24.39

TABLE 5.13 Mark, Pack, and Ship Data

Cost at 99 percent yield (%)	=Z=(Y*0.01)	0.24
Total fabrication cost per unit ($)	=AA=Y+Z	24.63

TABLE 5.14 Foreign Market Value (FMV) Formula Adders

R&D expense (15 percent)	=AB=0.15*AA	3.70
S,G&A expense (10 percent)	=AC=(AA+AB)*0.10	2.83
Profit (8 percent)	=AD=(AA+AB+AC)*0.08	2.49
Constructed FMV	=AE=(AA+AB+AC+AD)	33.66

5.20 REVIEW MATERIAL

1. Calculate that the final cost of a packaged Power PC603 chip is $45.51 and that the packed cost of a SuperSPARC/60 chip is 318.31. Assume the 1996 data of Patterson and Hennessy (1996b).
2. What manufacturing improvements have occurred since 1996 that will lower this cost?

CHAPTER 6

COMPUTER MANUFACTURING

6.1 INTRODUCTION

Chapter 5 reviewed the transistor, the integrated circuit, and the techniques of front-end semiconductor manufacturing. The "journey along the product development path with emphasis on the fabrication techniques" now continues from a packaged IC to the manufacture of a finished computer. The case study at the end of this chapter considers the specific design and manufacture of a portable, wireless computer aimed at an approximate selling price of $300.

In Chapter 6, components are mounted on printed circuit boards (PCBs) and assembled with the disc drive and other pieces of hardware into a fully connected computer system. The components must also be placed in a physical structure to anchor and protect them. The physical structure could still be in prototype form (Chapter 4) or mass-produced by polymer injection (Chapter 8) from a machined mold (Chapter 7). In Figure 6.1 and Table 6.1, key features include:

- The motherboard with main microprocessor, including the control, the data paths for the integer and floating point parts of the processor, and the cache memory
- The main memory boards on the motherboard
- Secondary memory in the form of floppy or hard discs
- The input and output devices

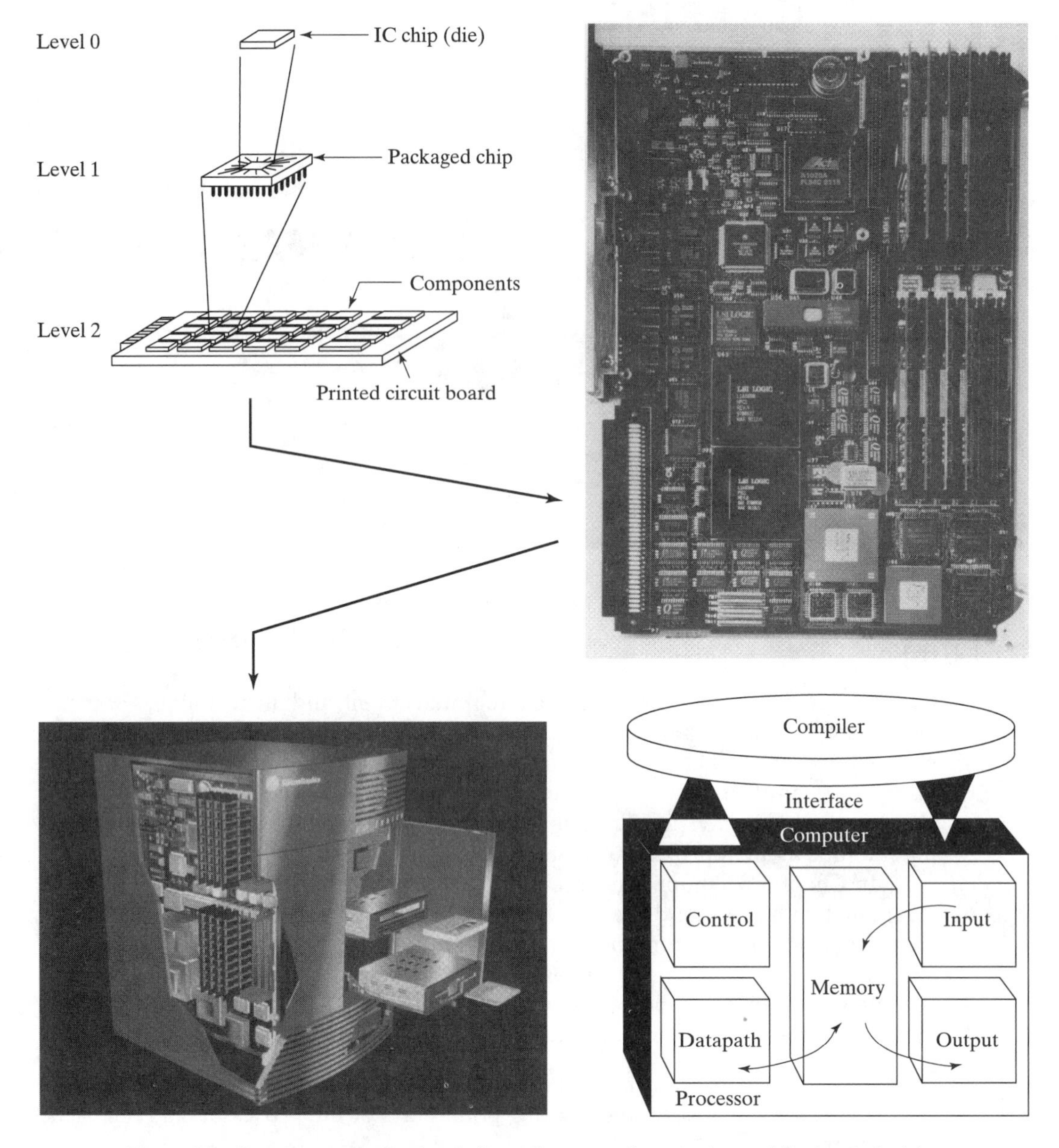

Figure 6.1 Computer packaging levels (from *Computer Organization and Design,* 2nd ed. by David Patterson and John Hennessy, © 1996. Morgan Kaufmann Publishers.). From the top left from the previous chapter, packaged integrated circuits (ICs) (e.g., the main processor and memory chips) are first assembled onto PCBs (the motherboard and the eight main memory boards, assembled vertically on the motherboard). System level packaging on the lower left also shows the secondary memory (floppy and hard drive). The schematic shows the main functional abstractions of the physical devices.

TABLE 6.1 Key Functional Abstractions of a Computer System

Component	Function
1. Processor (CPU) data path	Performs arithmetic operations
2. Processor (CPU) control	Sends signals that determine the operation and sequencing of data paths, memory, and use of the input/output devices
3. Memory	(a) Primary (main) memory: volatile memory of programs or data being used by the processor (b) Secondary (floppy or hard drive) memory: nonvolatile memory or storage of programs
4. Input	Includes keyboard, mouse, voice activation, digital camera, incoming e-mail, fax, and so forth
5. Output	Includes screen, printer, outgoing e-mail, fax, and the like

6.2 PRINTED CIRCUIT BOARD MANUFACTURING

6.2.1 Introduction

Printed circuit boards (PCBs) provide the foundation for the interconnections among subcomponents. The interconnections are provided by copper *tracks* that are applied to the circuit board in a series of additive or subtractive processing steps similar to those used in the fabrication of integrated circuits (ICs). Copper *lands* are also applied for the connections to individual ICs and components. Colloquially speaking, the PCB is a "subway map" of circuit tracks that connect the ICs and other devices located at the stations.

The board itself also provides the rigid structure that holds chips and other fragile system components in place and allows for the physical connections to the outside world of input/output devices such as the hard drive, monitor, keyboard, and mouse. The earliest circuit-laying methods used screen printing techniques, from which the term *printed* circuit board or *printed* wiring board developed. Photolithography is now the preferred method for circuitizing boards. There are three general types of PCBs, depicted in Figure 6.2:

- Single-sided boards have copper tracks on only one side of an insulating substrate.
- Double-sided boards consist of copper tracks on both sides of the insulating layer.
- Multilayer boards are constructed from alternating copper and insulating layers.

6.2.2 "Starting Board" Construction

Starting boards are so-called because the circuit patterns have not yet been applied. A double-sided PCB is a flat laminated "sandwich." A thin substrate (0.25 to 3 mm thick) of insulating material is sandwiched between thin copper foil (0.02 to 0.04 mm thick) on both sides. Epoxy resin is the most commonly used insulating polymer for the inner substrate, usually reinforced with an epoxy/glass fiber called e-glass. The insulating substrate is formed by taking multiple sheets of thin glass fiber impregnated with partially

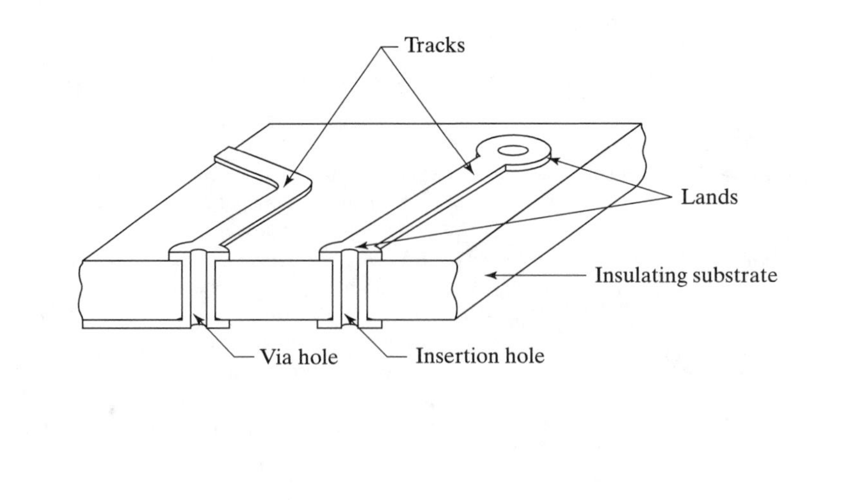

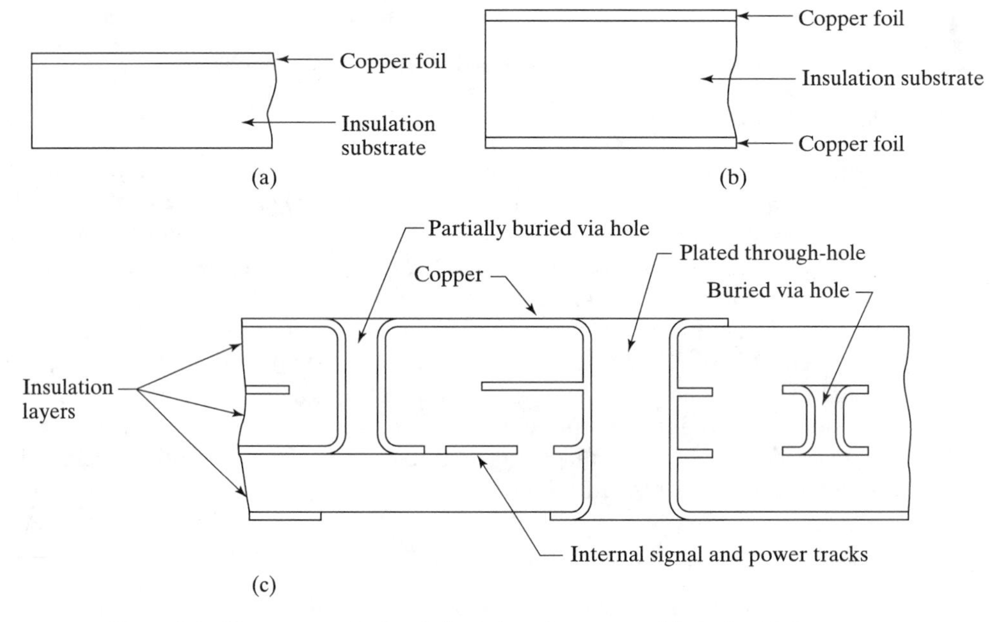

Figure 6.2 Three types of printed circuit board structures: (a) single-sided, (b) double-sided, and (c) multilayer showing vias and pathways between layers (courtesy of Groover, 1996).

cured epoxy. The boards are then pressed together between hot plates or rolls. The heat and pressure cure and harden the laminates, creating a strong and rigid board that is heat and warp resistant.

6.2.3 Board Preparation

The starting board must be prepared for further processing through a variety of shaping procedures. First, shearing operations are carried out to obtain the desired size for the final computer/electronic equipment. Second, tooling or alignment holes, typically 3 mm in diameter, are drilled or punched into the corners of the board. The tooling holes are used to precisely align the boards as they move from one machine to another in the sequence of fabrication steps. The board may be bar coded at this point to facilitate identification. The final step in this preparation phase is to carefully clean and degrease the surfaces. While board making does not require the stringent cleanliness standards of chip making, a fairly high level of cleanliness is essential to minimize defects.

6.2.4 Hole Drilling, Punching, and Plating

Additional holes are then created in the board. Automatic hole punchers or CNC drilling machines are used. For large batch sizes punching is the most efficient, but CNC drills can also drill a stack of several panels, thereby increasing productivity. The holes allow conducting paths, called *vias,* between the two sides of a double-sided board. Other *insertion* holes are for any pin-in-hole (PIH) components. Additional holes provide anchoring *locations* for heat sinks and connectors.

These holes, or vias, drilled through any insulative layers are nonconducting. Therefore, conductive pathways must be created between the sides of a double-sided board. These pathways are typically formed by electroless plating. This process is tailored to the deposition of copper onto the epoxy/glass fiber surface of the through-holes. Regular electroplating will not work because the surfaces are nonconducting. Electroless plating takes place chemically in an aqueous solution containing copper ions, but without any anode/cathode action. Specific details of these reactions are given by Nakahara (1996) and Duffek (1996).

6.2.5 Circuit Lithography

In this important step, a circuit pattern is transferred to the board's copper surface(s) using selective photolithography and etching. PCB industries may use a *subtractive* method. In Figure 6.3, the starting board's surface is already the thin copper foil. It is first coated with a polymer resist, sprayed on in liquid form or rolled out over the board from a spool of dry film (see Clark, 1985, p. 175). Ultraviolet (UV) lithography then exposes the resist in areas that are not wanted for circuits. The exposed resist is then strip-washed away; next, the now-unprotected copper areas are chemically etched with any of the following solutions: ammonium persulphate, ammonium hydroxide, cupric chloride, or ferric chloride. The remaining, unattacked copper areas constitute the board's circuitry or the lands. Alternatively, an *additive* process of PCB circuitizing begins with an unclad board, namely, the insulating material without a copper foil. Photoresist is spread on the board and then exposed in the pattern of the desired tracks. This exposed photoresist is then strip-washed away. Thus, at this stage, the board exhibits the exact pattern of the desired tracks except that no copper has been laid down yet into these "valleys." During electroplating, the board is shielded under the "hills" of remaining photoresist. Meanwhile, the copper is added—that is—electroplated, into the exposed "valleys," creating the desired circuits and lands (Figure 6.4).

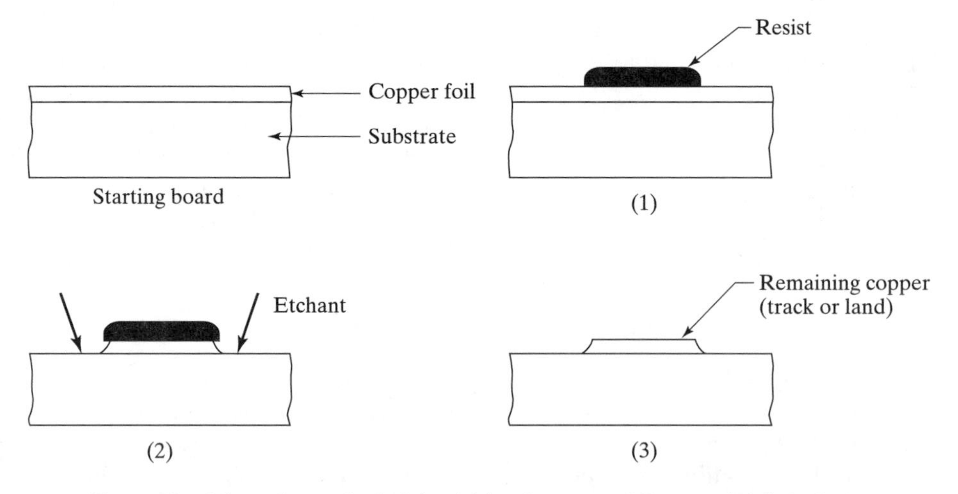

Figure 6.3 Subtractive method of circuitizing (courtesy of Groover, 1996). In the subtractive process, the copper foil is protected where the circuits and lands are needed. Lithography exposes the resist in areas that are not ultimately required. Once the exposed resist has been removed, that part of the copper is etched away. The final sketch shows the desired layout.

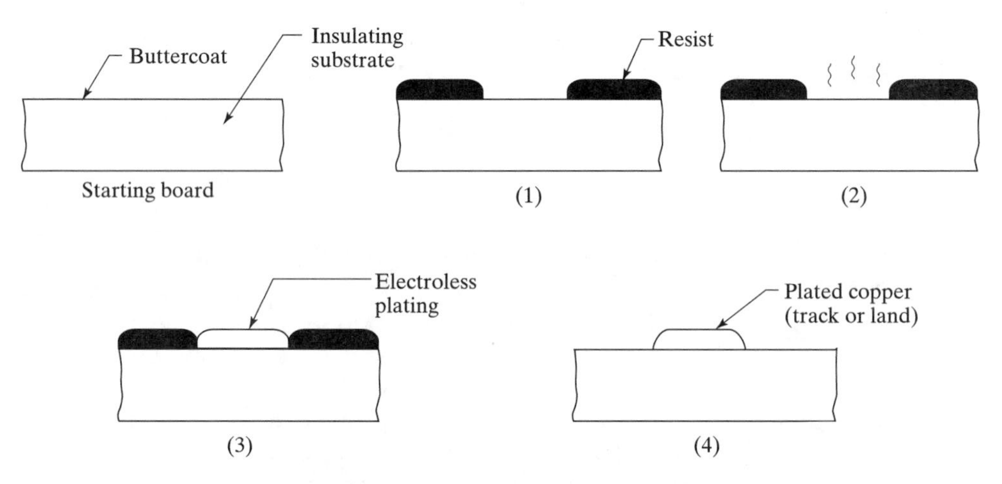

Figure 6.4 Additive method of circuit manufacture (courtesy of Groover, 1996). Photoresist is spread on an unclad board and then exposed in the pattern of the desired tracks. This exposed photoresist is then strip-washed away. During electroplating, copper is electroplated into these exposed "valleys," creating the desired circuits and lands.

6.2.6 Multilayer Board Fabrication

In multilayer board fabrication, the circuit designs are first applied to individual boards. Once the layers have been integrated, a multilayer board resembles a double-sided board from the outside, but it will have several integrated middle units with

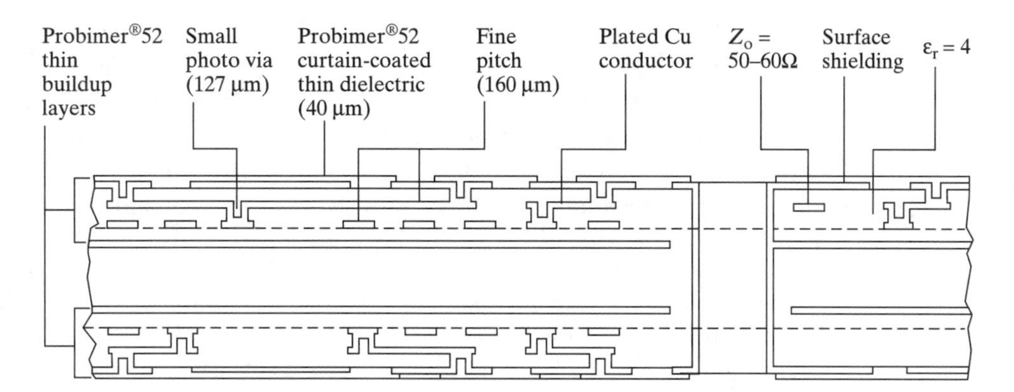

Figure 6.5 Surface laminar circuits created for blind and buried via holes in multilayer boards.

copper patterns on both sides. Precise alignment between each layer is obviously essential and is achieved by the alignment pins that fit tightly into the tooling holes.

It may have already occurred to the reader that creating vias and connections among the inner boards involves a special manufacturing challenge. In particular, the creation of buried and blind vias deserves special attention. Surface laminar circuits (SLCs) are created using intermediate photolithography methods on the inner boards (Figure 6.5). Inner layer patterns for the ground and power distributions are first created on the inner boards, and then the board is oxidized. Next, an insulating photosensitive resin is coated over the panel. The desired via locations are formed by photoexposure, development, and strip-washing. These via hole surfaces are coated with copper by either direct metallization or electroless plating. Further inner connections with thicker layers of copper are also added. With the constant push for miniaturization, higher speeds, and the use of surface-mount components on both outer surfaces of a board, these technologies will be more in demand.

The final phase in the fabrication of a printed circuit board is to test and finish the circuitry. Both visual and electronic test methods are used to check the functionality of the copper wiring. Detailed information on testing is given by Andrade (1996). The last step is to screen print a legend guiding the placement of components on the board, as well as a bar code. The finished board is now ready to have electronic and mechanical components attached to it to form a final PCB assembly.

6.3 PRINTED CIRCUIT BOARD ASSEMBLY

6.3.1 Overview

A multilayer PCB will probably contain hundreds of individual components ranging from sophisticated ICs to rather ordinary heat sinks and perimeter connections. As a result several, or perhaps all, of the following assembly styles will be used on any given board. The list that follows gives a summary, and the details are shown in Figures 6.6 to 6.13.

- Pin-in-hole (PIH) is the older classical method. It involves inserting the leads of standard components into holes drilled in the board, then clipping and soldering the leads into place on the opposite side of the board.
- Surface mount technology (SMT) is now the preference in industry because it allows greater packing densities. The SMT method directly solders component leads to copper lands on the same side of the board. This approach greatly reduces the surface area needed to fit components (requiring 40% to 80% less space than PIH), making it possible to build smaller and higher performance circuit boards. The leads used at the edge of a surface mounted IC typically have the "gull wing" or a "J-lead" shape shown in the diagrams toward the end of Chapter 5.[1]
- Multichip modules (MCM) consist of several SMT chips all mounted side by side inside one larger outer package. These have the following advantages: closer packing densities; reduced routing needs in the PCB, hence reducing the number of layers needed in a multilayer board; reduced power consumption; higher performance due to tighter noise margins, smaller output drivers, and smaller die sizes; and lower overall packaging costs. An excellent review may be found in Green (1996).
- Ball grid array (BGA) is a development of individual SMT components, where the connections are made underneath the chip instead of on the perimeter. Small balls of solder make the connections between the chip's underside and the PCB.
- Flip chip technology (FCT) extends SMT/BGA for even greater packing density. In this case, the IC is turned over and placed face down on the board. As earlier, solder balls and a perimeter solder ring create the circuit connections to the board. Additional mechanical bonding with epoxy is required.

Just as SMT has gradually replaced PIH for many applications due to the increased packing densities it offers, BGA and FTC have been growing in popularity in comparison to standard SMT. The costs of these newer methods are of course higher but can be justified in certain devices such as cellular phones where miniaturization is key to market leadership. Figure 6.6 shows many of these trends.

All these assembly methods involve similar basic processing steps. Components are first soldered into place on the board, and then the whole assembly is cleaned, tested, and, if necessary, reworked. The key differences lie in the method for placing and soldering components on the board; there are also some differences in the subsequent testing and reworking steps. Most SMT components also share the "real estate" on a multilayer board with PIH components. This complicates the assembly sequence, but the basic processing steps do not change.

[1]Back-end packaging was already introduced in Chapter 5. However, with continuing miniaturization, it is hard to differentiate where the IC package ends and where the PCB begins, and so some further discussion is warranted from a PCB perspective.

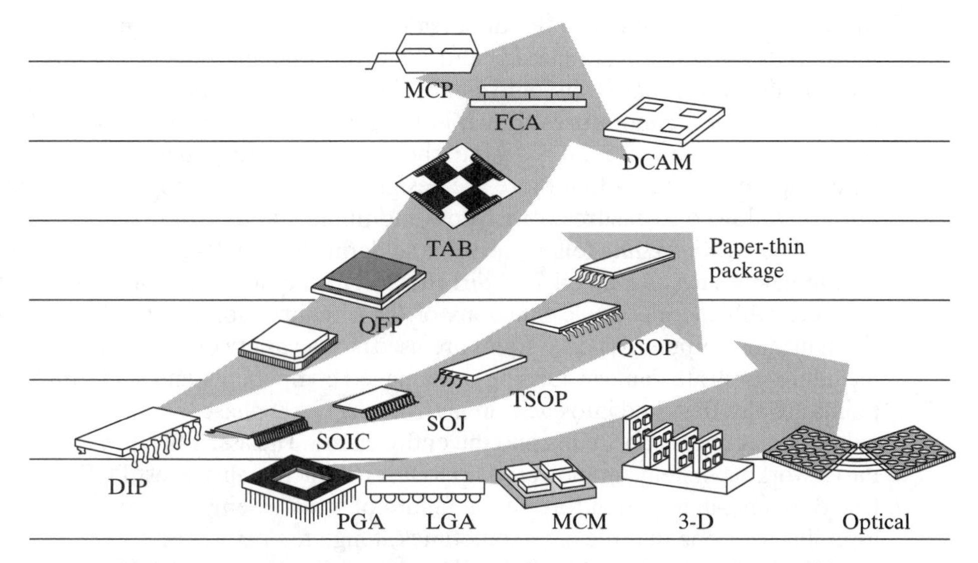

Figure 6.6 IC packaging families and trends (from *Printed Circuits Handbook* by Clyde F. Coombs, © 1996. Reprinted by permission of the McGraw-Hill Companies).

6.3.2 Fabricating with Pin-in-Hole Technology (PIH)

Insertion is the first step in the "old classic" PIH process. This involves inserting the leads of each component into the holes that have been predrilled in the board during fabrication. The insertion method depends on the type of component. For example, axial components—commonly including resistors, capacitors, and diodes—are cylindrical in shape, and their leads project from each end; the leads must be bent at right angles to be inserted in the board. Preforming is thus required so that component leads, which are straight, are bent into a U shape (Figure 6.7). Light-emitting diodes and fuse holders are common radial lead components, with parallel leads radiating from the component body, and require a different type of work head and preforming.

Wave soldering is the next major step in manufacturing. For example, a PCB with inserted PIH components is passed over a standing wave of molten solder such that the solder just touches the bent leads on the underside of the board. Figure 6.8a

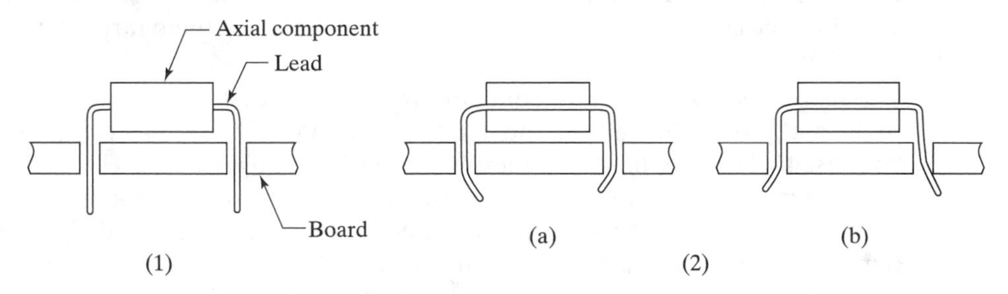

Figure 6.7 Affixing a component to a PCB with the "old classic" PIH method: (1) an axial component is first inserted; (2) bending and cropping inward (a) and outward (b) (courtesy of Groover, 1996).

shows that flux is applied to the underside of the board at the beginning of the conveyor. After preheating, the board and the projecting leads of the components meet the agitation wave that "wets" and cleans the surfaces. The final laminar wave creates the joints at the temperatures shown in Figure 6.8b. This process forms solder joints by forcing the liquid solder to flow into the clearances between the leads and through-holes. Figure 6.8c shows that there are design rules (layout rules) for this process that must be followed to ensure correct flow and filling and to avoid "shadowing."

Cleaning and testing follow the wave soldering. The PCBs are degreased to remove contaminants such as flux, oil, and dirt that might chemically degrade the assembly or interfere with the electronic functions of the circuitry. Boards are visually inspected (human and computer vision systems are used) for a variety of potential quality defects, including substrate damage, missing or damaged components, and soldering faults. Test points are also designed into each circuit from the CAD phase. Contact probes test individual components, subcircuits, and the entire circuit. The assembly may also be plugged into a working system and powered up to test its functionality. Most PCBs are also subjected to burn-in tests that force early failure of weak assemblies; this test operates the assemblies for one to three days, sometimes at high temperatures.

Rework is the final step that will commonly be seen in any factory tour of a subcontract board assembly operation. Because of the high value of electronic components, as well as the cost of board fabrication and assembly, it is economically more feasible to repair defects than to discard the entire board. Rework is always a skilled manual operation involving manual solder touch-up, replacement of defective or missing components, or repair of the copper substrate.

6.3.3 Fabricating with Surface Mount Technology (SMT)

As mentioned, surface mount technology uses an assembly method in which component leads are soldered to lands on the surface of the board rather than into holes running through the board. There are two primary methods shown in Figures 6.9 and 6.10:

For *adhesive bonding and wave soldering,* epoxy or acrylic is first dispensed through a stencil onto the desired locations on the board. Components are then automatically placed on the board surface by a computer-controlled "onsertion" machine at a rate of up to several components a second. The adhesive is cured with heat, UV, and/or infrared radiation to bond the components to the PCB surface. The board is then wave soldered as described in the PIH method. The difference is that in SMT assembly, the components are first shielded before passing them through the molten solder wave.

The reflow method is a more common method that first stencils down the solder paste and a flux binder on the lands of the PCB. Next, the components are "onserted." The flux binder is then baked at low temperatures. The final step, to create strong adhesion, is to heat the solder paste in a solder reflow oven. Boards move on conveyors through heated chambers under controlled conditions. This step remelts the solder sufficiently to form a high-quality mechanical and electrical joint between the component leads and the board's circuit lands. Finally, whichever attachment process is used, the board is put through the standard test/inspection/rework operations described earlier.

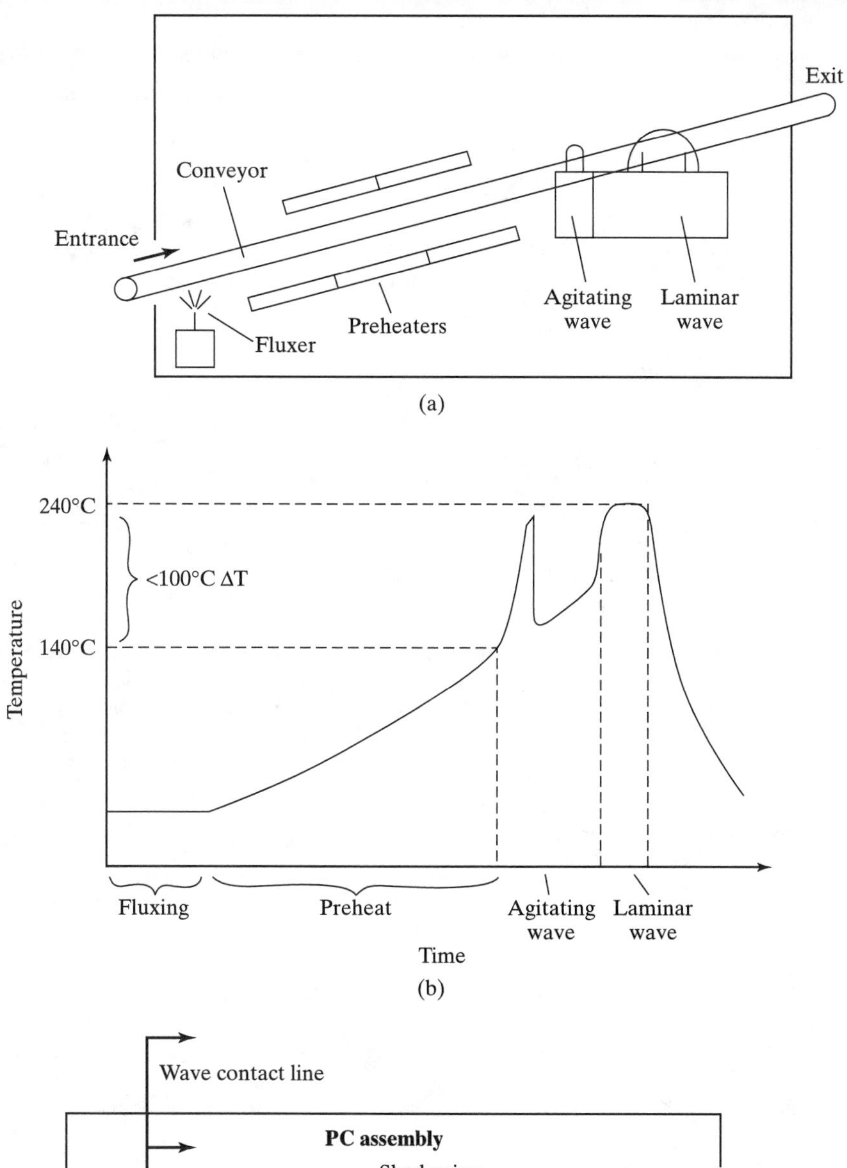

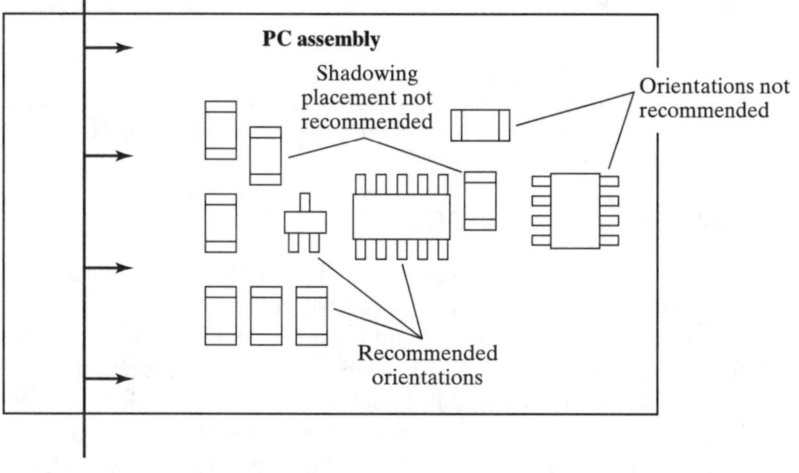

Figure 6.8 (a) Wave soldering equipment, (b) temperature profile, and (c) design rules.

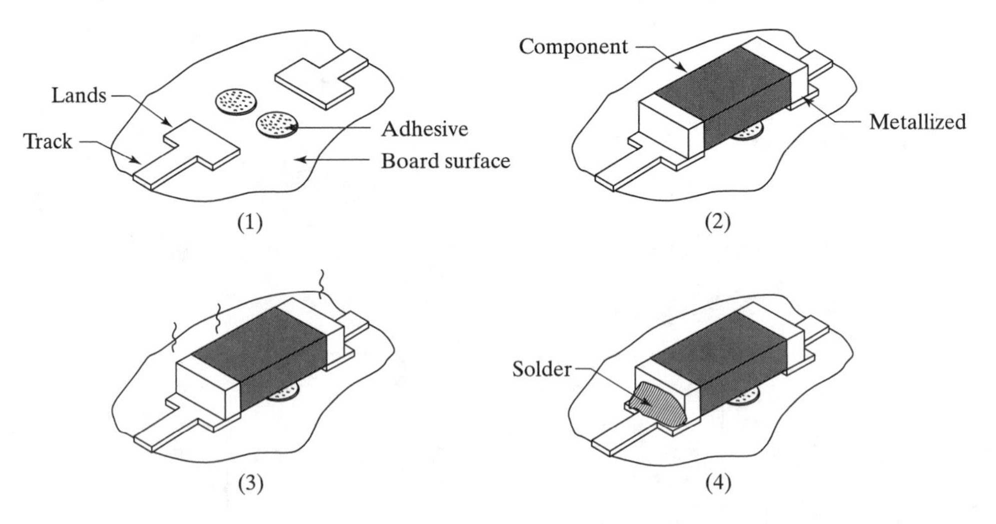

Figure 6.9 Adhesive bonding and wave soldering: (1) adhesive applied to board where components will be placed; (2) components placed on adhesive-coated areas; (3) adhesive cured; and (4) solder joints formed by wave soldering (courtesy of Groover, 1996).

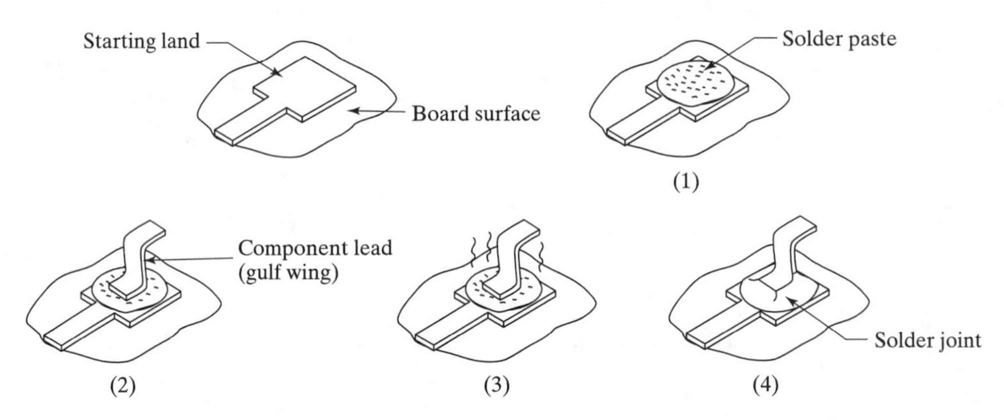

Figure 6.10 Solder paste and reflow method: (1) solder paste applied to land areas, (2) components placed on board, (3) paste baked, and (4) solder reflow (courtesy of Groover, 1996).

6.3.4 Ball Grid Arrays (BGA)

Faster, more versatile computing with miniaturized devices drives the demand for more input/output (I/O) counts between the functional blocks of the circuit. This means there is a greater demand for a product in which the leads have less spacing between them. This may be referred to as fine pitch technology (FPT), which allows, say, the standard size quad flat pack IC with many "gull wing" or "J-leads" to be surface mounted onto the PCB. Unfortunately, there is a natural limit to how close these leads can be placed without a solder short occurring between the finely pitched leads. Also the delicate, fine pitched leads are vulnerable to bending distortions.

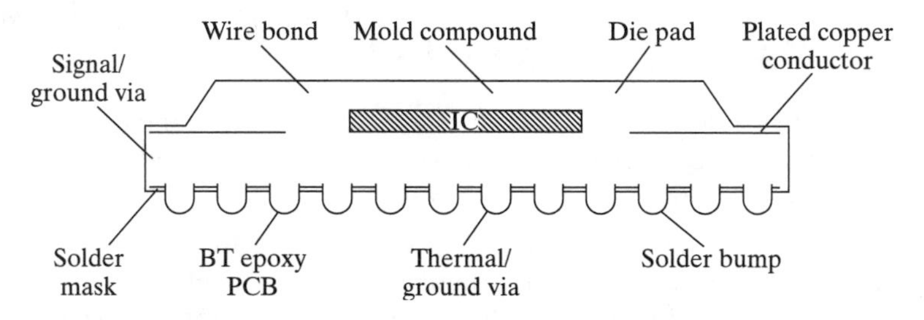

Figure 6.11 The ball grid array (from *Printed Circuits Handbook* by Clyde F. Coombs, © 1996. Reprinted by permission of the McGraw-Hill Companies).

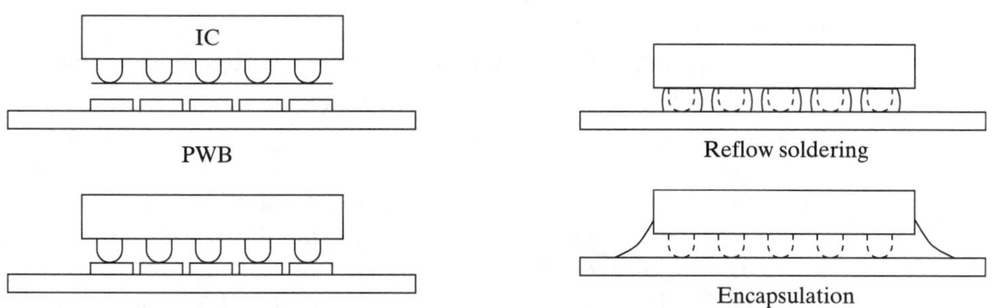

Figure 6.12 In flip chip technology, the solder balls on the chip's face make direct contact with the PC(W)B when the IC is "flipped" onto the board (upper figure). The flux/solder mix (second figure) holds the IC in place until it is reflow soldered onto the PCB (third figure). Nevertheless, the epoxy encapsulation is needed as a stress distribution layer in order to compensate for the different thermal expansion of the silicon IC and the board.

The ball grid array (BGA) and flip chip technology both overcome these difficulties (Figures 6.11 and 6.12). Here, the leads that would otherwise be sticking out from the side of the quad flat pack (QFP) are replaced by an array of solder balls on the underside surface. These are originally made by silk screening, and then the solder is reflowed to create the balls. Leicht (1995) predicts that soon BGA ICs will largely replace QFP ICs with side leads. The trade literature indicates that some manufacturers are presenting their standard quotations only in the BGA format. Mitt and associates (1995) describe the future need for micro-BGAs with an example of a BGA that has a 125 micron pitch and 576 locations for pins in a package that is only 21 × 21 millimeters in size. Evidently, a number of rival techniques will continue to emerge to address this issue.

6.3.5 Direct Chip Attachment Methods, Including Flip Chip Technology (FCT)

It is understandable that products like today's cellular phones, PDAs, and thin-format laptops relentlessly push for small chips and smaller, more efficient PCBs. The result is this next family of direct chip-on-board (COB) methodologies, which takes BGA one step further. The following three methods are extensively used for mounting a bare chip directly onto a PCB:

- *Direct wire bonding* attaches the chip to the lands on the board. The wire bonding requires more assembly costs, but the reliability is high once the manufacturing processes are established.
- *Tape automated bonding* (TAB) also attaches the chip directly to the lands on the board. The chips are first held by a metal lead frame that is printed onto a polyamide film that looks like a roll of photographic film with sprockets along its edges. The wires (left side of Figure 6.13f) are thermocompression-bonded to gold-tin "bumps" on the I/Os of the IC (center Figure 6.13f). In later assembly, this prefabricated tape indexes passed the desired locations, and the outer leads of the TAB-bonded die are thermocompression-bonded with a thermode to pretinned pads on the interconnects of the PCB. The thermode is shown in Figure 6.13f.
- *Flip chip technology* (FCT) represents today's culmination of methods that package ICs and provide connections to the PCB. Pads on the top of the chip are directly bonded to the substrate by flipping the chip and placing it with its pad array downward onto the corresponding pads on the PCB. The literature shows that IBM tried these methods in the 1960s (see Gilleo et al., 1996), together with a few peripheral contacts between the chip and the board. A schematic of the FCT process is shown in Figure 6.12. It is emphasized that the epoxy underfill in the fourth sketch is used not just to fix the chip on more strongly but to distribute thermal stresses more evenly. Without this stress-distribution layer, the difference in coefficient of thermal expansion between silicon chip and copper coated PCB can cause damage to the individual contacts shown in the upper sketches. Rabaey (1996) points out that power consumption and clock speed may also improve with the flip chip style because the copper or gold interconnect materials on the PCB substrate are typically of better quality than the aluminum interconnects on a chip.

6.3.6 Summary

Figure 6.13 summarizes the major designs. The trend to SMT using BGA is well established, and some observers (see Leicht, 1995) see them becoming the major attachment method in these next few years. FCT is the next generation after BGA and indeed is a fast-growing category. However, Messner (1996) observes that they still represent only a small percentage of all ICs used and, for economic reasons, are used only in situations in which a large I/O count or small pitches are needed. Messner postulates that around the year 2000, the ICs with over 100 I/Os (the type where FCT comes into play) amount to only 10% of the total worldwide IC consumption. Of course as FCT matures, a range of techniques is being investigated to make them easier to use and more cost effective. For example, new attachment methods will increase reliability and circuit efficiency. These include deposited metal, mechanical attachment, and conductive anisotropic adhesive (see Palmer et al., 1997).

A final note on flexible circuits—made from rolled copper, polyamide, or polyester film—is added for completeness. Figure 6.14 illustrates four of the typical folding configurations reviewed in more detail by Sheldahl Inc. (1996).

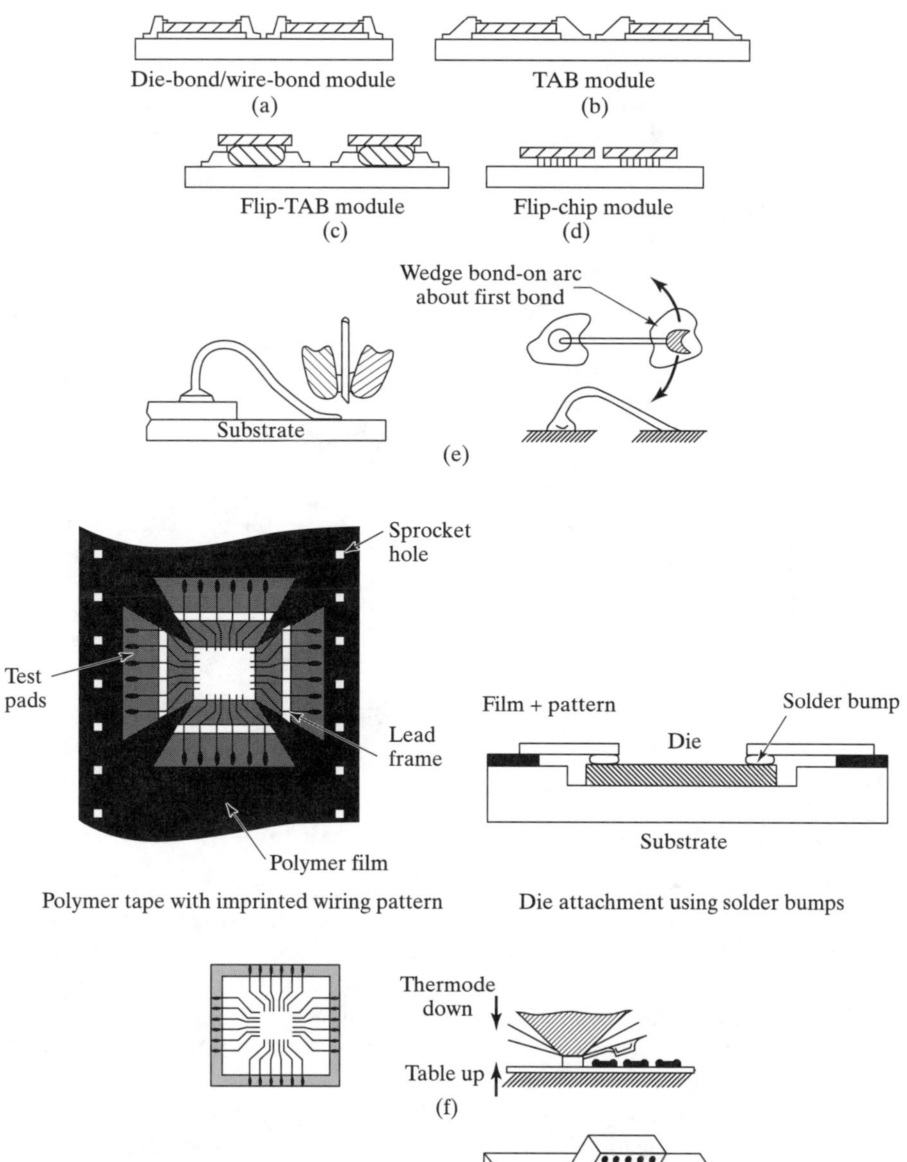

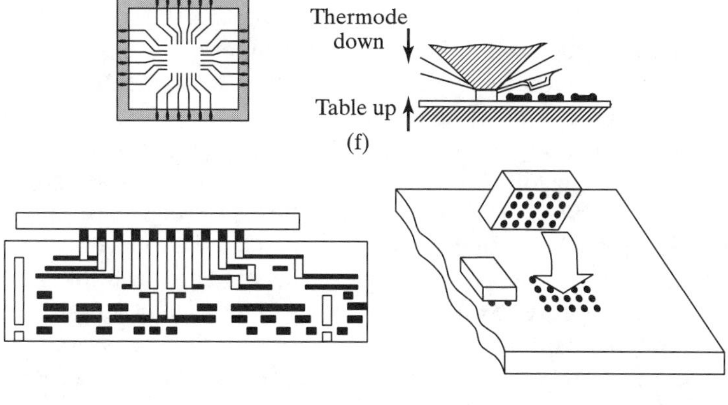

Figure 6.13 Die bonding, tape automated bonding, and flip chip technology (from *Printed Circuits Handbook* by Clyde F. Coombs, © 1996. Reprinted by permission of the McGraw-Hill Companies).

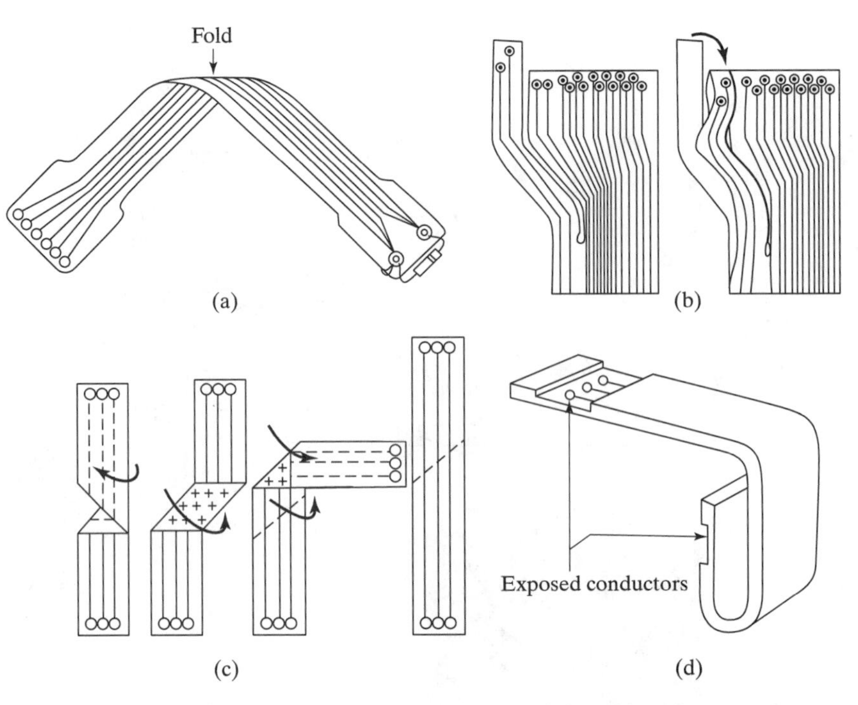

Figure 6.14 Folding configurations for flexible circuits (adapted from literature of Sheldahl Inc., 1996).

6.4 HARD DRIVE MANUFACTURING

6.4.1 Introduction

At the beginning of Chapter 5 the IC, especially the central processor, was colloquially referred to as the "brains" of the system. Pushing the analogy a little further, the primary memory (the cache and dynamic random access memory [DRAM]) might be thought of as short-term memory. By contrast, the secondary memory (the floppy or hard drive) might be thought of as permanent or long-term memory. In any event, this long-term memory needs to be kept safe even when a computer is shut off. Thus, it needs to be kept in a nonvolatile format.

Since 1965, magnetic discs (or disks) with surfaces that can be magnetically recorded over (just like audiotapes) have been used for permanent memory. In addition, since the 1980s, computer manufacturers have incorporated thin-film coils (usually called heads) into their hard disc drives for the recording heads. These extremely small electromagnetic coils are one of the standard technologies that can be used to write data to, and read data from, the hard disc on a computer. They are the critical link between a computer's virtual desktop and the files stored more permanently on the hard disc. As a user opens, closes, saves, and transfers files, the gentle whirring sounds that can be heard come from these heads and discs doing their job.

The markets for such miniaturized heads and discs are driven by the associated markets for small, high-capacity storage devices for personal computers (PCs), workstations, and file servers. At the time of this writing, a modest PC might use one or two 3.5 inch discs storing 4 gigabytes of data together, all accessible in less than 20 milliseconds. Note, though, that the access time for DRAMs is 50 to 150 nanoseconds.

These numbers, and the data in Figures 6.15 and 6.16, suggest that the design of small disc drives, and hence the design and demand for disc drive heads, will greatly escalate during the next few years. It will make the given data seem comically old-fashioned with each year that passes.

The fabrication and assembly of these miniature heads—on the order of 1.7 millimeters along the longest edge for the so-called nanosliders—are extremely

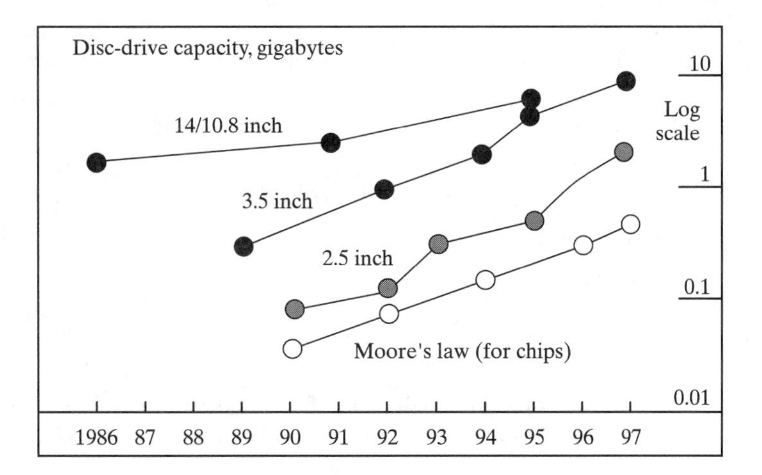

Figure 6.15 Disc drive capacity over the last decade.

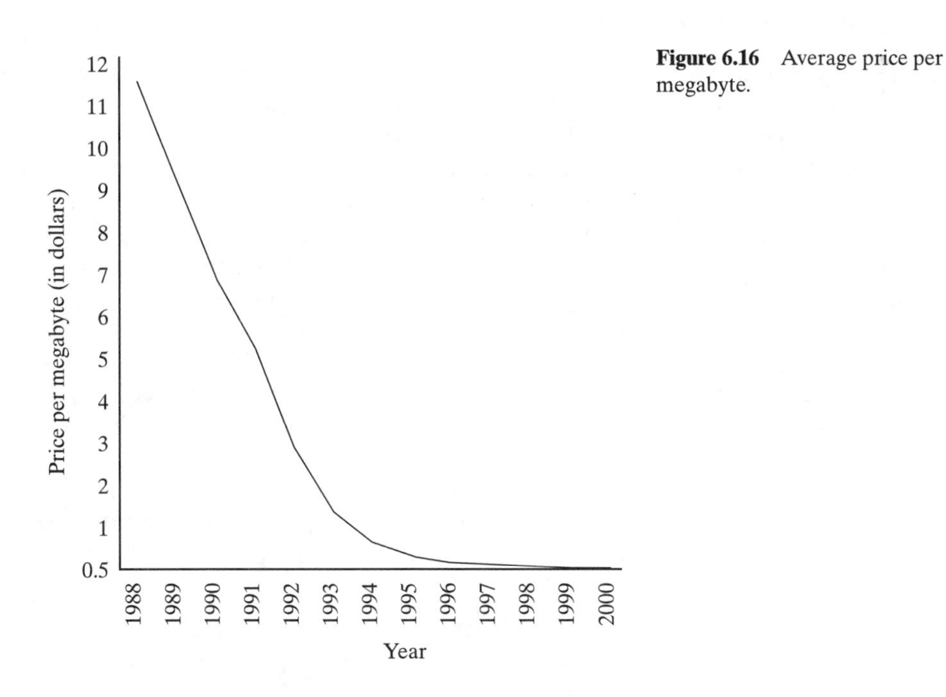

Figure 6.16 Average price per megabyte.

challenging. In some firms it is performed by human operators viewing the components through stereomicroscopes. The assembly of the sliders onto the arms is a painstaking and awkward process requiring great dexterity—also warranting the use of inexpensive labor in developing countries. But as assembly workers look through stereomicroscopes and guide the head assemblies into custom-made jigs and fixtures with tweezers, even well-trained people with great dexterity can damage components and thereby reduce yield. Additionally, many overseas manufacturing sites are difficult to monitor in real time, and changes in product specifications may not be efficiently communicated to foreign assembly plants.

Other manufacturing companies are thus moving in the direction of low-batch, high-precision, flexible assembly work cells for some of the steps described. This requires high investment of capital and is usually feasible only for the largest companies. It also demands robotic assembly systems that are "agile" and able to be retooled or fixtured for a new product design "six months down the road." The pressures in this industry are similar to those faced by the IC and PCB industries:

- Increased miniaturization
- Reduced batch sizes
- Demands for higher quality

Consumer trends such as video-on-demand both in the household and in the transpacific airline seat are some of the biggest drivers. Once again these remarks illustrate the observations described in Chapter 2, that *consumer pull* and *technology push* fuel each other in a spiral of growth.

6.4.2 Design and Manufacturing

The inner components of a hard drive resemble a miniature version of an old-fashioned juke box, with memory discs instead of records. The magnetic coil, or head, shown on the left of Figure 6.17 is held at the end of a small arm. Commonly there are several discs in a stack and so several arms and heads. The arms and heads track the disc, just as if they were playing Elvis's old 45s. The difference is there is no "pickup needle": the arm/head assembly skims over the disc surface. In the figure, the first bit is "1" and written as (← →). The second bit is differentiated from the first by opposite polarity (→). Since it is going to be a "0," the bit is written (→ →).

In the factories that make disc drives, these heads are called thin-film heads, read-write heads, or sliders. Actually, at the speed at which they operate in a disc drive, the heads literally fly over the surface. The majority of practicing engineers who are familiar with this read-write technology share a general sense of awe regarding the level of miniaturization that has been achieved and is still being advanced upon. They describe the read-write mechanism as a 747 airliner skimming over a forest of tree tops with perfect precision. Perhaps more remarkably, disc drives mostly keep on working even after people drop their laptops on the floor or get dust in them.

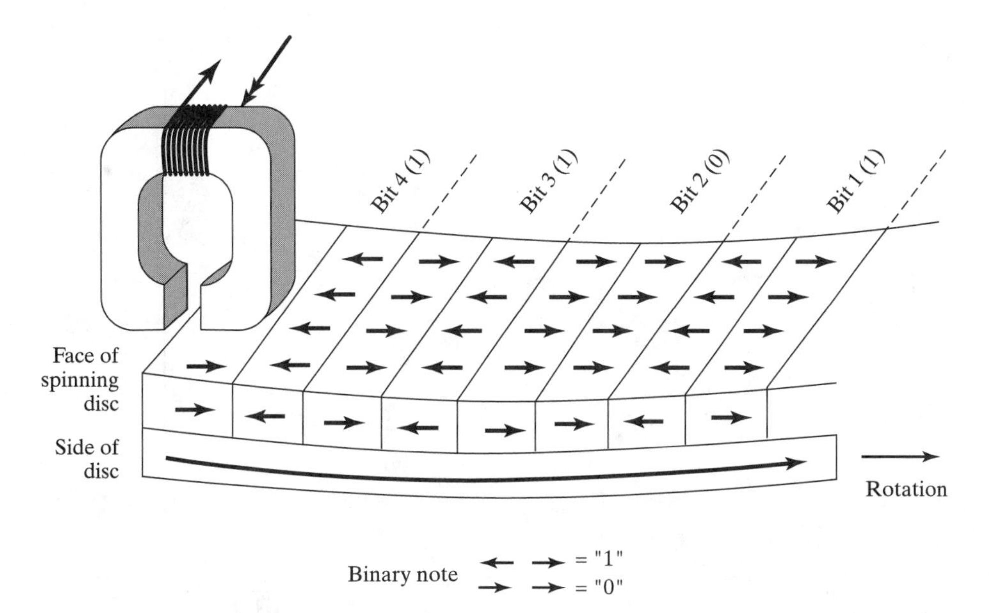

Figure 6.17 Coil reading data to a disc with the convention that "1" bits have opposite polarity and "0" bits have like polarity. The first bit is "1" and written as (← →). The second bit is differentiated from the first by opposite polarity (→). Since it is going to be a "0," the bit is written (→ →). The third bit is going to be a "1." It starts with (←) to differentiate it from the second bit and continues (→ ←) to maintain the convention that "1" bits have opposite polarity and "0" bits have like polarity. The fourth bit is going to be "1" and stays with (→ ←).

6.4.3 Fabrication Steps of Standard Read-Write Heads

Figure 6.18 illustrates the basic steps for head fabrication used by Read-Rite (1997).

- *Wafer fabrication:* Wafers are fabricated and polished flat from raw alumina/titanium (rather than silicon). The front-end fabrication processes are similar to semiconductor wafer fab: photolithography, electrodeposition (plating), thin-film deposition, and ion beam processes are used to fabricate a matrix of heads on the wafer.
- *Coil/circuit fabrication:* The circuitry of the heads is created using electroplating and sputtering processes onto the raw alumina/titanium wafers. Electroplating creates copper coils and tips shown in Figure 6.18, whereas sputtering is used to lay down gold conducting pads. The detailed steps are as follows:
  (a) Lithography is used to selectively create the patterns where the coil's pole tips and yoke will be built up by subsequent electroplating. A similar process to the one shown in Figure 5.22 for lithography is thus used to define the shape of the coil shown in Figure 6.14, but at least 20 layers of copper are needed to build up to obtain the required thickness. Steppers

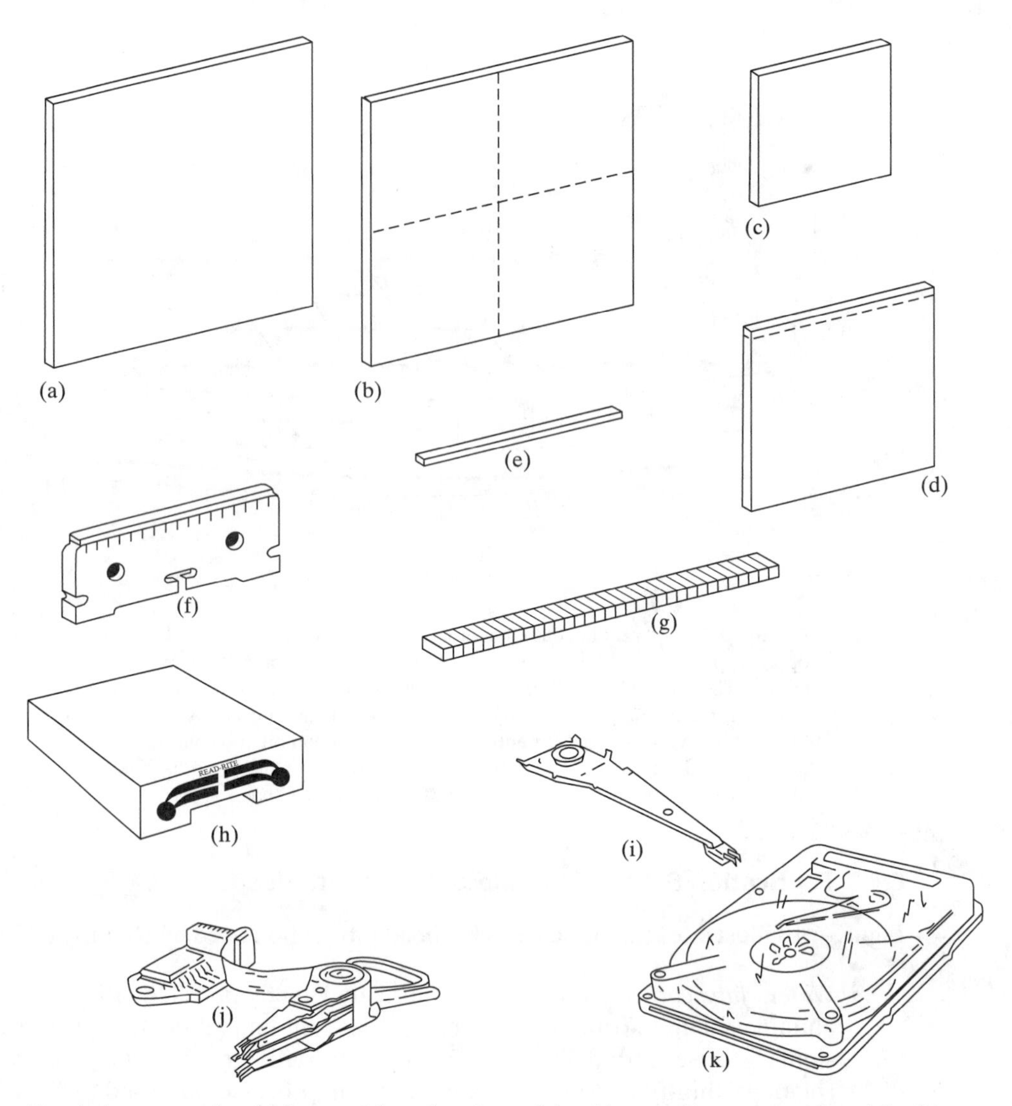

Four inch square wafers (a) have serial numbers for each head laser written on one side and the thin film elements are built up on the other side. The wafers are scribed and sawed into four "quads" (b and c) for further machining. Each quad (d) is sawed into rows of sliders (e), which are then attached to tooling bars (f) for further machining. After machining, the heads are individual sliders (g and h) and are removed from the tooling bar. The sliders are assembled into head gimbal assemblies (i), or HGAs. The HGAs are assembled into head stack assemblies (j), or HSAs. The HSAs are shipped ready for installation into the customers's disc drive (k).

Figure 6.18 Steps in the fabrication of magnetic read and write heads (courtesy of Read-Rite Corporation).

that provide accuracies in the order of 0.2 micron are thus needed to maintain the "verticality" of the walls of the coil.

(b) Electroplating takes place in wet acid plating baths where a strong current drives the copper onto the magnetic wafer in these selected regions.

(c) Sputtering takes place in an argon filled vacuum chamber. A strong electric field ionizes the argon, and a stream of heavy ions bombards a target material to create gold ions that are deposited on the charged wafer. In the manufacture of the magnetoresistive heads in the next section, thin nickel-iron layers that form the read elements are also deposited by sputtering.

The wafers are probed to ensure that each potential head meets specifications. In the design of nanoslider wafers, there are as many as 12,000 individual elements on the wafer that require testing.

- *Slider fabrication:* The wafers are sliced (each die contains the circuitry for one head) and polished down to extremely fine tolerances. As part of this process, strips or "rows" of heads are cut from the wafer and attached to tooling bars for handling and mechanical processing. The rows mounted on the bars are removed and reduced to individual heads for additional processing, testing, and further assembly.
- *Head gimbal assembly (HGA):* The heads are attached to an arm fixture or "flexure," which holds the head in place over the rotating disc. This is one of the most painstaking aspects of the manufacturing cycle. Small electrical wire assemblies (consisting of a twisted pair of wires, covering tube and contact pad) are attached to the slider, the slider is glued to the arm, and then protective packaging and general wiring are added.
- *Head stack assembly (HSA):* HSAs consist of up to 30 total parts. The HGAs, the actuator coil, and a flexible printed circuit cable are mounted on the "E-block" such that the heads can be positioned within the disc drive. The HSA also includes a read/write preamplifier and head selection circuit and will include other miscellaneous parts such as bearings, a head-retraction coil, and a connector, depending on the design of the disc drive.

6.4.4 Fabrication of Magnetoresistive Heads

A noticeable "jump" in feasible storage density occurred around 1991 with IBM's commercial introduction of magnetoresistive (MR) heads.

Figure 6.19 shows how the MR head differs. Although a standard copper coil (on the top) is still used for the "write" function, the "read" function is now carried out by the MR sensor in the center of the figure. The commercial significance is that this "merged MR head" is much smaller, and the magnetic domain on the disc can be correspondingly smaller—hence increasing areal density.

Although the processing operations are still similar, the design of the MR head involves a separate read and write element formed on top of each other and sharing the same material layers.

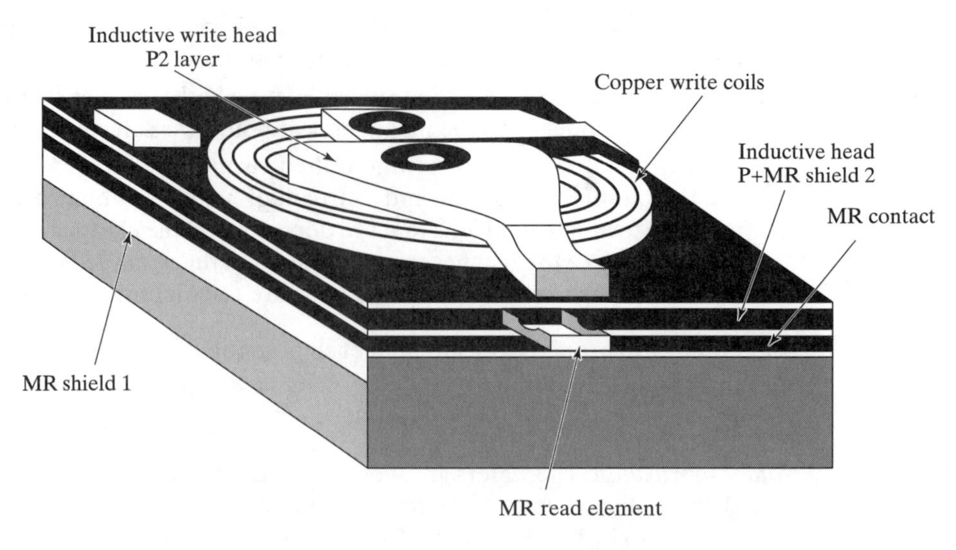

Figure 6.19 Magnetoresistive head operation.

This multilayer sandwich arrangement is shown in Figure 6.19. The copper write coil is shown at the top of this figure, and the MR read element is further down the stack.

- The write coil is copper. It is laid down by the same photolithography and plating steps as described earlier. However, now that it only has to perform the write function, it is much thinner and easier to make than a head that must perform both the write and the read functions. The head requires fewer copper coils, material layers, photolithography steps, and tolerance controls. This in itself also leads to fewer complications in manufacturing and a better yield.
- The read element is nickel/iron and is laid down by sputtering. The NiFe alloy exhibits a change in resistance as it passes over a magnetic field: this is the MR effect. Shielding layers protect the element from other magnetic fields. One shield is shown to the left of the MR sensor in Figure 6.20. The other shield is actually merged with and part of the write coil. Further developments in the material itself lead to the "giant MR effect." The pickup sensor is a multilayer sandwich with a thin metal interlayer able to respond to even smaller magnetic fields.

As before, an inductive write element writes bits of information to magnetically biased regions within radially concentric tracks on the disc. When it is necessary to read the data back, the MR sensor is used rather than another inductive head. The MR sensor picks up this magnetic transition (or flux reversal) between bits, causing the magnetization in the MR sensor to rotate. The read current is indicated in Figure 6.20. This rotation is detected as a resistance change by a precision amplifier, which then produces a stronger signal to the disc drive output.

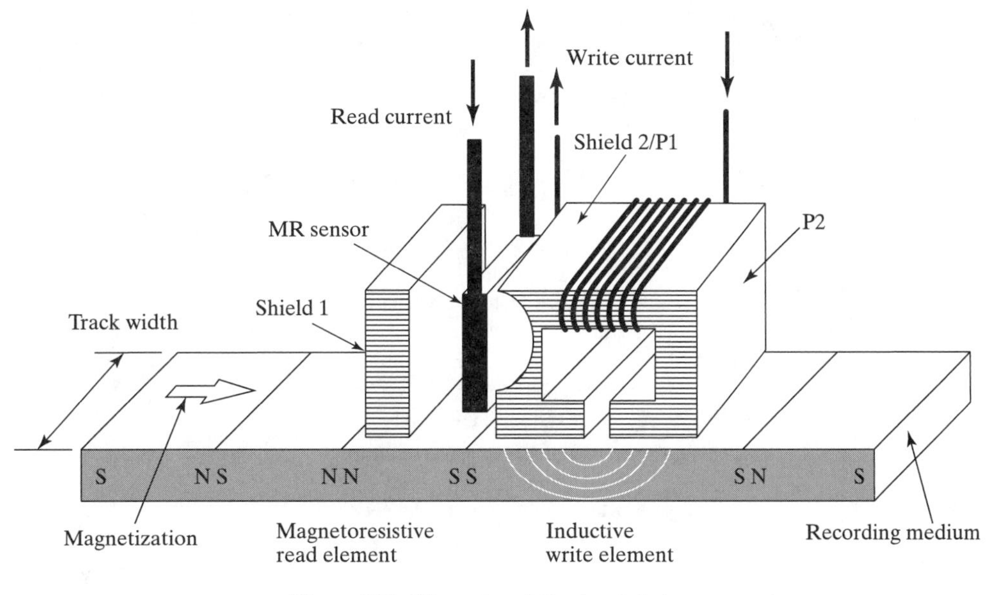

Figure 6.20 Magnetoresistive head design.

6.5 MANAGEMENT OF TECHNOLOGY

6.5.1 The Culture and History of the Computer Industry

The evolution of the electronic computer over the past six decades from a rare, highly specialized item to a commodity has repeatedly reshaped the computer industry (Stern, 1980; Bell, 1984). Table 6.2 shows some of the main milestones.

As one example, in 1953, having assembled transistors with other discrete components, IBM decided to make a prototype and investigate the market for the Type 650 magnetic drum calculator. The Type 650's computing power was roughly equivalent to a modern VCR and rented for $3,250 per month, equivalent to $20,000 in today's dollars. IBM was then a large, slow-moving corporation with an appropriately conservative marketing group. The market for a commercial computer was estimated to be small. But when the stalwart Type 650 was finally withdrawn from the market in 1962, several thousand had been sold. In this situation, IBM successfully *crossed the chasm* described in Chapter 2 and created a unique and viable product: one of the world's first mass-produced computers. By 1964, IBM launched the /360 series of machines, which incorporated a wide range of possible products for a wide variety of users. But it should be stressed that at that time, the users of this computing world were the academic/scientific community on the one hand and commercial/business corporations on the other. The average consumer and the average family certainly did not sit down in the evening to write a letter on a computer, let alone read e-mail or surf the Web.

For this average consumer, the most significant breakthrough in computing did not occur until the development of the microprocessor. Throughout the 1960s,

TABLE 6.2 Some of the Milestones in the Development of the Electronic Computer

Decade	Prototype	Comments	Figure(s)
1940s	ENIAC	Vacuum tubes of the electronic numerical integrator and calculator	Eckert and Mauchly
	EDVAC	Stored programs of the electronic discrete variable automatic computer	von Neumann
	EDSAC	Stored programs of the electronic delay storage automatic calculator	Wilkes
	Transistor	Semiconductors: eventually replace vacuum tubes	Shockley, Brattain, and Bardeen
1950s	UNIVAC1	Universal automatic computer launched as the first commercial electronic computer	Eckert and Mauchly
	650 and 701	First systems launched by IBM	IBM
	IC	Combined functions on one chip	Kilby
1960s	/360 Series	First "family" of machines of widely different capability and cost	IBM
	PDP8	Minicomputer sold for <$20,000	DEC
	CDC 6600	First supercomputer	Cray
1970s	4004	First microprocessor	Hoff and Intel
	Apple II	First personal computer	Jobs and Wozniak
	Internet	Interconnectivity for the academic community	DARPA
1980s	IBM PC	Best-selling personal computer + multimedia	IBM, Intel, and Microsoft
	The Web	Networked computing	CERN/Berners-Lee
1990s	Compaq	The PC as a commodity	Pfeiffer
	Mosaic	Interconnectivity for mass consumption	NCSA/Andreesen (Netscape)
	Java	"Write once run anywhere" computing	Sun
	PDAs	Handheld devices and network computers (NC)	Palm Pilot
~2000	Wireless	Merging of computing/wireless/information appliances into wide array of consumer products (sometimes called a "post-PC" phase)	Nokia and Motorola

advances in IC design and fabrication methods set the stage for the first microprocessor, the 4004 (shown at the beginning of Chapter 5), developed at Intel by a group led by Ted Hoff. It was based on two key innovations:

- All logic was on one chip.
- The device was programmable by software.

The microprocessor made possible a huge *middle ground of general-purpose machines,* most notably the personal computer (PC) and high-performance workstations. The distinction between these categories is now blurred, particularly as powerful networked PCs rival the functionality of professional workstations.

By 1977, Jobs and Wozniak built and sold the Apple II as a commercial product, and by 1981, the IBM-PC was announced, using the Intel 80 × 86 microprocessor. Although its PC was a tremendous success, IBM did not fully capitalize on its brilliant new product. It was not appreciated at the time that *modest computing power on everyone's desk would be more attractive than greater computing power centralized on mainframes.* As a result, two historic choices were made by IBM:

- It subcontracted the PC operating system to Microsoft.
- It subcontracted the IC fabrication to Intel.

Many professional and amateur analysts now look back and criticize these two choices, which, at first glance, appear to have been shortsighted. At the time, however, these subcontracting arrangements were considered sensible because they minimized research and development (R&D) investment in areas that were not IBM strengths. And as time has gone on, there has in fact been an increasing trend in all industries toward subcontracting. While it is easy to look back and criticize IBM for not getting the maximum benefit from the first PC, on balance, the decision made at the time was consistent with today's conventional wisdom—namely, to focus on one's core strengths and main markets while outsourcing all other operations. Today, for most of the well-known, brand-name computer companies (e.g., Compaq, Dell, Hewlett-Packard,) the trend is toward more outsourcing to companies that provide specialized manufacturing services (e.g., Solectron, Flextronics, SCI systems).

Nevertheless the commercial significance of the computer operating system and main microprocessor is now obvious to anyone who uses what has become known as the "Wintel" de facto standard. The significance of keeping pace with IC technology is illustrated in Figure 6.21, which shows the tremendous increases in IC complexity and power over the last three decades.

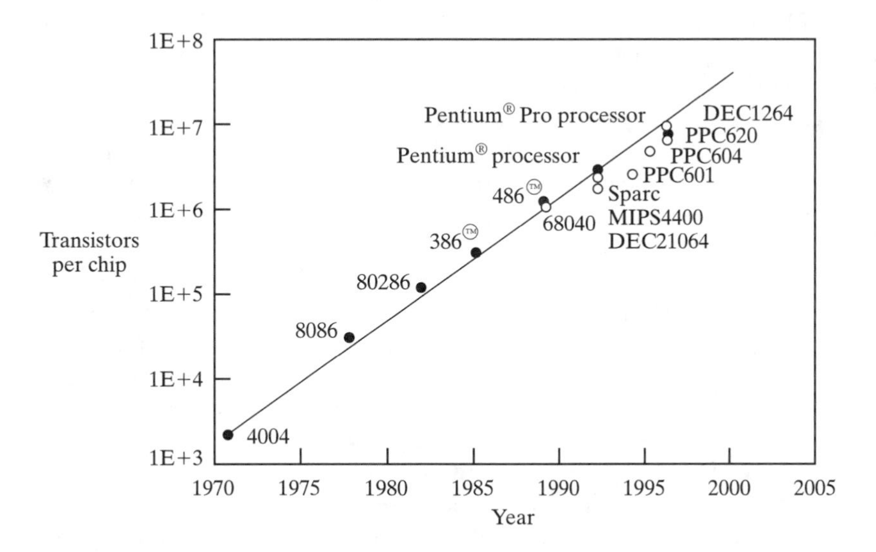

Figure 6.21 The transistor count in the central processing unit (CPU) from 1970 and projected to 2005.

6.5.2 The Present

From a management of technology point of view, the computer industry is about coping—or failing to cope—with *change.* With new technologies and applications emerging all the time, the dust barely settles after one revolution before another is well under way. In this environment of constant change, staying in business requires mastery of all aspects of the technology management process, from research and development to design, manufacturing, and marketing. The recent history and some of the present situations in the computer industry illustrate how difficult this can be, even for some top performers.

- IBM and DEC suffered during the early 1990s because they were large conservative organizations, overcommitted to a mainframe philosophy.
- Apple lost market share throughout the 1990s. Analysts and economic observers cite many possible reasons, which usually include a closed proprietary operating system that deters third party software development (see the comments on VHS versus Betamax in the preface), low supplies of the best-priced products at critical selling times during the year, and a neglect of design for assembly manufacturing (DFA/M). Apple's future still remains uncertain at the time of this writing despite the captivating aesthetic designs of recent products.

In contrast, Dell, Compaq, and Gateway have boosted profitability and captured the market lead by (a) redesigning their products to aggressively cut costs and (b) improving their manufacturing productivity with DFA/M. To hold on to their present lead, such companies have also introduced major innovations in their supply chains.

Dell in particular has become famous for the "direct sales" model. Capitalizing on a well-organized Website, the company builds each PC to order. This has the benefit of eliminating the middleman—the computer dealer. This direct sales approach delays commitment on the final product configuration until the last possible moment. In this way, unnecessary inventories do not build up, and subcomponents can be selectively stocked based on the most popular configurations. Other examples of this strategy are the European assembly plants in the Netherlands, which build computers for the multilingual European market. Again, by delaying commitment to the very last minute, a company does not get stuck with too many keyboards or software applications in the wrong language.

Dell also minimizes working capital and maximizes the return on it by using a technique known as a *negative cash conversion cycle.* This means that a consumer pays Dell for the assembled computer and FedEx shipping costs long before Dell pays its subsuppliers for the parts. This has a double benefit given that the price of subcomponents is constantly tumbling. Curry and Kenney (1999) describe the loss-of-value dynamics of critical subcomponents in the PC industry. They report that many of Dell's competitors continue to lose market share because they are unable to manage time as effectively and consequently buy subsupplies at higher prices than they can later package and sell them for. By delaying payment to their subsuppliers until the PC is sold to the consumer, Dell effectively buys the subcomponents at the last-minute (actually postminute) market price.

6.5.3 The Future

In summary, the versatility and the power of PCs have now made them the workhorses of the information age. All professionals now depend on their desktop, laptop, or handheld computers for word processing, e-mailing, and access to Web-based information and services.

Despite this range of possibilities, fewer and fewer of us are in any way overwhelmed by computers. Or if so, we try not to let it show! The real news is that computers are not just smaller and faster; they are also a lot cheaper. The basic computer, especially in the form of the PC, is a common commodity, well on its way up the market adoption curve in Chapter 2. The PC is not quite on a par with pork bellies on the Chicago Stock Exchange, but it is getting there. The "sub-$1,000 machine" has become the center of today's consumer market, and price performance has become the main focus for the manufacturers. Handheld, networked wireless devices are more recent additions to the marketplace. These take advantage of the wireless application protocol (WAP), which allows PDAs to easily access the Internet.

In the eyes of many analysts, these are heralding a "post-PC age" of convenient devices that boot up directly to specific applications, rather than have the user stumble through the icons on a PC desktop just to get to the Internet.[2]

As a result, multimedia and communications technologies are merging with computers, and industry boundaries are dissolving. Newly formed business alliances—sometimes called "virtual corporations"—are clashing for technical and market leadership. These new alliances usually consist of *two or more* from the following list of constituents:

- PC and PDA makers of Silicon Valley and other high-tech regions
- Chip makers, with Intel being the obvious giant
- Operating system and software developers, dominated by Microsoft
- Telephone companies and network suppliers
- Television and cable TV companies
- Hollywood studios, backed up by special effects companies

At first glance, it is difficult to tell which type of alliance will come out on top, *and what the computer of the future will look like and what it will do.* But based on the history of the IC, the microprocessor, the PCB, and especially the computer itself, it is certain that rapid dramatic changes are in store.

Consider, for example, the following thought exercise. Choose the most likely scenario for the next several years:

Option 1: "WebTV" will offer even more powerful set-top boxes and smart keyboards for "interactivation." Television will have so much more interactivity and

[2]For example, Alan Kessler, president of 3Com's Palm Computing, is quoted as follows in *U.S. News and World Report,* December 13, 1999, p. 52: The new mantra is "give them [consumers] just what they need when they need it" rather than respond to the "old" consumer demand of: "Give me more memory. Give me more power. Give me more complex software."

bidirectional communication that the standard desktop PC will be made redundant.

Option 2: The Web-based PC will become a high-resolution "information furnace" (a buzzword courtesy of Avram Miller of Intel). Voiceover modems, video-telephone links, live concerts by musicians, MP3, and high-quality video images will make TV obsolete.

Option 3: Neither TVs nor PCs will diminish in popularity. Rather, consumers will continue to have high-quality TV entertainment in the living room and high-quality information processing in the home office.

Option 4: The PC in its current instantiation will disappear, and its central microprocessor will essentially be absorbed internally as the central *information motor* into all such *information appliances.* Norman (1998) and other observers make the analogy that a stand-alone *electric motor* was once, in the 1920s, a consumer product in and of itself. It was advertised in the Sears catalog as something "every home should have," connectable to washing machines, refrigerators, and hair dryers. Now, in the passing of time, the electric motor is of course just as important, but it is not seen as an external stand-alone device; rather it is buried deep inside consumer appliances and taken for granted. So this may be the future of today's PC. It will be "reduced to a powerful microprocessor" and just be the central "information motor" for TVs, PDAs, communication devices, and information appliances. This idea is now a recurring theme in the popular magazines of the computer industry, such as *Wired, PC Computing,* and *Red Herring* (1998).

6.5.4 Philosophy

Archaeologists and historians traditionally view the growth of civilization in terms of the predominant technology of particular eras. Chapter 1 mentioned the Stone Age, the Bronze Age, the Iron Age, and the Steel Age of the industrial revolution. Observers of the history of computing also try to document the chronological "eras" that summarize the rise of the computer from the early mechanical computers, to the vacuum tube era, to the IC, and to the microprocessor (e.g., see Stern, 1980; Bell, 1984; Patterson and Hennessy, 1996a; *Economist,* 1996).

Partially based on these other writings, the present text hypothesizes that the history and anticipated future of commercial computers may be divided into four distinct phases. Note that these commercial developments could not have been launched without some truly revolutionary scientific research discoveries, such as the transistor and the planar transistor in the period beginning in 1947. Usually, the *commercial development phase* is 5 to 10 years behind the *scientific discovery phase* and prototype use by the academic community. This is certainly true of the World Wide Web (see Berners-Lee, 1989). Actually, this particular gap is 25 years if today's "dot-com-fever" is measured from the beginning of the DARPAnet and its use in the academic community.

6.5.4.1 The Iron Age (1953 to 1980)

The reign of the mainframe computer.

6.5.4.2 The Desktop PC Age (1981 to 1991)

The age of stand-alone desktop personal computers, augmented by CD-ROM.

6.5.4.3 The World Wide Web Age (1992 to 2001)

The age of multimedia applications carried to a global communication level well beyond the limits of an individual user's desktop PC. It involved the merging of the World Wide Web, CD-ROM, TV, telephone, workstation, and wireless communications technology.

6.5.4.4 The Integrated Man-Machine Age (2002 to 2020 and Beyond)

For 1999, The *Economist* (1999) states that U.S. consumers purchased 16.9 million PCs—17% more than in 1998—raising household penetration to 52%. However, the same and other observers indicate a possible reduction over the next few years due to several factors: (a) overcapacity; (b) reducing demand for upgrades—many users have "powerful enough" machines; and (c) the rise of PDAs, smart cellular phones, and networked computers (see *Red Herring,* 1998).

Obviously the PC "ruled" in the desktop age (1981–1991) and was the key workhorse or platform for the World Wide Web age (1992–2001). However, in the new age of man-machine devices, distinctions and interfaces between human beings and their communication devices are now blurred. As a result, the monolithic PC solution to life will fade, just as the monolithic mainframe faded.

Today's wireless-handheld combination of a cellular phone and PDA is only the beginning of a new age of man-machine devices. *Wearable computers* are already established devices in advanced applications. Weiss (1999) provides a popular review. Akella and associates (1992), Smailagic and Siewiorek (1993), and Finger and colleagues (1996) provide more scientific details. Extrapolating from these existing prototypes, how might the following list of technical developments influence future products?

- Assuming success with the developments in x-ray lithography and so forth described in Chapter 5, it is reasonable to assume that more than a billion transistors will be packaged on a logic chip in the near future, opening up a whole new range of computing capabilities at a scale never before possible.
- With billion-transistor chips, all the technologies of the World Wide Web age might well be packaged into a voice-activated, hearing-aid-sized device that can be worn at all times.
- Beyond 2020, with advances in engineering biologically compatible materials, it might well be possible to embed such a tiny but powerful electronic device in the lining of the scalp; a subcutaneous radio modem would be a realistic option.

Several decades ago, the philosopher and physicist Heisenberg was one of the first people to futurize about such possibilities. He used the following metaphor to conjecture about our future. Snails, crabs, and similar creatures can exist and live without their protective helmets (shells) but not very effectively. Is it possible, Heisenberg then asked, that human beings are living beneath our full potential? If we were equipped with a kind of "information helmet"—using these technologies of the integrated man-machine age—then we would dramatically increase information access and expand the effectiveness of our lives.

When these ideas are discussed in a lecture, many people squirm at the thought of embedding a microprocessor and a radio modem under their skin. People seem to accept and welcome external devices like hearing aids and pacemakers, but it does seem a threatening "jump" to go to internal devices. However, other philosophers have postulated that humankind could have discovered the wheel more quickly if our thinking patterns were not blocked off by observing the rest of nature where no wheel-like devices are found.

Perhaps we are also blocking the thought of internally embedded electronic devices for the same reason—namely, that they do not appear anywhere else in nature. If we can look beyond this threatening jump—to devices that will improve our personal communication networks, our ability to compensate for injury, and our general health and immunity support functions—then perhaps the IC and the microprocessor will indeed reach out to an even wider range of tasks.

6.6 GLOSSARY

6.6.1 Ball Grid Array (BGA)

Development of individual SMT components, where the connections are made underneath the chip instead of on the perimeter. The term *ball grid* refers to the small balls of solder used to make the connections.

6.6.2 Bus

The bus connects the microprocessor, disc drive controller, memory, input/output ports, and other parts of the system.

6.6.3 Central Processing Unit (CPU)

The main arithmetic and control units plus working memory.

6.6.4 Compiler

A program that translates from high-level problem-oriented computer languages to machine-oriented instructions.

6.6.5 Design for Assembly and Manufacturing (DFA/DFM)

Strategy of lowering cost by aiming at lowering assembly time and reducing the number of subcomponents. Design for assembly involves three key ideas: reducing the number of subcomponents, increasing their quality, and simplifying the assembly operations between subcomponents.

6.6.6 Flip Chip Technology (FCT)

Extension of SMT/BGA that offers even greater packing density. The IC is turned over and placed face down on the board before creating the circuit connections.

6.6.7 Head Gimbal Assembly (HGA)

An assembly of a read/write head on an arm. This holds the head in place over the rotating disc.

6.6.8 Head Stack Assembly (HSA)

An assembly of the HGAs (above), the actuator coil, and a flexible printed circuit cable. The HSA also includes a read/write preamplifier, a head selection circuit, and other miscellaneous parts.

6.6.9 Interconnection

The process of mechanically joining devices together to complete an electrical circuit. Also, the conductive path needed to connect one circuit element to another or to the rest of the circuit system. Interconnections may be leads, soldered joints, wires, or another joining system.

6.6.10 Known Good Die (KGD)

A semiconductor die that has been tested and is known to function properly according to specifications.

6.6.11 Lands

Small solder lands/regions of the PCB that provide for the connection of individual ICs and components.

6.6.12 Multichip Modules (MCM)

A device containing two or more packaged ICs mounted and interconnected on a substrate.

6.6.13 Pin-in-Hole (PIH)

A PCB assembly method that involves inserting the leads of components into holes in the board, clipping, and soldering the leads into place.

6.6.14 Printed Circuit Board (PCB)

Also called a printed wiring board (PWB), this is a rigid insulating substrate with conductors etched on the external and/or internal layers. PCBs include single-sided, double-sided, and multilayer boards. A raw "starter board" is a PCB without components attached to it.

6.6.15 Printed Circuit Board Assembly (or Printed Wiring Assembly [PWA]

A PCB with all components mounted and interconnected on it.

6.6.16 Sliders

The heads, or magnetic coils, of the read/write unit of the disc.

6.6.17 Substrate

On a PCB, the base material that provides a supporting surface for etching circuit patterns, as well as attaching components.

6.6.18 Surface Mount Technology (SMT)

The process of attaching components directly to the surface of a PCB. Increasingly, SMT is replacing the older pin-in-hole method.

6.6.19 Tape Automated Bonding

A process in which precisely etched leads (supported on a flexible tape or plastic carrier) are interconnected to the chip or a substrate by a heated pressure head. This process simultaneously creates a bond for all leads at once.

6.6.20 Test Coupon

A preset pattern of copper pads and/or holes for testing during manufacture.

6.6.21 Tracks

Parts of the PCB that provide the interconnections among components and various ICs.

6.6.22 Wave Soldering

Technique of collectively soldering the components to the printed circuit board by passing the board over a standing wave of molten solder that fixes the leads of the components.

6.7 REFERENCES

ACIS Technical Overview. 1999. *ACIS geometric modeler. Programming manual.* Boulder, CO: Spatial Technology Inc.

Akella, J., A. Dutoit, and D. P. Seiwiorek. 1992. A prototyping case study. In *Proceedings of the 3rd IEEE International Workshop on Rapid System Prototyping.* Research Triangle Park, NC: IEEE.

Allen, W., D. Rosenthal, and K. Fiduk. 1991. The MCC CAD framework methodology management system. In *Proceedings of the 28th ACM/IEEE Design Automation Conference,* 694–698.

Amir E., H. Balakrishnan, S. Seshan, and R. Katz. 1995. Efficient TCP over networks with wireless links. In *Proceedings of the Fifth Workshop on Hot Topics in Operating Systems.* Orcas Island, WA.

Andrade, A. D. 1996. Acceptability of fabricated circuits. In *Printed circuits handbook,* 4th ed., edited by C. F. Coombs Jr., 35.3–35.41. New York: McGraw-Hill.

Barnes, T., D. Harrison, A. Newton, and R. Spickelmier. 1992. *Electronic CAD frameworks.* Kluwer Academic Publishers.

Bell, C. G. 1984. The mini and micro industries. *IEEE Computer* 17 (10): 14–30.

Berners-Lee, T. 1989. Information management: A proposal. CERN Internal Proposal, March.

Bohr, M. 1998. Silicon trends and limits for advanced microprocessors. *Communications of the ACM* 41 (3): 80–87.

Brodersen, R. W. 1997. The network computer and its future. In *Proceedings of the IEEE International Solid-State Circuits Conference.* San Francisco, CA.

Burstein, A., A. C. Long, S. Narayanaswamy, et al. 1995. The InfoPad user interface. In *COMPCON '95,* 159–162.

Cho, T., G. Chien, F. Brianti, and P. R. Gray. 1996. A power-optimized CMOS baseband channel filter and ADC for cordless applications. *VLSI Circuit Conference Digest 96,* June.

Clark, R. H. 1985. *Handbook of printed circuit manufacturing.* New York: Van Nostrand Reinhold.

Cole, R. E. 1999. *Managing quality fads: How American business learned to play the quality game.* New York and Oxford: Oxford University Press.

Curry, J., and M. Kenney. 1999. Beating the clock: Corporate responses to rapid changes in the PC industry. *California Management Review* 42 (1): 8–36.

Duffek, E. F. 1996. Plating. In *Printed circuits handbook,* 4th ed, edited by C. F. Coombs Jr., 19.1–19.55. New York: McGraw-Hill.

Economist. 1994. A survey of the computer industry, 17 (suppl.): 1–22.

Economist. 1999. A bad business, July, 53–54.

Finger, S., J. Stivoric, and C. Amon, et al. 1996. Reflection on a concurrent design methodology: A case study in wearable computer design. *Computer Aided Design* 28 (5): 393–404.

Fulton, R. E. 1987. A framework for innovation. *Computers in Mechanical Engineering,* March.

Gilleo, K., T. Cinque, and A. Silva. 1996. Flip chip 1, 2, 3: Bump bond and fill. *Circuits Assembly,* June, 32–34.

Green, H. D. 1996. Multichip modules. In *Printed circuits handbook,* 4th ed., edited by C. F. Coombs Jr., 6.1–6.31. New York: McGraw-Hill.

Groover, M. P. 1996. *Fundamentals of modern manufacturing,* 878–906. Prentice-Hall.

Guerra, M., M. Potkonjak, and J. Rabaey. 1994. System-level design guidance using algorithm properties. In *VLSI Signal Processing VII,* 73–82: IEEE Press.

Gupta, R., et. al. 1989. An object-oriented VLSI CAD framework: A case study in rapid prototyping. *IEEE Computer* 22 (5): 28–37.

Guy, E. T. 1992. An introduction to the CAD framework initiative. In *Electro 1992 Conference Record.* Boston, MA.

Inside Read-Rite Corporation. 1993. (Informational brochure. Milpitas, CA: Read-Rite Corporation.

Keller, K. H. 1984. An electronic circuit CAD framework. Ph.D Thesis, Department of Electrical Engineering and Computer Science, University of California, Berkeley.

Lao, A., J. Reason, and D. Messerschmitt. 1994. Asynchronous video coding for wireless transport. *IEEE Workshop on Mobile Computing,* December, Santa Cruz, CA.

Le, M. T., F. Burghardt, S. Seshan, and J. Rabaey. 1995. InfoNet: The networking infrastructure of InfoPad. In *Proceedings of Compcon '95.*

Leicht, H. W. 1995. Reflow soldering and repair of BGAs. In *10th European Microelectronics Conference,* 508–520.

Long, A. C., S. Narayanaswamy, A. Burstein, R. Han, K. Lutz, B. Richards, S. Sheng, R. W. Brodersen, and J. Rabaey. 1995. A prototype user interface for a mobile multimedia terminal. In *Proceedings of the 1995 Computer Human Interface Conference.*

Mead, C., and L. Conway. 1980. *Introduction to VLSI systems.* Reading, MA: Addison Wesley.

Messner, G. 1996. Electronic packaging and interconnectivity. In *Printed circuits handbook,* 4th ed. edited by C. F. Coombs Jr., 1.3–1.22. New York: McGraw-Hill.

Mitt, M., G. Murakami, T. Kumakura, and N. Okabe. 1995. Advanced interconnect and low cost micro stud BGA. In *The 1995 IEEE/CPMT Electronics Manufacturing Symposium,* 428–521.

Nakahara, H. 1996. Types of printed wiring boards. In *Printed circuits handbook,* 4th ed., edited by C. F. Coombs Jr., 3.1–3.14. New York: McGraw-Hill.

Narayanaswamy, S., S. Seshan, E. Brewer, R. Brodersen, F. Burghardt, A. Burstein, Y.-C. Chang, A. Fox, J. Gilbert, R. Han, R. Katz, A. C. Long, D. Messerschmitt, J. Rabaey. 1996. Application and network support for InfoPad. *IEEE Personal Communications Magazine,* March.

Norman, D. A. 1998. *The invisible computer.* Cambridge, MA: MIT Press.

Palmer, P. J., D. J. Williams, and C. Hughes. 1996. Assembly and packaging of conventional electronics. *Process Group Technical Report No. 96/13.1.* England: Loughborough University.

Patterson, D. A., and J. L. Hennessy. 1996a. *Computer architecture: A quantitative approach.* San Francisco, CA: Morgan Kaufmann Publishers.

Patterson, D. A., and J. L. Hennessy. 1996b. *Computer organization and design: The hardware/software interface.* San Francisco, CA: Morgan Kaufmann Publishers.

Rabaey, J., L. Guerra, and R. Mehra. 1995. Design guidance in the power dimension. Paper presented at the International Conference on Acoustic, Speech and Signal Processing.

Read-Rite Corporation. 1997. Technical literature available from 345 Los Coches St., Milpitas, CA.

Red Herring. 1998. The post-PC world, December, 50–66.

Sarma S. E., S. Schofield, J. A. Stori, J. MacFarlane and P. K. Wright. 1996. Rapid product realization from detail design. *Computer-Aided Design* 28, (5): 383–392.

Sheldahl Technical Staff. 1996. In *Printed circuits handbook,* 4th ed. edited by C. F. Coombs Jr., 40.1–40.31. New York: McGraw-Hill.

Sheng, S., R. Allmon, L. Lynn, I. O'Donnell, K. Stone, and R. W. Brodersen. 1994. A monolithic CMOS radio system for wideband CDMA communications. In *Wireless '94 Conference Proceedings.* Calgary, Canada.

Smailagic, A., and D. P. Siewiorek. 1993. A case study in embedded-system design: The VuMan 2 Wearable Computer. *IEEE Design and Test of Computers,* September, 56–67.

Stafford, J. W. 1996. Semiconductor packaging technology. In *Printed circuits handbook,* 4th ed., edited by C. F. Coombs Jr., 2.1–2.16. New York: McGraw-Hill.

Stern, N. 1980. Who invented the first electronic judicial computer? *Annals of the History of Computing* 2 (4): 375–376.

Sturges, R. H., and P. K. Wright. 1989. A quantification of dexterity. *Journal of Robotics and Computer Aided Manufacturing* 6 (1): 3–14.

Wang, F.-C., B. Richards, and P. K. Wright. 1996. A multidisciplinary concurrent design environment for consumer electronic product design. *Concurrent Engineering: Research and Applications* 4 (4): 347–359.

Wang, F.-C., P. K. Wright, B. A. Barsky, and D. C. H. Yang. 1999. Approximately arc-length parametrized C3 quintic interpolatory splines. *Transactions of the ASME, Journal of Mechanical Design* 121 (3): 430–439.

Weiss, P. 1999. Smart outfit. *Science News* 156 (21): 330–332.

Yeh, C. P. 1992. An integrated information framework for multidisciplinary PWB design. Ph.D Thesis, Georgia Institute of Technology.

Yeh, C. P., R. E. Fulton, and R. S. Peak. 1991. A prototype information integration framework for electronic packaging. Paper presented at the ASME 1991 Winter Annual Meeting. Atlanta, GA.

6.8 CASE STUDY ON COMPUTER MANUFACTURING[3]

6.8.1 Overview

This case study presents the product development process of the InfoPad shown in Figure 6.22. The InfoPad is a portable, wireless computer aimed at an approximate selling price of $300. It provides text and graphics, pen input, limited speech input,

Figure 6.22 The InfoPad—a wireless "information appliance" (see <**www.eecs.bwrc.berkeley.edu**>).

[3]The InfoPad was a large collaborative project, and particular acknowledgments are made to the following colleagues: Professor Robert Brodersen, Professor Jan Rabaey, Dr. Frank Wang, Brian Richards, Susan Mellers, and many students at the Berkeley Wireless Research Center.

audio output, and full-motion color video. In a restricted classroom or home environment it can be used as a mobile communication device and a sketch pad (Brodersen, 1997). Twenty prototypes were produced as evaluation kits for marketing purposes and user testing in a college classroom.

6.8.2 Goals of the Case Study

Some key points that may be learned in the case study include the following:

- Designing and fabricating a complex system like the InfoPad require collaboration between many engineering disciplines. Specifically, most consumer electronic products are electromechanical systems. They consist of mechanical components such as structures, enclosures, and mechanisms, combined with electrical components such as printed circuit boards, power supplies, wires (harnesses), and switches. In spite of the advancements *within* each field—namely, the electrical CAD tools (ECAD) shown in Figure 5.9 and the mechanical CAD tools (MCAD) shown in Table 3.2—a gap still exists today for good communication *between* ECAD and MCAD. The cartoon of Figure 6.23 captures this struggle.
- An environment called the domain unified computer aided design environment (DUCADE) has thus been developed to address this need. It is a concurrent engineering system for ECAD/MCAD. The links from (a) conceptual design to (b) detail design to (c) fabrication are smooth and deterministic, creating a fast link from the initial design to a fabricated product. This integration improves product quality and time-to-market.

Figure 6.23 DUCADE has the goal of reducing the wall between ECAD and MCAD.

- A specific focus is on *constraint resolution* between electrical and mechanical issues. A central virtual-white-board environment is created to share and communicate *coupled design issues* during the design process.
- Various electrical and mechanical subsystems can be designed for modularity and reused in successive design generations. This further accelerates the design to production time.

6.8.3 Conceptual Design

The conceptual design phase of the InfoPad involved a functional requirement tree. Figure 6.24 is a global overview of the product design space. Design constraints were specified for each functional requirement as hard (cannot be changed) or soft (can be changed) constraints. For example, a hard design constraint was the desired weight of less than 2 pounds, under the functional requirement of "portable."

6.8.4 Concurrent Detailed Design Using DUCADE

The detailed designs for the InfoPad subsystems were conducted by various design teams. These included the pad group, the wireless communication (radios) group, the multimedia network group, the user interface group, and the mechanical design group.

Each group used its own domain design tools to perform specific design tasks. However, at certain critical junctures, predetermined design data within each team's domain were shared with other teams in a collaborative way. For example, Figure 6.25 illustrates the collaboration between the "pad group" and "mechanical design group." The PCB of the pad was designed using the Racal PCB layout tools, and the InfoPad casing was designed using the MSC/ARIES mechanical design package.

The domain unified computer aided design environment (DUCADE) then provided concurrent access to all the design tools of each team. Importantly, at critical junctures, it specifically provided online checking and verification of the design issues that were predetermined as being coupled between design teams.

Commercial CAD packages that were encapsulated in the DUCADE system included four MCAD packages and two ECAD packages: *MSC/ARIES,*[4] *AutoCAD,*[5] *ProEngineer,*[6] *ACIS,*[7] *Finesse,*[8] and *Racal/Visula.*[9] ARIES and AutoCAD/ProEngineer were primarily used for mechanical component design and mechanical analyses such as interference and thermal analyses. ACIS was the solid modeling kernel and package for solid modeling. Racal/ Visula was the major electrical design tool for PCB layout design.

[4]MSC/ARIES™ is a trademark of MacNeal Schwendler Corporation.

[5]AutoCAD™ is a trademark of Autodesk Inc.

[6]ProEngineer™ is a trademark of Parametric Technology Corporation.

[7]ACIS™ is a trademark of Spatial Technology Inc.

[8]Finesse™ is a trademark of Harris EDA Inc.

[9]Racal/Visual™ is a trademark of Racal-Redac.

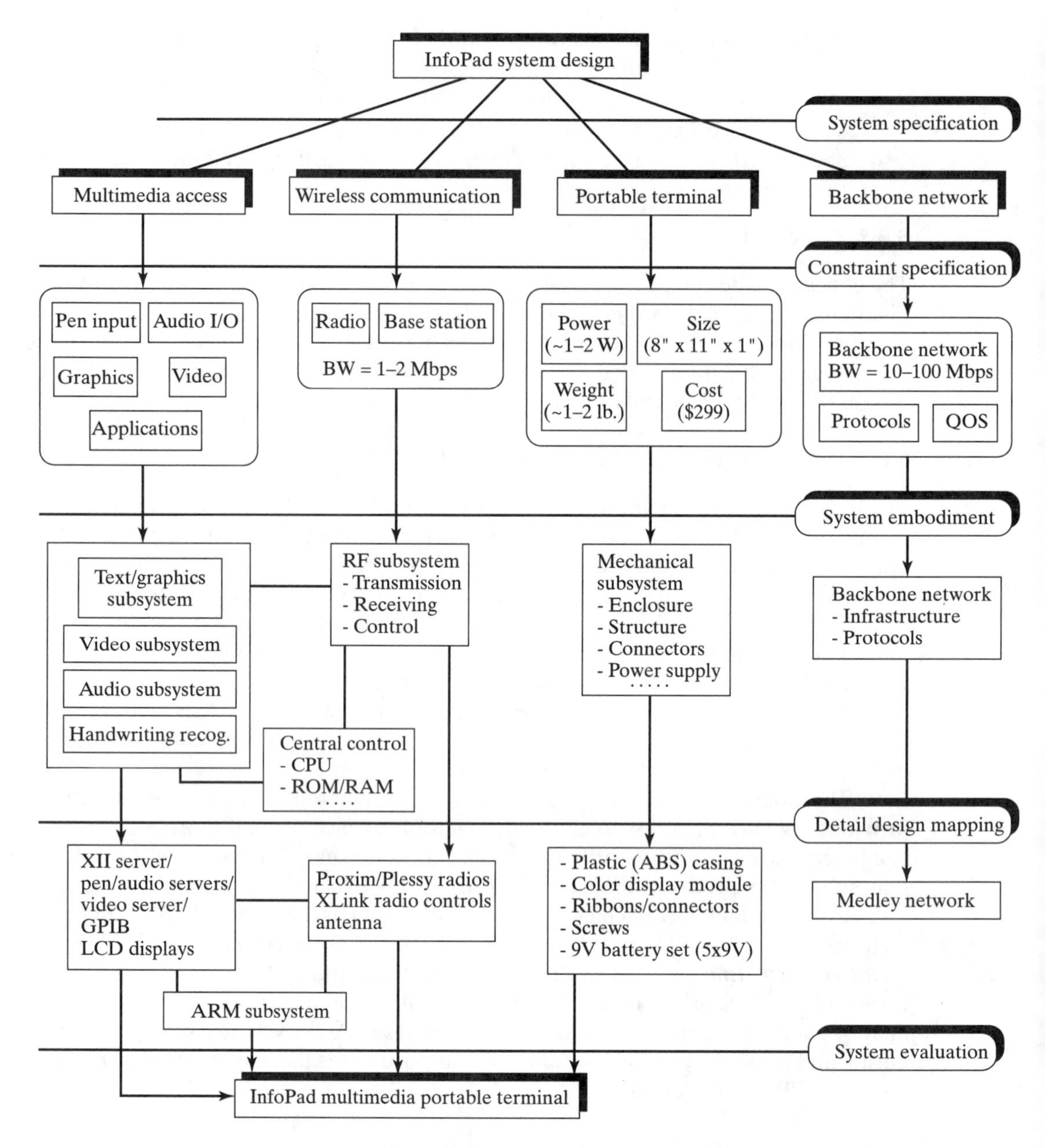

Figure 6.24 Design architecture of the InfoPad system.

Readers who are interested in the electrical system design can refer to the literature for radios in wireless communication development (Sheng et al., 1994; Cho et al., 1996), mobile multimedia networking and applications (Le et al., 1995; Amir et al., 1995; Narayanaswamy et al., 1996), video and graphic transport (Lao et al., 1994), user

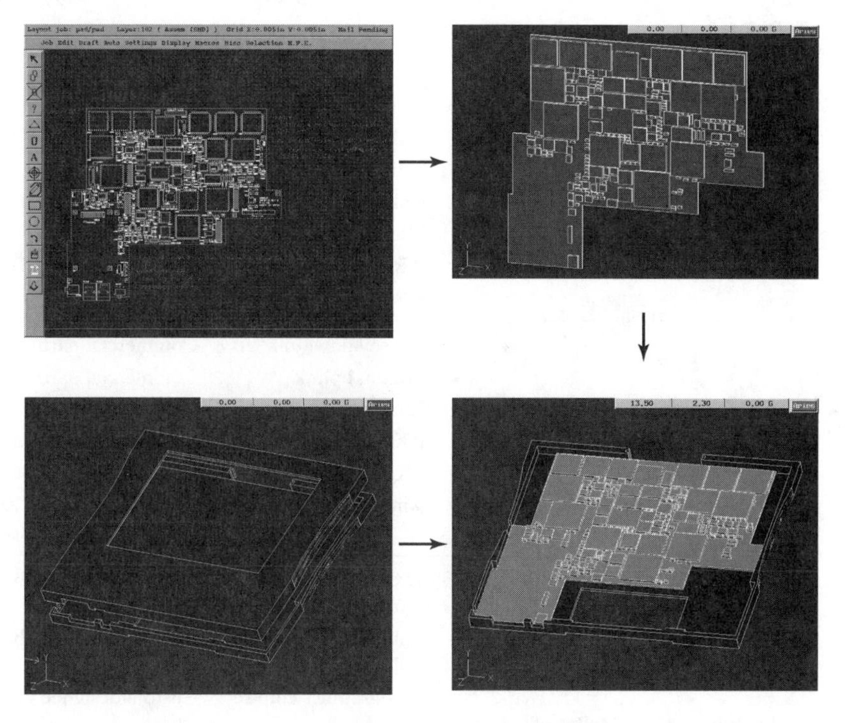

Figure 6.25 Detailed design of mechanical casing and PCB.

interface applications (Long et al., 1995), and design tools and framework (Guerra et al., 1994; Rabaey et al., 1995; Wang et al., 1996). Table 6.3 summarizes the implementation.

6.8.5 Coupled Design Constraints

Figure 6.26 shows some of the coupling constraints that occurred between the mechanical and electrical domains. For clarity, only one arrow is shown for the coupling relationship between the locations/orientations of the electronic components and the enclosure's shape. This coupling relationship usually required iterative design tasks with close communication between mechanical and electrical designers to achieve compact packaging. Because the InfoPad was designed to be a low-power device, many constraints focused on the compact packaging of ICs, devices, and displays.

6.8.6 Coupling Constraints Originating in the Mechanical Domain ($C_{m>e}$)

Several constraints originated on the "mechanical side" and had to be accommodated on the "electrical side." They are given the symbol $C_{m>e}$. They were as follows:

- Given the $8 \times 11 \times 1$ format—deliberately chosen to mimic the familiar engineers' clipboard—the available space inside the terminal enclosure was limited. It thus impacted the available area as well as the height for the display and the PCBs and their components.

TABLE 6.3 Major Electrical Subsystems and Components of InfoPad

Major subsystems	Functionality	Major parts	Source/part No.
Arm subsystem	Central control	PAL	Commercial part/ATV 2500L
		EPROM	Commercial part/AM27C010
		Octal buffer	Commercial part/HCT574
		SRAM	Commercial part/TC551001 BFL-85
		ARM60	Commercial part/GPS-P60ARMPR
		ARM interface chip	Custom designed and fabricated
Radio subsystem	Wireless communications	Plessey downlink	Commercial part/GEC-DE6003
		Proxim uplink	Commercial part/RDA-100/200
		Xilinix	Commercial part/XC-4008
		RX chip	Custom designed and fabricated
		TX SRAM	Commercial part/TC551001 BFL-85
		Antenna (× 2)	Commercial part/EXC-VHF 902 SM/EXC-UHF 2400
Multimedia subsystem	Multimedia I/O	Text/graphics LCD display	Commercial part/Sharp LM64k83
		Color video LCD display	Commercial part/Sharp LQ4RA01
		Text/graphics chip set (× 5)	Custom designed and fabricated
		Color video chip set (× 5)	Custom designed and fabricated
		Audio control chip set (× 5)	Custom designed and fabricated
I/O subsystem		Gazelle pen board	Commercial part
		Codec	Commercial part/MC145554
		Speaker	Commercial part
Power subsystem	Power supply	9V battery × 5	Commercial part

- Given the standard mechanical/UI features on the terminal casing—for example, the window for the LCD display on the top case, the access window for the battery set—again, the shapes, dimensions, and positions of the PCBs and their components were limited to certain values.
- Given the fact that the casing needed mechanical supporting structures, ventilation grids, and antenna positioning relative to the user's body, certain restrictions on placement of ICs were inevitable.

To display these for the electrical designers, DUCADE provided a simple "layered" view of the ($C_{m>e}$) mechanical constraints. This was because the electrical design teams were familiar with 2.5-D layout tools (rather than 3-D tools) for ICs and PCBs. Figure 6.27 shows the internal layout of the bottom casing. The maximum

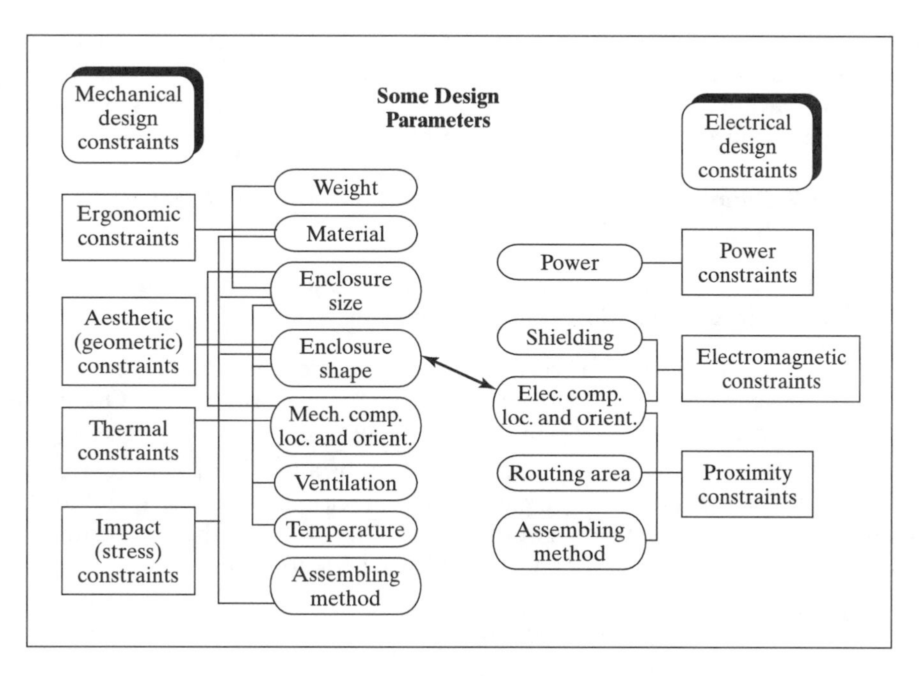

Figure 6.26 Coupling mechanical/electrical design constraints.

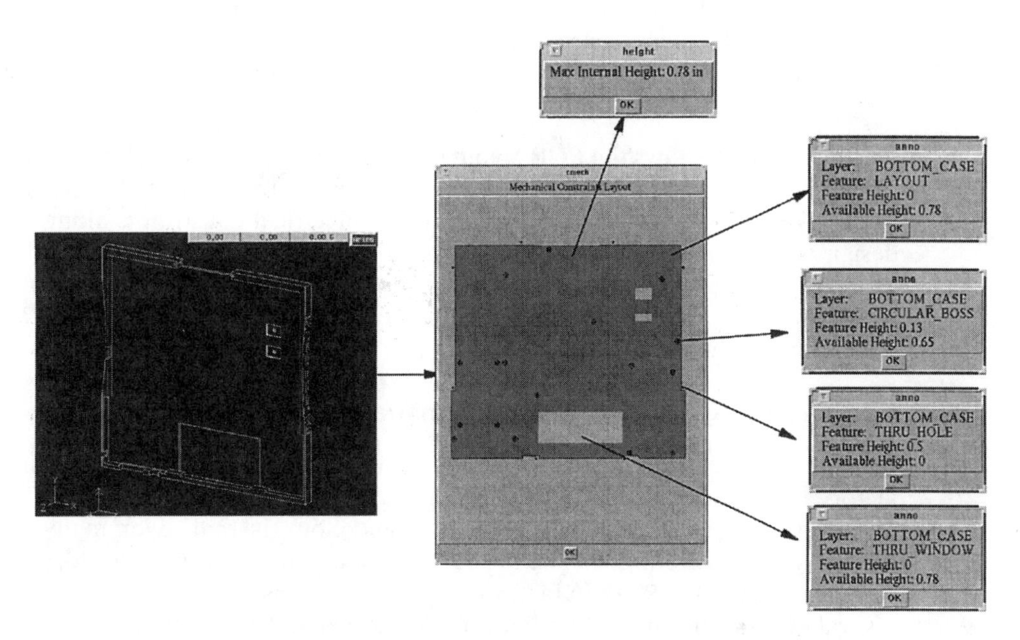

Figure 6.27 Layout of the mechanical constraints.

internal height for the electrical subsystems is 0.78 inch. Special features on the bottom case are the circular bosses for mounting the PCB, the through window for the battery set, and various mechanical dimensions and shapes of electrical switches. Such features needed to be compatible with the work of the electrical design team.

6.8.7 Coupling Constraints Originating in the Electrical Domain ($C_{e>m}$)

Other constraints originated on the "electrical side" and had to be accommodated on the "mechanical side." They are given the symbol $C_{e>m}$. They were as follows:

- The "z-height" of certain large devices on the PCBs were flagged and conveyed to the mechanical designers for consideration. Often this exchange resulted in more favorable packaging and overall space saving.
- Various switches, volume control knobs, and other I/O were of course mounted on the side of PCBs and their precise dimensions were conveyed to the mechanical team. Ultimately, this meant that they protruded through the plastic casing's boundary with user-friendly dimensions. These components were a power connector, a power switch, serial/parallel ports, audio I/O jacks, a keyboard jack, two volume control knobs, and a connector for an attachable color display.
- Certain devices, such as the power supply, required electromagnetic shielding. Once the sites for these devices were determined their dimensions were shared with other teams.
- The locations and sizes of the mounting holes that attached the PCB to the bottom casing were also the product of a dialogue between electrical and mechanical designers.

Figure 6.28 shows the PCB layout of the InfoPad terminal with these electrical constraint features highlighted. To present these electrical constraints to mechanical designers in the DUCADE environment, electrical constraints along with the design were transformed into mechanical representation (3-D solid models) for analysis. The approach captured only the geometric aspects of the electrical design constraints. Methods for presenting other types of electrical constraints are still under development.

6.8.8 Resolving the Coupling and Constraints ($C_{e<>m}$)

After several iterations between the mechanical and electrical design teams, satisfactory compromises were achieved. They are given the symbol $C_{e<>m}$. The evidence in firms such as Hewlett-Packard and Sony is that, over many product revisions, these couplings can be further refined and the product gets more efficient and compact (Cole, 1999). In fact, an interesting exercise for students is to "dissect" several generations of the familiar Sony Walkman and see how the engineers have invented new ways of resolving the constraints ($C_{e<>m}$). For a first prototype there

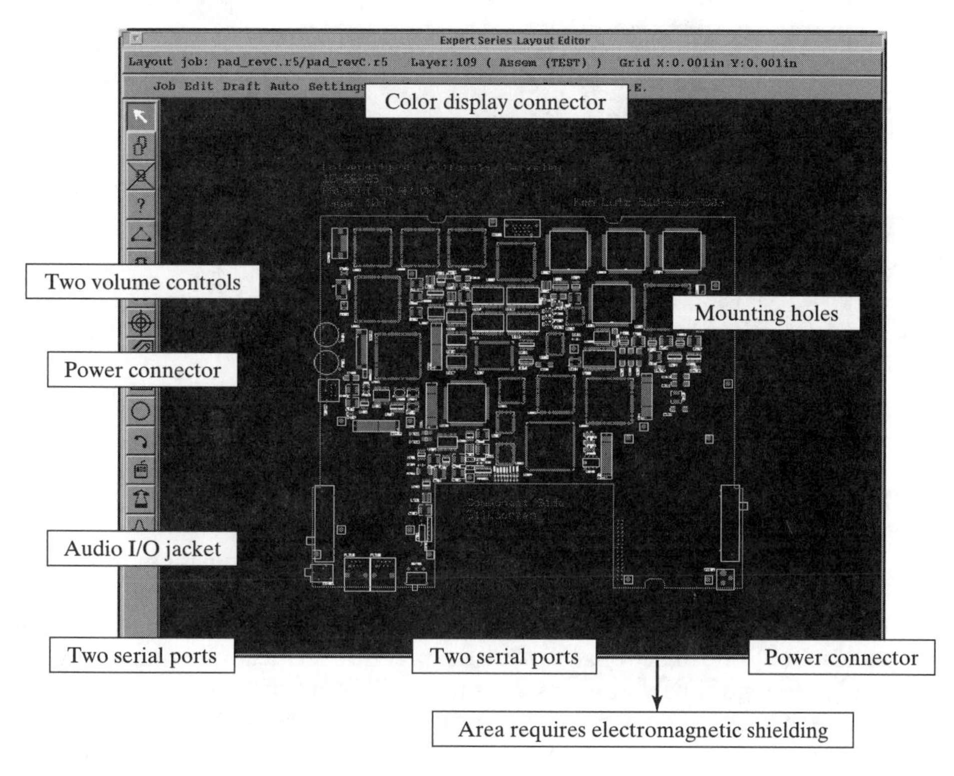

Figure 6.28 PCB layout and some electrical constraints.

is a limit to how much can be achieved in the first few iterations. Nevertheless for the InfoPad the main constraints were resolved, leading to the following new aspects of the "final" design:

- The PCB design was changed to a customized "U shape." It provided a smaller form factor for the new specification and allowed access to the selected battery set.
- Some important electrical components were redesigned using different IC packaging and different surface mounting techniques. This reduced the device sizes and allowed them to fit into the smaller interior space of the mechanical casing.
- The boundary features of the mechanical casing were changed for the added audio, keyboard jacks, two serial ports, and the color video unit.
- The interior contour of the mechanical casing was modified according to the shape of the new PCB. Circular bosses were added and modified on the bottom casing for mounting the newly designed PCB.
- Curved, spline features were added to the two sides of the terminal casing to improve the ergonomic and aesthetic design (Figure 6.29).

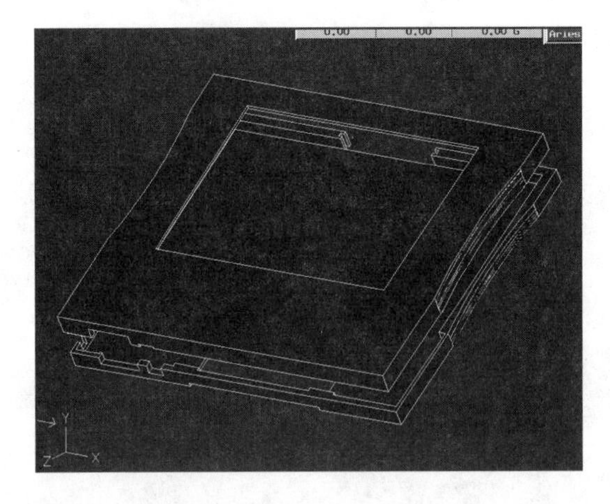

Figure 6.29 CAD solid model of the InfoPad.

6.8.9 Fabrication

The MOSIS service was used for the various ICs <**www.mosis.org**>, the PCBs were "fabbed" at a local bureau <**http://sierraprotoexpress.com**>, and the molds were fabricated on the CyberCut service <**cybercut.berkeley.edu**>. The design of the molds for plastic injections involved the precise specification of taper angles for separating the mold halves, shrink factors for different materials, core design, running gate design, and parting plane specification. Prototype molds and casings are shown in Figure 6.30.

Figure 6.30 Top: the aluminum mold halves; bottom: the injection molded plastic casings.

CHAPTER 7

METAL-PRODUCTS MANUFACTURING

7.1 INTRODUCTION

7.1.1 The "Garage" Shop at www.start-up-company.com

The sepia photograph above, taken around 1917, shows William Woodland, one of my grandfathers. He was an aircraft mechanic. He is sitting in the cockpit of a Vickers Vimmy biplane. My other grandfather, Browett Wright, was a railroad engineer; specifically he was a "knocker." He walked around the stockyards at Watford Junction, on the outskirts of London, carrying a small hammer. By lightly knocking on the wheels and axles of wagons, and listening to the resulting "ring," his well-trained ear could detect the absence or presence of potentially dangerous fatigue cracks.

As a hobby, these men and their friends often set up small machine shops in their garages or basements where they would fabricate small personal projects. Sometimes, around the Christmas consumer season, they even did *small-batch manufacturing* runs for local suppliers/wholesalers.

The garage shop allowed autonomy, custom fabrication, and reasonably good delivery times. A wide variety of projects could be accomplished with a well-equipped tool chest and just two main machines: a small lathe and a medium-sized three-axis milling machine. A drill press was an inexpensive addition that saved these men from setting up the entire mill just to drill a hole. If a few extras were added—such as a bench grinder, a small sheet-metal press, and a welding set—then they were really in business.

Many of today's famous figures in Silicon Valley and elsewhere also began their start-ups in a humble garage. It is most likely that many of these garages contained these basic tools in their early days.

The most important point is this: these simple metal-cutting machines allow a range of parts to be made. The same set of cutting tools can be used to make a wide variety of parts of rather complex geometry. Machining is the most important metal-forming operation for this reason, despite the more glamorous appeal of FDM and other SFF processes. In the last few decades, machining has endured a bad reputation for being wasteful and for being a little slow, but it always bounces back in fashion for these reasons of flexibility and good accuracy.

7.1.2 The Origins of the Basic Machine Shop

From these humble origins emerges the job shop or machine shop—really just a larger version of the garage shop but with bigger machines, more specialized machines, and perhaps several machinists. However, in the end it all boils down to flexibility and specialized service with reasonable delivery.

Machining is a powerful way of producing the outer structures of weird prototypes that are just beginning the product development cycle or perhaps just entering the market: products located in the lower left corner of Figure 2.3. Also the job shop in a university or research laboratory is a place where machinists turn the new designs of mad inventors into reality.

All over the world, unusual start-up companies are being formed: computer start-ups, biotech firms, in fact any new-product companies. At some point these companies will need a prototype, often to show the bank for a loan or to impress potential investors. Most likely, the company will need to go to a machine shop to get the prototype made. For example, the first few Infopad casings—see Chapter 6—were readily machined with good tolerances of +/−0.002 inch (~50 microns). The fabrication time was approximately three days from the moment the design was fixed until a finished casing was fabricated by milling.

7.1.3 The Tool and Die Shop—Machining and EDM

Metal cutting is more ubiquitous in industrial society than it may appear from the preceding description of the traditional machine shop. All forgings, for example those used in cars and trucks, and many sheet metal products, for example those used in steel furniture and filing cabinets, are formed in *dies* that have been machined. In fact, most of today's electronic products are packaged in a plastic casing that has been injection-molded into a die: cellular phones, computers, music systems, the Walkman, and all such products thus depend on the *cutting of tools and dies for other processes* that follow metal cutting. While the initial prototype of a new cellular phone might well be created with one of the newer rapid prototyping processes that emerged after 1987—stereolithography (SLA), selective laser sintering (SLS), or fused deposition modeling (FDM)—the final plastic products, made in batch sizes in the thousands or millions (see Chapter 8), will be injection-molded in a die that has been cut in metal with great precision and surface finish constraints.

The machining of such dies, usually from highly alloyed steels, requires some of the most exacting precisions and surface qualities in metal-cutting technology. It prompts the need for new cutting tool designs, novel manufacturing software that can predict and correct for tool deflections and deleterious burrs, and new CAD/CAM procedures that incorporate the physics and knowledge bases of machining into the basic geometrical design of a component. At the same time it should be noted that if the die is a complex shape, a two-step machining process may be needed. First, the "roughing cuts" will be done with conventional end milling as shown in Section 7.2.4. Second, the "finishing operations" will be created by the electrodischarge machining (EDM) process. In this process, a shaped carbon electrode slowly "sinks" into the metal mold. Electric arcing between the electrode and die surface takes place in a dielectric bath. Material removal takes place by surface melting of the die surface and flushing the debris away with the dielectric fluid.

7.1.4 Full-Scale Production Using Machining Operations

The previous sections focus on small-batch manufacturing operations that generate a small number of parts, or even one-of-a-kind dies. In other sectors of industry, *large-batch manufacturing* is more the normal situation. Thus machining in all its forms—turning, milling, drilling, and the like—can also be seen as a large-scale manufacturing process. It supports mass-production manufacturing such as the auto and steel industries, both positioned well along the market adoption curve in Figure 2.2.

To summarize this economic importance, the cost of machining amounts to more than 15% of the value of all manufactured products in all industrialized countries. Metcut Research Associates in Cincinnati, Ohio, estimates that in the United States the annual labor and overhead costs of machining are about $300 billion per year (this excludes work materials and tools). U.S. consumption of new machine tools (CNC lathes, milling machines, etc.) is about $7.5 billion per year. Consumable cutting tool materials have U.S. sales of about $2 to $2.5 billion per year. For comparison purposes, it is of interest to note a ratio of $300 *to*

\$7.5 *to* \$2.5 billion for labor costs *to* fixed machinery investments *to* disposable cutting tools.

7.1.5 Full-Scale Production Using Other Metal-Processing Operations

Sheet rolling is also a large-batch manufacturing process in which a rolling mill continuously produces flat strip in coils. Such strip is sold to a secondary producer, who will shear it into smaller blanks that are then pressed into an ordinary soup can or filing cabinet. The *sheet-metal forming* of single discrete items, such as the hood of a car, is also a large-batch manufacturing process because many similar parts are produced on a continuous basis from one very expensive die. This chapter considers these other examples of large-batch manufacturing, focusing as an example on sheet-metal forming.

The machines and dies for all these processes are *extremely* expensive, so the analysis of the forces on the machines and dies is crucial. As managers of technology, a very valuable service is created if these force predictions lead to sensible machinery purchases—that is, machines that are powerful enough to deform the typical materials and products being created by a company, but not overly powerful, hence wasting capability and investment costs. At the same time, the understanding of which factors affect quality assurance and the properties of the deformed material is equally important.

7.2 BASIC MACHINING OPERATIONS

7.2.1 Planing or Shaping

A cutting tool moves through a steel block and removes a layer from its top surface (Figure 7.1). The discarded layer is called the chip. The mental image that seems to work well even in sunny California is that of a handheld snow shovel being pushed along a snow-covered sidewalk. The shovel is analogous to the tool, and the chip that rises up the face of the tool—actually called the *rake face*—is analogous to the layer of snow.

The chip then curls away from the face at some distance, called the chip-tool contact length, and falls away onto the surface being machined or to one side. If the shovel face made a perfect right angle with the sidewalk, the rake angle would be 0 degrees. Of course, a snow shovel is usually tilted back to a rake angle of about 20 or 30 degrees. In metal machining the *rake angle* for today's tools is often 6 degrees. It is also quite common in metal machining for the rake face to be at a slight negative angle of -6 degrees. This gives added strength to the delicate cutting edge.

Wood planing has a similar *geometry* to metal cutting by planing. However, this is not altogether the best analogy because in wood cutting the *physics* of the process is different: the wood ahead of the sharp cutting edge is split, and a long crack runs ahead of the tool to make for rather modest cutting forces.

In the machining of metals, although a ductile crack of microscopic proportions is obviously formed right down near the tip of the tool (otherwise the two surfaces would not separate), most of the work done is related to the shearing of metal in the *shear plane*. The shear plane is shown as *OD* in Figure 7.2.

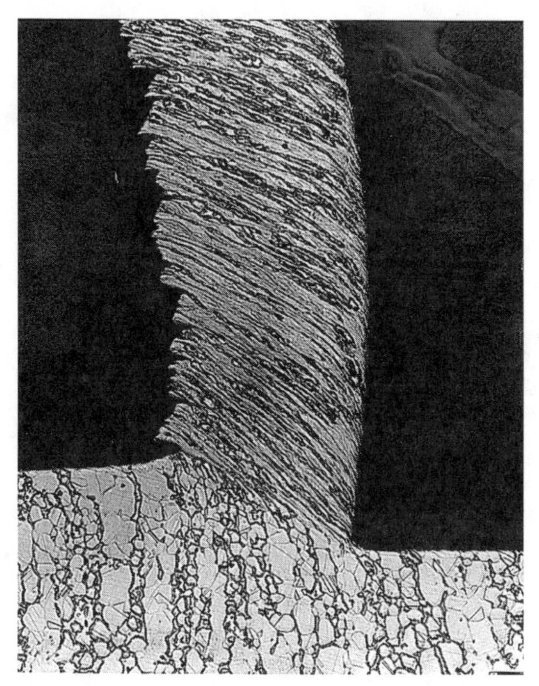

Figure 7.1 Micrograph of chip formation.

In Figure 7.2 and accompanying equations, the shear plane angle is related to the *undeformed* chip thickness t and the *deformed* chip thickness t_c. It is common to define a chip thickness ratio ($r = t/t_c$). Since the chip slows down by frictional drag on the tool, t_c is always greater than t. Thus r is always less than unity.

In the special case in which the rake angle, α, is set to zero, it can be seen that the tangent of the shear plane angle is just the undeformed chip thickness divided by the deformed chip thickness. If the main cutting forces (F_C and F_T) are measured with a force dynamometer, they can be resolved onto the all-important forces on the cutting edge of the tool. Why are these other forces F_N and F_R important? The answer is that they govern the life of the tool. Large values of F_R will create high shear stresses and temperatures in this region. A high value of F_N will be associated with a high normal pressure on the delicate cutting edge. Such high forces will create high friction and wear of the rake face. A large value of F_R will also tend to lift the tool away from the surface being machined and make the surface finish irregular. It is therefore worth paying for lubricants and diamond-coated tools that minimize this force.

7.2.2 Turning

The basic operation of turning (also called semiorthogonal cutting in the research laboratory) is also the one most commonly employed in experimental work on metal cutting. The work material is held in the chuck of a lathe and rotated. The tool is held rigidly in a tool post and moved at a constant rate along the axis of the bar, cutting

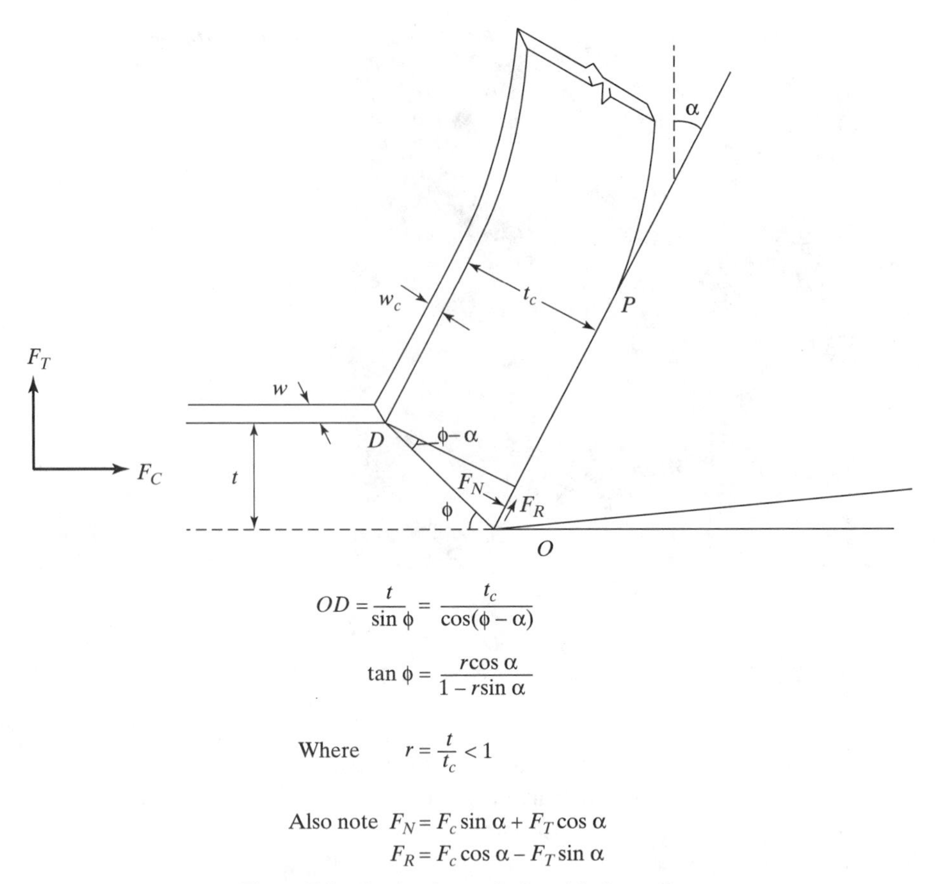

Figure 7.2 Cutting forces during chip formation.

away a layer of metal to form a cylinder or a surface of more complex profile. This is shown diagrammatically in Figure 7.3.

The cutting speed (V) is the rate at which the uncut surface of the work passes the cutting edge of the tool, usually expressed in units of ft/min or m min^{-1}. The *feed* (f) is the distance moved by the tool in an axial direction at each revolution of the work.

The *depth-of-cut (w)* is the thickness of metal removed from the bar, measured in a radial direction.[1] The product of these three gives the rate of metal removal, a parameter often used in measuring the efficiency of a cutting operation.

$$V f w = \text{rate of removal} \tag{7.1}$$

[1]The feed rate (f) during turning is also called the undeformed chip thickness (t). The depth-of-cut, w, in turning is also referred to as the undeformed chip width. Since machining has developed from a practitioner's viewpoint, the terminology is not really consistent from one operation to another. For example, in the diagrams for end milling, the term *depth-of-cut* is used in a different way. This is unfortunately confusing for a new student of the field. Perhaps the best way to accommodate these inconsistencies is to always view the "slice" of material being removed as the "undeformed chip thickness" (t) and the direction normal to this (into the plane of the paper) as the "undeformed chip width" (w).

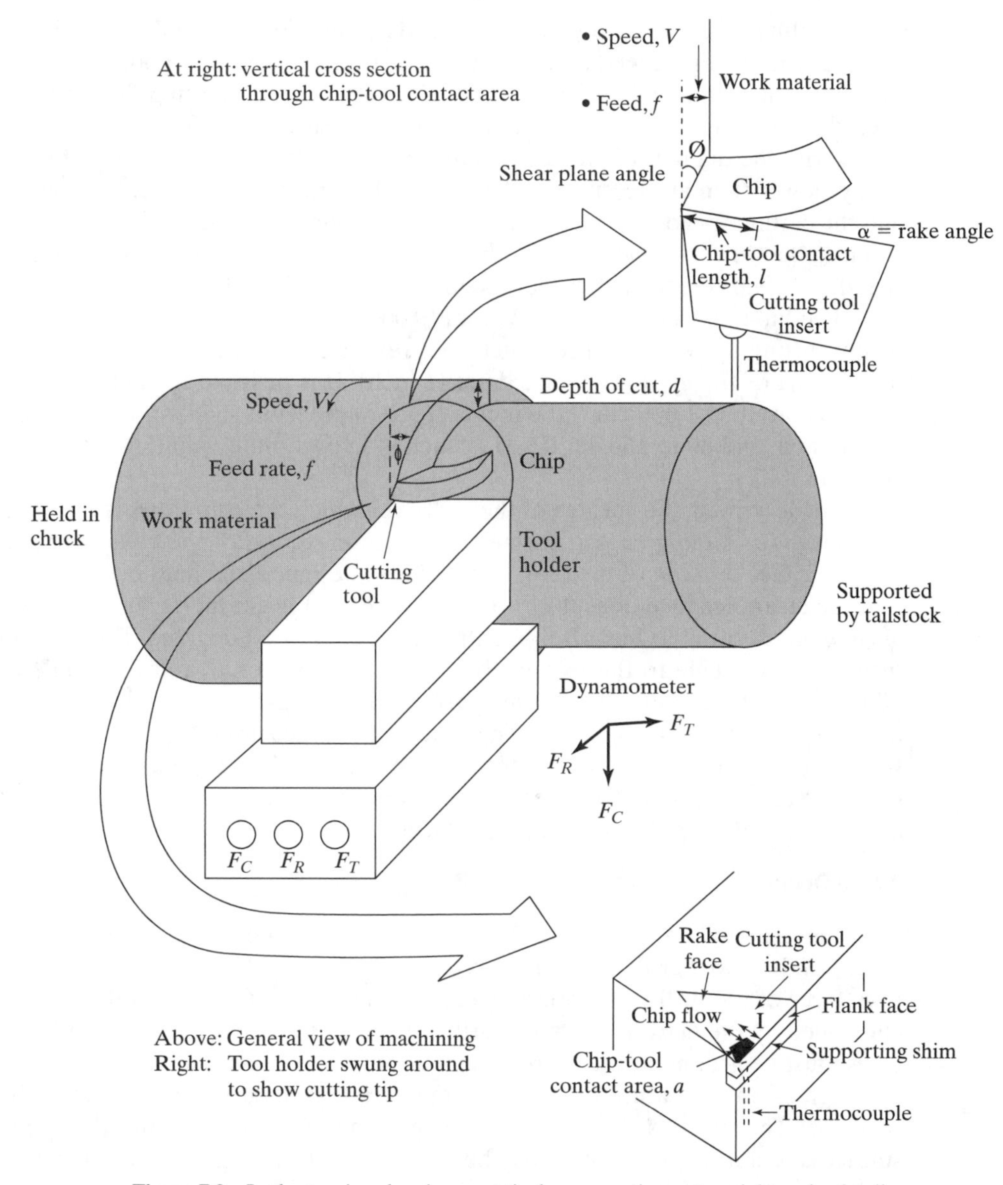

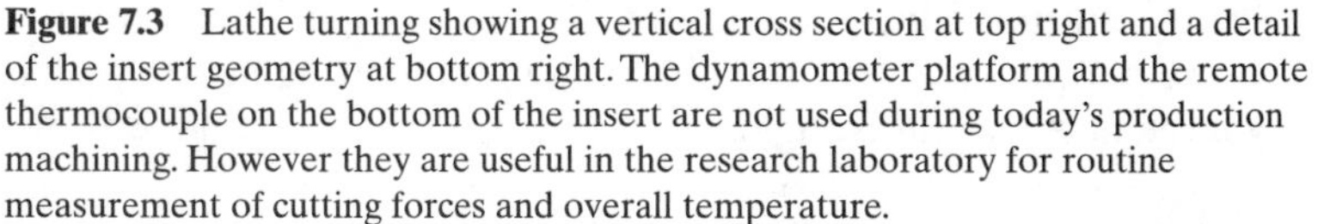

Figure 7.3 Lathe turning showing a vertical cross section at top right and a detail of the insert geometry at bottom right. The dynamometer platform and the remote thermocouple on the bottom of the insert are not used during today's production machining. However they are useful in the research laboratory for routine measurement of cutting forces and overall temperature.

The cutting speed and the feed are the two most important parameters that can be adjusted by the operator or programmer to achieve optimum cutting conditions. The depth of cut is often fixed by the initial size of the bar and the required size of the product.

Cutting speed is usually between 3 and 200 m min^{-1} (10 and 600 ft/min). However, in modern high-speed machining, speeds may be as high as 3,500 m min^{-1} when machining aluminum alloys. The rotational speed (rpm) of the spindle is usually constant during a single operation, so that when cutting a complex form the cutting speed varies with the diameter being cut at any instant. At the nose of the tool the speed is always lower than at the outer surface of the bar, but the difference is usually small and the cutting speed is considered as constant along the tool edge in turning. Recent computer-controlled machine tools have the capacity to maintain a constant cutting speed, V, by varying the rotational speed as the workpiece diameter changes.

Feed may be as low as 0.0125 mm (0.0005 inch) per revolution and with very heavy cutting up to 2.5 mm (0.1 inch) per revolution. Depth of cut may vary from 0 over part of the cycle to over 25 mm (1 inch). It is possible to remove metal at a rate of more than 1,600 cm^3 (100 inches^3) per minute, but such a rate would be very uncommon, and 80 to 160 cm^3 (5 to 10 inches^3) per minute would normally be considered rapid.

As described, the surface of the tool over which the chip flows is known as the rake face. *The cutting edge* is formed by the intersection of the rake face with the *clearance face* or *flank* of the tool. The tool is so designed and held in such a position that the clearance face does not rub against the freshly cut metal surface. The *clearance angle* is variable but often on the order of 6 to 10 degrees. The rake face is inclined at an angle to the axis of the bar of work material, and this angle can be adjusted to achieve optimum cutting performance for particular tool materials, work materials, and cutting conditions. The rake angle is measured from a line parallel to the axis of rotation of the workpiece (Figures 7.2 and 7.3). The *nose* of the tool is at the intersection of all three faces and may be sharp, but more frequently there is a *nose radius* between the two clearance faces.

7.2.3 Drilling

In drilling, carried out on a lathe or a drilling machine, the tool most commonly used is the familiar twist drill. The "business end" of a twist drill has two cutting edges. The rake faces of the drill are formed by part of each of the *flutes,* the rake angle being controlled by the *helix angle* of the drill. The chips slide up the flutes, while the end faces must be ground at the correct angle to form the clearance face.

An essential feature of drilling is the variation in cutting speed along the cutting edge. The speed is a maximum at the periphery, which generates the cylindrical surface and approaches zero near the centerline of the drill, the *web* (Figure 7.4), where the cutting edge is blended to a chisel shape. The rake angle also decreases from the periphery, and at the chisel edge the cutting action is that of a tool with a very large negative rake angle.

7.2.4 Milling

Both grooves and flat surfaces—for example, the faces of a car cylinder block—are generated by milling. In this operation the cutting action is achieved by rotating the tool while the work is clamped on a table, and the feed action is obtained by moving it under the cutter (Figure 7.5).

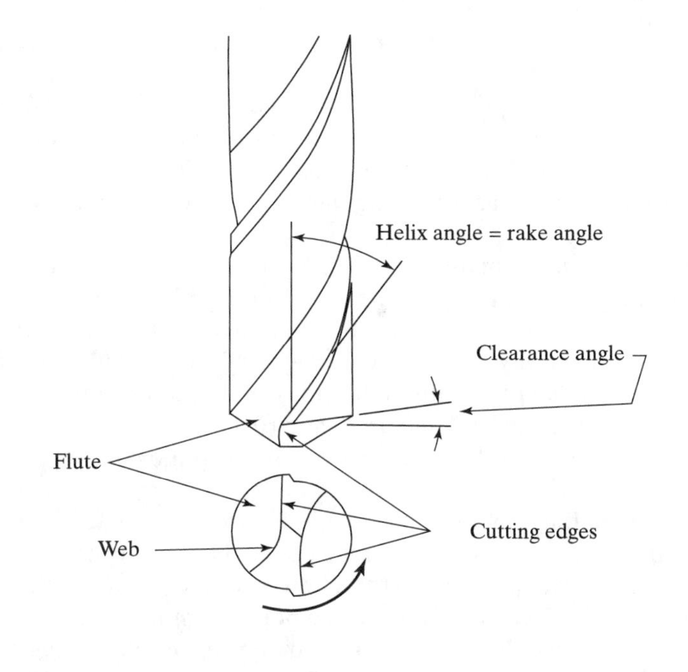

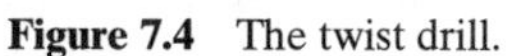
Figure 7.4 The twist drill.

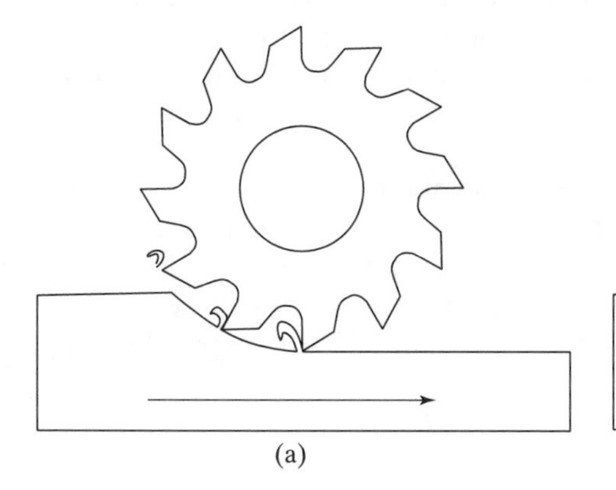

(a)

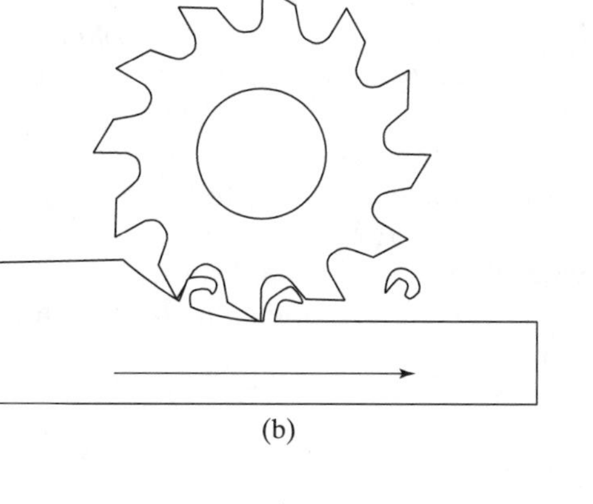

(b)

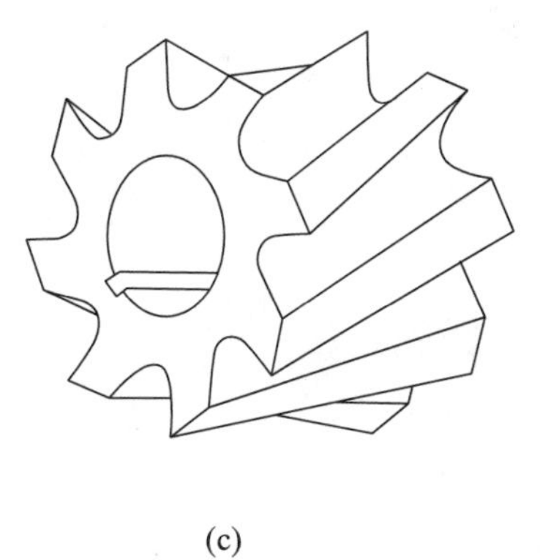

(c)

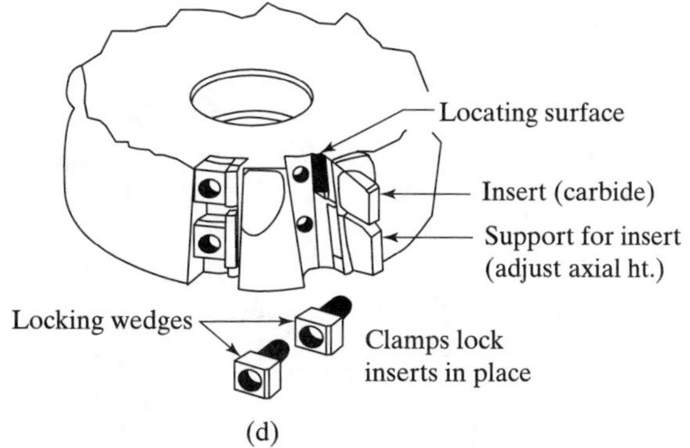

(d)

Figure 7.5 Milling cutters: (a) up milling, (b) climb milling, (c) edge milling, and (d) detail of face milling with indexable inserts—the inserts protrude below the tool holder.

Milling is used also for the production of curved shapes, while end mills (Figure 7.6), which are larger and more robust versions of the dentist's drill, are employed in the production of hollow shapes such as die cavities. The end mill can be used to machine a feature such as a rectangular pocket with (ideally) vertical walls.

There are a very large number of different shapes of milling cutter for different applications. Single toothed cutters are possible, but typical milling cutters have a number of teeth (cutting edges), which may vary from 2 to over 100. The new surface is generated as each tooth cuts away an arc-shaped segment, the thickness of which is the *feed* or *tooth load.* Feeds are usually light, not often greater than 0.25 mm (0.01 inch) per tooth, and frequently less than 0.025 mm (0.001 inch) per tooth. However, because of the large number of teeth, the rate of metal removal is often high. The feed often varies through the cutting part of the cycle.

In the orthodox side milling operation shown in Figure 7.5a, the feed on each tooth is very small at first and reaches a maximum where the tooth breaks contact with the work surface. If the cutter is designed to "climb mill" and rotate in the opposite direction (Figure 7.5b), the feed is greatest at the point of initial contact.

In practice it is observed that the sideways pressure on an end cutter can deflect it away from the vertical surface, as shown in Figure 7.7. Since the slender end mill is like a cantilever beam, the deflection is more pronounced down at its tip. Instead of being vertical, the walls of the pocket begin to have a "ski-slope" cross section—reasonably vertical at the top, but sloping out near the bottom where the tool is most deflected.

Even this is a somewhat simplified view, and as shown in the detail of Figure 7.8, the nonverticality—the form error—shows an S-shaped undulation at the top of the wall, before the ski-slope effect gives rise to the maximum error at the bottom of the wall. Stori (1998) has shown that the precise geometry is related to the depth of the pocket versus how many of the flutes shown in Figure 7.6 are engaged in chip removal.

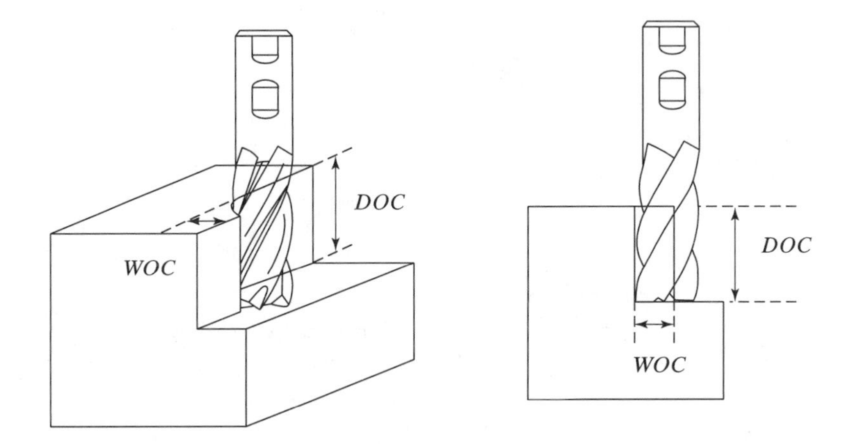

Figure 7.6 Ideal cutting conditions of an end mill: *WOC* = width of cut, *DOC* = depth of cut (courtesy of Dr. James Stori).

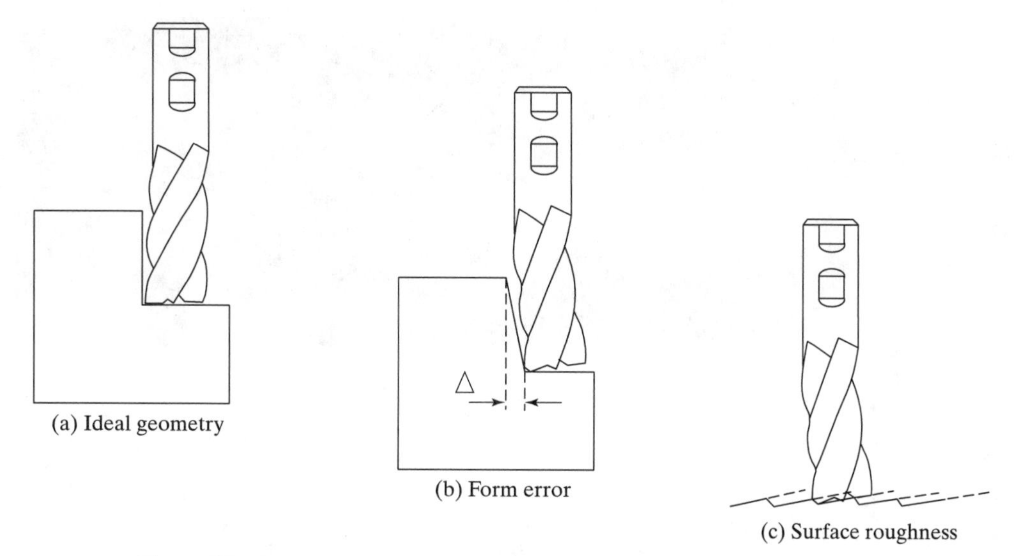

Figure 7.7 Effect of tool deflection on form error and surface roughness (courtesy of Dr. James Stori).

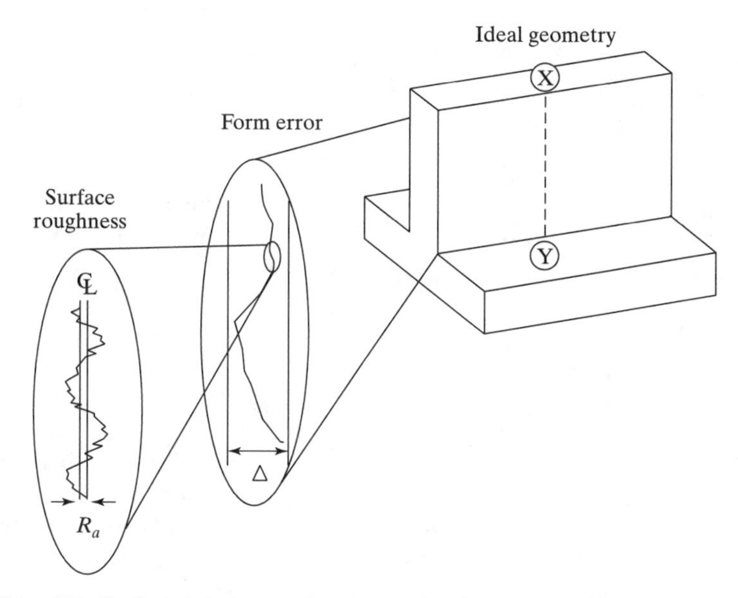

Figure 7.8 Deviations in form and surface quality (courtesy of Dr. James Stori).

7.2.5 Feature Creation by Milling Using "G and M Codes"

The link from any of the electronic representations in Chapter 3—whether in wire-frame, CSG, or DSG—to a physical component is critical to the success of quickly producing a quality part. Ideally, the interchange from a design format to a manufacturing format should preserve all the design information, plus it should be easy to

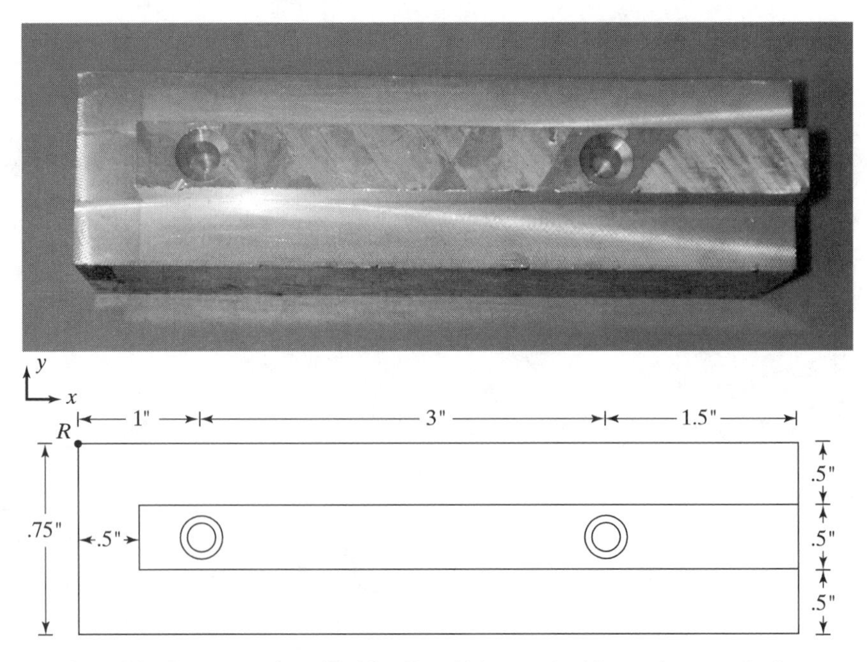

Figure 7.9 Test part to be milled for G and M example. Above: photograph of machined block. Below: the dimensions of the three DSG shoulders and two holes. The screw head chamfers are applied with a spot drill prior to blind hole drilling (Table 7.1).

interpret from a manufacturing point of view. This interpretation work is known as process planning. It involves fixture selection, tool selection, and tool path creation.

T1 shell mill (3 inch diameter)

T2 end mill (1.5 inch diameter)

T3 spot drill

The example maps the CAD features into physical features onto the piece of aluminum, using a computer numerically controlled (CNC) machine tool. The movements of these standard machine tools are controlled by *G and M codes* (Figure 7.9). These painstaking cutting tool movements were worked out in the 1950s. The "G" codes are the commands that create machine moves such as *G1 = linear feed.* The "M" commands are associated with machinery actions such as *M3 = turn on the spindle clockwise.* Even though these methods have been automated, they still form the communication routines for the low-level communication to today's CNC machines. G and M codes were also recently integrated into the RS274 standard.

The G and M codes were enhanced by APT (automatically programmed tool) as early as 1955–1960. APT is a higher level language than G and M codes that treats each line in Figure 7.9 above as an "object." Methods based on APT are still used for machine tool programming in today's factories, and again they compile into G and M codes. However, many other high-level programming systems that break down

TABLE 7.1 G and M Program for Machining the Shape in Figure 7.9

"O" blocks are alignment blocks that allow the machine to start. They also permit restart if there is an error. The machine cannot from "N" block dimensions.

O1	G0 T1 M6	G0 = rapid travel; M6 = call tool T1 into spindle
O5	*x*-2.00 *y*-.75 *z*0 s3000 M3	s = spindle speed (rpm); M3 = turn on clockwise
N10	G1 *x*7.50 F50	G1 = linear feed; *x* = new coord.; F = feed rate (in/min)
O15	G0 T2 M6	Call tool T2 into spindle
O20	*x*-1.0 *y*.26 *z*-.49 s1700 M3	−1″ off clearance; *y* =.26 and *z*=-.49 = rough cut
N25	G1 *x*6.26 F50	Linear cut to *x*=6.26; creates first shoulder
N30	*y*-1.76	
N35	*x*-.26	Stays in linear feed (G1)
N40	*y*.26	
O45	G0 *x*-1.0 *y*.25 *z*-.50 s2000	Rapid travel to starting point for finish cut
N50	G1 *x*6.25 F70	Finish cut
N55	*y*-1.75	
N60	*x*-.25	
N65	*y*.25	
O70	G0 T3 M6	Tool 3 spot drill
N75	G81 *x*1.0 *y*-.75 [*z*-.156 R0] F10 s4000 M3	G81 drill code canned cycle—down to depth and rapid retract
N80	*x*4.0	Other hole done (G81)
O85	G0 T4 M6	Drill T4
N90	G81 *x*1.0 *y*-.75 [*z*-.575 R0] F15 s3000 M3	R0 is a subdatum, 100 mils (.1″) above part
N95	*x*4.0	
N100	M2	End of program

Notes: Please note the difference between "O" and "0".
The first three lines clean up the rough stock to make a smooth surface.

CAD features into individual tool paths are also available. Note that without these higher level programming environments, all the painstaking details have to be specified, concerned with tool offsets and ordering the roughing and finishing cuts.

In chapter 3 several URLs were given to high-level CAD systems such as Pro-Engineer and SDRC's IDEAS. These systems are usually augmented by manufacturing planning environments for fixture planning, tool selection, and creating cutter paths that can be postprocessed to G and M codes.

7.3 CONTROLLING THE MACHINING PROCESS

The ideal situation from a practitioner's viewpoint would be to purchase and set up the lathe or the milling machine in a nice, clean laboratory-like environment, acquire a set of new cutting tools, and then reliably and quickly zip out parts on an NC machine with high accuracy and a mirror finish.

Initial questions might include: Which tools should be used? How fast should the machine run? How should the convex shape of this casting be clamped during finish machining? Which cutting fluid should be used?

Later on in the same day, other, less polite questions might arise: Why does this drill keep snapping off and getting stuck in the expensive casting? Why do the chips look like continuous ribbons and keep snarling around the spindle and stalling the mill? Why does the surface finish look like images from the Mars Rover? Will the operators get sick from all the cutting fluid pouring all over the place?

More seriously, the following major points are the issues that interest the practitioner and that any R&D activity ought to address if it is going to be useful:

- Prediction of the accuracy of the component being machined
- Prediction of the surface finish on the component being machined
- Prediction of tool life
- Prediction of chip control
- Prediction of the loads on the tool, and/or workpiece, and/or fixtures

Of course machining is a difficult process to predict. The upper diagram in Figure 7.10 represents the metal extrusion process—akin to squeezing out toothpaste. The diameter of the undeformed material on the left is known, as is the diameter of the deformed material on the right, because it is all constrained to flow through a die of prescribed diameter. However, in the lower sketch of Figure 7.10, the free edge of the chip is unconstrained. The thickness of the undeformed material on the left is known, but the thickness of the chip is not known because there is nothing like a die to restrain the free outer edge of the chip. As a result, the angle ϕ is not known a priori because it is not constrained, unlike processes such as extrusion.

The value of ϕ governs the calculation of the following factors:

- The surface finish on the component
- The stress σ on the tool face, and forces on the fixtures and on the machine tool
- The temperature T of the tool edge, the tool rake face, and the component
- The power that needs to be exerted by the machine tool

7.3.1 Predicting the Shear Plane Angle

Ernst and Merchant (1940–1945) studied the physics of the shear plane in great depth. Photomicrographs revealed a band of "intense shear occurring at a clearly

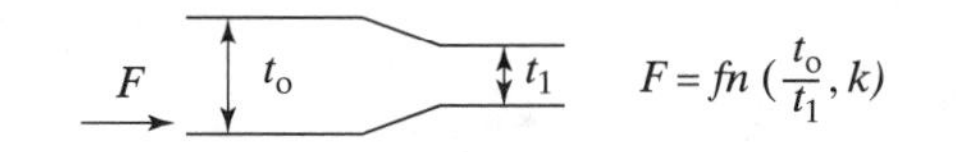

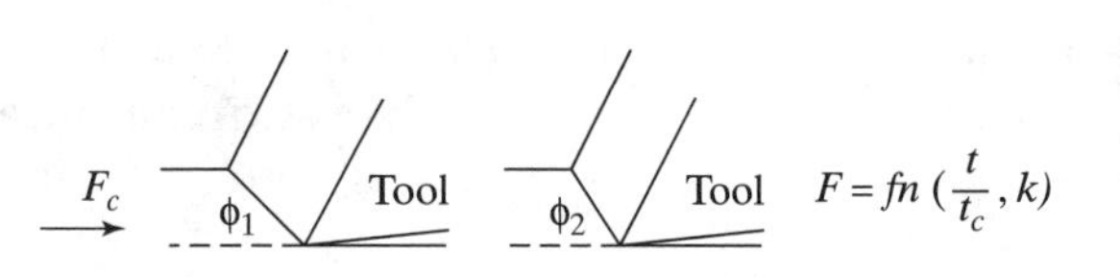

Figure 7.10 Comparisons between extrusion and machining (the forces F are a function of yield strength in shear, k; the undeformed and deformed extrusion/machining dimensions; and, in the case of machining, the shear plane angle ϕ).

defined angle" (Figure 7.1). Their observations set the stage for the prediction of the shear plane angle. Their classic "minimum energy" approach was developed between 1940 and 1945, around the time that the transistor was on the horizon at Bell Labs.

This section of the book now considers the prediction of the shear plane angle. The thickness of the chip is not constrained by the tool, and the question is, What *does* determine whether there is a thick chip with a small shear plane angle and high cutting force, or a thin chip with large shear plane angle and minimum cutting force?

In the last 60 years there have been many attempts to answer this question and to devise equations that will predict quantitatively the behavior of work materials during cutting from a knowledge of their properties. Colloquially speaking, one could even say that the prediction of the shear plane angle has become a sort of "holy grail" preoccupation in the machining research community!

In the pioneering work of Ernst and Merchant (1940–1945) a model of the cutting process was used in which the shear in chip formation was confined to the shear plane, and movement of the chip over the tool occurred by classic sliding friction, defined by an average friction angle λ.

This approach did not produce equations from which satisfactory predictions could be made of the influence of parameters, such as cutting speed, on the behavior of materials in machining. The inappropriate use of friction relationships relevant only to sliding conditions was probably mainly responsible for the weakness of this analysis. With this model the important area of contact between tool and work was not regarded as significant, and no attempt was made to measure or to calculate it.

Nevertheless, Merchant's force circle, shown in Figure 7.11, remains an important milestone in metal-cutting theory.

The forces can be found from Figure 7.11 as the following two equations:

$$F_c = \frac{twk\cos(\lambda - \alpha)}{\sin\phi\cos(\phi + \lambda - \alpha)} \tag{7.2}$$

$$F_T = \frac{twk\sin(\lambda - \alpha)}{\sin\phi\cos(\phi + \lambda - \alpha)} \tag{7.3}$$

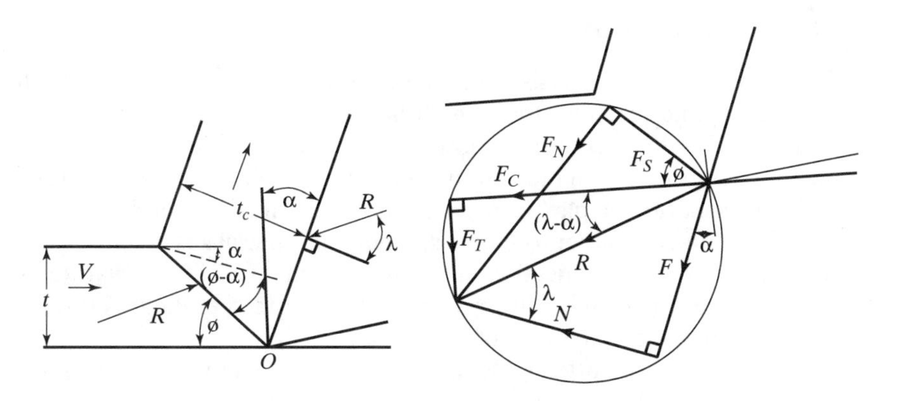

Figure 7.11 Merchant's force circle.

where

t = undeformed chip thickness

w = undeformed chip width

k = the shear yield strength of the metal being machined

λ = the friction angle on the tool's rake face

Differentiating the first equation with respect to the shear plane angle gives

$$\frac{dF_c}{d\phi} = \frac{twk\cos(\lambda - \alpha)\cos(2\phi + \lambda - \alpha)}{\sin^2\phi\cos^2(\phi + \lambda - \alpha)} = 0 \tag{7.4}$$

and the Merchant equation as

$$\phi = \frac{\pi}{4} - \frac{1}{2}(\lambda - \alpha) \tag{7.5}$$

After this expression has been found, the cutting forces can then be written. For the main cutting force acting in the tool/work direction

$$F_c = \frac{wtk\cos(\lambda - \alpha)}{\sin[(\pi/4) - \frac{1}{2}(\lambda - \alpha)]\cos[(\pi/4) + \frac{1}{2}(\lambda - \alpha)]} \tag{7.6}$$

$$\text{and} \qquad F_c = 2wtk\cot\phi \tag{7.7}$$

For the feed (or tangential) force acting normal to the main cutting force

$$F_T = \frac{wtk\sin(\lambda - \alpha)}{\sin[(\pi/4) - \frac{1}{2}(\lambda - \alpha)]\cos[(\pi/4) + \frac{1}{2}(\lambda - \alpha)]} \tag{7.8}$$

$$\text{and} \qquad F_T = wtk(\cot^2\phi - 1) \tag{7.9}$$

These forces allow the calculation of the power needed by the machine tool and the temperatures in the tool.

7.3.2 Significance to Power Needed

The first use of the analysis from an economic viewpoint is to calculate the power required by the machine tool. Small lathes and mills, which have a spindle in the range of 3.75 kilowatts (5 horsepower), may cost only $40,000 and may be quite adequate for a modest machine shop preparing the aluminum molds for rapid prototyping by small-batch plastic injection molding (see Chapter 8).

By contrast, a larger tool-and-die shop that specializes in full-scale production of steel dies for long-run injection molding might need a 22.5 kilowatt (30 horsepower) spindle. Such a machine will probably have a correspondingly larger bed and tool changer and cost on the order of $250,000.

In industry, too often educated guesses are used to make such purchasing decisions. Formal analysis of the type shown here is recommended.

Having predicted F_C for the typical size of the cuts and knowing the typical material strengths, the power, P, can be found (in SI units) from

$$P = VF_c \text{ watts} \tag{7.10}$$

If in this form of the equation, F_c is measured in newtons and V in m sec^{-1}.

7.3.3 Significance to Temperatures, Tool Selection, and Tool Wear

Heat is generated in the three regions shown in Figure 7.12, leading to the contours shown in the top of the figure. These contours were calculated with a finite-element program and checked against an experimental technique that examined the tempering in the tool (Stevenson et al., 1983).

This calculation of the temperatures may also be done in a simple way, carrying on from the Ernst and Merchant analysis. The temperature rise (ΔT_P) in the primary shear zone is given by

$$\Delta T_P = \frac{(F_c\cos\phi - F_T\cos\phi)\cos\alpha}{\rho c w t \cos(\phi - \alpha)} \tag{7.11}$$

In this equation, the density, ρ, and specific heat, c, are included. The analysis can also be extended to calculate the tool temperatures both at the delicate cutting edge and further along the rake face. When cutting steels at standard conditions, temperatures of 1,000°C are common toward the end of the contact length at position P in Figure 7.2 (see Trent and Wright, 2000).

High-speed steels, cemented carbide, and polycrystalline alumina-based materials are today's three main families of cutting tool material. The first two families can easily be coated with thin layers of other inert materials such as titanium nitride, reducing friction on the rake face. The beneficial effects are lower tool temperatures and a smoother surface finish. Trent and Wright (2000) describe the properties of these cutting tool materials in great detail.

The high-speed steel family has the highest toughness, or impact resistance, and it is favored for drills and many milling cutters in the small-batch machining shop. But for faster production, the carbide family resists high temperatures better and is therefore used as much as possible as the standard industry tool material. The cemented carbide tools are manufactured by sintering. They can therefore be made into inserts that clamp into the end of a toolholder. The inserts have six or eight cutting edges and can be quickly indexed around by the operator when an edge becomes worn. Cemented carbide inserts thus represent a highly efficient tooling solution.

Diamond-coated cemented carbide inserts are also being used for high-speed turning of aluminum-silicon alloys, and they provide an excellent surface finish because the seizure between chip and tool is minimized with such an inert material at the interface. Such coatings can be applied with the CVD process.

In spite of their high strength and hardness at high temperature, diamonds are not used for machining steel because tool wear is very rapid. The tools are smoothly worn by a mechanism that appears to involve transformation of diamond to a graphitic form and/or interaction between diamond and iron and the atmosphere.

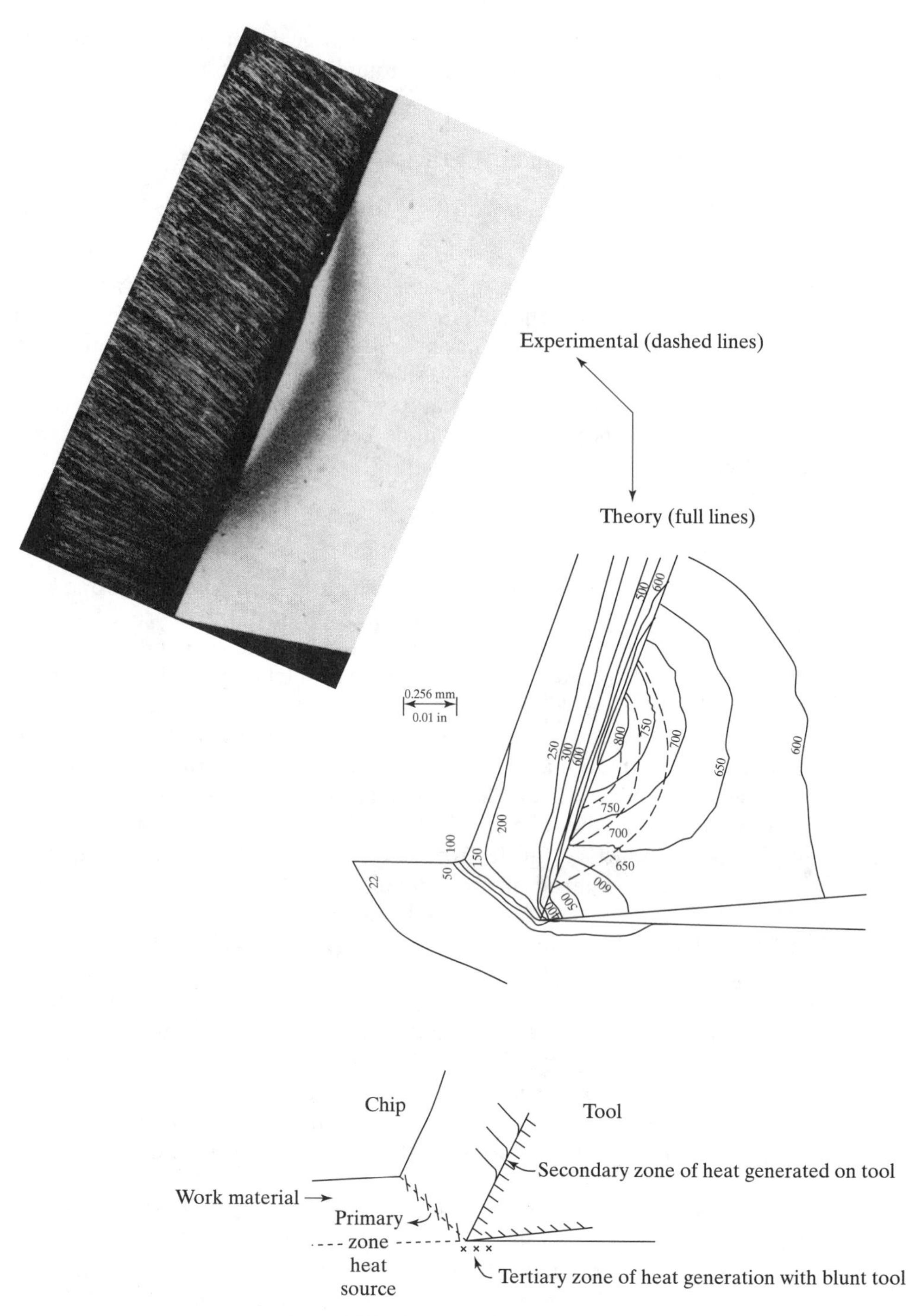

Figure 7.12 Regions of heat generation.

Diamond is not the stable form of carbon at atmospheric pressure. Fortunately, it does not revert to the graphitic form in the absence of air at temperatures below 1,500°C. In contact with iron, however, graphitization begins just over 730°C, and oxygen begins to etch a diamond surface at about 830°C.

It is also disappointing that diamond tools are rapidly worn when cutting nickel and aerospace alloys. Generally, they have not been recommended for machining high-melting-point metals and alloys where high temperatures are generated at the interface.

The family with the highest hot hardness is the alumina-based (Al_2O_3) group, and these are favored for high-speed facing of cast iron. Cast iron machines with a well-controlled "shower" of short chips that facilitate high-speed cutting. However, the Al_2O_3 -based materials are also very brittle, and they have limited use for cutting steels.

Empirically, it can be shown that tool life decreases with increases in cutting speed, as shown in Figure 7.13.

It turns out that the prolific F. W. Taylor also took great interest in this topic. The optimization of cutting speeds fell in naturally with his interests in the principles of scientific management. By the time the results of his *Taylor equation* were applied to the Midvale Steelworks, a productivity gain of 200% to 300% was achieved on the machine tools, which also created a 25% to 100% increase in the wages of the machinists. Taylor found that if the data are replotted on log-log axes, a straight line is obtained for most tool-work combinations.

This observation led to a wide series of plots of the type shown in Figure 7.14. The famous "Taylor equation" relates the cutting speed, V, and tool life, T, to the constants n and C, particular to each tool work combination.

$$VT^n = C \tag{7.12}$$

$$\log T = \frac{1}{n}\log V + \frac{1}{n}\log C \tag{7.13}$$

$$T = \left(\frac{C}{V}\right)^{1/n}_{f \text{ and } d \text{ held constant}} \tag{7.14}$$

Tool life, T, is also sensitive to feed rate, f (with V and d held constant), and to depth-of-cut (with V and f held constant), see Figure 7.15.

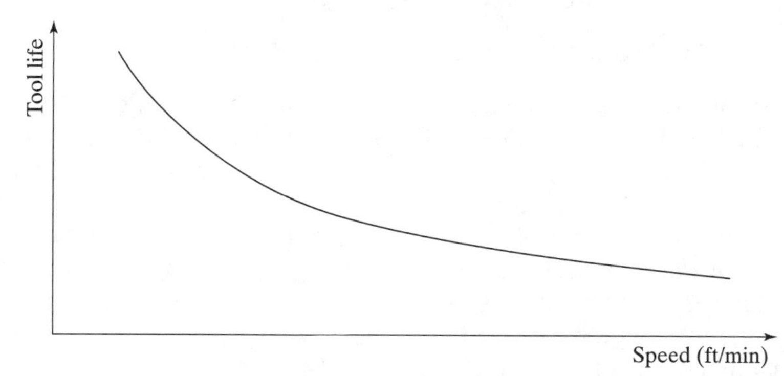

Figure 7.13 Tool life versus cutting speed.

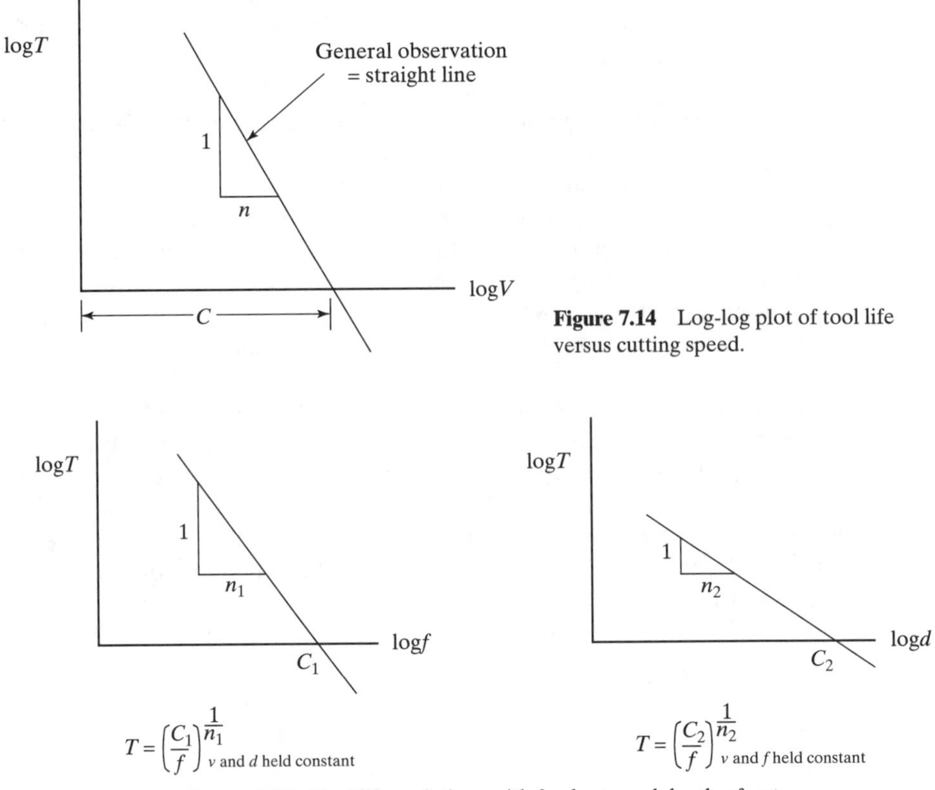

Figure 7.14 Log-log plot of tool life versus cutting speed.

Figure 7.15 Tool life variations with feed rate and depth-of-cut.

However, it is found that

$$\frac{1}{n_2} < \frac{1}{n_1} < \frac{1}{n} \tag{7.15}$$

This physically means that with $(n_2 > n_1 > n)$, changes in cutting speed, rather than feed rate or depth-of-cut, will result in greater amounts of tool wear.

7.3.4 Significance to Work Holding and Fixturing

The forces F_C and F_T generated during milling or turning are resisted by a family of *work-holding* devices called—depending on the context and the specific machining process—fixtures, jigs, clamps, vises, and chucks. The accuracy that can be obtained in a particular machining process is directly related to the reliability of these work-holding devices that allow standard manufacturing machines to process specific parts. *Fixtures* are a subset of work-holding units designed to facilitate the setup and holding of a particular part. The fixture must conform to specific surfaces on the part so that all 6 degrees of freedom are stabilized. Forces and vibrations inherent in the manufacturing process must be resisted by the fixture. A *jig* supports the work like a fixture while also guiding a tool into the workpiece. A jig for drilling, for instance, might contain a hardened bushing to guide the drill to a precise location on the part being processed.

Both fixtures and jigs are usually custom configured to suit the part being manufactured. Hence tooling engineers have endeavored to give these devices flexibility and modularity so that they can be applied to the greatest possible set of part styles. Such flexibility is even more important today, since the trend in manufacturing is toward production in small batch sizes (Miller, 1985). Batch production represents 50% to 75% of all manufacturing, with 85% of the batches consisting of fewer than 50 pieces (Grippo et al., 1988). As the batch size for a particular part decreases, modularizing fixtures and jigs can help to minimize the setup costs per unit produced. Developments in microprocessor-based controllers, sensors, and holding devices in the last decade have made this goal more feasible.

Today's fixture designers depend on heuristics such as the "3-2-1 rule," which states that a part will be immobilized when it is rigidly contacting six points (Hoffman, 1985). Three points define a plane called the primary datum, and two additional points create the secondary datum. The tertiary datum consists of a single point contact. These six locations fix the part position relative to the cutter motions (see Figure 7.16).

If friction is considered, fewer contacts can be used, so long as the applied cutting forces are not excessive. The choice of these datum points is often left up to the fixture designer. However, workpieces used in demanding applications can have their datums explicitly stated in the part drawing. These datums are also used to specify geometric relationships between part features such as perpendicularity, flatness, or concentricity. Information on tolerancing can be found in Hoffman (1985).

Once a suitable set of contact locations on the part has been determined, a rigid structure must be devised to hold these contacts in space. Also, the contact type must be selected. Finally, a set of clamps is chosen that apply forces to the part so that it will remain secured. For complex parts, the final fixture will be a custom designed device that only works for that part with minor variations.

A fixture is composed of *active elements* that apply clamping forces and *passive elements* that locate or support the part. For simple parts a custom designed fixture is not needed. Instead, simpler setups are built that use at least one active element and optional mechanical stops. In the absence of stops, the part can be manually located. Since the loaded position of each part of the same type must be measured,

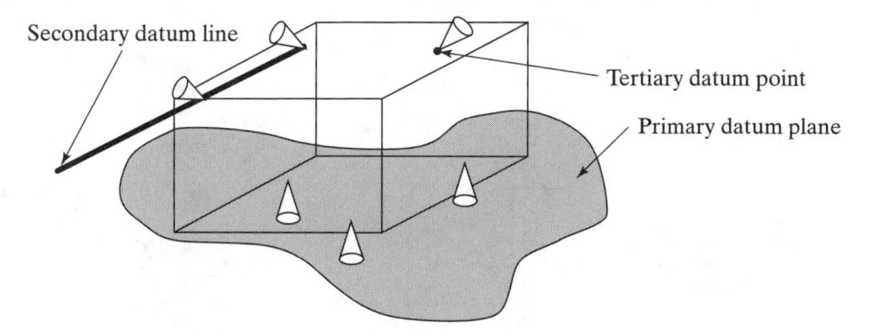

Figure 7.16 The "3-2-1" rule on the primary datum plane.

the time cost of using a simpler setup balances against the cost of building a special fixture.

Figure 7.17 shows some common *passive fixturing* elements. The primary datum can be defined by a subplate that is fixed to the machine tool. When angled features are called for in the part drawing, a *sine plate* may be used. It can reorient the primary datum to any angle from 0 to 90 degrees. They are usually set manually. Angle blocks or plates perform the same function but are not adjustable. Parallel and riser blocks can lift the part up a precise amount. Fixed parallels can be used as a "fence" to prevent motion in the horizontal plane.

Vee blocks give two line contacts so that cylindrical parts can be fixtured. Spherical and shoulder locators are used to establish a vertical or horizontal position. The spherical locator more closely approximates a point contact. This is desirable when the surface being clamped is wavy or when datums are explicitly defined in the part drawing.

The parallel-sided *machining vise* is a versatile tool capable of both *active* clamping and locating prismatic workpieces (Figure 7.18). Special jaws can be inserted that conform to irregular part shapes. The vise consists of two halves, one that is fixed and one that moves toward the fixed portion of the vise. When the vise

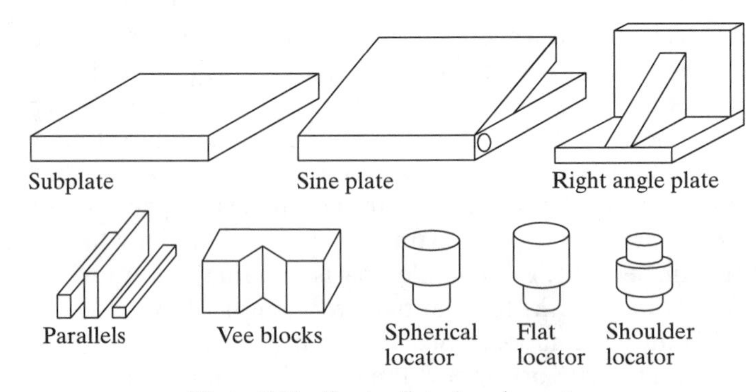

Figure 7.17 Passive fixturing elements.

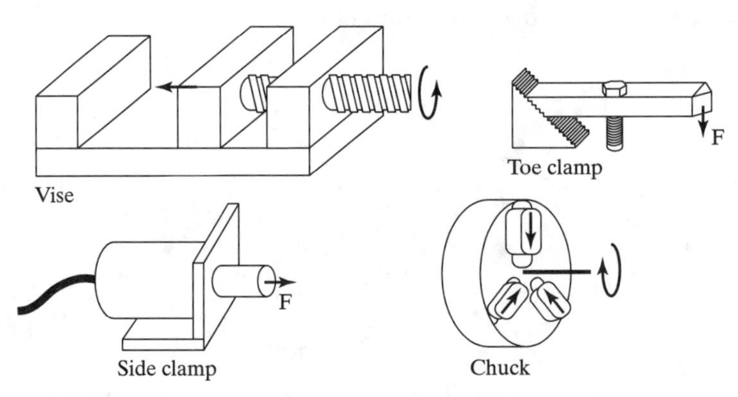

Figure 7.18 Active fixture elements including the standard parallel-sided vise.

jaws have a shoulder and one additional stop, all degrees of freedom are eliminated. Under light machining loads, these additional locators may not be necessary.

Chucks provide an analogous function for rotationally symmetric parts. They have multiple jaws that move radially and, in some cases, independently. A chuck is used in Figure 7.3 to locate and clamp the part. Although such three-jaw chucks have limited accuracy due to finite rigidity and clearances similar to the vise, their flexibility makes them the standard lathe fitting.

Toe clamps and side clamps provide a smaller area of contact and do not locate the part. Toe clamps exert vertical forces on the workpiece and are often used when large or irregular parts, such as castings or flat plates, are being machined. Side clamps provide supplemental horizontal forces that support the part against stops. For safety reasons, they are rarely used alone since the part may become dislodged.

The nature of the contact between the part and the fixture or chuck establishes the maximum clamping force that can be exerted on the part without crushing it and the number of degrees of freedom effectively removed. A greater area of contact means that the clamping forces can be lower. One area of research has been in developing conformable fixtures that increase the area of contact for irregular workpiece shapes. Line contact and point contact induce greater stresses in the material but provide a more precise workpiece location. Large area clamps can also hinder tool accessibility to the component being machined. This is a measure of how many faces of the part are exposed in a given setup and how easy it is to load the workpiece in the tool. The capacity of the fixture to handle different part shapes is a measure of its reconfigurability. Other important qualities for fixtures are reliability, precision, and rigidity.

The development of *new workpiece fixturing devices* is an important area of research. As a first example, *modular tooling sets* (Figure 7.19) are used extensively in industry and represent the state of the art in fixturing as practiced on the factory floor. They were first invented in Germany in the 1940s.

The basic concept of "modular" fixturing is well known: these systems typically include a square lattice of tapped and doweled holes with spacing toleranced to 0.0002 inch (0.005 mm) and an assortment of precision locating and clamping elements that can be rigidly attached to the lattice using dowel pins or expanding mandrels. The tooling's base can be rapidly loaded onto a machining center. This is then fitted out with a complement of active and passive fixturing elements and fasteners.

The elements are assembled in "Erector set" fashion, using standard parts. Extraordinary part shapes might require special elements to be machined. Use of these sets can speed the design and construction of fixtures for small batch sizes. The sets can also reduce the cost of storing old fixtures, since they can be disassembled and reused. The setups can be rapidly replicated, once they have been recorded with photographs and notes. In order to achieve sufficient precision in the assembled fixture, all component surfaces are hardened and ground.

When using modular fixturing, there is a general need for systematic algorithms for automatically designing fixtures based on CAD part models. Although the lattice and set of modules greatly reduce the number of alternatives, designing a suitable fixture currently requires human intuition and trial and error. Furthermore, if the set of alternatives is not systematically explored, the designer may settle upon a suboptimal design or fail to find any acceptable designs.

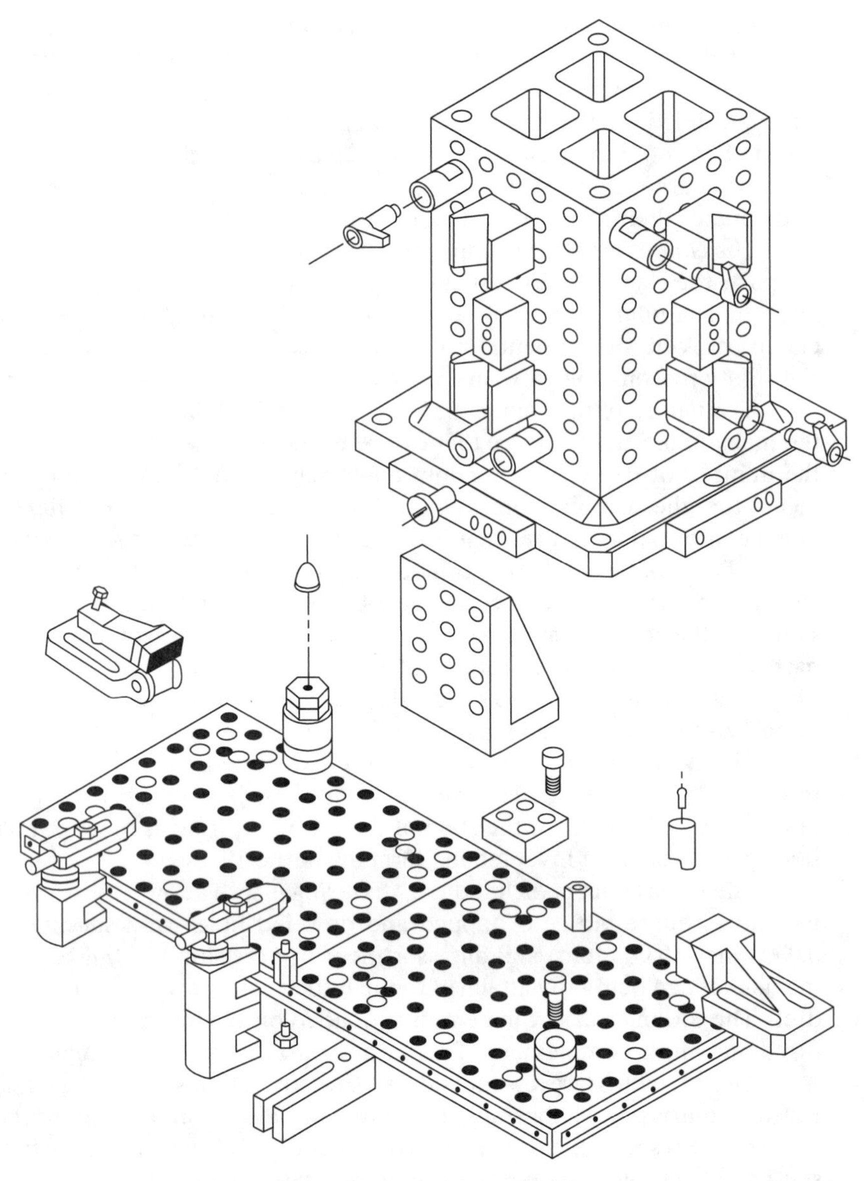

Figure 7.19 Modular tooling kit.

Goldberg and colleagues (Wagner et al., 1997) have thus considered a class of modular fixtures that prevent a part from translating and rotating in the plane. The implementation is based on three round locators, each centered on a lattice point, and one translating clamp that must be attached to the lattice via a pair of unit-spaced holes, thus allowing contact at a variable distance along the principal axes of the lattice. World Wide Web users may now use any browser to "design" a polygonal part. Goldberg's FixtureNet returns a set of solutions, sorted by quality metric, along with images showing the part as the fixture will hold it in form closure for each solution.

The current version of FixtureNet is described in Section 7.12. The links on the Website include an online manual and documentation. This initial service provides an algorithm that accepts part geometry as input and synthesizes the set of all fixture designs in this class that achieve form closure for the given part. This is one of the first fixture synthesis algorithms that is complete, in the sense that it guarantees finding an admissible fixture if one exists. Planning agents can call upon FixtureNet directly and explore the existence of solutions, practical extensions to three dimensions, and issues of fixture loading.

As a second example, *quick change tooling* is helpful in factories that use extensive automated material handling. It can also reduce the setup time at the machining workstation. For instance, the automated pallet changer receives pallets of standard size and connections, carrying a diverse array of part shapes. It can act as the tool base for a modular work-holding system. In this way, a part can travel from a lathe to a mill with no refixturing time, potentially on material handling equipment with this same receiver. Standard connections to the equipment can be made in seconds. In flexible manufacturing systems (FMS), these pallets are built up and loaded off-line at manual workstations.

As a third example, *hydraulic clamping systems* have been developed to replace manually actuated active elements. The oil charged cylinders provide a much more compact and controllable source of clamping power. Hydraulic circuits can be created that result in self-leveling supports, sequenced clamping order, and precise clamping forces. When accumulators are used, the hydraulic power source can be disconnected without a reduction in clamping force.

As a fourth example, the *automatically reconfiguring fixture system* described by Asada and colleagues (1985) is intended for sheet-metal drilling operations. The tool base has a number of tee slots into which a cartesian assembly robot inserts vertical supports. The supports feature a lock mechanism that permits them to be assembled with one "hand." The act of grasping the clamp unlocks it, after which it can be slid into position along the tee slot. The height of the locators can also be set by the robot. An operator selects contact points on a 3-D wireframe model of the part, and the system decomposes this into a series of manipulation tasks.

As a final example, the *reference free part encapsulation* (RFPE) system is designed to "free up" the design space and greatly expand the possible range of the parts that can be designed and then machined (Sarma and Wright, 1997). RFPE allows the machining of parts with thin spars and narrow cross sections. RFPE uses a biphase material (Rigidax) to totally encapsulate a workpiece and provide support during the machining process (Figure 7.20).

After the first side of a component has been machined, the Rigidax is poured around the features, returning the stock to the encapsulated, prismatic, bricklike appearance that can be easily reclamped. Machining then continues on the other sides. This iterative process at the manufacturing level of abstraction (encapsulate/machine side-1/repour-to-reencapsulate/reposition/machine side 2, etc.) has a dramatic "deconstraining" effect on the designer. The RFPE fixturing rules are described by a smaller set than those for conventional fixturing.

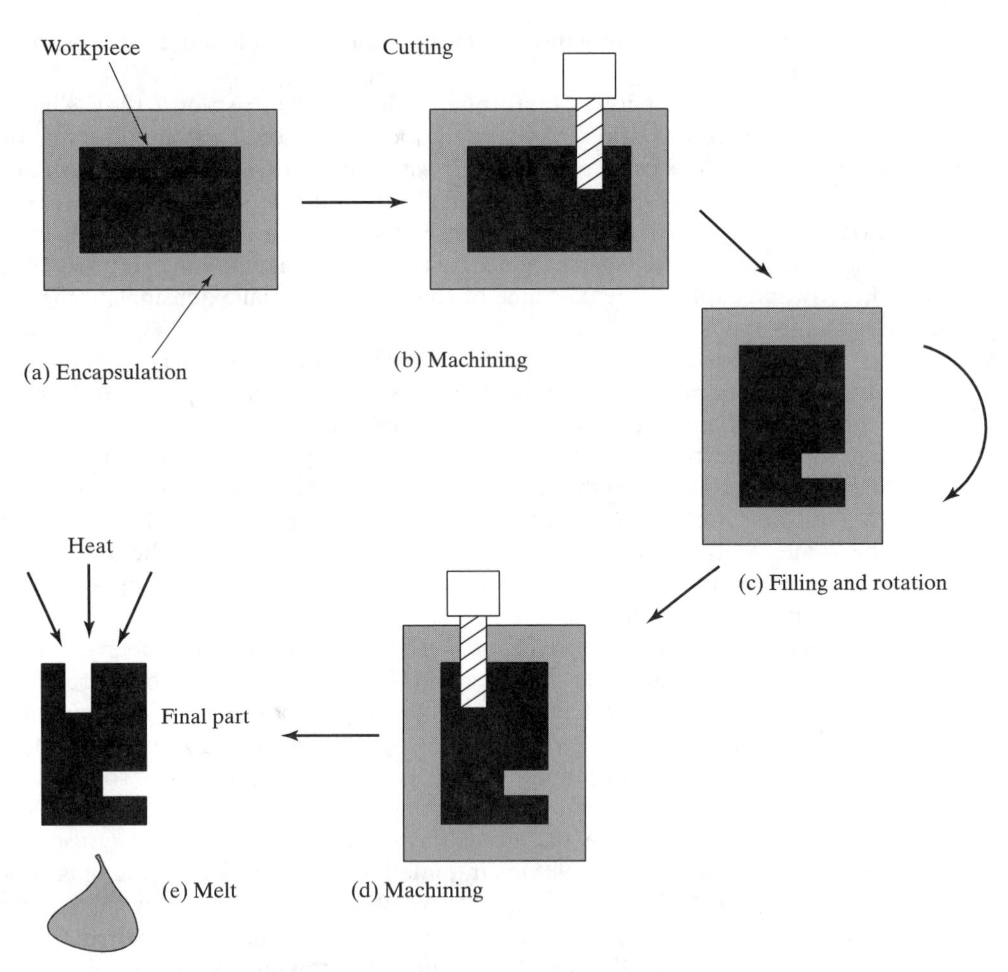

Figure 7.20 Reference free part encapsulation (RFPE) "deconstrains the design space" during fixturing for machining.

The use of RFPE does decrease the achievable tolerances to some degree. Without RFPE a machine tool offers a daily accuracy of +/−0.001 inch (0.025 mm). Also Mueller and colleagues (1997) have used simulation packages prior to cutting, and sensors during the machining process, to obtain tolerances down to +/−0.0002 inch (0.005 mm). During fabrication with RFPE, typical tolerances average +/−0.003 inch (0.075 mm). Ongoing research will aim to improve the machining accuracy using RFPE techniques.

7.4 THE ECONOMICS OF MACHINING

7.4.1 Introduction

A method is now introduced to optimize the costs of operating the machine tools in a production shop. Actually, the general method is applicable to many variable cost analyses in manufacturing. A detailed treatment of this topic is therefore generally relevant to shop-floor microeconomics. The general goals are to achieve one of the following:

- Minimize the production cost per component
- Minimize the production time per component
- Maximize the profit rate

The symbols shown in Table 7.2 are needed for the analysis.

7.4.2 Production Cost per Component

The cost to produce each component in a batch is given by

$$C_{\text{PER PART}} = WT_L + WT_M + WT_R\left[\frac{T_M}{T}\right] + y\left[\frac{T_M}{T}\right] \tag{7.16}$$

In this equation, the symbols include

W = the machine operator's wage plus the overhead cost of the machine.
WT_L = "nonproductive" costs, which vary depending on loading and fixturing.
WT_M = actual costs of cutting metal.
WT_R = the tool replacement cost shared by all the components machined. This cost is divided among all the components because each one uses up T_M minutes of total tool life, T, and is allocated of T_M/T of WT_R.

Using the same logic, all components use their share T_M/T of the tool cost, y.

TABLE 7.2 Symbols and Explanations for the Analysis on the Economics of Machining

Symbol	Explanation	Usual Units	Usual Units (SI)
V	Cutting speed	ft/min	m/min
f	Feed rate for the turning operation in Figure 7.3. It has been found empirically that speed is much more damaging to the tool than either feed rate or depth-of-cut. Thus V appears in the analysis more than f or d.	inches/rev	mm/rev
d	Depth-of-cut in the turning operation	inches	millimeters
T	Tool life	minutes	minutes
T_M	Time cutting metal	minutes	minutes
T_R	Replacement time of a worn tool	minutes	minutes
T_L	Part loading time, which includes (loading + fixturing + advancing + overrun + tool withdrawal + part unloading)	minutes	minutes
W	Average cost per minute of operating the machine plus the operator's wage	$/minute	$/minute
y	Cost of the cutting edge of the tool. For a cemented carbide indexable insert the cost of a single edge is the cost of the insert divided by the number of edges (usually 3, 4, 6, or 8)	$	$

Today's turning tools are usually cemented carbide indexable inserts, and there are three, four, six, or eight edges that are available for use on any individual insert. The number depends on whether the insert is triangular or square and whether it is positive or negative rake angle. Positive rake tools yield only three of four edges. An economic reason for using negative rake tools is that both faces of the insert can be used to give the six or eight available edges. In general the cost y = cost of insert divided by the number of usable edges (three, four, six, or eight).

7.4.3 Production Time per Component

The time to produce each component in a batch is given by

$$\text{Total time} = T_L + T_M + T_R\left(\frac{T_M}{T}\right)$$

In the event that time is more important than money, perhaps to accommodate a valued customer, this equation should be optimized rather than the previous one.

7.4.4 Profit Rate

The third consideration might be the profit rate, given by the following equation:

$$\text{Profit rate} = \frac{\text{income per component} - \text{cost per component}}{\text{time per component}}$$

7.4.5 Minimum Cost versus Minimum Time

It is possible to calculate either the recommended speed for the minimum cost, $V_{\text{opt 1}}$, or the recommended speed for the minimum time, $V_{\text{opt 2}}$. The calculations are essentially the same except that the time-oriented analysis ignores tooling costs (though not the tool replacement time). Sacrificing the tooling cost, perhaps to please a valued customer, creates the higher value for the optimum cutting speed shown on the x axis in Figure 7.21. In either case, though, as speed V, feed rate f, or depth-of-cut d, is increased, the tool is stressed more.

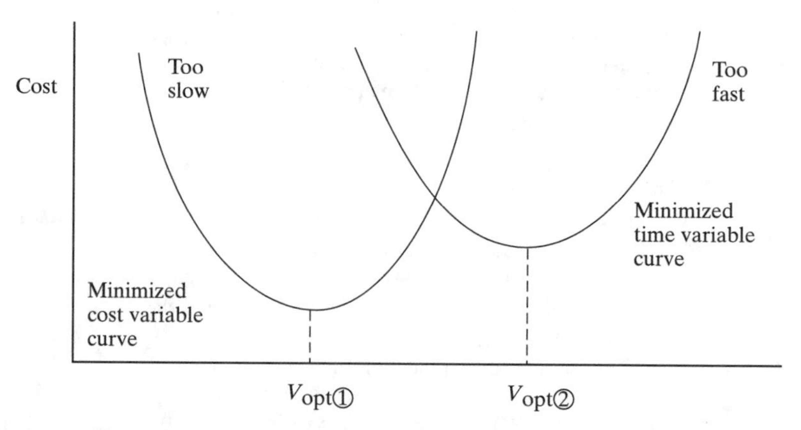

Figure 7.21 Optimal cutting speeds for minimized cost and time.

- Thus on the one hand, if V is too low, then the machining time T_M will be too high.
- On the other hand, if V is too high, then T will be too low, and T_R and y will be too high.

This trade-off between machining time on the one hand and tool life on the other hand creates the minimums and recommended optimum speeds in Figure 7.21.

7.4.6 Analysis of Minimum Costs

Limiting the analysis to turning, rather than milling, it can be shown that the time taken to machine the bar in Figure 7.3 is

$$T_M = (\pi dl)/1000fV \tag{7.17}$$

This is the expression for the time to machine the round bar in the lathe of Figure 7.3, where the length of the bar is (l), its diameter is (d), the feed rate is (f), and the cutting speed is (V).

Units are peculiar to the standard industrial ways of expressing speed in meters per minute and feed in terms of millimeters. Length and diameter are also in millimeters. To make all the units compatible, the meters per minute are multiplied by 1,000.

It is possible to calculate the optimum cost per component with respect to cutting speed. Essentially, the idea is to *differentiate Equation 7.16 with respect to V* and find the minimum in the curve in Figure 7.21. The following steps are taken:

Step 1: Maximize the feed rate, f, for a desirable surface finish. Section 7.7.24 describes how surface finish (R_a) is measured by the arithmetic mean of the surface undulations. In Equation 7.18, (R) is the nose radius of the lathe tool.

$$R_a = 0.0321(f^2/R) \tag{7.18}$$

Step 2: Perform the differentiation of Equation 7.16 using the Taylor Equation (Equation 7.12) and machining time (Equation 7.17) to isolate the parameter V. The expressions are rather cumbersome. Detailed analyses are presented in other machining textbooks such as Cook (1966) or Armarego and Brown (1969). Only the final equations are given here. The value of T appearing in the following equations is the value of tool life that will give minimum cost with variations in V. The cutting speed, V, at the minimum cost is also shown in these equations and in Figure 7.21 as V_{opt}. At this optimum set of values, all the variable parameters (f, V, T, etc.) are denoted with an asterisk (*).

Step 3: Generate the Taylor constants n, n_1, and K. Also calculate $\Re$ in Equation 7.19:

- First, Taylor equations of (T versus V) and (T versus f) are needed. Recall that increases in feed rate are "less damaging" to the tool life than increases in speed. Values of n and of n_1 appear in the equations that follow.
- Second, since the Taylor equations are now a function of both V and f, the constant (C) is replaced by the constant (K), which combines both the feed and speed constants. This is also shown in Equation 7.21.

- Third, to account for the variables in the main Equation 7.16 that are not directly related to change in speed, another cost-related constant ($\Re$) is formed that combines the tool cost, y, the (operator + machine cost) = W, and the tool replacement time, T_R. Here, y/W is a constant without units and is added to a value of T_R in minutes.

In the following equations, all times are measured in minutes, and all costs are in cents. The values of n, n_1, K, and $\Re$ are constants.

$$\Re = T_R + (y/(W)) \tag{7.19}$$

$$T^* = \Re\left(\frac{1}{n} - 1\right) \tag{7.20}$$

$$T^* = K(V^*)^{-1/n}\,(f)^{*-\frac{1}{n_1}} \tag{7.21}$$

$$V^* = \left(\frac{K}{T^*(f^*)\frac{1}{n_1}}\right)^n \tag{7.22}$$

$$\left(C^* = W\left(T_L + \frac{T_M^*}{1-n}\right)\right) \text{ – or – } \left(C^* = W\left(T_L + T_M^*\left(1 + \frac{\Re}{T^*}\right)\right)\right) \tag{7.23}$$

In summary, the preceding equations relate the optimized tool life, T^*, the recommended cutting speed, V^*, and the recommended feed rate, f^*, to get the minimum in the parabolic graph shown earlier. Equation 7.17 gives $T_M{}^*$.

7.5 SHEET METAL FORMING

7.5.1 Deformation Modes in Sheet Forming

The wide variety of sheet metal parts for both the automobile and electronics industries is produced by numerous forming processes that fall into the generic category of "sheet-metal forming." Sheet-metal forming (also called stamping or pressing) is often carried out in large facilities hundreds of yards long.

It is hard to imagine the scope and cost of these facilities without visiting an automobile factory, standing next to the gigantic machines, feeling the floor vibrate, and watching heavy duty robotic manipulators move the parts from one machine to another. Certainly, a videotape or television special cannot convey the scale of today's automobile stamping lines. Another factor that one sees standing next to such lines is the number of different sheet-forming operations that automobile panels go through. Blanks are created by simple shearing, but from then on a wide variety of bending, drawing, stretching, cropping, and trimming takes place, each requiring a special, custom-made die.

Despite this wide variety of subprocesses, in each case the desired shapes are achieved by the modes of deformation known as drawing, stretching, and bending. The three modes can be illustrated by considering the deformation of small sheet elements

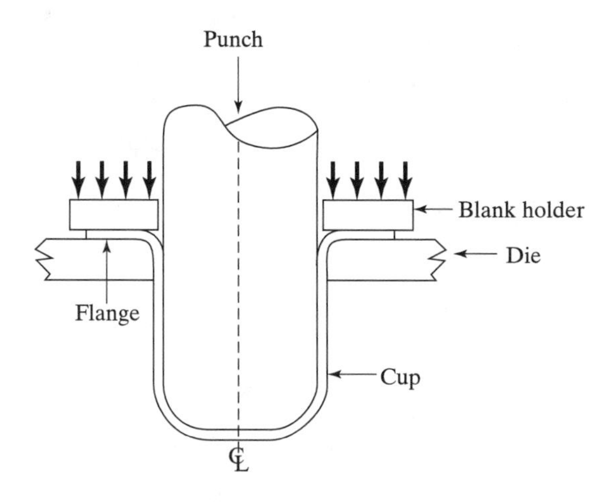

Figure 7.22 Sheet forming a simple cup.

subjected to various states of stress in the plane of the sheet. Figure 7.22 considers a simple forming process in which a cylindrical cup is produced from a circular blank.

1. *Drawing* is observed in the blank flange as it is being drawn horizontally through the die by the downward action of the punch. A sheet element in the flange is made to elongate in the radial direction and contract in the circumferential direction, the sheet thickness remaining approximately constant (see top right of Figure 7.23).
2. *Stretching* is the term usually used to describe the deformation in which an element of sheet material is made to elongate in two perpendicular directions in the sheet plane. A special form of stretching, which is encountered in most forming operations, is *plane strain stretching.* In this case, a sheet element is made to stretch in one direction only, with no change in dimension in the direction normal to the direction of elongation but a definite change in thickness, that is, thinning.
3. *Bending* is the mode of deformation observed when the sheet material is made to go over a die or punch radius, thus suffering a change in orientation. The deformation is an example of *plane strain elongation and contraction.*

7.5.2 Materials Selection to Avoid Failure during Stretching

In the stretching operation shown at the bottom of Figure 7.23, fracture may often occur by local thinning (i.e., "*necking*") near one of the corners of the sheet. The combination of the stretching at the dome of the punch and the bending near the corners creates the highest strain in the deforming metal. It follows, then, in stretch forming that if localized thinning is to be prevented, materials with an ability to increase in strength during deformation should be selected.

At the start of a process, a metal becomes stronger in the deformed region and the strain is transferred to another location. This process of "shifting the next increment of strain to adjacent weaker material" continues. However, eventually, the strain-hardening capacity of a local region is exhausted and necking starts. The

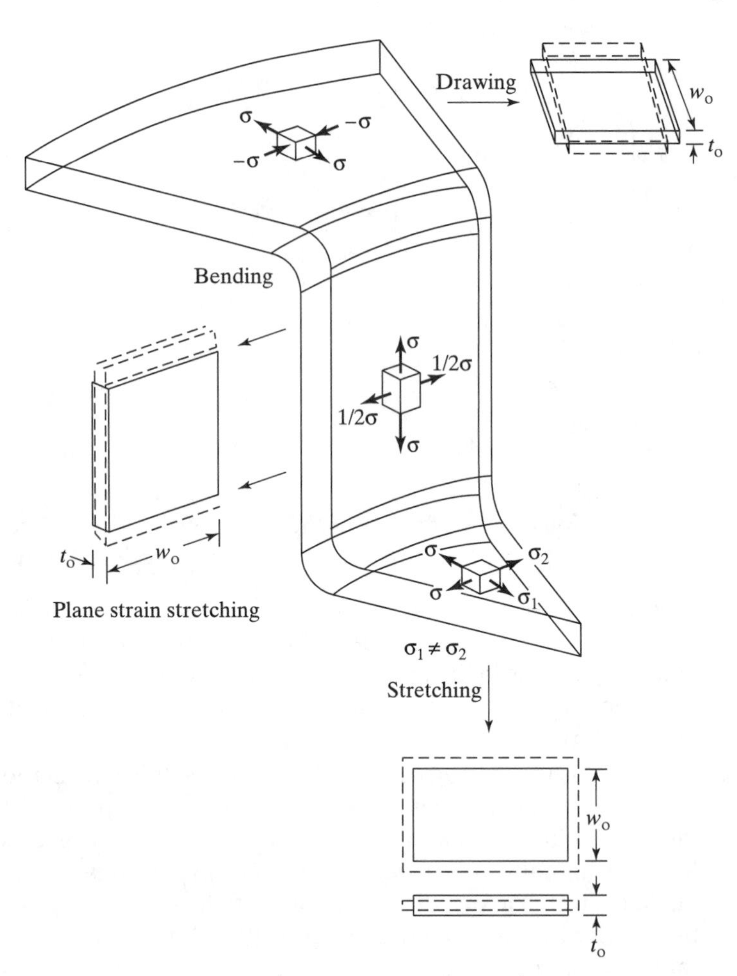

Figure 7.23 Modes of sheet forming.

strain-hardening characteristics of sheet materials are usually described by the exponent n in the true stress–true strain relationship:

$$\sigma = K\varepsilon^n$$

where σ = true stress

K = a material constant

ε = true strain

n = strain-hardening exponent

Figure 7.24 shows the standard plot of true stress versus true strain (see Rowe, 1977). On a log-log plot, this usually gives a straight line for the n value. High values of n are desirable in materials subjected to stretching operations because they lead to a more uniform distribution of strain, that is, less localized strain.

Figure 7.25 illustrates the influence of n in a set of bulging tests. The data were obtained by Meyer and Newby (1968) by bulging circular blanks of three dif-

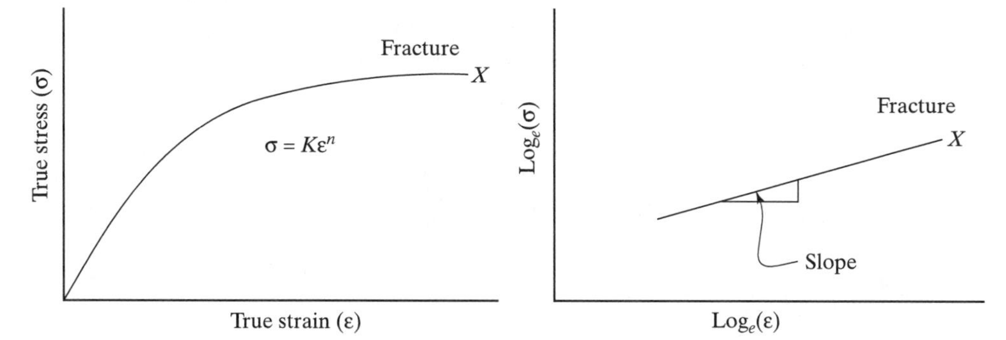

Figure 7.24 The stress–strain curve plotted on a log-log scale gives a straight line for n.

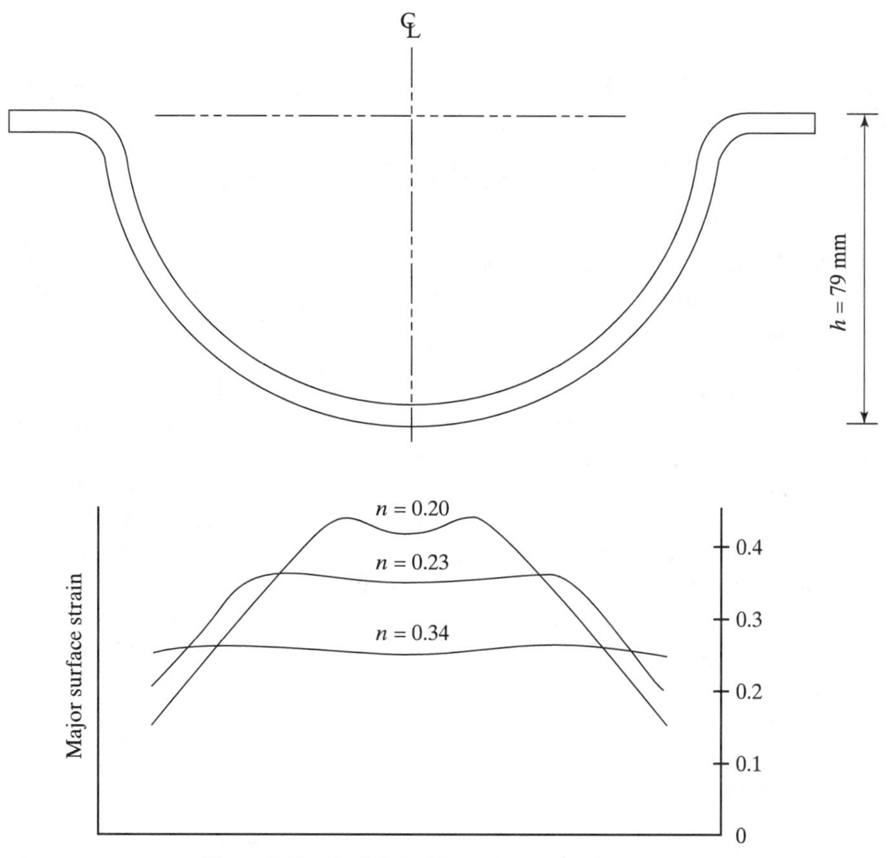

Figure 7.25 Radial strain in a hemispherical dome.

ferent materials to the same height (79 mm) with a hemispherical punch. The material with the higher n value exhibited a much lower strain at the top center of the dome because more of the deformation had been transferred to the peripheral regions.

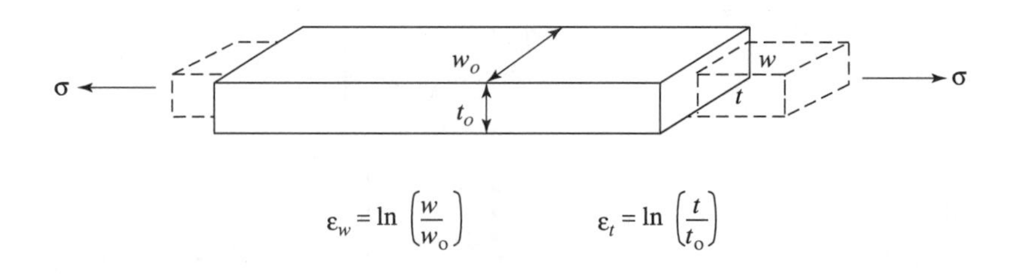

$$\varepsilon_w = \ln\left(\frac{w}{w_o}\right) \qquad \varepsilon_t = \ln\left(\frac{t}{t_o}\right)$$

$$\text{Normal anisotropy strain ratio, } R = \frac{\varepsilon_w}{\varepsilon_t}$$

Figure 7.26 Deformation of a tensile specimen to find the R value.

Some typical n values for various materials are shown below:

Mild steel (capped, Al-killed, rimmed), $n = 0.22\text{–}0.23$
Austenitic stainless steels, $n = 0.48\text{–}0.54$
Ferritic stainless steel, $n = 0.18\text{–}0.20$
70/30 brass (annealed), $n = 0.48\text{–}0.50$
Aluminum alloys, $n = 0.15\text{–}0.24$

7.5.3 Materials Selection to Avoid Failure during Drawing Operations

While the previous stretching modes require ductile materials with good strain-hardening properties, drawing operations require materials with strong normal anisotropy, that is, stronger in the through-thickness direction than in the sheet plane. (In the following, the goal is to have a low value of strain in the through-thickness and a high value in the plane, hence a high value of the parameter R.)

Resistance to thinning in the through-thickness is measured by the plastic anisotropy parameter, R, which is defined as the ratio of the plastic strain in the plane of the sheet to the plastic strain in the thickness direction (Figure 7.26).

A high value of R indicates good drawability because the value of ε_w will be greater than ε_t. Actually, sheet materials nearly always exhibit marked crystalline anisotropy, meaning that the rolled strip has different properties in the "rolling direction," "directly across," and at "any angle across the sheet."

As shown in Figure 7.27, an average value of R is determined from four specimens cut so that the tension axes are, respectively, 0, +/-45, and 90 to the rolling direction. The average value is then evaluated to give R_m.

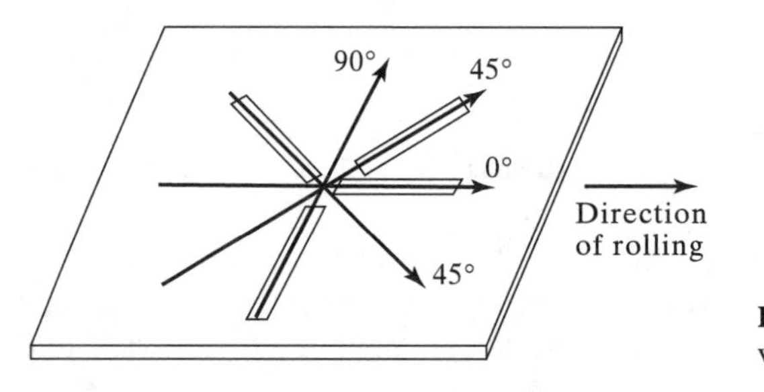

Figure 7.27 Obtaining the mean R value from four different specimens.

Some typical R values are shown below:

Mild steel, $R_0 = 0.90–1.60$; $R_{45} = 0.95–1.20$; $R_{90} = 0.98–1.90$; $R_m = 0.98–1.50$
Aluminum alloys, $R_m = 0.6–0.8$
Austenitic stainless steels, $R_m = 0.90–1.00$
Ferritic stainless steel, $R_m = 1.00–1.20$
70/30 brass (annealed), $R_m = 0.80–0.92$
Titanium, $R_m \cong 3.8$
Alpha-titanium alloys, $R_m = 3.0–5.0$
Zircaloy: 2 sheet (cold rolled), $R_m \cong 7.5$

Drawbeads are often introduced in practice to avoid failure around the top of the sheet as it flows into the die wall. The drawbeads resemble "bumps" that are machined or inserted into the surface of the die. They hold on to the sheet as it flows toward the die zone, as shown in Figure 7.28.

7.5.4 Testing Methods

A range of specialized tests has been developed to assist in simulating each aspect of forming. Two examples of such tests are outlined here. The first measures stretchability, and the second drawability.

The Erichsen test. In this test the stretchability limits of sheet materials are established under conditions of balanced biaxial tension. A specimen 90 mm wide is clamped tightly against a 27-mm diameter die, and a spherical punch of 20 mm diameter is pressed against the specimen until fracture occurs. The bulge that forms is almost entirely due to stretching, and the depth of the bulge at fracture is then taken

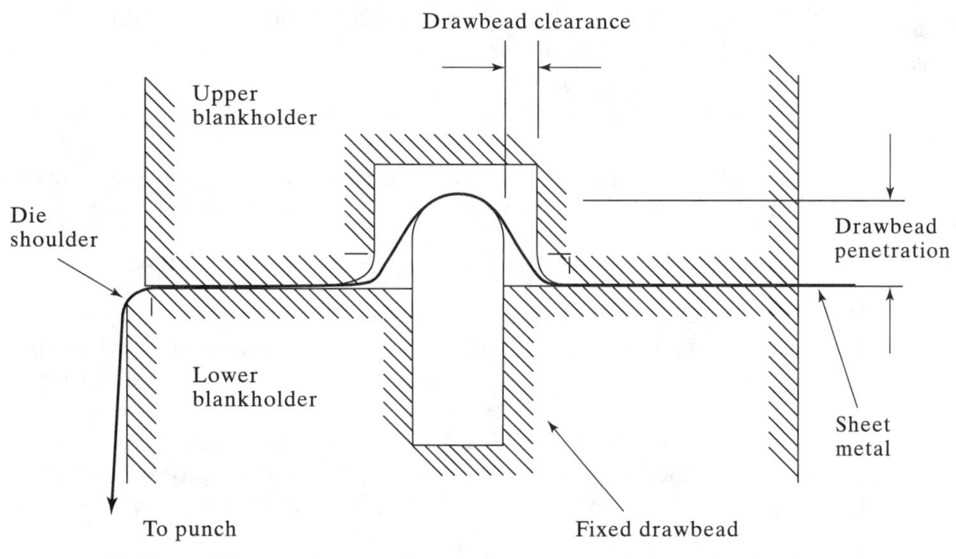

Figure 7.28 Drawbead configuration to restrain material during drawing.

as the limit of stretching for the material. This test measures stretchability but does not assess drawability.

The Swift test. In this test, flat-bottomed cups are drawn from a series of circular blanks having slightly different diameters until a blank size is found above which all cups fracture. If this blank diameter is divided by the punch diameter, the limiting drawing ratio (LDR) is obtained. The Swift test is obviously applicable to drawing operations but is of little value in assessing stretchability.

7.5.5 The Forming Limit Diagram

The Erichsen and Swift tests are useful in providing some guidance to the die setter in practice. However, because of their restricted nature, they cannot be used to establish the forming limits for complex processing in which both drawing and stretching modes of deformation occur simultaneously.

The *forming limit diagram* therefore provides a more comprehensive graphical description of the various surface-strain combinations that lead to failure in a generalized forming operation. The first diagrams, introduced by Backofen and associates (1972) and Goodwin (1968), were determined by empirical methods that involved a large number of simulative tests, similar to the two described previously. Such forming limit diagrams indicate the failure strains (i.e., at necking and fracture) in a given material for various *combinations* of the maximum (e_1) and minimum (e_2) strain components in the sheet plane.

As an example, consider a case when a sheet is stretched in such a way that the two surface-strain components are equal in magnitude and direction at all times (i.e., $e_1 = e_2$). This represents a balanced biaxial tension stress situation and corresponds to that obtained in the Erichsen test. This situation is represented by the line on the far right of Figure 7.29. Various additional tests—for example, with $e_1 = 2e_2$, $e_2 = 0$, $e_1 = -e_2$, and so on—can be performed on the same material and the strain values at failure determined. The locus of all such failure conditions at points x in Figure 7.29 can then be drawn. *This locus is termed the forming limit diagram.*

With reference to Figure 7.29, it can be readily appreciated that the region in which e_2 is negative (i.e., compressive) describes the deformation conditions encountered during a drawing operation, while the region in which e_2 is positive (i.e., tensile) represents stretching. The particular case when $e_2 = 0$ describes the plane strain stretching mode of deformation.

7.5.6 Usefulness of the Forming Limit Diagram in Practice

It is of interest to note that $e_2 = 0$ represents the least favorable combination of surface strains in any forming operation. Therefore, increasing *or* decreasing the strain e_2 in a critical region of a pressing permits a greater amount of deformation to take place before failure occurs.

To further study the formability of an automobile panel, a common practice is to first imprint the blank with a grid pattern. It is possible to use an etching process that creates a grid of small circles with a diameter of 2 to 3 mm. The dimensions of the circles are then measured after pressing (see Figure 7.30).

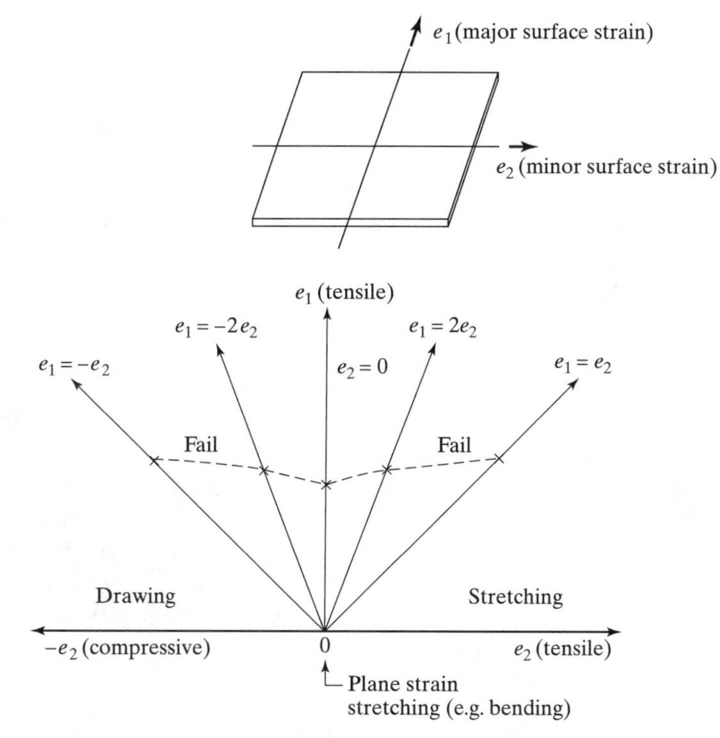

Figure 7.29 The basic forming limit diagram.

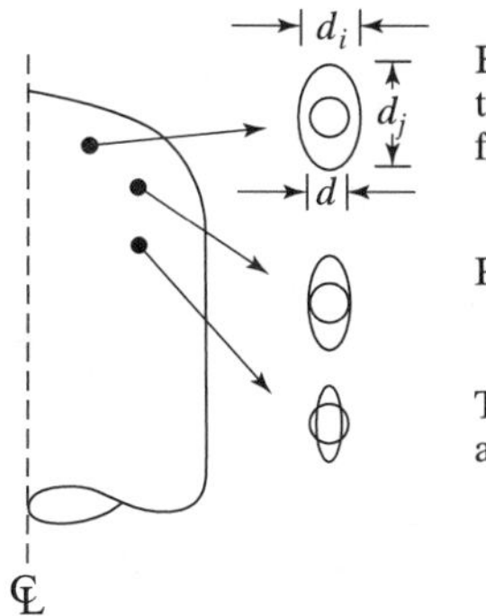

Figure 7.30 Strain states in a formed shell: small circles etched onto the upward formed shell can be used to study the strain distributions in practice.

The grid circles are deformed during pressing into ellipses, and the mutually perpendicular major and minor axes of these ellipses define the principal surface true strains ε_1 and ε_2.

From the geometry of circles and ellipses:

$$\varepsilon_1 = \log_e \frac{d_j}{d} \qquad \text{and} \qquad \varepsilon_2 = \log_e \frac{d_i}{d}$$

where

$$d = \text{grid circle diameter before pressing}$$

$$d_j = \text{major diameter of ellipse after pressing}$$

$$d_i = \text{minor diameter of ellipse after pressing}$$

It is frequently more convenient to express the resulting principal strains in terms of engineering rather than true or natural strain definition—that is,

$$e_1 = \frac{d_j - d}{d} \quad \text{and} \quad e_2 = \frac{d_i - d}{d}$$

If the analysis is made for the circles immediately adjacent to necked or fractured regions, a plot of major strain (e_1) against minor strain (e_2) will yield a curve that separates the strain conditions for successful pressings from those that result in weakness or fracture (i.e., forming limits will be established).

Thus, as indicated in the experimental press shop data of Figure 7.31, "safe" regions and "fail" regions for different materials and thicknesses can be established.

It is informative to consider some day-to-day uses of the forming limit diagram. The die setter can quickly determine, from a single pressing on a gridded sheet, whether a new component with its given set of tools is going to be easy, hard, or

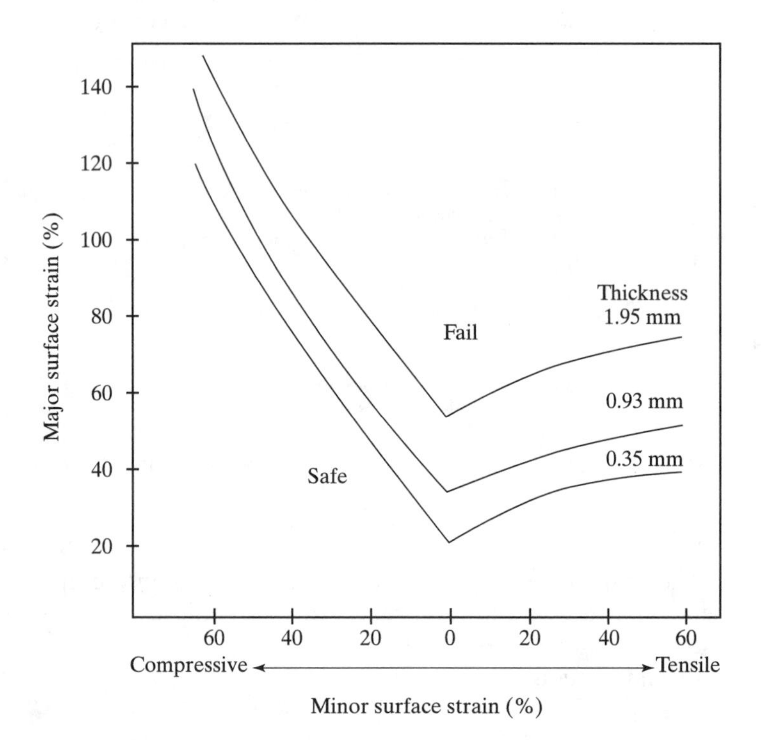

Figure 7.31 Forming limit diagram. Different sheet thicknesses shown on right (data from Rose, 1974).

impossible. Then it might be possible to argue quantitatively for a design modification or a change of material. Or, as discussed previously, it might be possible to move from failure to success by increasing one of the strain components through a slight change in die geometry.

In many cases, special lubricants containing molybdenum disulfide can be locally applied to critical areas of the pressing, just to change the strain distribution near potential thinning and fracture points. In a different scenario it might be found that the material being used is too good and that a cheaper grade of sheet could be introduced.

By keeping a satisfactory reference pressing, the die setter can also locate a source of trouble if, later on in the production run, the press gets out of adjustment or the properties of the sheet change. Finally, the training of press operators and die setters can be made considerably easier and quicker if gridded blanks are available for reference and demonstration.

7.6 MANAGEMENT OF TECHNOLOGY

7.6.1 Precision Manufacturing Services and Their Clients

Today's clients for "high-tech" machine shops and metal fabrication shops include the medical industry, the biotechnology industry, mold-making industries that create the plastic casings of electromechanical consumer products, the aerospace industry, and the special-effects companies for Hollywood's movie industry.

Small-batch high-precision machine shops are also the key suppliers of equipment for the semiconductor industry (Chapter 5) and PCB industries (Chapter 6). The stepper machines that increment the masks in photolithography are an excellent example of $1 to $2 million machines that are initially fabricated by metal machining.

This review of the "client base" for machining brings out a key historical observation that will hold true in the future. Namely, *the machine tool industry is a key building block for industrial society, since it provides the base upon which other industries perform their production.* This fact was especially true in the decades that followed the first industrial revolution (approximately 1780 to 1820). Throughout the period to the 1920s, the machine tool industry was the foundation for the shipbuilding, railroad, gun-making, construction, automobile, and early aircraft industry.

Since then it has also become the foundation for semiconductor fabrication and all forms of consumer product manufacturing. As these devices become more specialized and miniaturized, the construction of equally specialized equipment will still be performed at these "high-tech" machine shops and metal fabrication shops.

Given this range of services, it is not surprising that a new awareness of precision and optimization is emerging. Also, processes such as laser machining that were once regarded as specialized are now being used on a day-to-day basis for precision hole drilling (Chryssolouris, 1991).

Overall, best practices include rapid links from design to G and M codes as indicated in Section 7.2.5, highly tuned economically operated machine tools as indicated in Section 7.4, and a greater appreciation for tooling design as indicated in Section 7.5.6. Deeper understandings of the physics of sheet-metal forming and machining—for example, the prediction of the cutting forces in Section 7.3.1—result in sensible investments in machine tools, forming machines, and rolling mills.

7.6.2 Open-Architecture Manufacturing

At the same time, more sophisticated control of the metal (cutting and metal) forming machinery is allowing these more traditional processes to keep pace with the SFF technologies described in Chapter 4. Some new developments in the last decade that have given more flexibility to CNC machinery controllers are now described.

Today, factory-floor CNC machines are supplied by the machine tool companies with "closed controller architectures." Fanuc, Mazak, and Cincinnati-Milacron are some of the most often seen controllers. Specifically this means that a user or programmer is constrained to work with the predefined library of G and M codes (now the RS 274 standard) that are supplied with each machine tool company's vendor-specific controller. This results in limited library functions, written in local formats. These are adequate today for routine production machining but they are not "open" to any arbitrary third-party software developers able, say, to supply C-based routines for new CAD geometries or new machining sensors coming onto the market.

A broader "openness" to any outside third-party developer is one of the design goals of several U.S. government projects (Schofield and Wright, 1998; Greenfeld et al., 1989). The aim is to improve the productivity of the U.S. machine tool industry, not just by focusing on machine tool companies alone but also by expanding market opportunities for CAD companies, sensor companies, diagnostic software developers, and all ancillary product suppliers. The paradigm is the vastly expanding PC industry. It is anticipated that by using generic products and open systems, a large number of third-party products will be supported commercially, hence increasing the productivity of standard CNC machines and flexible manufacturing systems.

"Open-architecture" machinery control (Figure 7.32) will allow faster access between high-level computer aided design (CAD), computer aided process planning (CAPP), and computer aided manufacturing (CAM).

- As a first example, especially for mold making and some aerospace parts, it is crucial to be able to take interesting, highly complex geometries from CAD and convert them into cutting tool motions. For example, a particular goal of the work by Hillaire and associates (1998) is the ability to take NURBS (nonuniform rational B-spline) curved surfaces from CAD and execute them on a standard three-axis milling machine. By contrast, with "closed architectures" it is likely that the user would be confined to the geometries and standard interpolations in the machine tool company's library.
- As a second example, open architectures allow a machine tool to automatically compensate for errors in the positioning of the workpiece and make possible the active control of the machining process by accepting inputs from external sensors—something that the previous generation of controllers could not do. This results in faster production, more flexibility, and more opportunity for on-machine inspection and quality control. More flexibility in sheet forming can also be created with controllable die surfaces (Walczyk and Hardt, 1998).

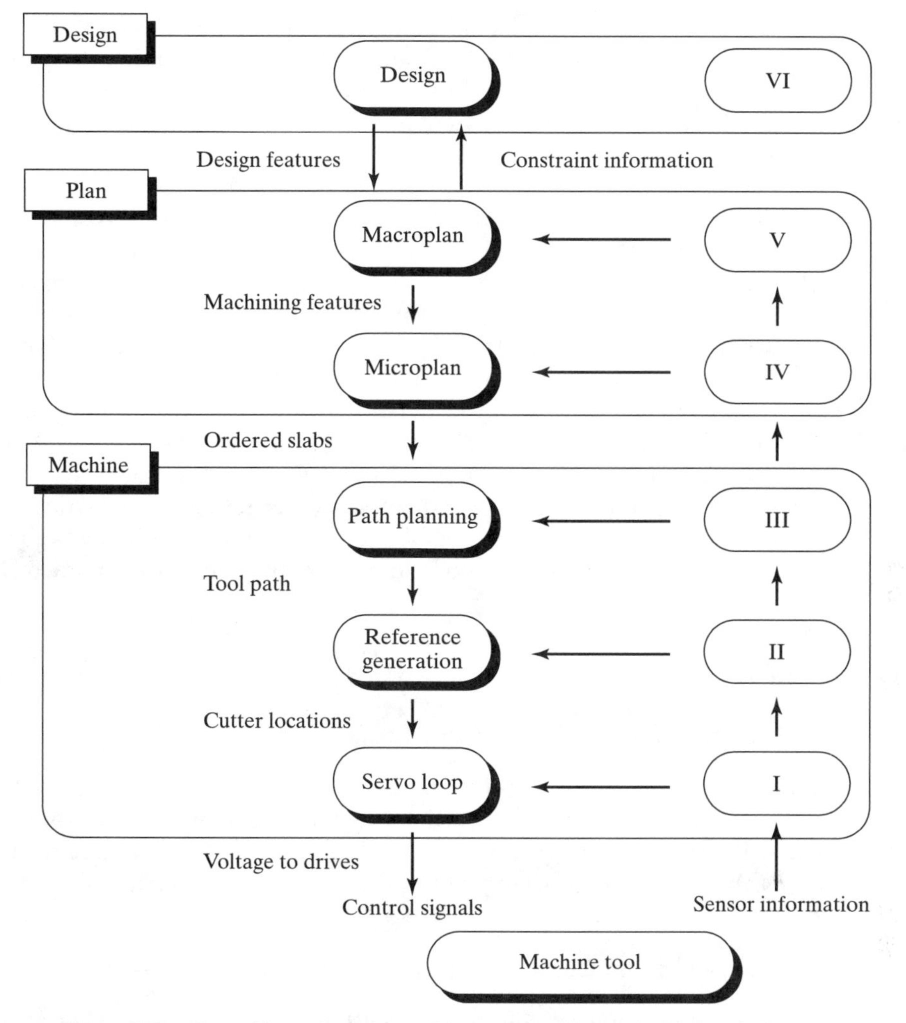

Figure 7.32 Control loops for open-architecture machine tools. Sensors and feedback are shown at six levels on the right (I–VI): (I) at the lowest level, vibration and force sensors monitor levels, and changes in speed or feed can be made "on the fly"; (II) at the next level, "on-the-machine touch probing after machining" can suggest changes in the cutter locations to compensate for form errors in the verticality of pocket walls; (III) at the next level, undesirable burrs can be compensated for by changing the entry and exit angles of the milling cutters; (IV) at the next level—now within the process planing domain—it is often desirable to reallocate the proportion of roughing versus finishing cuts so that the last "slab" milled into the bottom of a pocket creates the desired surface finish; (V) at the next level—still within the process planning domain—it is often desirable to reorder the sequence in which the several features of the part are cut, in order to improve accuracy or fixturability; (VI) at the design level, NURBS and new graphics routines can be directly sent to the open-architecture machine.

Since the mid-1990s, open-architecture machine tool controllers have thus been commercially launched by some industrial companies including Hewlett-Packard, Allen-Bradley, Delta Tau, and Aerotech (e.g., see Delta Tau, 1994). Such new products are usually PC based, use either the UNIX or NT operating systems, and are open to third-party suppliers of sensors, diagnostic systems, programming interfaces, and software tools.

Targeted at sophisticated users in industries such as aerospace, these open-architecture machine tools will be very useful as stand-alone machines, *and* they will provide powerful, networked-based machines for agile manufacturing. As individual systems they will be capable of producing small lot sizes of components with high accuracy. They will also be the factory-floor building blocks of systems in which machines can communicate "bidirectionally" with the rest of the factory. Cole (1999) has emphasized that such bidirectional knowledge exchange is the key to implementing TQM, JIT, and 6 sigma procedures, for the total integration of quality in the factory.

To close with an analogy, a banking machine or a telephone is useful not only because the machine itself is sophisticated but because it has been designed to allow bidirectional knowledge exchange all over a global network. Access to the network and all its services is actually more important than the local characteristics of the machine itself.

7.7 GLOSSARY

7.7.1 Cemented Carbide Cutting Tools

A family of sintered cutting tools that use a few percent of cobalt as the binder phase and a variety of hard carbide particles as the high-temperature, abrasion resistant phase. These may be tungsten carbide, titanium carbide, or tantalum carbide. Commonly, these cemented carbide materials are additionally coated with thin, abrasion-resistant layers.

7.7.2 Ceramic and Cubic Boron Nitride (CBN) Cutting Tools

A family of hard, nonmetallic sintered cutting tools that have higher abrasion resistance than carbides but relatively low toughness.

7.7.3 Chatter

A machine tool vibration initiated by resonance with a machine tool element but worsened as the part surface becomes undulated and regenerative chatter occurs.

7.7.4 Chuck

The clamping device in a lathe.

7.7.5 Cup

The test part shape in methods that assess the stretching (Erichsen) and drawing (Swift) characteristics of sheets of metal.

7.7.6 Deep Drawing

Essentially the same as drawing but often related to processes that use several repeated drawing operations so that long products can be formed.

7.7.7 Drawing

The general term for sheet-metal forming and more specifically the behavior of material in the flange of a product that gets drawn into the die wall.

7.7.8 Deformed/Undeformed Chip Thickness

The geometry of machining can be described by the chip dimension (t_c) and the uncut dimension (t). If the rake angle is zero, then $\tan \phi = t/t_c$.

7.7.9 Depth-of-Cut (d)

In turning operations, the depth-of-cut is measured radially into the bar being machined. In milling it is the vertical depth into the block.

7.7.10 Feed Rate (f)

In turning, the feed rate (f) is measured longitudinally along the bar, usually in millimeters or inches per revolution of the bar. In milling, the feed rate is usually the table speed in millimeters or inches per minute, so that it represents the relative motion between the tool and part in the plane being machined.

7.7.11 Fixture

A work-holding device that supports, clamps, and resists the cutting forces between tool and work.

7.7.12 Flank Face/Flank Angle

On a turning tool, the face is given a clearance angle at the side of a tool. This prevents it from rubbing on the shoulder being cut (usually to the left of the tool).

7.7.13 Forces

The main cutting force is F_c, acting on the tool face from the advancing tool in milling or the advancing work in turning. The tangential force, F_T, acts normal to the main cutting force.

7.7.14 Form Error

Ideally the walls of a milled pocket should be vertical. However, form errors often occur because of fixture deflections, part deflections, or tool deflections. In the latter case, the walls often exhibit a "ski-slope" appearance related to the tool deflection shape. Similar form errors can occur in turning if the bar is slender and pushes away from the tool.

7.7.15 Forming Limit Diagram

A plot of minor strain, on a $+/-x$ axis, and major strain, on a y axis. The strains are measured from small circles that are etched onto a sheet prior to the test. The circles might become ellipses or bigger circles depending on the deformation that occurs. The diagram also notes at which combination of major and minor strain failure by tearing of the sheet occurs. This locus of failure points is the forming limit curve or diagram.

7.7.16 G Codes

The standard low-end machine tool command set that gives motion; for example, G1 = linear feed.

7.7.17 Jig

A modified work-holding device or fixture that additionally guides the cutting tool into a desired location on the surface of the part.

7.7.18 Machinability

A relative term that judges the ease of machining of different materials. Usually, the tool wear or tool life is the main objective function that appraises relative machinability.

7.7.19 M-Codes

The standard low-end command set for machine tool operations that are not related to x, y, or z motion of the axes; for example, M6 = call tool into spindle.

7.7.20 Milling

A machining process suited to prismatic parts.

7.7.21 *n* Value (the Work-Hardening Coefficient)

Defines the slope of the stress–strain curve plotted on log axes. Physically, large n values occur with materials that work-harden a great deal during deformation. Austenitic stainless steels are in that category.

7.7.22 Power

The power supplied by the lathe or mill is usually measured by the product of the main cutting force and cutting velocity.

7.7.23 Rake Angle

The rake angle is measured from the face of the tool to the normal to the surface being cut.

7.7.24 *R* Value

Defines a ratio between the strain in the plane of a sheet and the strain in the thickness of the sheet. Large R values indicate good drawability because the material will extend and draw down without thinning in the thickness direction.

7.7.25 Roll Gap

The area between the rolls where plastic deformation of the strip is occurring.

7.7.26 Roll Load

The force between the rolls related to the deformation of the strip.

7.7.27 Shear Plane Angle, ϕ

The shear angle ϕ is not a single plane but a narrow zone identifiable in micrographs. The shear angle is then measured between this zone and the direction of the tool/work velocity factor.

- Primary shear: the main shear process that creates the chip
- Secondary shear: the shear zone between the bottom of the chip and the tool face

7.7.28 Strain

Defined as the extension divided by the original length:

- Engineering strain: the extension divided by the original length
- True strain: the extension divided by the current length as deformation increases

7.7.29 Stress

Defined as the load divided by the area of contact of the two opposing load bearing elements:

- Engineering stress: the load divided by the original area
- True stress: the load divided by the current area as deformation increases

7.7.30 Stretching

The deformation mode in sheet-metal forming in which an original square element of the sheet surface is deformed in both the x and y dimensions to become larger in all directions.

7.7.31 Surface Finish, Surface Roughness

Cutting tools leave distinctive markings on the surface that are a function of feed rate and the nose radius of the cutting tool edge (see Armarego and Brown, 1969). A profilometer can be used to trace over the surface and measure the roughness. (Imagine the stylus of an old-style record player being dragged across the tracks rather than following the tracks.)

The surface roughness can be measured by the arithmetic mean value (R_a)—which used to be known as the centerline average—or the root-mean-square average (R_q). To obtain these values, imagine that a cross section is like a rough or uneven sine wave about a centerline datum. The R_a value is found by taking a large number,

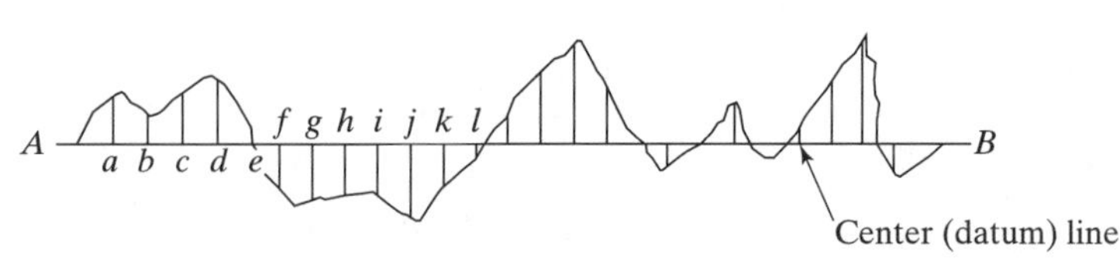

Figure 7.33 Surface finish.

n, of amplitude or ordinate values of the rough sine wave $(a + b + c + \ldots + n)$ and dividing them by n. The R_q value is obtained by taking the square root of $[(a^2 + b^2 + c^2 + \ldots + n^2)/n]$. Typical values of R_a might be 125 microinches for standard surfaces and 60 to 80 microinches for smoother, well-finished surfaces (Figure 7.33).

7.7.32 Taylor Equation ($VT^n = C$)

The result of replotting cutting speed (V) against the tool life data (T) on log-log axes.

7.7.33 Tool Life (T)

Usually defined by 0.75 millimeter (0.03 inch) of flank wear.

7.7.34 Turning

A machining process suited to axisymmetrical parts.

7.7.35 Wear Mechanisms

Tool wear by abrasion, attrition, and fracture occurs at lower cutting speeds. At higher speeds diffusion occurs especially at the rake face, where high-temperature conditions exist.

7.8 REFERENCES

Armarego, E. J. A., and R. H. Brown. 1969. *The machining of metals.* Englewood Cliffs, NJ: Prentice-Hall.

Asada, H., and A. Fields. 1985. Design of flexible fixtures reconfigured by robot manipulators. In *Proceedings of the Robotics and Manufacturing Automation ASME Winter Annual Meeting,* 251–257.

Backofen, W. A. 1972. *Deformation processing.* Reading, MA: Addison Wesley.

Chryssolouris, G. 1991. *Laser machining.* New York: Springer-Verlag.

Cole, R. E. 1999. *Managing quality fads: How American business learned to play the quality game.* New York and Oxford: Oxford University Press.

Cook, N. H. 1966. *Manufacturing analysis.* Reading, MA: Addison-Wesley.

Delta Tau Data Systems Inc. 1994. *Product Literature: "PMAC-NC."* Northridge, CA.

Ernst, H., and M. E. Merchant. 1940–1945. In particular see M. E. Merchant. 1945. The mechanics of the metal cutting process. *Journal of Applied Physics 16:* 267–275.

Goodwin, G. M. 1968. Application of strain analysis to sheet metal forming problems in the press shop. In *Proceedings of the Fifth Biennial Congress I.D.D.R.G.,* Torino, Italy.

Greenfeld, I., F. B. Hansen, and P. K. Wright. 1989. Self-sustaining, open-system machine tools. In *Proceedings of the 17th North American Manufacturing Research Institution Conference,* 17: 281–292.

Grippo, P. M., B. S.Thompson, and M. V. Ghandi. 1988. A review of flexible fixturing systems for computer integrated manufacturing. *International Journal of Computer Integrated Manufacturing* 1 (2): 124–135.

Hill, R. 1956. *The mathematical theory of plasticity.* New York and Oxford: Oxford University Press.

Hillaire, R., L. Marchetti, and P. K.Wright. 1998. Geometry for precision manufacturing on an open architecture machine tool (MOSAIC-PC). In *Proceedings of the ASME International Mechanical Engineering Congress and Exposition,* 8: 605–610.

Hoffman, E. G. 1985. *Jig and fixture design.* Albany, New York: Delmar.

Johnson, W., and P. B. Mellor. 1973. *Engineering plasticity.* London: Van Nostrand Reinhold.

Lu, L., and S. Akella. 1999. Folding cartons with fixtures: A motion planning algorithm. In *IEEE Conference on Robotics and Automation.* Detroit.

Meyer, R. H., and J. R. Newby. 1968. Effect of mechanical properties of bi-axial stretchability on low carbon steel. Paper presented at the *SAE Automotive Engineering Congress.* Paper No. 680094.

Michler, J. R., M. L. Bohn, A. R. Kashani, and K. J. Weinmann 1995. Feedback control of the sheet metal forming process using drawbead penetration as the control variable. In *Proceedings of the North American Manufacturing Research Institution,* 23: 71–78.

Miller, S. M. 1985. Impacts of robotics and flexible manufacturing technologies on manufacturing cost and employment. In *The Management of Productivity and Technology in Management,* edited by P. R. Kleindorfer, 73–110. New York: Plenum Press.

Mueller, M. E., R. E. DeVor, and P. K. Wright. 1997. The physics of end-milling: Comparisons between simulations (EMSIM) and new experimental results from touch probed features. In *Transactions of the 25th North American Manufacturing Research Institution,* 25: 123–128. See <http://mtamri.me.uiuc.edu>.

Rose, F. A. 1974. Grid strain analysis technique for determining the press performance of sheet metal blanks. In *International Conference on Production Technology.* Melbourne. Institution of Engineers.

Rowe, G. W. 1977. *Principles of industrial metalworking processes.* London, Arnold.

Sarma, S., and P. K. Wright. 1997. Algorithms for the minimization of setups and tool changes in 'simply fixturable' components in milling. *Journal of Manufacturing Systems* 15 (2): 95–112.

Schofield, S. M., and P. K. Wright. 1998. Open architecture controllers for machine tools, part I: Design principles. *ASME Journal of Manufacturing Science and Engineering,* 120: 425–432.

Stevenson, M. G., P. K. Wright, and J. G. Chow. 1983. Further developments in applying the finite element method to the calculation of temperature distribution in machining and comparisons with experiment. Transactions of the *ASME, Journal of Engineering for Industry* 105: 149–154.

Stori, J. A. 1998. Machining operation planning based on process simulation and the mechanics of milling. Ph.D. dissertation, University of California, Berkeley.

Trent, E. M., and P. K. Wright. 2000. *Metal cutting, 4th ed.* Boston and Oxford: Butterworths.

Wagner, R., G. Castanotto, and K. Goldberg. 1997. FixtureNet: Interactive computer aided design via the WWW. *International Journal on Human-Computer Studies* 46: 773–788.

Walczyk, D. F., and D. E. Hardt. 1998. Design and analysis of reconfigurable discrete dies for sheet metal forming. *Journal of Manufacturing Systems* 17 (6): 436–454.

7.9 BIBLIOGRAPHY

Bammann, D. J., M. L.Chiesa, and J. C. Johnson. 1995. Modeling large deformation anisotropy in sheet metal forming. In *Simulation of materials processing: Theory, methods, and applications,* 657–660. edited by Shen and Dawson, Rotterdam: Balkema.

DeVries, W. R. 1992. *Analysis of material removal processes.* New York: Springer-Verlag.

Klamecki, B. E., and K. J. Weinmann. 1990. Fundamental issues in machining. In *Proceedings of the Winter Annual Meeting of ASME in Dallas Texas,* 43: New York: American Society of Mechanical Engineers.

Kobayashi, S., S-I. Oh, and T. Altan. 1989. *Metal forming and the finite element method.* New York and Oxford: Oxford University Press.

Komanduri, R. 1997. Tool materials. In *The Kirk-Othmer Encyclopedia of Chemical Technology,* 4th ed., 24. New York: John Wiley and Sons.

Oxley, P. L. B. 1989. *The mechanics of machining: An analytical approach to assessing machinability.* New York: Halsted Press.

Pittman, J. T., R. D. Wood, J. M. Alexander, and O. C. Zienkiewicz. 1982. *Numerical methods in industrial forming operations.* Swansea, U.K.: Pineridge Press.

Shaw, M. C. 1991. *Metal cutting principles.* Oxford Series on Advanced Manufacturing, Vol. 3. Oxford: Oxford Science Publications, Clarendon Press.

Stephenson, D. A., and R. Stevenson. 1996. *Materials issues in machining III and the physics of machining processes III.* Warrendale, PA: TMS Press (Minerals, Metals, and Materials Society).

Wang, C. H. 1997. *Manufacturability-driven decomposition of sheet metal products.* Robotics Institute Technical Report CMU-RI-TR-97-35. Pittsburgh, PA: Carnegie Mellon University.

7.10 URLS OF INTEREST

A collection of sites for machining planning and automation can be found at <**http://kingkong.me.berkeley.edu/html/contact/mach_software.html**>. A site for metal products in general is <**www.commerceone.com**>.

7.11 INTERACTIVE FURTHER WORK 1: THE SHEAR PLANE ANGLE

Use Netscape with Java capability to access <**http://cybercut.berkeley.edu/ merchant**>. Dr. Sandstrom of The Boeing Company has built an interesting Java applet that investigates the variables in the Ernst and Merchant theory of metal cutting.

Complete the table for the following 12 cases:

Rake angle: α (degrees)	Friction coefficient: μ (0 to 1)	Friction angle: *Write in* (degrees)	Shear angle: *Write in* (degrees)
0	0		
+45	0		
+45	0.5		
+45	1.0		
+6	0		
+6	0.5		
+6	1.0		
−6	0		
−6	0.5		
−6	1.0		
−42	0		
−42	1.0		

7.12 INTERACTIVE FURTHER WORK 2: "FIXTURENET"

Modular fixturing on the World Wide Web is by Dr. Kenneth Goldberg and his students. The URL to use is <**http://riot.ieor.berkeley.edu**>, and then click on FixtureNet.

Brost, R., and K. Goldberg. 1996. A complete algorithm for designing modular fixtures using modular components. *IEEE Transactions on Robotics and Automation* 12 (1).

Wagner, R., G. Castanotto, and K. Goldberg. 1997. FixtureNet: Interactive computer aided design via the WWW. *International Journal on Human-Computer Studies* 46: 773–788.

A modular fixture consists of a metal lattice with holes spaced at even intervals (Figure 7.34a), three locators (Figure 7.34b), and a clamp (Figure 7.34c), which make four contacts and hold objects in "form closure." Figure 7.34d is a photograph of their use.

Figure 7.35 shows a part on the World Wide Web with three locators and clamp in form closure. The three locators fit into the fixed lattice and are positioned in such a way that they are touching three edges of the part. The clamp must also be placed on the lattice so that clamp motion is horizontal or vertical. The clamp can be positioned to push against the object.

An admissible fixture is an arrangement of the three locators and clamp on the lattice that holds the part in form closure. The conservative assumption is made that there is no friction. Two fixture arrangements are equivalent if one can be transformed into the other using rigid rotation and translation.

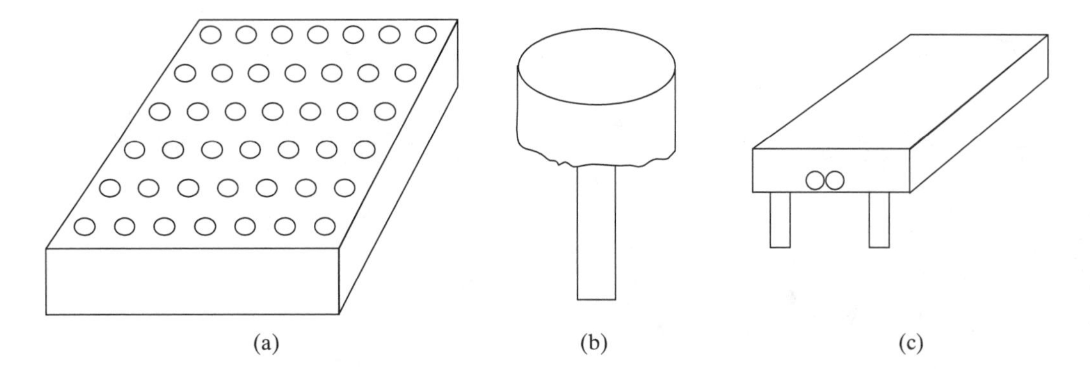

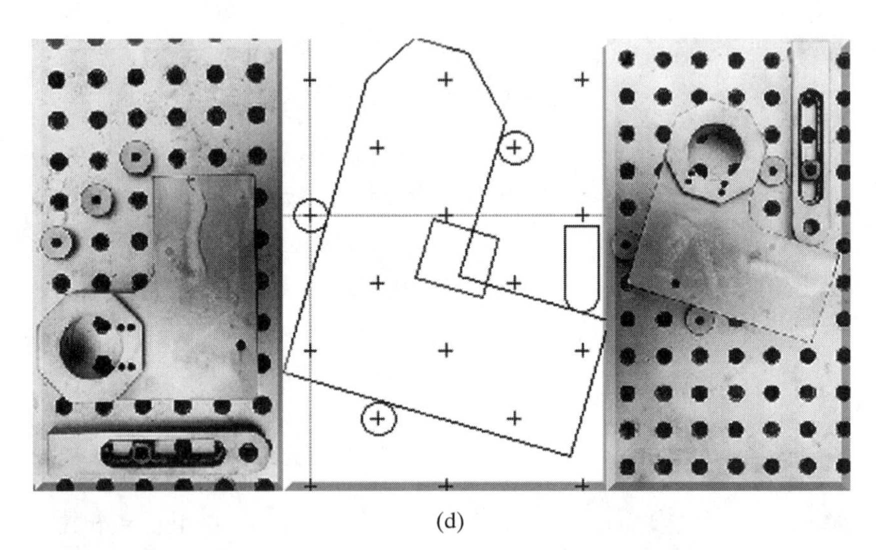

Figure 7.34 (a) Modular lattice, (b) locator, (c) clamp, and (d) physical setup.

The general problem is: Given a polygonal part boundary, find all admissible fixtures (if any). *The Algorithm is:*

Step 1: Grow the part by the radius (r) of the locators, and shrink the locators to a point. Curved portions are eliminated because we assume that locators and clamps have to be placed on straight surfaces and locators may also damage corners of delicate parts.

Step 2: Label each edge $1, 2, 3, \ldots, n$ in a counterclockwise fashion.

Step 3: Consider all combinations of triplets in counterclockwise increasing order—for example, 1, 2, 3 or 1, 2, 4 or 1, 2, 5 or 1, 3, 4 or 2, 3, 4. For each triplet, call the edges *a, b, c.* This will give us all possible arrangements of the three locators in contact with the three edges of the part.

Step 4: Without the loss of generality, assume edge a is in contact with a locator at $(0, 0)$. Find all possible positions for L2 in contact with edge b. First trans-

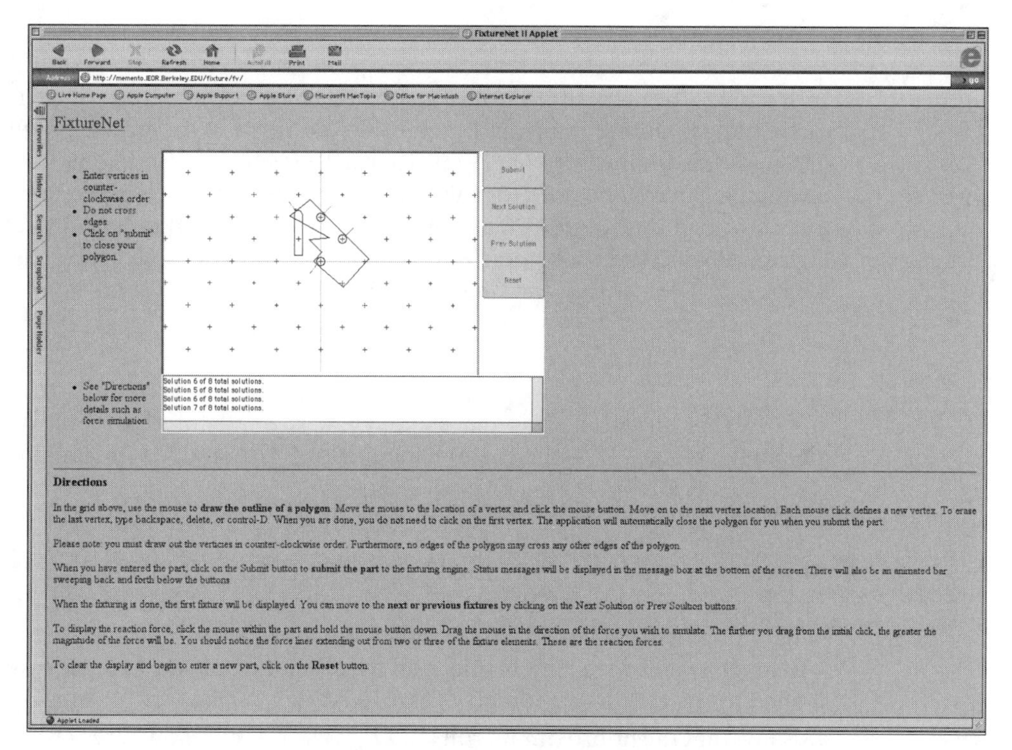

Figure 7.35 Screen dump from the Web site.

late and then rotate the part while maintaining contact between edge a and the locator. Consider only one quadrant of all possible locations that the second locator (L2) can be placed, because the other three quadrants would be reflections about the origin.

Step 5: For each choice of L2, find all possible choices of L3. Given (a-L1, b-L2), solve for c. Or (very fast) solve for (a-L1, c-L3) an annulus and (b-L2, c-L3) another annulus. Intersect to find all consistent choices for (c-L3).

Step 6: For each triplet of locator-edge matches (two acceptable designs in this example), find all possible clamp arrangements. Use the Reuleaux rotation center construction.

Step 7: Repeat Steps 3 to 6 for each triplet of edges.

Step 8: Output all possible solutions. The time order of this algorithm is $O(n5d5)$, where n is the number of edges and d is the diameter of the part in lattice units.

Question: Can all polygonal parts be fixtured?

Specific assignment: Use FixtureNet to design two modular alternative fixtures for a part. Compare these and explain why one might be preferable.

7.13 REVIEW QUESTIONS

1. In forming, forging, and extrusion operations, a popular technique for predicting approximate loads and metal flow patterns is the upper bound technique (Johnson and Mellor, 1973; Rowe, 1977; Hill, 1956). The upper bound technique can also be used to make an estimate for the force necessary to form the chip in metal cutting. The analysis first enlarges the center section in Figure 7.36 and then considers the complete shear band *OD*, which has a total length of (s). Show that the final result for the force F_c is found as:

$$F_c = \frac{k \cdot V_s \cdot s}{V} \tag{7.24}$$

In this equation, k is the shear yield strength of the metal, V is the incoming velocity, V_s is the shear velocity along *OD*, and s is the length of *OD*.

2. The basic rolling operation creates a wide flat strip in a coil. This strip is sold to a secondary processor, who carries out the sheet-metal forming operation. Automobiles, washing machines, office furniture, filing cabinets, and the inside casings that carry the PCBs in a computer all start as rolled product. The secondary processor takes the large coils that come off a rolling mill, shears them into much smaller starting blanks, and then sheet forms them in a pressing die shaped to the required geometry (Figure 7.37).

a. Show that the approximate roll load $P = w \cdot \overline{Y} \cdot \sqrt{R\,dh}$ where w is the width of the strip, Y is the average yield strength of the material as it goes through the roll gap, R is the radius of the rolls, and dh is the reduction.

b. Figure 7.38 shows a strip being pushed from left to right and into the roll gap. The top edge of the strip (E) is shown meeting the roll. It experiences two counteracting forces: one that tries to push it out, and another, due to friction, that tries to suck it in. The conditions that allow the strip to go in and be rolled require that the friction component be greater than the pushing-out component.

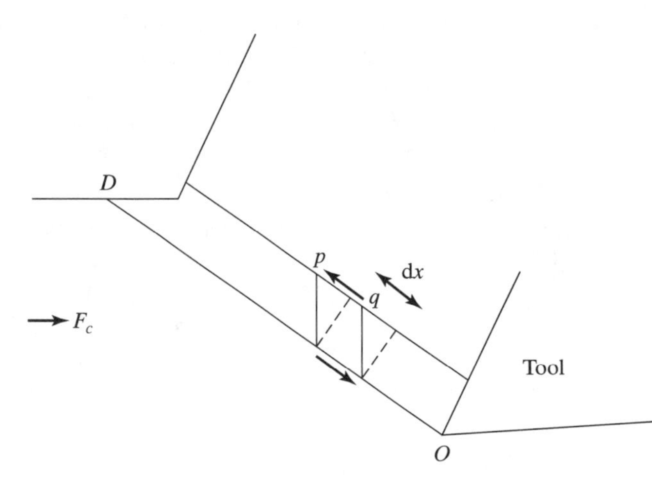

Figure 7.36 Stress element at the shear plane.

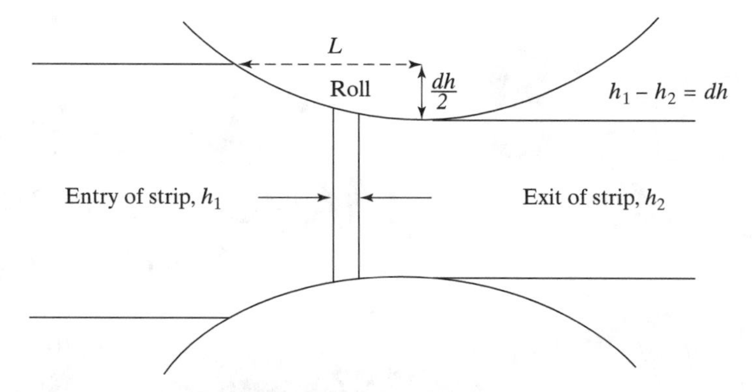

Figure 7.37 Sheet rolling: material on the left enters the roll gap and is plastically deformed by an amount $(h_1 - h_2 = dh)$.

Show that because of the balance between the friction that "pulls in" the sheet and the roll angle that "pushes out" the sheet, the maximum reduction in one pass is given by

$$\mu \geq \tan\beta \geq \sqrt{\frac{dh}{R}}$$

$$(dh)_{MAX} = \mu^2 R \tag{7.25}$$

The basic physics of friction, and the roll radius, control the maximum reduction in one pass. These mechanical analyses show why ultraexpensive multiple stands are needed at the standard steel mill to produce flat rolled strip for consumer products.

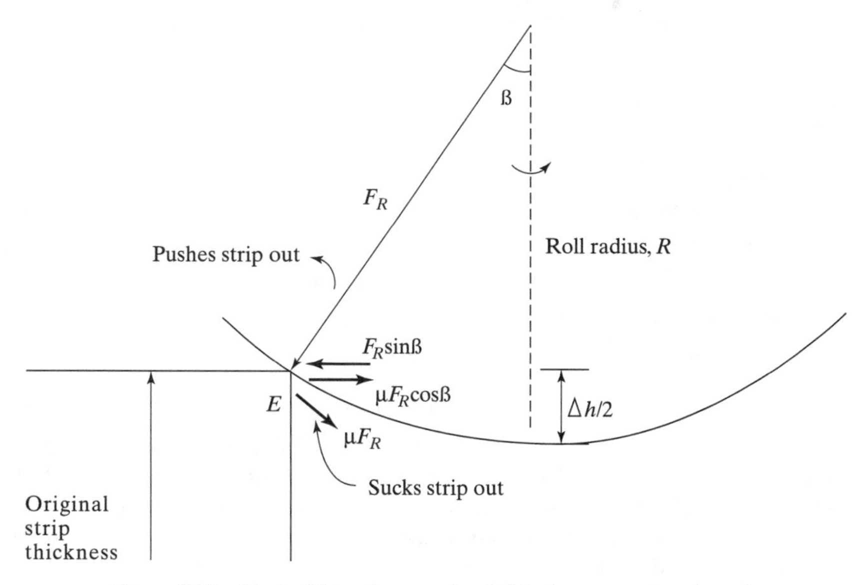

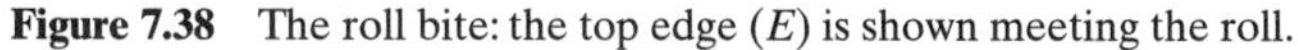
Figure 7.38 The roll bite: the top edge (E) is shown meeting the roll.

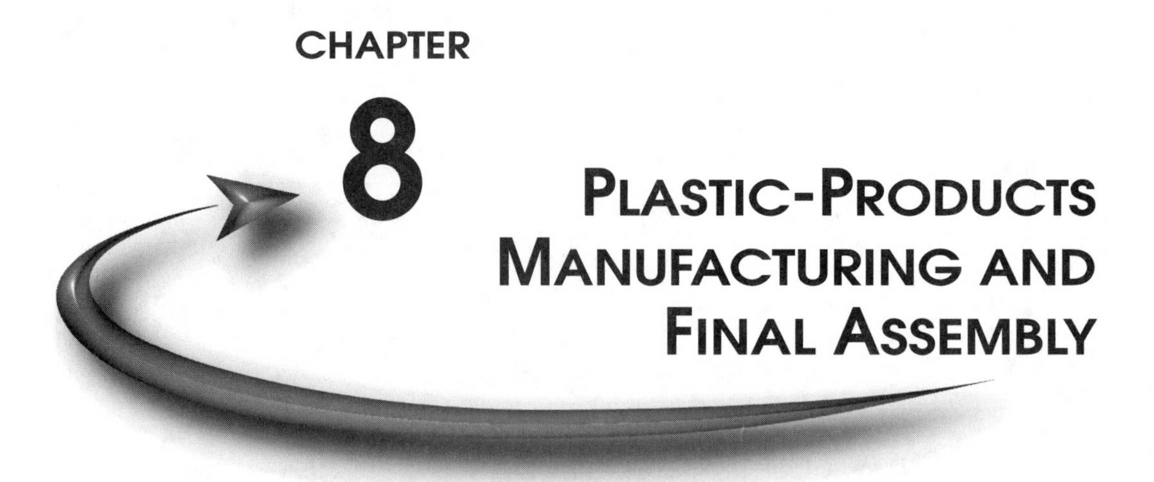

8.1 INTRODUCTION

University students arrive on campus wearing in-line skates, listening to a Walkman, and sipping designer water from a plastic bottle. The manufacturing processes for these three products are reviewed in this chapter. In particular, injection molding is presented in detail because it creates the packages for consumer electronics products. The outer bodies of these products must be lightweight, protective, and cheap. They must also offer the aesthetic impact to give the product shelf appeal at the nearby mall or dot.com site.

It is also important to look back on our "journey along the product development path." As shown on the clock-face diagram in Figure 2.1, the various steps were:

- Design of the product (Chapter 3)
- Prototyping of the product (Chapter 4)
- Making the inner brains (Chapter 5)
- Assembling the inner system (Chapter 6)
- Machining a mold (Chapter 7)
- Injection into this mold (Chapter 8)

As a result, plastic injection molding and product assembly can be seen as a culmination of the processes and devices in all the previous chapters, arriving at the production of millions of units ready for the consumer. At the same time it should be recalled from the case study in Chapter 2 on the fingerprint recognition device that injection molds can be machined from aluminum, allowing small batches of only 200 units to be made for early customer testing or for evaluation kits.

8.2 PROPERTIES OF PLASTICS

Plastics, or polymers, have different properties than the metals presented in Chapter 7, and it is important to review these before moving on to injection molding or blow molding. In fact the molecular and thermal properties of polymers govern many, if not all, of the part design and equipment design issues shown in subsequent figures.

As in Chapter 7, it is assumed that the reader has enjoyed a freshman class in material science and recalls that polymers fall into two broad classes:

- Thermosetting molding materials. These include the melamine-formaldehyde used in hard plastic tableware and the epoxy resins used for glues and reinforced cast products such as kayaks and tennis racket frames. Thermosetting products are heated until they become a viscous liquid, poured or injected into a mold, and then allowed to solidify. Chemical cross-linking occurs to create an irreversible, infusible mass.
- Thermoplastic molding materials. These include polymers such as acrylonitrile-butadiene-styrene (ABS) and polycarbonate (Lexan is a common brand) used for toys, consumer electronic products, and more flexible kitchenware products. The key feature is that these plastics can be heated to a viscous fluid, molded, and cooled in a reversible, time-and-time-again manner. As a result, they are perfect for the routine injection molding processes described later. They are therefore reviewed in more detail in the next section.

8.2.1 Properties of Thermoplastics

Which particular polymer should be used for a given component? The answer depends on how that polymer behaves at the operating temperature of the device. All thermoplastic polymers go through the generic transition described in Table 8.1 for polystyrene, but they do so at different temperatures.

At low temperatures, the polystyrene's structure is glassy and it has a high stiffness as measured by Young's modulus, E. The stiffness can also be increased by increasing the molecular weight of a polymer, by increasing the branching of the polymer chains, by creating specific crystallization patterns in which the chains are folded against each other, and by adding elements that cross-link the chains. Speaking colloquially, the mechanical properties at low temperatures can be viewed as being comparable with metals and involving bond stretching, but at higher temperatures the molecular chains of the polystyrene slide over each other like cooked spaghetti.

TABLE 8.1 General Characteristics of Thermoplastic Materials Related to Polystyrene

°C	Macroview	Microview
<90	Glassy	Bond stretching as in metals
90–120	Transition leathery	Chain bending/uncoiling
120–140	Rubbery plateau	Chain slipping
>140	Viscous liquid	Chain sliding

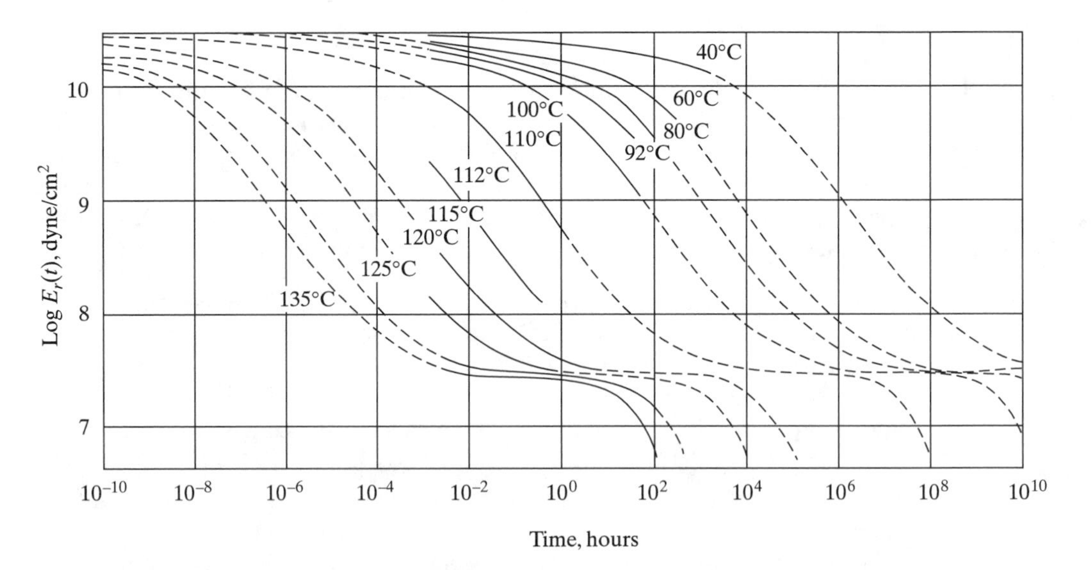

Figure 8.1 Stress relaxation of PMMA between 40 and 155 degrees. Dashed lines are extrapolations (adapted from McLoughlin and Tobolsky, 1952).

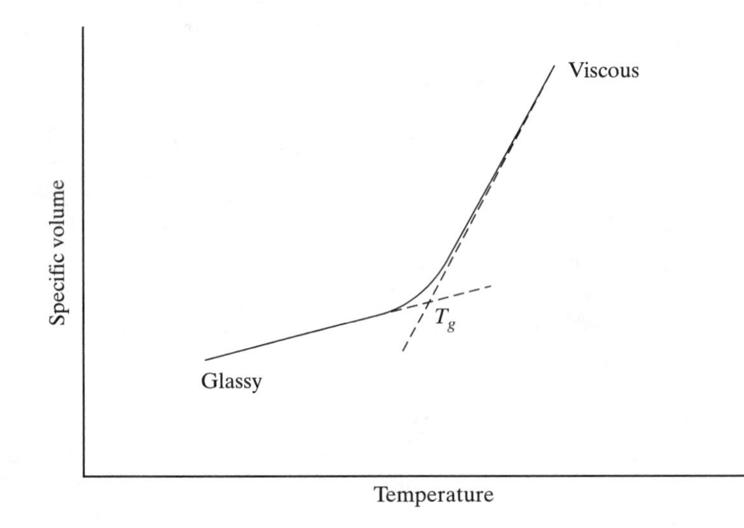

Figure 8.2 Schematic curve to show the glass transition temperature. Specific volume versus temperature.

This use of Young's modulus, E, is too simplistic for thermoplastics because deformation is both temperature and time dependent. The stress-relaxation modulus, E_r, is thus used. Plastic specimens at several temperatures, T, are tested by imposing a selected elongation, or strain ε_1. The test measures how the imposed stress decays away over time.

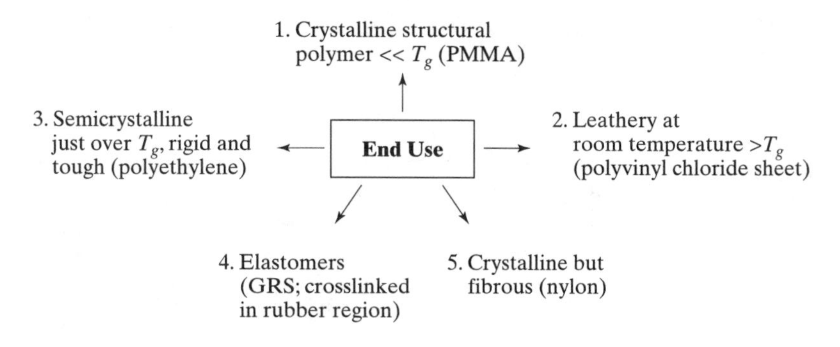

Figure 8.3 Design choices with polymers.

The stress-relaxation modulus is then given by,

$$E_r(t, T) = \frac{\sigma_{(t,T)}}{\varepsilon_1} \tag{8.1}$$

Typical results (McLoughlin and Tobolsky, 1952) are shown in Figure 8.1.

These tests are for polymethyl methacrylate (PMMA), commonly called Plexiglass (in the United States) or Perspex (in the United Kingdom). At 40°C, the material remains rigid for long periods, but with increasing temperature, the material becomes leathery above temperatures of 135°C and eventually viscous.

Another key concept is the glass transition temperature, at which a thermoplastic transitions from its glassy to leathery behavior. In Figure 8.2 the specific volume of polyvinyl acetate is plotted against temperature. The value of the glass transition temperature is found by extrapolating the glassy region and the leathery region to the intersection point at $T_g = 26°C$ in this case.

8.2.2 The Influence of Properties on High-Level Design

From a design perspective, the strategy is to pick a polymer that displays the desired characteristics at the operating temperature of the product, most often room temperature. Figure 8.3 shows this design strategy, which includes:

- Polymethyl methacrylate, which is a rigid-structured material at room temperature, considerably below T_g.
- Polyethylene and acrylonitrile-butadiene-styrene (ABS), which are just over T_g at room temperature but considerably below the melting point and therefore rigid and tough. These are suitable for toys, car parts, and electronics packaging.
- Polyvinyl chloride sheet, which is leathery at room temperature and suitable for some forms of clothing and imitation leather products.

This background sets the scene for the injection molding of ABS to create devices like the fingerprint recognition unit and the InfoPad. At the conceptual level, the ABS is heated into the highly viscous state, pumped into a die cavity, and then allowed to cool into the desired product. The details of the process, with some of its more challenging aspects, are described next.

8.3 PROCESSING OF PLASTICS I: THE INJECTION MOLDING METHOD

8.3.1 Overview

Injection molding is a key production method for the casings of consumer products. It is cheap, reliable, and reduces the device's weight compared with metals. The ST Microelectronics' TouchChip™ (the case study in Chapter 2) and the InfoPad casings (the case study in Chapter 6) were made in this manner. A very wide array of consumer products ranging from toys to telephones to automobile parts is also made through this process.

In Figure 8.4, some general features of the mold are shown. A simple bucket or cuplike component could be made in the gap between the core and the cavity shown at the bottom right. In this case the *parting plane,* shown on the bottom left, could be at the lip of the bucket. For other products the parting plane might be less conveniently located. Thus, a ridge is often visible around plastic toys and simple appliances where the parting plane has been located. Further finishing by hand might be desirable for such parts.

Hand finishing might also be needed (a) for the injection's "gate marks" that may be visible as tiny "pimples" on the surface of a part and (b) for the "ejector-pin" marks that for a relatively small object like a cellular phone casing are usually about 6 mm (about 0.25 inch) in diameter. The reader might be interested in picking up any familiar plastic consumer product and searching for these inevitable markings. However, these are often located in noncosmetic areas of the part, because of the cost involved in finishing operations. Not surprisingly, they are often found on the bottom or base of the object.

Injection molding is shown in Figure 8.5. Pellets of the desired thermoplastic are loaded into the hopper on the right side and heated as they are pushed by a screw

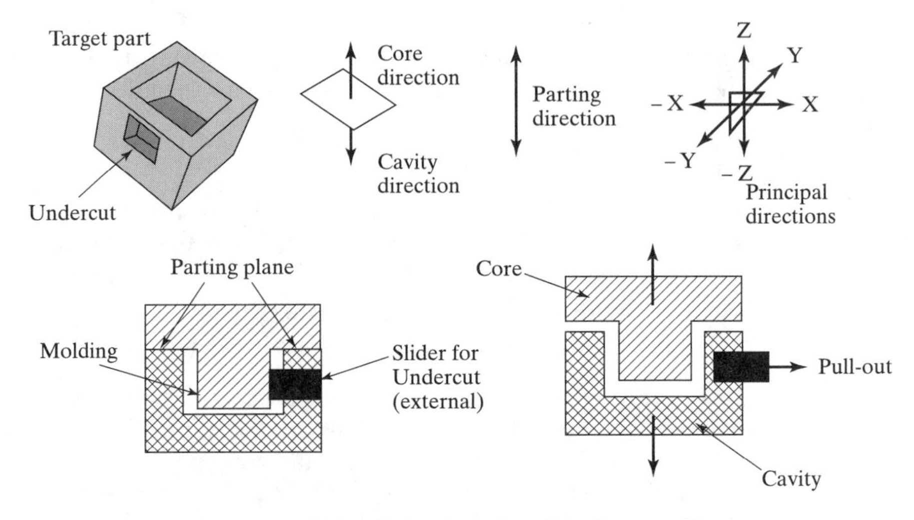

Figure 8.4 Basic mold description, including slider for an undercut.

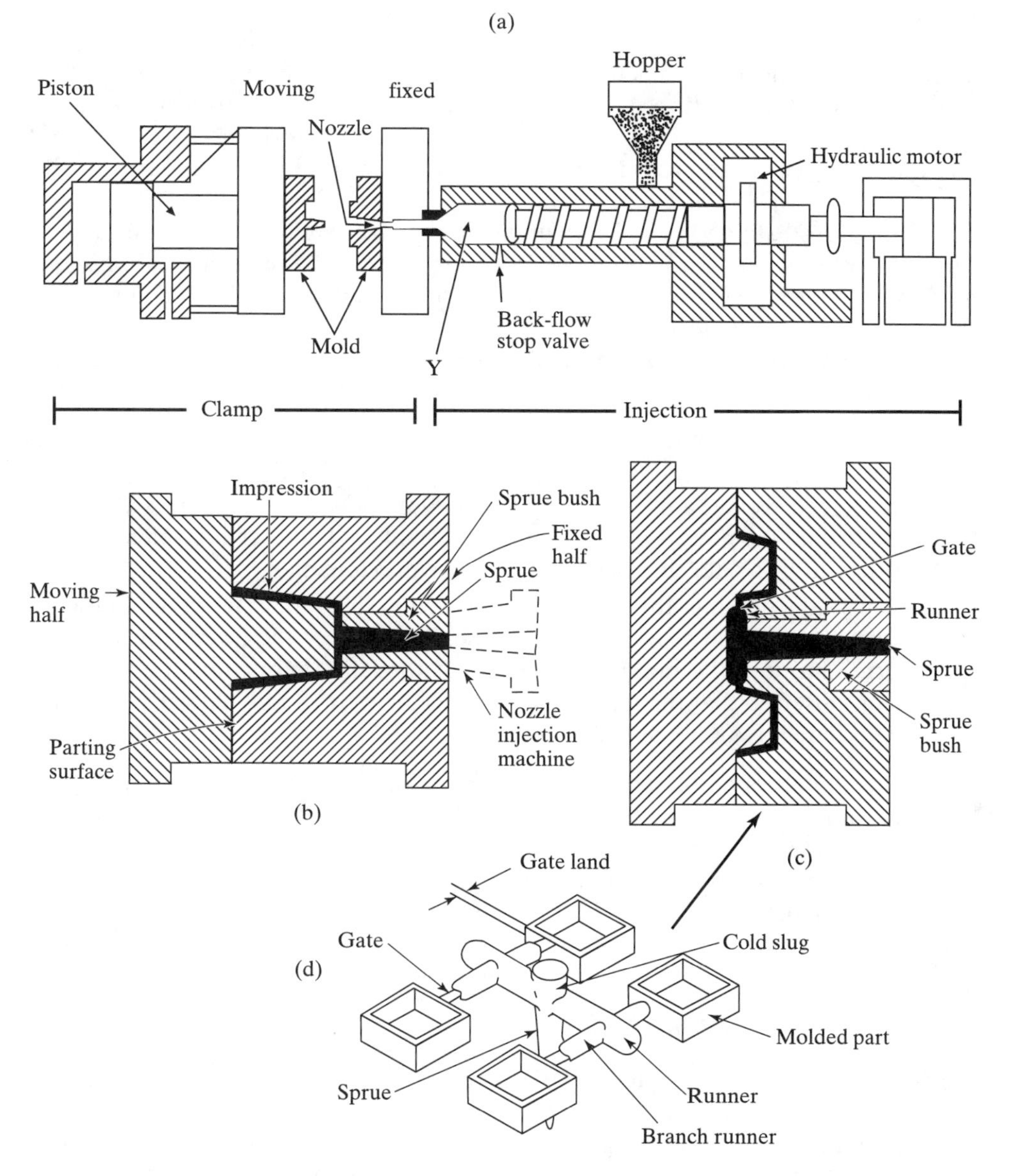

Figure 8.5 Injection molding with reciprocating-screw machine.

through heating zones. This melted and mixed material is forced through a nozzle into the mold by a hydraulic ram. The thermoplastic then cools, sets, and hardens in the mold. Once the two halves of the mold have been separated, the ejector pins facilitate "popping the part" out of the lower part of the mold. The process is now reviewed in more detail.

8.3.2 The Reciprocating-Screw Machine

The reciprocating-screw machine is the most used machine in industry. It is shown in the open position in Figure 8.5a. Below the machine is a single-cavity mold and a multiple-cavity mold. Note that the machine is horizontally constructed and operated. Thus the mold is effectively laying on its side in comparison with Figure 8.4.

In Figures 8.5b and 8.5c the cavity is on the right and the core is on the left. Molten plastic is shot into the mold from the nozzle of the reciprocating-screw machine and into the sprue of the mold. If there are multiple cavities (Figure 8.5d), the sprue feeds the runners and gates.

The barrel of the machine is heated, and as the screw pushes the pellets forward, there is additional heating from the mechanical pulverizing effect. In fact the machine is designed to operate in two distinct phases:

- Step 1, plasticizing the thermoplastic: the combined action of the heaters and screw feed creates a metered volume of liquid polymer that arrives at area **Y,** just behind the nozzle. Since the nozzle is small in diameter, it is sealed shut at this point by a cold slug of plastic from the previous shot.
- Step 2, injection into the mold: the screw action stops, and the whole stationary screw is now used as a ram to force the liquid out of the nozzle, through the sprue, and into the mold. Immediately before the ram action, the mold has been closed so that the liquid polymer fills the impressions shown as the dark areas in Figures 8.5b and 8.5c.

It should also be noted that a nonreturn valve just behind the ram head prevents the liquid from moving backward into the screw channels. Following this two-step process several things happen in unison. The part begins to cool down, and after a sufficient waiting period, the mold can be opened to eject the part. To reduce the cooling period, the mold is actively cooled by water lines. But during the cooldown, the screw can begin to turn again to collect its next shot of polymer pellets and move back to create space in position **Y** for the next shot.

8.3.3 Computer Aided Manufacturing

McCrum, Buckley, and Bucknall (1997) describe the research into equipment and control design in recent years. The screw has the triple function of transporting the pellets, compressing and melting them with help from the heater zones, and then, during the ramming action, having enough strength to pump the melt through the nozzle into the die. The clearance between the barrel and the flat lands of the screw flights is only 10^{-2} mm, demanding that the machine screw and the barrel be made from hardened steel with wear resistant coatings. Internal pressures of 100 Mpa are typical.

CNC controllers are used to monitor various sensors in the system and to adjust the various parameters shown in Figure 8.6. The key features include:

- Thermocouples for the temperatures of the barrel, nozzle, and mold
- Pressure sensors on the screw/ram

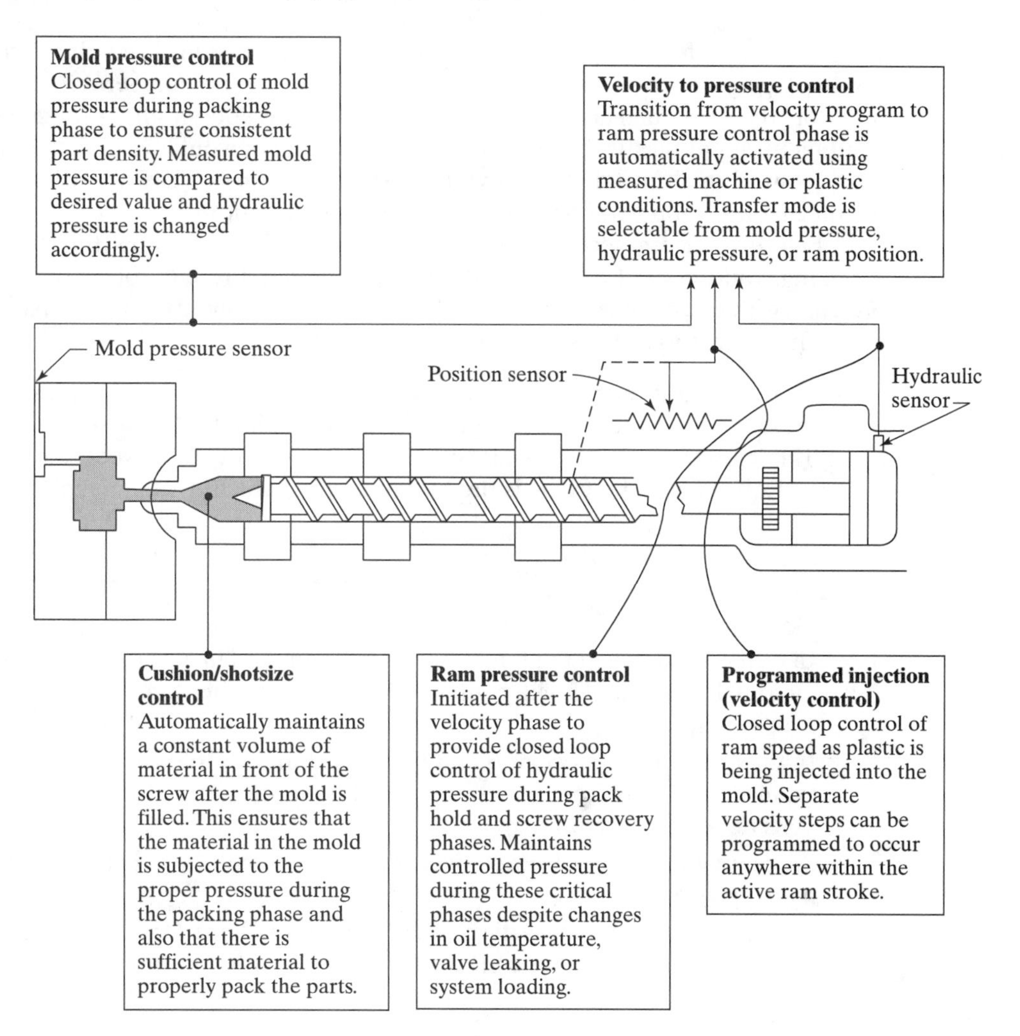

Figure 8.6 Control of injection molding (courtesy of Barber-Colman DYNA Products).

- Pressure sensors monitoring the pressure in the mold
- A screw position sensor (potentiometer or LVDT), which also measures velocity during the ramming step

The controller is used to optimize the cycle and to pack and hold the polymer in the mold under the continuing pressure of the ram. Two methods for doing this are described in Section 8.3.5.

8.3.4 Behavior of the Polymer inside the Mold during the Filling Stage

When a polymer cools in the die cavity, it shrinks dramatically and grips the sides of the core walls. During the milling operations (Chapter 7) to create the molds, these walls must therefore be tapered to allow easier part ejection. The volume contractions of a polymer between its liquid temperature and room temperature are of the order of 10% at normal atmospheric pressure. This presents another problem for the manufacturing of plastics because voids and sink marks would occur if the polymer were allowed to shrink this much without special controls. These voids can reduce the mechanical properties and cause cosmetic imperfections.

Figure 8.7 from McCrum and colleagues (1997) shows how this volumetric shrinking is addressed. Lines *a, b,* and *c* in the diagram represent isobars of increasing pressure, with line *c* being the highest. The details are explained in the next paragraph.

The cycle is as follows:

- At the highest temperature and specific volume (V_A) the liquid plastic is first pressurized, in the mold, to a pressure of approximately $P = 10^3$ atmospheres. This leads to a volume decrease of about 10% (i.e., A to B between lines a and c). In Equation 8.2 below, K is the bulk modulus, which is 1 GPa, typical of most liquids:

$$\frac{\Delta V}{V} = \frac{P}{K} = \frac{10^3 \times 1.01 \times 10^5}{10^9} = 0.101 \tag{8.2}$$

- In other words when the liquid polymer is rammed in at this pressure, and the mold is 100% filled, it will contain 10% more polymer than if it were filled at 1 atmosphere.
- In the next phase, between B and C, the part in the mold cools under pressure (value, C), and the ram keeps additional liquid flowing into the mold to compensate for contractions.
- At point C, the gate freezes over, sealing the mold and preventing further packing.
- Between C and D, the plastic cools at constant volume (V_D) under decreasing pressure until atmospheric pressure returns again at position D.

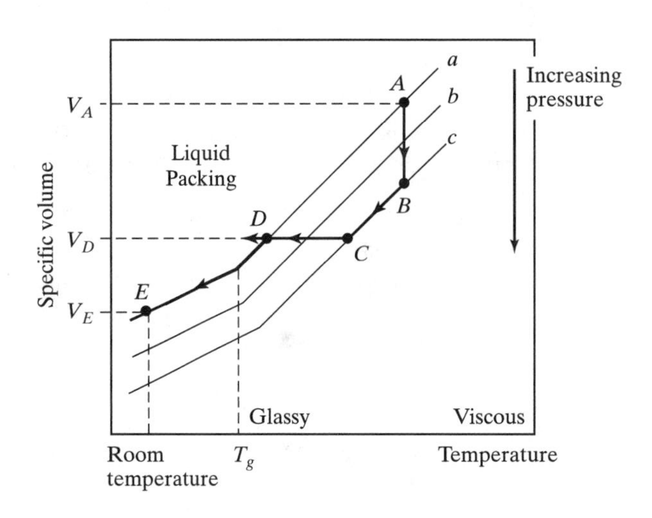

Figure 8.7 Liquid polymer follows the path A to B as a pressure of approximately 10^3 atmospheres is applied. The liquid is "packed" between B and C. At point C, the gate freezes over. Between C and D, the plastic cools at constant volume (V_D) (adapted from McCrum, Buckley, and Bucknall, 1997).

- From D to E the mold cools naturally and undergoes normal thermal contraction. If the polymer had been allowed to cool naturally, the free thermal shrinkage would have been $= (1 - V_E/V_A) = 10\%$ to 15%. However, because of the pressurized cooling cycle, the contraction is $= (1 - V_E/V_D) = 1\%$.

The maximum available clamping force is one of the key specifications in designing and costing equipment. Other main specifications include: (a) shot size, which is expressed as a mass equivalent volume of polystyrene, and (b) ejection stroke, which limits the maximum part depth that can be molded.

8.3.5 Controlling the Polymer Cooldown

There are two main methods to control the pressurized cooldown in Figure 8.7. These are summarized in the next two diagrams from the Barber-Colman Company. Actually, in both cases, Step 1, the mold filling with the ram, is similar:

- Step 1, for filling: a sequential velocity profile is programmed in, shown by the steps on the left of Figure 8.8. The velocity is high at first to accelerate the liquid polymer into the mold, but then it decreases as filling gets completed.

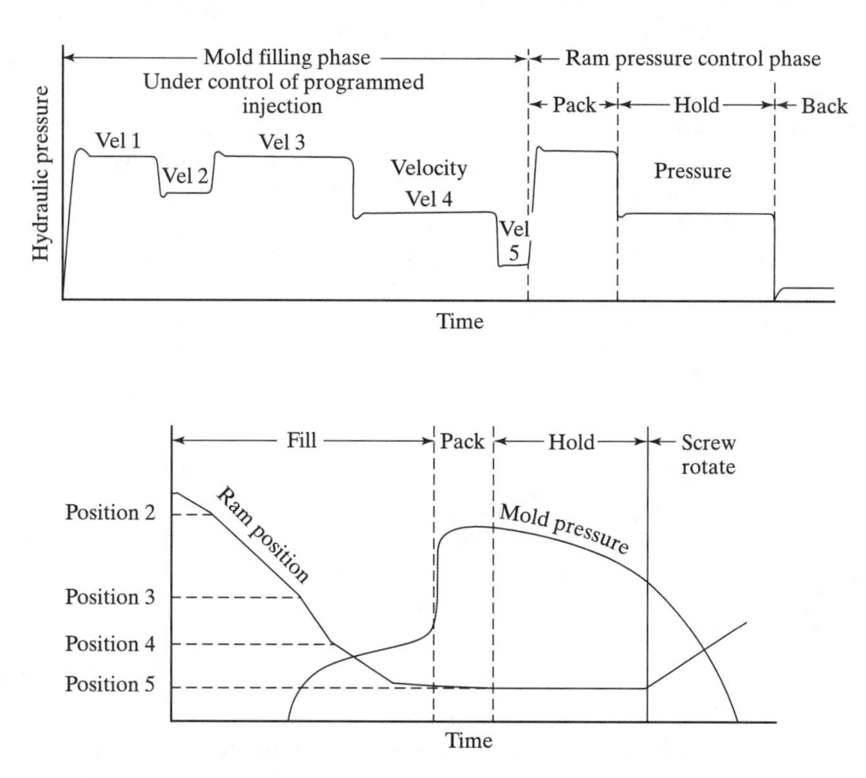

Figure 8.8 Injection molding cycles. The upper figure is ram pressure controlled, and the lower figure is mold pressure controlled (courtesy of Barber-Colman DYNA Products).

- Step 2, alternative 1 to packing, for the packing and holding phase (along *A* to *B* to *C* in Figure 8.7): the ram pressure is used to control the process. The pressure sensor on the ram is used to monitor and control the highest pressure during packing the mold to its desired mass. The system then transfers to the lower holding pressure until the gate freezes, and then after a set time the screw/ram rotates and moves back, shown on the extreme right of the figure.
- Step 2, alternative 2 to packing: a more direct method is to install a pressure sensor in the mold cavity. Figure 8.8 shows that this allows a more graceful release of the mold pressure and a direct measure of the processes inside the cooling mold.

8.3.6 Ejection and Resetting Stage

In the ejection and resetting stage, the cavity and core are separated as the mold is opened, and the ejector pins "pop" the part away from the die surface. The schematic of a "two-plate mold" shown in Figure 8.9 is from Glanvill and Denton (1965). It indicates that real production molds are rather more substantial than might be suggested by the earlier figures showing general nomenclature. In fact, even Figure 8.9 is a more simplified view of a mold. Classical texts in this area, such as that by Pye (1983) and by Glanvill and Denton (1965), show that many extra alignment pins and backing plates are needed beyond those shown in the schematic. The extra bulk, support blocks, ejector plates, and clamping plates reflect the facts that molding pressures are high and production runs are long (e.g., 10,000 to 10,000,000 parts). Thus a substantial device is needed so that the ejector pins and moving parts do not wear out. The downside is that molds open relatively slowly with hydraulic or mechanical actuators, and it takes care to lift the parts out of the cavity after the ejector pins have pushed it off the core plate (center of Figure 8.9). As much ejection area as possible is desirable to minimize part distortions and to minimize cycle time.

8.3.7 Positioning the Gate

Figure 8.10a shows some typical gate designs, and Figure 8.10c indicates the preferred sprue/gate for a cup shape. The gate is a very small orifice, and as the liquid is forced through, additional shearing causes a temperature rise of about 20°C, which is beneficial in mold filling. At a later stage, the small orifice is beneficial again because the polymer first freezes at the gate, thereby separating the molded part from the material in the barrel. The correct gate design is also the key to efficient mold filling, as shown on the right of Figure 8.10c.

Finally, the gate should be placed in a position that creates an acceptable part. Key design issues include:

- Positioning the gate on a surface that is noncosmetic. For consumer electronics this usually means it will be on the inside of the casing. In other situations it should be inexpensive to remove or present no finishing problems.
- Avoiding placing the gate near a stress concentration point of the product.

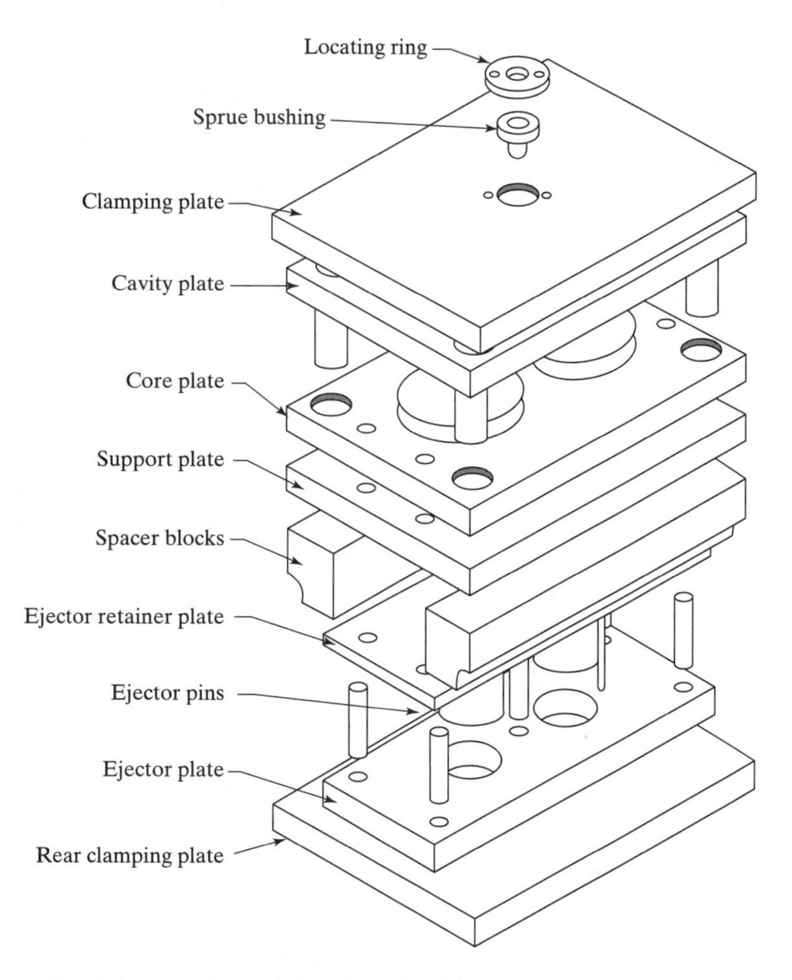

Figure 8.9 Two-plate mold based on Glanvill and Denton (1965). Schematic courtesy of Boothroyd et al. (1994).

- Positioning the gate where it will facilitate air expulsion as the injected liquid cools down.
- Avoiding the formation weld lines at cosmetic or stress critical areas.

The reader might be wondering how the air escapes as the ram fills the mold. In general, minute vents are installed on the parting plane, allowing the air to escape. Air can also be made to escape around the ejector pins. The position of the gate relative to these minute vents is important: there should be an easy flow of air from the gate-location side of the mold to the vent positions. If the air is not properly vented, it will become superheated, because of the high pressures involved, and scorch the part.

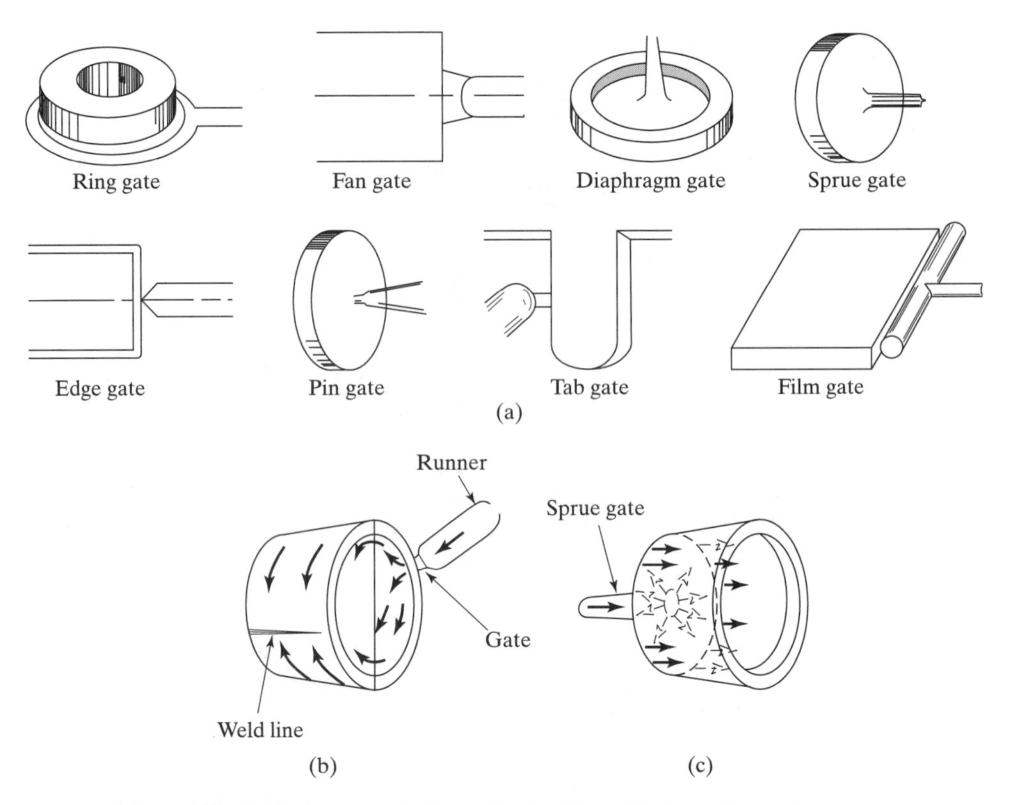

Figure 8.10 (a) Various gate designs in the top figure. The lower figure shows (b) an inefficient gating method and (c) the preferred method (on the right) to fill a cup (adapted from ICI and Pye).

8.3.8 Design Guides for the Part to Be Molded

The design of the mold requires great experience to account for the shrinkage[1] of the thermoplastic during cooling, the draft angles that must be added to vertical walls so that the part can be ejected, and the location of the parting plane.

Some other key design guides are shown in Figure 8.11. This figure is taken from Magrab (1997) and is based on the guides by Bralla (1998) and by Niebel, Draper, and Wysk (1989). The guides recommend:

- The avoidance of undercuts for simplification of ejection
- Uniform wall thickness for obtaining uniform shrinkage and avoiding warpage
- Dual diameters for deep holes to stiffen core pins under pressure and for cooling

[1]Typical shrinkages between the CAD geometry and the final part are 1% to 2%. A typical draft angle of 1 to 5 degrees is needed on a CAD design for a part that will be injection molded. These values will depend on the plastic and the texture used. Rather than quote values that may date quickly, for extensive listings, the reader is referred to <**www.metalcast.com**>.

Representative Design Guidelines for Injection Molding

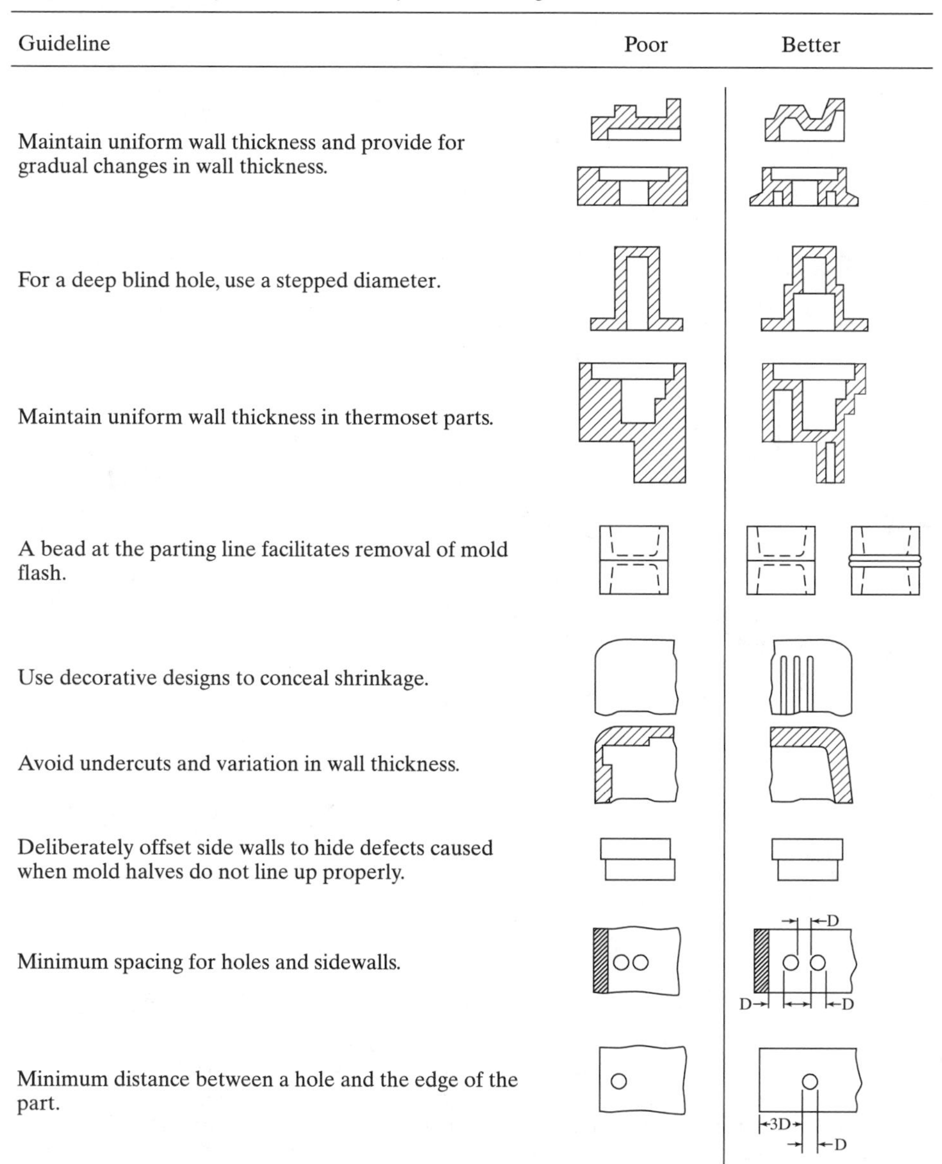

Guideline	Poor	Better
Maintain uniform wall thickness and provide for gradual changes in wall thickness.		
For a deep blind hole, use a stepped diameter.		
Maintain uniform wall thickness in thermoset parts.		
A bead at the parting line facilitates removal of mold flash.		
Use decorative designs to conceal shrinkage.		
Avoid undercuts and variation in wall thickness.		
Deliberately offset side walls to hide defects caused when mold halves do not line up properly.		
Minimum spacing for holes and sidewalls.		
Minimum distance between a hole and the edge of the part.		

Figure 8.11 Design guides for plastic injection molds. Reprinted with permission from *Integrated Product and Process Design and Development* by E. B. Magrab. Copyright CRC Press, Boca Raton, Florida.

- Minimum spacing distances from holes to holes and from holes to walls to ensure that the mold fills correctly and uniformly in the spaces between features
- Cavity shapes that reduce "flash" at the parting line
- The use of decorations to divert the eye from difficult-to-mold areas (One "trick," for example, is to scribe shallow circles around the gate position. After

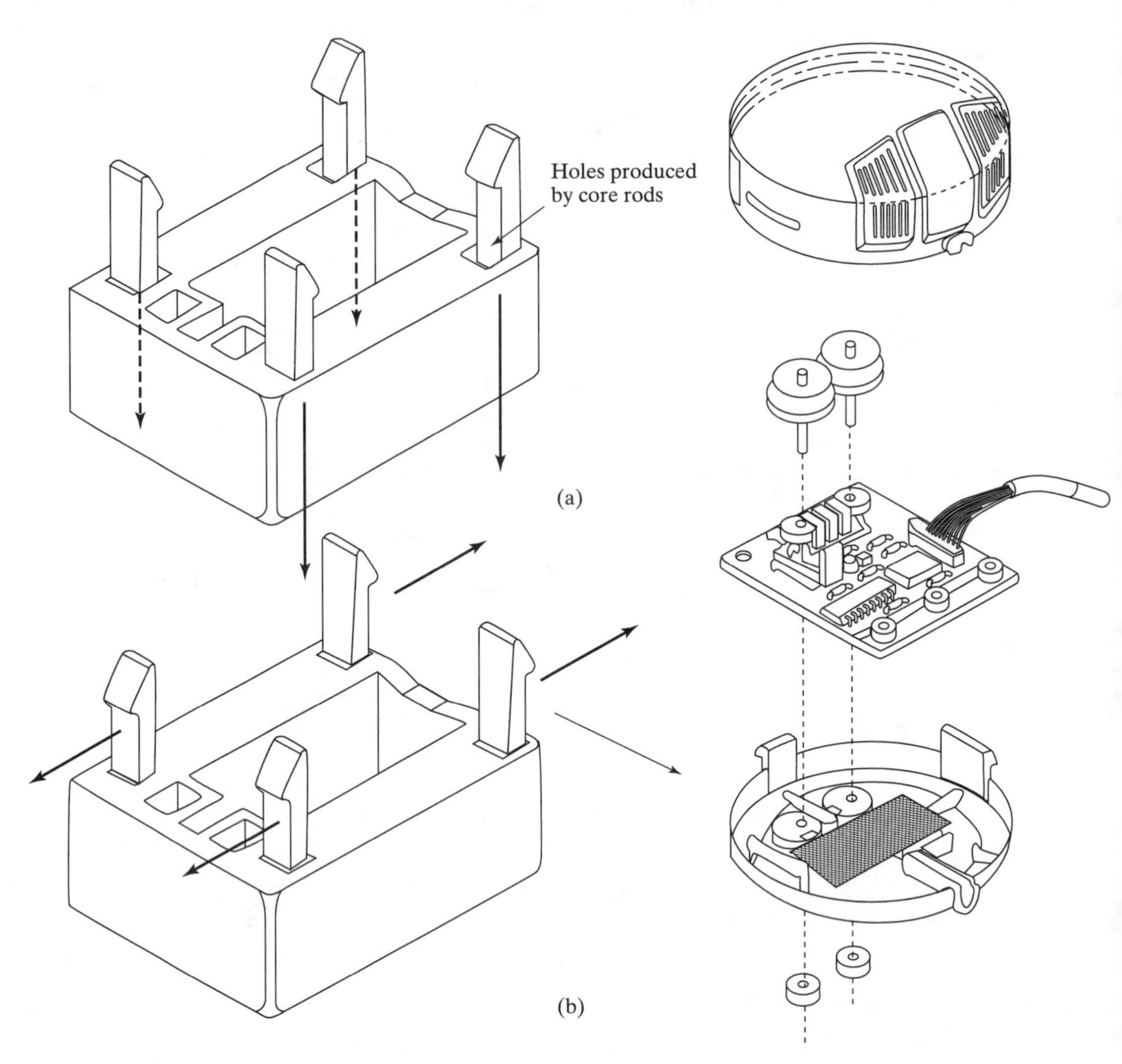

Figure 8.12 Cantilever snap-fit assembly made possible in the injection molding process: (a) the direction of the arrows shows where core rods would be pulled down after molding to create the required undercuts; (b) the arrows show the direction of side pullouts. On the right is a Digital Corporation mouse that was redesigned with the Boothroyd and associates DFA methods—the snap fits are shown in the lower subassembly (both diagrams courtesy of Boothroyd et al., 1994).

molding the user/consumer sees the kind of pattern seen around a stone thrown in a pond. It seems a deliberate cosmetic design feature to the unsuspecting eye, but actually it camouflages the rougher gate nipple.)

- The use of textured molds to create a "fine-orange-peel–like" surface that camouflages shrinkage and other blemishes

- Increasing offsets at parting lines rather than trying to make them line up perfectly (This is a well-known trick in carpentry—a bigger "setback" looks deliberate, whereas a small one looks like a mistake.)

The relatively low Young's modulus of plastics compared with metals also facilitates design with snap fits. In Section 8.7, the famous story of IBM's ProPrinter redesign will be mentioned. IBM carried out a full redesign of its standard printer, reduced the "part count," and used snap fits as much as possible rather than screws. This had a great impact on the part count, which was reduced from 152 to 32. Consequently the assembly time, also related greatly to the thermoplastic snap fits, was reduced from 30 to 3 minutes (see Dewhurst and Boothroyd, 1987). Figure 8.12 shows mold designs that create such snap-fit cantilevers. In the upper diagram, core rods in the mold are movable and leave the plastic projections freestanding and thus able to exhibit flexure. In the lower diagram, side-pullout pieces of metal insert create the undercuts beneath the plastic projections.

8.4 PROCESSING OF PLASTICS II: POLYMER EXTRUSION

In the bulk manufacturing of plastic products simple extrusion is the most common process. The products include tubes, plastic pipes for plumbing, window frame moldings, and insulated wire for electrical use. The polymer passes through the barrel's heated zones and is also plasticized by the screwing action. At the die mouth, liquid product is extruded through the preshaped die shown at the left of Figure 8.13.

Note also that the internal diameter of the screw is less at the back end near the hopper than at the front end. This is because compressed solid pellets pack irregularly and have a larger volume than the same mass of liquid. However, since the rate of mass flow must be constant at any plane in the screw mechanism, the area of the screw channel must decrease as the polymer transforms from solid pellets to liquid.

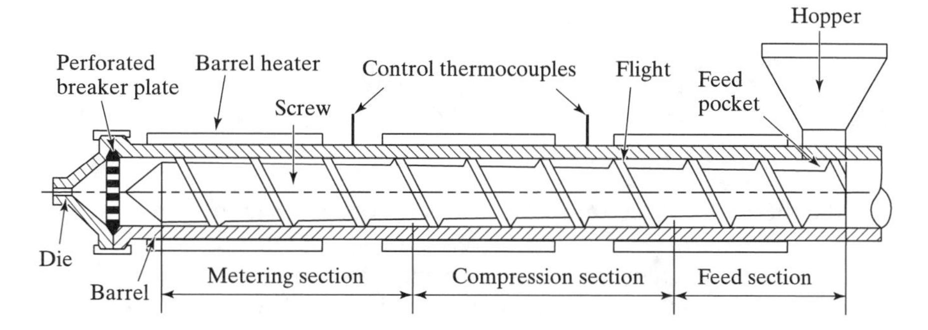

Figure 8.13 Extrusion with the die on the left.

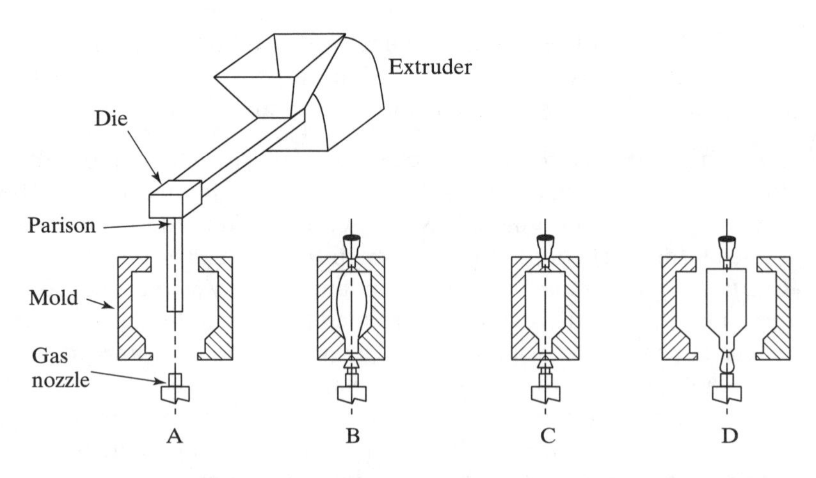

Figure 8.14 Blow molding of plastic bottles.

8.5 PROCESSING OF PLASTICS III: BLOW MOLDING

The blow-molding process produces the ubiquitous designer water bottle and similar consumer products. Extrusion blow molding is shown in Figure 8.14, and similar processes include injection blow molding and stretch blow molding.

The extruder first extrudes a heated dangling hollow tube, called a parison, into the open mold. The mold closes and pinches off the upper tube end: this usually corresponds to the bottom of the bottle. The reader might glance at any plastic drinking bottle and see the rough nipple that gets created. In some designs a crease may be used. Either way, a new problem is created for the part designer! The rough projection on the bottom will make the bottle unstable as it sits on a user's desk. Another glance at the bottom of any bottle will thus show a carefully designed convex hull or some feetlike projections around the bottle's perimeter.

After the mold closes, the parison is inflated and blown out to adopt the shape of the mold. Note that the extruded parison is already heated and the mold opening is warmed so that the leathery polymer easily takes up its molded shape. Subsequently, the now formed bottle is cooled and removed from the mold. A new parison is then extruded into the open mold.

8.6 PROCESSING OF PLASTICS IV: THERMOFORMING OF THIN SHEETS

8.6.1 Introduction

The thermoforming (or blow-forming) process has many similarities in geometry to the metal-forming operation described in Chapter 7. The difference is that plastic blanks are heated and then deformed by air pressure rather than a die. The expanding plastic dome can be unrestrained or blown against a sculptured die to create a desired shape. Whether the dome is free-formed or restrained, it is desirable to minimize the thinning, similar to sheet-metal forming. The next analysis considers the necessary properties for the plastic sheet being formed.

8.6.2 Analysis: Maintaining Uniform Sheet Thickness

The formed product—whether it is made from ABS, PMMA, HIPS, or polyethylene—invariably displays a nonuniform thickness distribution in the sheet. In other words, the strain experienced by the material at the pole of the free-formed dome is much higher than that experienced at the clamped edge.

This should sound familiar: it was the same problem faced earlier when a material with a high n value was needed in sheet-metal forming. The main difference now is that plastics undergo time-dependent creep at the typical temperatures of thermoforming, and this is added into the equations below.

The stress-strain-time (t) relationship can be represented by the following expression:

$$\sigma = kt^{m'}\varepsilon^{n} \tag{8.3}$$

where σ = stress

ε = strain after time t

m' = index of strain rate (time) sensitivity

n = index of strain-hardening sensitivity

k = a constant

The thermoforming behavior of different materials can be compared and explained by the m' and n values in the stress-strain-time relationship. From work on ABS, PMMA (polymethyl methacrylate), and HIPS (high-impact polystyrene), it has been found that more uniform thickness distribution is found in situations where the material has high n and low m' values.

From considering a unit width of the dome across any centerline, one obtains the following relationship. Standard mechanical engineering textbooks on machine design or stress analysis give such equations for pressurized pipes, cylinders, and shells.

$$\sigma \cdot s = P \cdot R$$

or

$$\sigma = \frac{PR}{s} \tag{8.4}$$

where σ = stress

P = thermoforming pressure

R = the inner radius of the dome

s = the thickness of the sheet as shown in Figure 8.15

Equating 8.3 and 8.4 yields

$$\frac{PR}{s} = kt^{m'}\varepsilon^{n}$$

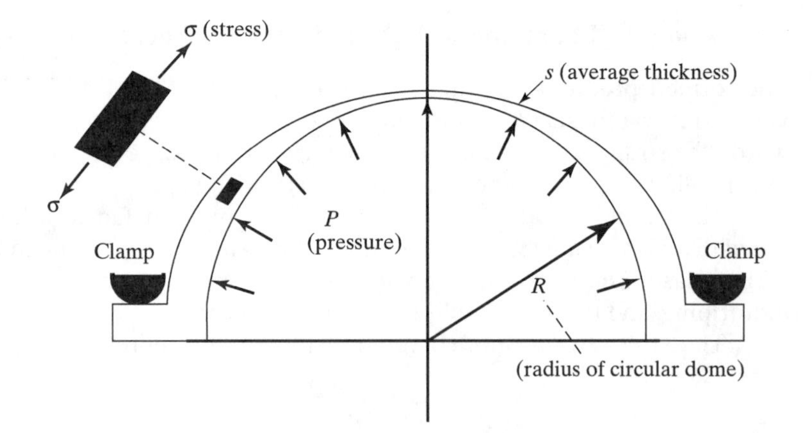

Figure 8.15 Dome section showing various parameters.

or

$$n \cdot \ln(\varepsilon) = \ln(P) + \ln(R) - \ln(s) - (m' \cdot \ln(t)) + \ln k$$

Differentiating this expression gives the following:

$$n \cdot \frac{\delta\varepsilon}{\varepsilon} = \frac{\delta P}{P} + \frac{\delta R}{R} - \frac{\delta S}{s} - m'\frac{\delta t}{t}$$

where δ, δ*P*, δ*R*, and δ*S* are incremental changes with respect to time (δ*t*). Now for fully plastic flow, the balanced biaxial strain is

$$d\varepsilon = \frac{\delta s}{s}$$

so that

$$n \cdot \frac{\delta\varepsilon}{\varepsilon} - \delta\varepsilon = \frac{\delta P}{P} + \frac{\delta R}{R} - m'\frac{\delta t}{t}$$

Therefore to obtain an even thickness distribution throughout the thermoformed dome, the strain at any two points must be the same, that is,

$$\Delta\left(\frac{\delta\varepsilon}{\varepsilon}\right) = \frac{1}{n - \varepsilon}\left[\frac{\delta P}{P} + \frac{\delta R}{R} - m'\frac{\delta t}{t}\right] \rightarrow 0 \quad (8.5)$$

This can be achieved by a high value of *n* and a low value of *m*′.

8.7 THE COMPUTER AS A COMMODITY: DESIGN FOR ASSEMBLY AND MANUFACTURING

8.7.1 System Configuration and Packaging in the Injection Molded Case

System assembly is the final step in the production of a finished computer, cellular phone, or any modern consumer electronics device. In Figure 6.1, the motherboard is a finished PCB containing the main microprocessor and eight additional memory boards. It is assembled into the computer casing as indicated on the right of Figure 6.1. The hard disc drive and other system components are put in place and connected to power supply cables. The unit is then powered up for testing. Once the functionality has been verified, the test cables are removed and the injection molded casing is enclosed. Similar steps were carried out for the ST Microelectronics' TouchChip™ and the InfoPad (case studies in Chapters 2 and 6), even though these devices did not have a hard drive.

8.7.2 Commercial Impact of DFA and DFM Techniques on the Computer Industry since the Early 1990s

A well-known (true) story . . .

> Once upon a time a big computer company called IBM decided to redesign its ProPrinter so that it could be assembled by robots. Because robots could not pick up little screws or perform fine assembly tasks, the engineers dramatically simplified the printer's design. They used snap rivets, for example, and they made sure that as many of the assembly operations as possible were coaxial and helped by gravity. By the time the redesign was done, the assembly process was so sleek that the humans could assemble the ProPrinter faster and more cheaply than the robots. So the robots were fired, and the people lived happily ever after.

This story illustrates the potential impact of design for assembly and manufacturing (DFA and DFM). Increasingly, the application of DFA and DFM has had a major influence in the computer industry, in terms of both the basic machines and all the important peripherals such as printers and scanners.

For example, Compaq[2] aggressively applied DFA and DFM starting in the period around 1991 when new management took over the company. Furthermore, it reduced net profit margins from above 40% to 20% to 25%. In the period from 1992 to 1994 the cost of Compaq's basic 486DX2/50 machine was reduced from over $3,000 to around $1,800. This "shook up" the other PC clone makers considerably and was a clear step in the direction of turning the PC into a commodity item.

The following list delineates Compaq's strategies in the early 1990s, while Figure 8.16 illustrates its success:

- The use of DFA to reduce assembly time and the number of subcomponents. On average, one-third fewer components were needed.
- The consequent reduction, by about one-third, in the space and rent needs of Compaq's main factory.
- The use of just-in-time (JIT) manufacturing with all subcontractors located within a 15-mile radius of the main factory.
- Investment in new assembly machines.
- Reduced profit margins—from 40% to 20% to 25%—in the plan's initial years.

[2] At the time of this writing (spring 2000), Dell Computing has installed a widely admired supply chain management system and provided consumers with a popular online ordering Website for customized computers. It has put Dell more in the public eye than Compaq and other consumer-related computer companies at present. However, the book remains with these well-known, early 1990s events at Compaq, not to endorse its products but to show the power of DFA/DFM.

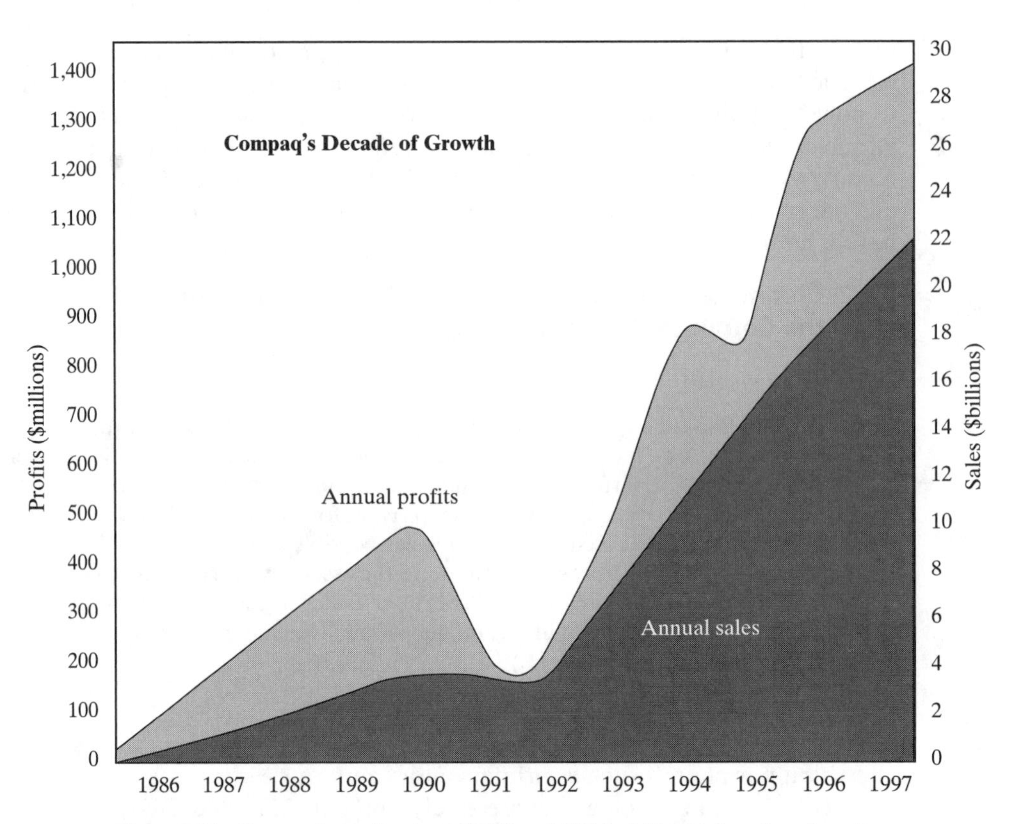

Figure 8.16 Compaq's application of DFA and JIT. It deliberately reduced profits in 1991 and 1992 but led to rapid growth for Compaq. It also made dramatic changes in the pricing structure of the standard PC throughout the world (data from Compaq).

The effects of this strategy can be seen in the dip in the 1991 figures, a temporary reduction that is more than balanced in subsequent years by dramatic increases in sales and profits.

8.7.3 Overview of Design for Assembly (DFA) Procedures

Design for assembly involves two key common sense ideas:

- *Individual subcomponent* quality must be high, and the number of subcomponents must be reduced as much as feasible.
- *Assembly operations* among subcomponents must be as simple as possible; for example, by keeping factory layouts orderly, by simplifying the shape of individual components, by altering certain design features that simplify the assembly of one component with another (Figure 8.17), and by not fighting gravity. (Anyone who has unscrewed a sump plug to change his or her oil will appreciate this!)

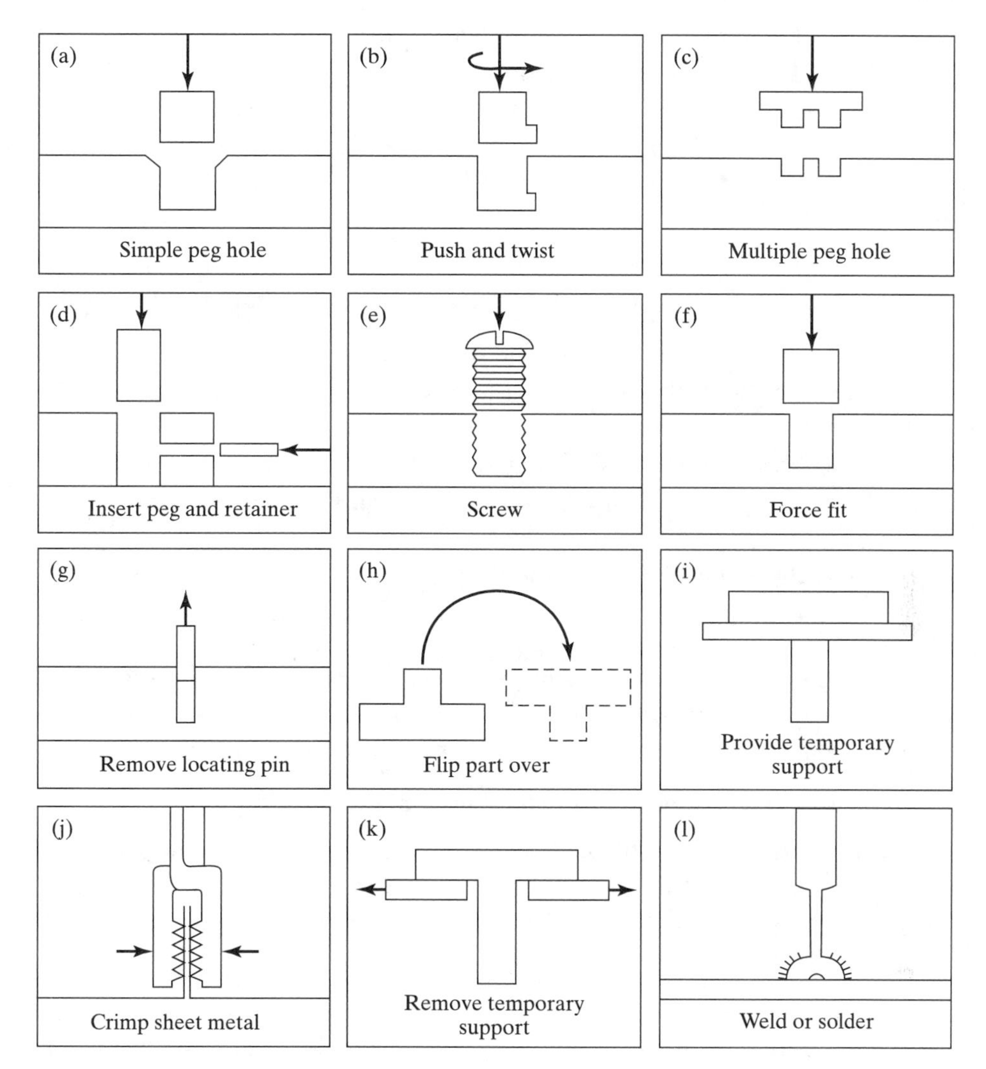

Figure 8.17 Typical assembly operations.

8.7.4 Design Checklist for Individual Subcomponents

Reducing the number of subcomponents and improving their quality involve the following steps.

First, the general geometry should be considered:

- Are the designs of subcomponents standardized so that different subcontractors can supply identical products?
- Are subcomponents symmetrical? Make them so if possible so that people or robots will not have to juggle them into a unique orientation.

- If asymmetrical, can they at least be oriented in a repeatable way?
- Have screws been eliminated as much as possible?
- Can lead-in chamfers be used as shown in Figure 8.17a?
- Will parts tangle, nest, or interlock and cause problems when people or machines try to pick them up?
- Will any parts "shingle" down onto each other and cause problems in picking them up?

Second, tolerances and mechanical properties should be reviewed:

- Are critical dimensions and tolerances clearly defined?
- Are tolerances as wide as possible?
- Have sharp corners been eliminated?
- Are burrs and flash eliminated?
- Are parts rigid enough to withstand buckling during push assembly as shown in Figures 8.17c, 8.17d, and 8.17f?
- Has weight been minimized?

Third, the design for manufacturability of the subcomponents should be analyzed:

- Have tooling requirements been minimized and standardized?
- Can the present shop equipment cope with all the assembly needs?
- Are subcontractors supplying parts according to 6 sigma quality?

8.7.5 Design Checklist for the Assembly Operation Itself

The early pioneers of industrial engineering such as Talyor and Gilbreth paved the way for efficient assembly work with their early time-and-motion studies. These studies involved a lot of common sense. They emphasized an orderly workplace with an ergonomic layout and well-designed delivery systems for subcomponents. Ideally, a person or a machine does not have to stretch too far to retrieve the components.

Keeping things orderly and close together can be related to Fitts's law. The speed and accuracy of hitting a pencil point into the windows in Figure 8.18 are given by Fitts's *index of difficulty* = the logarithm of **s/w.**

This index emphasizes that subcomponents should arrive in a tightly arranged workplace (small **s**), but that the actual assemblies should allow for margins of human or machine variability (large **w**). For example, the lead-in chamfer in Figure 8.17a creates more window (**w**) for the alignment of a pin-in-hole type assembly.

In summary, Fitts's overall recommendations are (a) to reduce the distance, **s,** that subcomponents have to be moved and (b) to increase the "give" in the assembly

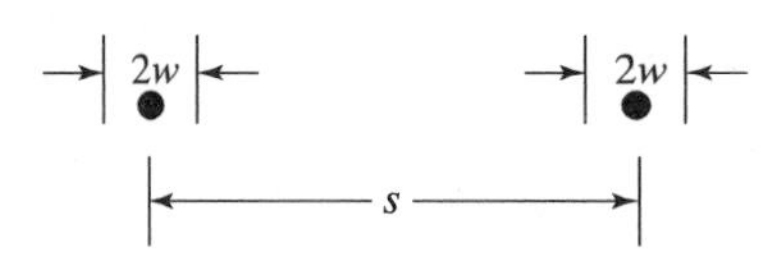

Figure 8.18 Fitts's tapping task.

with as large a value of **w** as can be permitted without compromising the functioning of the device.

8.7.6 Design Checklist for Electromechanical Devices

For the assembly of an electromechanical device, the following checklist arises:

- Is it possible to use a vertical stack and have gravity help the assembly? Figure 8.19 summarizes data from 10 industrial products (labeled **A** through **L**), showing that vertical assembly is the most common for these operations. Note that component **E** was the most complex to assemble given the number of direction 2 and 3 tasks.
- Does the design have a datum surface to aid referencing?
- Has unnecessary turning and handling been avoided?
- Have buried and inaccessible subparts been avoided?
- Is fixturing precise and helpful?

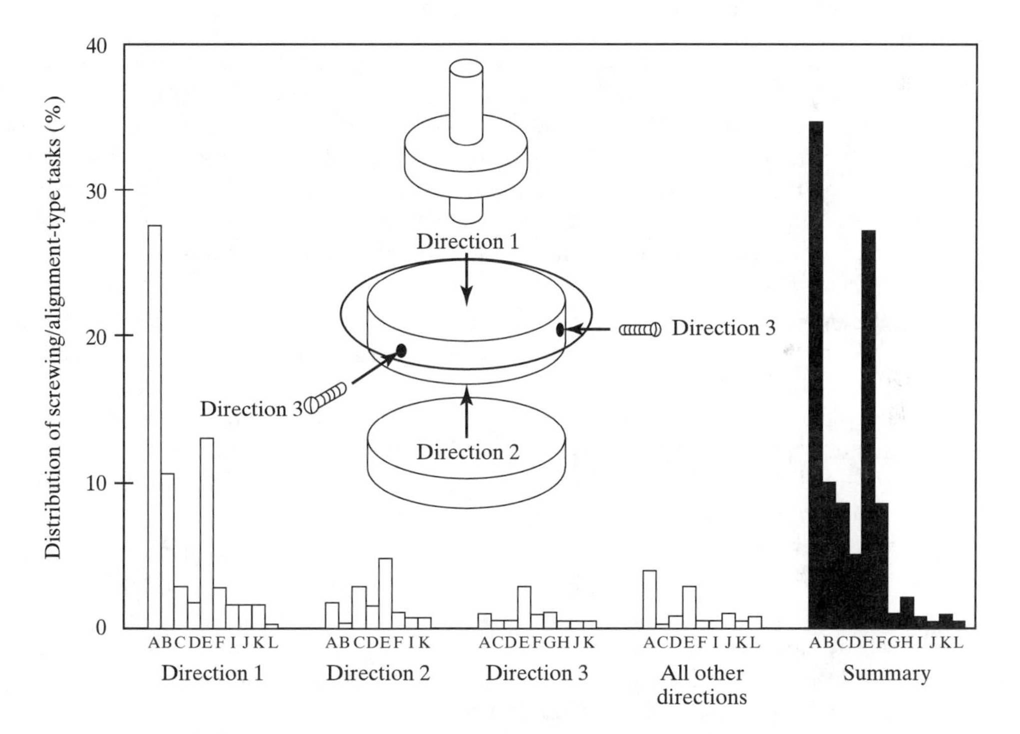

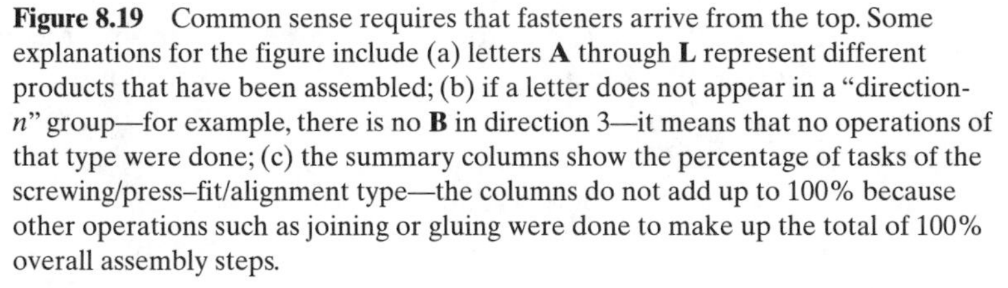
Figure 8.19 Common sense requires that fasteners arrive from the top. Some explanations for the figure include (a) letters **A** through **L** represent different products that have been assembled; (b) if a letter does not appear in a "direction-n" group—for example, there is no **B** in direction 3—it means that no operations of that type were done; (c) the summary columns show the percentage of tasks of the screwing/press–fit/alignment type—the columns do not add up to 100% because other operations such as joining or gluing were done to make up the total of 100% overall assembly steps.

- Have screws and other fasteners been minimized and reduced to snap fits?
- Can everything be done in an automatic assembly machine?
- Can a base part, base plate, or central axle be used, to which everything else can be assembled? This helps to orient everything toward a central assembly theme.
- Are subassemblies modular?
- Has group technology been used for the next part in the assembly process?
- Is the greatest value-adding task performed last? This is an important point in case something is damaged at the last minute.
- Has the required assembly dexterity been minimized?

8.7.7 Design Checklist for Welding, Brazing, Soldering, and Gluing

A brief description of joining methods is appropriate here. Kalpakjian (1995) and Bralla (1998) provide an excellent review of the physical chemistry and the DFA/DFM aspects.

Welding processes: Intense heat from the electric arc of a "welding-stick," a controlled plasma arc, or a "spot-welding" tool causes localized melting, mixing, and local resolidification of the surfaces of the two components being joined. This "micromelting/casting" needs to be done in a protective atmosphere; otherwise, the oxygen in the air forms local oxide deposits that damage the metallurgical integrity of the finished joint. For example, a consumable welding rod may serve to provide this atmosphere as it decomposes in the heat, thereby generating a covering shield of inert gases.

Brazing and soldering: A filler material is locally melted with a "soldering iron" (for soldered joints) or a flame (for a brazing operation) and made to flow between the two surfaces to be joined. In contrast to welding, the two main surfaces do not melt, but when the filler material resolidifies, a solid-state bond is created between each surface and the filler material. The filler material may be conventional electrical solder (tin-lead alloys) or brazing compound (silver or copper alloys). Brazing gives a higher strength than soldering.

Gluing methods: Epoxy resins and acrylic glues provide a chemical bond between the two surfaces to be joined. Clean surfaces devoid of grease and as much oxide as possible are the ideal conditions. Nevertheless, the bonds created are significantly lower in strength than the metallic bonds created by welding, brazing, or soldering. Glues are often susceptible over time to the ultraviolet rays in natural light. It is unwise to depend on a glued joint for long-term service.

During CAD, designers aim to create component geometries that enhance the structural integrity of a formed joint. Figure 8.20 shows some recommended joint geometries for soldering and brazing (Bralla, 1998). At the same time, for "downstream manufacturing," the accessibility of a manually operated welding torch or "spot-welding robot" should be considered. For example, on an automobile assembly line, the welding operations inside a car's trunk are done in tight quarters. The orig-

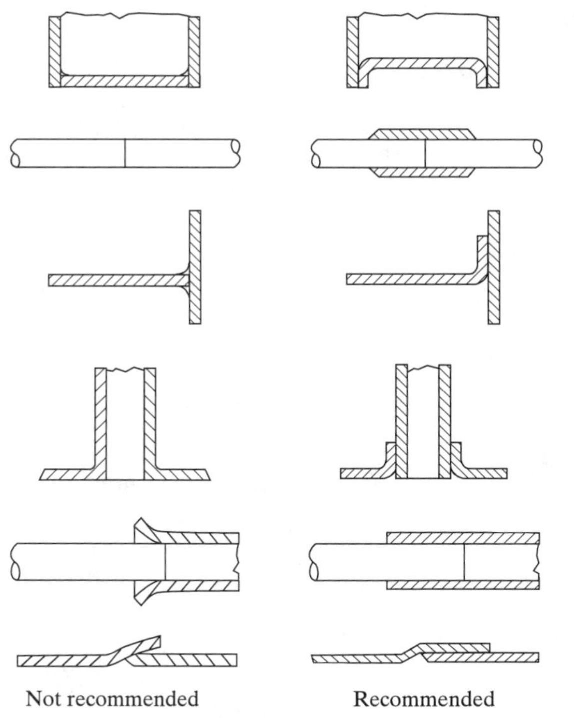

Figure 8.20 An example of design guides for soldering and brazing processes (from *Design Manufacturability Handbook* edited by J. G. Bralla, © 1998. Reprinted by permission of the McGraw-Hill Companies.)

inal designer, the process planners, and the fixturing engineers all play a role in making the process easy or hard to execute, which in turn affects the resulting quality of the weld. As with all assembly operations, a downward attack on the work surfaces is a main recommendation.

8.7.8 Formal Methods of Scoring Assemblies

Formal schemes are now being used by corporations—Chrysler and Compaq are two notables—in a variety of industries to quantify the preceding lists as much as possible. Obviously, the best practices for all companies should increasingly include these assembly evaluations.

8.7.8.1 Boothroyd and Dewhurst Method

The Boothroyd and Dewhurst evaluation method assigns scores to the following tasks:

- *Parts count:* this is simply counted, and where possible design changes are made to reduce the number of parts.
- *Symmetry:* axial symmetry is preferred and given the highest ranking.
- *Size of parts:* medium-sized parts that can be picked up easily by humans are given the highest ranking. Small read-write heads would be given a low ranking because they require stereomicroscopes and tweezers for assembly.

Heavy parts that might need hoists or human amplifiers to pick them up would also get a low ranking.

- *Shapes:* smooth shapes that avoid tangling are given a high ranking.
- *Quanta of difficulty:* finally, additional penalties are given to parts that are in any way awkward, slippery, or easily damaged. One way of calculating this ranking is to measure the time and level of skill needed to complete the task.

8.7.8.2 Xerox Corporation

The Xerox scoring method is similar to that of Boothroyd and Dewhurst. Quantitative scores are given in the following areas:

- Parts count (as before)
- Direction of assembly motion (as shown in Figure 8.19)
- Fixturing needs at each setup
- Fastening methods, with snap fits preferred over screws or joining methods

8.7.9 Maintaining a System Perspective—Closing Thoughts

In the design and prototyping of the ST Microelectronics' TouchChip™ (the case study in Chapter 2) several DFM/DFA strategies were employed. For example, snap fits were used to hold the PCB onto the upper lid. At the same time this made the aluminum molds more expensive to machine and more expensive to operate because of the undercuts needed. Thus, as a word of caution, the design team should certainly look at the "big picture." For the first run of 200 parts, it might not be worth the cost of the snap fit. But for millions of production parts, the extra time taken to machine the molds and also to operate the cores for the undercuts during molding will probably pay off in "downstream assembly costs."

This section concludes with an interesting success story (Prentice, 1997) from an extended "learning organization." During the redesign of a rather ordinary portable stereo, the question arose, What size should the outside plastic casing be? At first, it seemed that these dimensions could be rather arbitrary. But on closer analysis, a great benefit was gained by adjusting the size so that a certain number of stereos would completely fill cargo containers used in transpacific shipping. By adjusting the size of an individual stereo it was possible to fit a greater number into the containers. Furthermore, it was possible to minimize the packing materials somewhat because the perfect tightness of fit prevented the cargo from shifting and becoming damaged. It is perhaps unusual to be able to adjust a design based on a constraint so far away in the logistical chain, but the example does challenge an individual design engineer to think as broadly as possible in a learning organization.

8.8 MANAGEMENT OF TECHNOLOGY

8.8.1 Integrated Product and Process Design

Economic pressures, particularly related to the quality of manufactured goods and time-to-market, are forcing designers to think not only in terms of *product design* but also in terms of *integrated product and process design* and, finally, in terms of deterministic *manufacturing planning and control.*

As a result of these three high-level needs, there is now an even greater need for comprehensive models—of the type introduced in Chapters 7 and 8—that predict material behavior during a manufacturing process, the stresses and/or temperatures on associated tooling, and the final product integrity. The overall goal is to enrich a CAD/CAM environment with:

- Physically accurate finite element analyses (FEA) and visualizations of the manufacturing process
- Access to process planning modules that allow detailed cost estimates

For the polymer materials and mold making that have been the focus of this chapter, CAD/CAM-related URLs are given in Section 8.12.

8.8.2 Databases and Expert Systems

In addition to the FEA methods in manufacturing, *expert systems* (Barr and Feigenbaum, 1981) continue to be valuable. Expert systems formulate solutions to manufacturing concerns that cannot be solved directly with quantitative analysis. Since the early 1980s they have been useful in a wide variety of scheduling problems (see Adiga, 1993). Expertise is gathered by the formal questioning and recording process known as *knowledge engineering.* In this approach, engineers work with factory-floor personnel to compile records, tape recordings, and videotapes. These build up a qualitative model of the approaches needed for problem solving.

When carried out within a learning organization (see Chapter 2), it has been found that factory personnel and machinists react favorably to this approach—documenting the kinds of problems that often arise with production machinery and, similarly, documenting setup and monitoring procedures for individual machine tools (Wright and Bourne 1988). In the best situations the personnel are even flattered that their skills are valued and worth capturing for subsequent generations. The rules and qualitative knowledge of the experts are written down as a series of rules of the form "If . . . then. . . ." The qualitative parameters in these fields might be nonquantitative data such as colors or approximate percentages.

In other situations where manufacturing data is more quantitative, conventional relational databases or object-oriented databases are more useful (see Kamath, Pratt, and Mize, 1995). At a high level, such databases might describe the corporate history, meaning a history of the typical products, batch sizes, and general capabilities of the firm. At a more medium level of abstraction, particular capabilities of the factory-floor machinery might be described, with achievable tolerances, operational costs, and availability. At the lowest level, the databases might contain carefully documented procedures for lithography and etching times. In any industry, the immediate availability of accurate manufacturing parameters for machinery setup and diagnosis is very valuable. Such databases also facilitate incorporation of DFA and DFM data structures.

PDES/STEP has emerged as a worldwide scheme for developing a common informational framework for such databases and CAD/CAM systems. Its goal is to ensure that information on products and processes among different companies is compatible. Now that so many large firms rely on subcontractors and outside suppliers to create their supply chain, the need for a common interchange format is more important than ever (see Borrus and Zysman, 1997).

8.8.3 Economics of Large-Scale Manufacturing

Economically, the aims are to ensure a high-quality product and to reduce time-to-market by eliminating ambiguities and "rework" during CAM (Richmond, 1995). For example—as reported by Halpern (1998)—Grundig states that the dies for the front and back of its television casings cost approximately $300,000 each. The average cost to make a single change to one of these is typically 10% of the original die cost, or $30,000. Evidently, integrated CAD/CAM systems of the type described at the end of Chapter 6 are very important software tools for minimizing such rework during mold design, fabrication, and tryout.

Most *large-scale manufacturing* operations (in either metal or plastic) are by definition mature technologies that are well along the market adoption curve in Chapter 2. But customers deliberately choose these mature technologies because they are tried and true, giving reliable, predictable results. These basic processes in Chapters 7 and 8—such as machining, sheet-metal forming, injection molding, and thermoforming—may not have the glamour of stereolithography or selective laser sintering, but they remain central to many major industries and to the economy as a whole.

Nevertheless, to compete in global markets, all companies in these fields must apply creative methods and innovations. These clearly include new CAD/CAM techniques that reduce time-to-market, the use of sensor-based automation at the shop-floor level to reduce labor costs, and quality assurance techniques. To create customization, especially for Internet users, traditional manufacturing flows will need to be broken down into modular segments. Garment producers have considered such a change in order to address the custom tailoring market. The *Economist* (2000) argues that traditional manufacturers may well have to follow this example.

8.9 GLOSSARY

8.9.1 Blow molding

Various kinds of blow molding allow plastic tubes and plastic sheets to be inflated with air pressure against a mold. Plastic drinking bottles are the most obvious products made by these methods.

8.9.2 Branching

In these polymers, the side branches lock into adjacent chains and provide additional interlocking and stiffness.

8.9.3 Cross Linking

In these polymers, additional elements link one chain to another. The best example is the use of sulfur to cross-link elastomers to create automobile tires.

8.9.4 Crystallization

With mechanical processing, such as extrusion or rolling, polymer chains can be folded into explicit structures to give the material more stiffness.

8.9.5 Design for Assembly

Design for assembly involves reducing the number of components, keeping the quality of the components high so that they can be easily assembled, simplifying factory layout so that individual subcomponents come together easily, and ensuring that as many operations as possible can be done in a vertical direction. Vertical directions are shown in Figure 8.17.

8.9.6 Design Guides

A variety of heuristics that have been developed over time to aid the mapping from a part design to a mold design. Frequently a design guide relates to the elimination of sink and distortions.

8.9.7 Gate

The entrance to the mold cavity.

8.9.8 Glass Transition Temperature

The glass transition temperature is approximately halfway between the glass plateau and the leathery plateau shown in Figure 8.1. Also, by extrapolating the two curves shown in Figure 8.2, the glass transition temperature is the intersection of glassy behavior and viscous behavior.

8.9.9 Ejectors

These are typically pins used at the end of the cycle to lift the part from the mold.

8.9.10 Flash

If additional plastic is forced between the mold halves, because of a poor mold fit or wear, it is called flash. In general this is to be avoided and may require additional hand finishing if excessive.

8.9.11 Index of Strain-Hardening Sensitivity

Shown in Section 8.6, the strain-hardening sensitivity relates to the amount of strength increase with a given strain.

8.9.12 Index of Time Sensitivity

Shown in Section 8.6, the time sensitivity is the relaxation-related property of the material at a given temperature.

8.9.13 Injection Molding

Injection of plastic into a cavity of desired shape. The plastic is then cooled and ejected in its final form. Most consumer products such as telephones, computer casings, and CD players are injection molded.

8.9.14 Packing

The phase of injection molding where the ram holds the liquid mold at pressure. During this phase, approximately 10% more polymer is pumped into the mold cavity.

8.9.15 Parison

The dangling tube of plastic that is extruded into a heated mold for blow molding. It is subsequently pinched off at one end and inflated at the other during blow molding.

8.9.16 Parting Plane

The separation plane of the two mold halves.

8.9.17 Reciprocating-Screw Machine

The most used injection molding machine in industry: it combines the screwing action for the plasticization process and a ramming action for the injection process.

8.9.18 Runners

In a multipart mold, the runners extend from the sprue to the individual gates of each part.

8.9.19 Shrinkage

The amount of volume contraction of a polymer. Usually this is 1% to 2% given the reciprocating-screw process.

8.9.20 Snap Fit

Projections molded into a part that deflect to provide mechanical fastening with other parts.

8.9.21 Sprue

The runway between the injection machine's nozzle and the runners or the gate.

8.9.22 Thermoforming

In this process, plastic sheets are clamped around the edge, heated, and inflated with air pressure. The dome can be free-formed or formed against a mold to create surface impressions.

8.9.23 Thermoplastic Polymers

Polymers that undergo reversible changes between the glassy, leathery, viscous, leathery, glassy cycle.

8.9.24 Thermosetting Polymers

Polymers that undergo irreversible changes from the liquid to solid state, often by adding other chemicals such as epoxy resins.

8.9.25 Undercuts

"Sideways" recesses or projections of the molded part that prevent its removal from the mold along the parting direction. They can be accommodated by specialized mold design such as sliders.

8.9.26 Young's Modulus

Young's modulus defines the stiffness of a material and is given by stress divided by strain in the elastic region.

8.10 REFERENCES

Adiga, S. 1993. *Object-oriented software for manufacturing systems.* London: Chapman Hall.

Barr, A., and E. A. Feigenbaum. 1981. *The handbook of artificial intelligence: Volumes 1–3.* Los Altos, CA: William Kaufmann.

Beitz, W., and K. Grote. 1997. *Dubbel taschenbuch fúr den maschinenbau [Pocket book for mechanical engineering].* Berlin: Springer-Verlag.

Boothroyd, G., and P. Dewhurst. 1983. *Design and assembly handbook.* Amherst: University of Massachusetts.

Boothroyd, G., P. Dewhurst, and W. Knight. 1994. *Product design for manufacture and assembly.* New York: Marcel Dekker.

Borrus, M., and J. Zysman. 1997. Globalization with borders: The rise of wintelism as the future of industrial competition. *Industry and Innovation* 4 (2). Also see *Wintelism and the changing terms of global competition: Prototype of the future.* Work in Progress from Berkeley Roundtable on International Economy (BRIE).

Bralla, J. G., ed. 1998. *Design for manufacturability handbook,* 2d ed. New York: McGraw-Hill.

Dewhurst, P., and G. Boothroyd. 1987. *Design for assembly in action.* Assembly Engineering.

Economist. 2000. All yours. (April 1): 57–58.

GE Plastics. 2000. *GE engineering thermoplastics design guide.* Pittsfield, MA: General Electric Company. Also see **http://www.geplastics.com.**

Glanvill, A. B., and E. N. Denton. 1965. *Injection mold design fundamentals.* New York: Industrial Press.

Halpern, M. 1998. Pushing the design envelope with CAE. *Mechanical Engineering Magazine,* November, 66–71.

Hollis, R. L., and A. Quaid. 1995. An architecture for agile assembly. In *Proceedings of the American Society of Precision Engineers' 10th Annual Meeting.* Austin, TX.

Kalpakjian, S. 1995. *Manufacturing engineering and technology.* Menlo Park, CA: Addison Wesley. See in particular Chapters 27–30.

Kamath, M., J. Pratt, and J. Mize. 1995. A comprehensive modeling and analysis environment for manufacturing systems. In *4th Industrial Engineering Research Conference, Proceedings,* 759–768. Also see **http://www.okstate.edu/cocim.**

Magrab, E. B. 1997. *Integrated product and process design and development.* Boca Raton and New York: CRC Press.

McCrum, N. G., C. P Buckley, and C. B. Bucknall. 1997. *Principles of polymer engineering.* Oxford and New York: Oxford Science Publications.

McLoughlin, J. R., and A. V. Tobolsky. 1952. The viscoelastic behavior of polymethylmethacrylate. *Journal of Colloidal Science* 7: 555–568.

Niebel, B. W., A. B. Draper, and R. A. Wysk. 1989. *Modern manufacturing process engineering.* New York: McGraw-Hill.

Prentice, B. 1977. Re-engineering the logistics of grain handling: The container revolution. In *Managing enterprises: Stakeholders, engineering, logistics, and achievement,* 297–305. London: Mechanical Engineering Publications Limited.

Pye, R. G. W. 1983. *Injection mold design.* London: Godwin.

Richmond, O. 1995. Concurrent design of products and their manufacturing processes based upon models of evolving physicoeconomic state. In *Simulation of materials processing: Theory, methods, and applications*, edited by Shen and Dawson, 153–155. Rotterdam: Balkema.

Urabe, K., and P. K. Wright. 1997. Parting directions and parting planes for the CAD/CAM of plastic injection molds. Paper presented at the ASME Design Technical Conference, Sacramento, CA.

Wright, P. K., and D. A. Bourne. 1988. *Manufacturing intelligence.* Reading, MA: Addison Wesley.

8.11 BIBLIOGRAPHY

Modern Plastics Encyclopedia. New York: McGraw-Hill. Published annually.

8.12 URLS OF INTEREST

For mold design: **www.cmold.com**

General design with polymers: **www.IDESINC.com**

Bayer polymers division: **http://www.bayerus.com/polymers/**

Magics: **http://www.materialise.com/**

GE plastics: **http://www.ge.com/plastics/**

Society of Plastics Engineers: **http://www.4spe.org/**

Trading networks: **www.iprocure.com, www.memx.com,** and **www.commerceone.com.**

8.13 CASE STUDY ON ASSEMBLY

This case study invites the reader to think about how much investment in automation is needed to assemble a product. Batch size is a main consideration. Product revision is another: if designs change quickly, it may be difficult to justify automation if low-cost labor is available.

Referring to Figure 2.6, one helpful guide is to consider whether a company's current and future products and typical batch sizes are suited to (a) manual assembly, (b) human-assisted computerized assembly, (c) flexible robotic assembly, or (d) hard automation with less need for reprogrammability.

Manual assembly: This type of craftsmanship will dominate for one-of-a-kind machining/assembly of the kind seen in a university or the (R&D) model shop of a company. At the same time, manual assembly is likely to be the best choice for high-volume clothing and shoe manufacture. Since styles change quickly, the economics favor the use of intensely human assembly in countries that offer low wage rates. For example, shoe manufacturing in such countries is likely to be done more or less entirely by hand, by people sitting at simple gluing and sewing machines, or standing at simple transfer lines.[3]

Human-assisted computerized assembly: This is typically seen in U.S., Japanese, and European automobile factories for the final assembly of the seat units, fascia, and other internal finishes on the car. In this work, human dexterity is needed to carefully manipulate subcomponents into their proper places. This situation describes more of a middle ground of automation. The CIM system is installed to orchestrate the line flow and the delivery of subcomponents, *but human workers are very much part of the operation.* A similar situation can be observed at printed circuit board (PCB) assembly firms, many of which are subcontractors to the brand-name computer companies. These are the new service industries for the computer industry, delivering assembled PCBs with very little delay. Again, CIM systems orchestrate the flow lines, but a noticeable amount of human interaction is needed to load machines, monitor progress, and step in if there is a problem. The economics in this industry seems to justify U.S.-based assembly operations, perhaps because the batch sizes are smaller and communications between design and subcontractors are enhanced by proximity. These speciality PCB assembly firms are also able to buy large quantities of electronic devices in bulk and thus achieve economies of scale.

Flexible robotic assembly: Further along the spectrum, all the leading automobile companies in the United States and Japan have installed medium-cost robots and CIM systems to spot-weld and paint cars. The large batch sizes, heavy and/or unpleasant tasks, and a willingness to invest for the long haul have justified the investment in CIM. A tour of today's standard automobile line reveals that almost no shop personnel are needed to oversee such welding and painting operations. (Recall from earlier, however, that a great deal of personnel are needed to participate in final assembly.)

Hard automation: In Chapter 2 it was emphasized that for extremely large batch sizes, it might even be economical to revert to noncomputerized machines. Speaking colloquially, this batch size moves into the realm of "ketchup in bottles," where fixed conveyor lines pump out the same product day in, day out. As stated, this is often referred to as fixed or hard automation. In such factories, some basic computer controls and sensors are needed for monitoring and control, but reprogramming is not needed.

Chapter 6 reviewed the increasing miniaturization of disc-drive components and how difficult it is becoming to assemble them by hand with microscopes and tweezers. What are the considerations for automation? In the final analysis, will automating disc-drive assembly pay off? The batch sizes are large, but are they large enough? Today, overseas assembly workers can get the job done quickly and with sufficient reliability. Perhaps it is not worth risking a huge investment in automated

[3]It is a disheartening fact, but in today's civilization, some people are pleased to leave a rural environment to earn only $100 a month in an industrial setting, while others spend more than $100 at a shopping center on the purchase of just one pair of running shoes.

assembly systems. Whether this will always be true though, especially as components become even more miniaturized, remains to be seen.

To summarize, there is no question that an appropriate investment in CIM systems is important for U.S. and European firms. For large batches, unattended CIM systems are the only way for U.S. and European firms to compete globally. Research in this area to develop more agile systems is vital (see Hollis and Quaid, 1995). However, in cases where batch sizes are low and product designs change frequently, assembly may still be subcontracted to countries where labor costs are low. Finally, there are the intermediate cases where human-assisted low-cost CIM systems are the appropriate solution for both high-wage and low-wage operations.

In summary, while this might seem a frustratingly vague conclusion to an important topic, it is best left open because each case is special and warrants prudent analysis. By contrast, in the period around 1980, the U.S. research community and U.S. industry were not so prudent and were enthralled with the potential of robotics. Over time it has been more useful to think about robotics and automation with a different (although overlapping) emphasis:

- *Robotics* should encompass autonomous systems that emulate human capabilities and allow exploration or operation in environments that are too hazardous, tiring, or inappropriate for humans.
- *Automation of factories* should be analyzed strictly in context and provide an economic solution. This may range from intensely manual assembly through a spectrum of part-human/part-robotic CIM systems to hard automation.

8.14 INTERACTIVE FURTHER WORK

Visit the Metalcast Website and consider the following:

1. Find information on five different prototyping methods used to create the initial molds and describe, with diagrams where possible, the methods.
2. Make a table that lists the shrink rates for the following popular plastic materials: Allied Signal Capron, 8267 nylon, Amoco Polypropylene, Chevron (Poly)Styrene, Dow ABS, GE Lexan FL-410, Hoecht acetal, and Santoprene rubbers.
3. After clicking on "Data Exchange," list the six file formats that Metalcast can receive from customers.
4. View "Past Exhibits" and "Lost Core Manifold."

8.15 REVIEW MATERIAL

1. Based on the work of Kienzle, which became the German standard DIN 8580, manufacturing processes can be described in a two-dimensional taxonomy or framework (see Beitz and Grote, 1997). Six major groups are shown in Figure 8.21: (a) primary shaping, (b) forming (based on deformation), (c) dividing/separating, (d) bonding, (e) coating, and (f) changing of material properties. As a review activity for Chapters 2, 7, and 8, list five processing operations for each of the six categories.

2. The alternative Figure 8.22 shows an "ongoing three-dimensional taxonomy" that also includes rapid prototyping methods. Also, revisit the MAS and note that there are 20 manufacturing processes listed there.

3. As a review activity, using pens of different colors if needed, write on Figure 8.22 diagram to include all the other processes. SLA and SLS are done already. Casting and machining are perhaps done, but rethink the various types of machining and casting to make sure they fit OK. It might be interesting to add another layer above the diagram for the bulk shaping methods of forging and so forth.

<table>
<tr><th colspan="5">Coherence</th></tr>
<tr><td>Create...</td><td>Maintain...</td><td>Reduce...</td><td colspan="2">Increase...</td></tr>
<tr><td rowspan="2">(1) Primary shaping</td><td>(2) Bulk forming</td><td>(3) Dividing/ separating</td><td>(4) Bonding</td><td rowspan="2">(5) Coating</td></tr>
<tr><td colspan="3">(6) Changes to material properties by...
Moving... | Eliminating particles... | Adding...</td></tr>
</table>

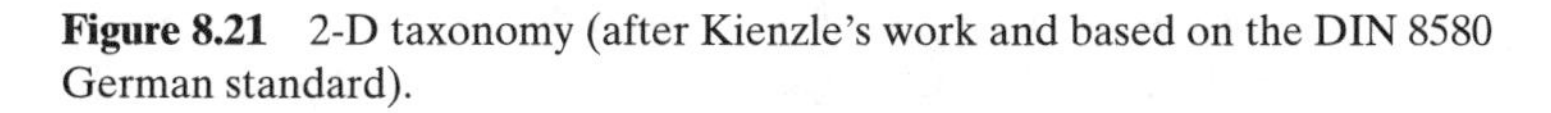

Figure 8.21 2-D taxonomy (after Kienzle's work and based on the DIN 8580 German standard).

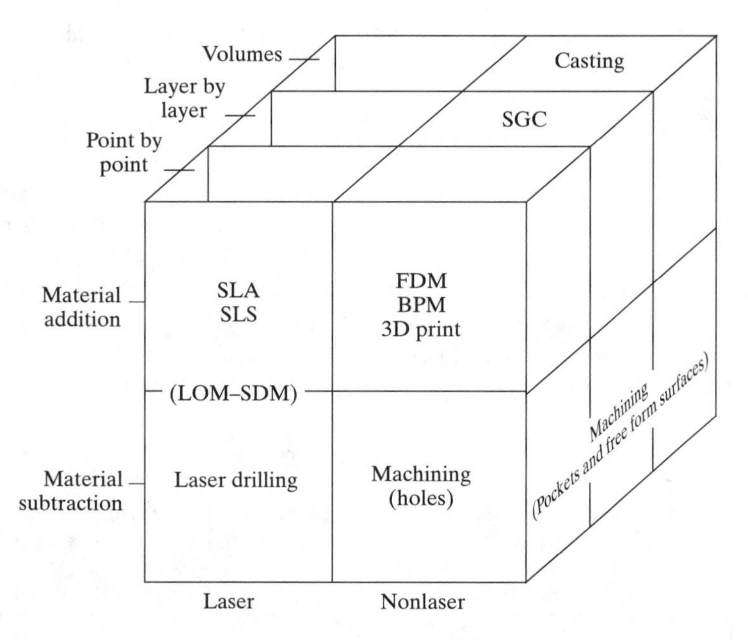

Figure 8.22 Three-dimensional taxonomy for manufacturing processes, including rapid prototyping.

CHAPTER 9

BIOTECHNOLOGY

9.1 INTRODUCTION

In the 1967 film *The Graduate* the young man played by Dustin Hoffman was offered this piece of advice: "One word . . . *plastics.*" If the script were written today, the family friend could just as easily advise "*biotechnology.*"

Biotechnology is defined for this book as the "utilization of biological processes to manufacture a desired product." However, the biotechnology industry draws on a cumulative base of scientific knowledge from a number of disciplines, including math, physics, chemistry, and biology. Thus, it may not be realistic to be too pedantic about definitions.

The biological techniques used in the biotechnology industry include recombinant DNA, cell fusion, and advanced processing techniques to grow or modify living organisms in order to produce useful products or processes. Since a recombinant gene was first used to clone human insulin in the 1970s, biotechnology has grown into a multibillion-dollar international industry. Along with microelectronics and computers, biotechnology is one of today's most technology-intensive industries. Biotechnology and bioengineering[1] could be considered the "fifth pillar" of the engineering world, joining civil, mechanical, chemical, and electrical engineering.

[1]How do *bioengineering* and *biomedical engineering* differ from *biotechnology*? Bioengineering could be defined as "the utilization of engineering analysis tools for the design and fabrication of devices that improve or augment humans; examples being artificial knee joints, heart valves, and tissue engineering"—also refer to Berger, Goldsmith, and Lewis (1996). Biomedical engineering could have a similar definition to that of bioengineering. However, it can be extended to include medical monitoring equipment and drug delivery systems. The considerable overlap between these three fields reemphasizes the caution concerning overly pedantic definitions.

Biotechnology is now an international business, but its roots can be traced to the San Francisco Bay Area. In the early 1970s, research on *gene splicing* and *cloning,* conducted at the University of California at San Francisco (UCSF) and at Stanford University, provided the basis for the creation of companies such as Genentech and Chiron as well as a plethora of smaller companies.

Colloquially described by the shortened word *biotech,* these industries expanded dramatically during the 1980s. Today's picture is that of a well-established worldwide industry looking for life sciences graduates, facility engineers, process development experts, information systems experts, and product support personnel.

The industry is hiring people with management, manufacturing, and marketing skills as well as technical skills and experience in the biological sciences. Venture capitalists, consulting companies, and patent law offices are eager to find people with knowledge of biotech products and processes.

There are also emerging prospects for the synergy of biotech and electronics. As an example, the genetic information for many bacteria has now been stored on inexpensive memory chips, and such information is of great importance. Such memory chips are useful in recombinant DNA procedures, gene cloning, and manufacturing (Campbell, 1998).

9.2 MODERN PRACTICE OF AN ANCIENT ART

Despite all the hype, biotechnology is nothing new. Over 10,000 years ago, in Sumeria, Babylon, and Egypt, yeast (a single-celled organism) was used to carry out one of the most fundamental industrial bioprocesses (fermentation) for the production of beer and wine. Over the centuries, the availability of bioengineered foodstuffs continued to grow. Besides alcohol, yeast was found to be useful for making bread. People learned to use rennin and mold to make cheese, bacteria to produce yogurt, and selective breeding to grow bigger crops and fatter livestock.

Also, in the 1850s, Louis Pasteur showed that microorganisms could be killed by the application of controlled heat. Such use of controlled heat became known as pasteurization and was used immediately to preserve food and drink, in particular milk. In related work his controlled processes were used to kill damaging microorganisms attacking silkworm eggs. Pasteur was credited at the time for saving both the wine and silk industries in France, and perhaps he can be regarded as one of the founders of modern biotechnology in its industrial applications. From a management of technology viewpoint, he certainly understood how to transfer the results from experimental science into industrial products and quickly "crossed the chasm" described in Chapter 2 (Figure 2.3).

In recent decades, the understanding of cellular and molecular biology has advanced to a remarkable degree. This new knowledge is opening a wide range of commercial opportunities as well as exciting possibilities for resolving many of today's great problems. The most widely visible applications are in the medical arena. Research in molecular and cell biology has played a critical role in the ability to understand and develop treatments for AIDS, certain types of cancer, multiple sclerosis, and other diseases. The tools and techniques of molecular biology have made it easier to diagnose or even anticipate an individual's risk of contracting specific diseases. Biotechnology researchers have synthesized products such as insulin and human growth hormones.

In addition to these medical advances, biotechnology has played a role in mitigating environmental problems and increasing global food supplies. Thus, beyond health care, biotechnology is creating tools and applications in such wide ranging fields as agriculture, genetics, energy, and environmental science.

Bioengineers have invented biodegradable plastics, organic pesticides, and microorganisms that break down oil and chemical spills. Improvements in crop productivity and resistance to disease will eventually help feed and clothe an increasingly populated world. And while it failed to impress the jury in the O. J. Simpson trial, genetic "fingerprinting" through DNA analysis has become a powerful forensic tool.

9.3 CAPTURING INTEREST

The manufacturing of biotechnology products will continue to accelerate and offer huge future employment opportunities for today's students. This section and the diagram in Figure 9.1 are thus aimed at capturing the imagination of any readers who might be tempted to ignore this chapter as not being relevant to "traditional manufacturing."

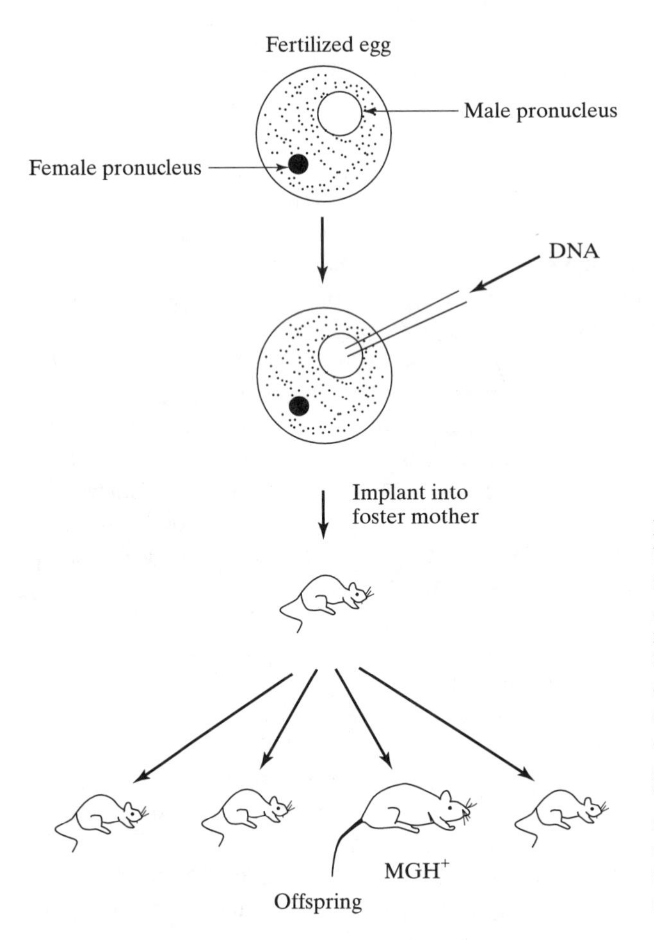

Figure 9.1 Creating a "supermouse" (diagram based on the experiments of Palmiter et al., 1982 and reprinted from *An Introduction to Genetic Engineering* by Desmond S. T. Nicholl, © 1994; reprinted with the permission of Cambridge University Press). In the top sketch, fertilized eggs were first removed from a female. In the second sketch, DNA carrying MGH—a fusion of the mouse genetic information to the rat growth hormone—was injected into the eggs. The eggs were then implanted in a foster mother. In the lower sketch, one of the mouse offspring expresses the MGH construct—that is, the offspring is MGH^+—and grows to an abnormal size.

The way in which DNA and RNA direct the "manufacture" of proteins in the human body is a remarkable process, in and of itself worthy of study. The new advances in genetic engineering are even more remarkable. Examples include:

- The production of useful proteins such as insulin for diabetic people.
- The creation of transgenic plants for increased production (see *Economist,* 1998).
- "Supermouse," which is an example of a transgenic animal. Supermouse was the outcome of experiments in which the growth hormone gene of a rat was combined with DNA from a mouse. Such combinations of different types of DNA are called *recombinant DNA* methods. The new DNA was then injected into fertilized eggs and implanted in female mice. In some offspring the rat growth hormone was expressed, leading to rapid growth and a large mouse (Palmiter et al., 1982).

9.4 MILESTONES IN BIOTECHNOLOGY HISTORY

9.4.1 Evolution, Genetics, and Biochemistry

Evolution

In the early 19th century, Charles Darwin proposed that the evolution of plants and animals occurred by "the survival of the fittest"; that is, the fittest of the species pass on their favorable traits to their young, whereas unfavorable traits eventually die out. On balance, the best-adapted individuals with desirable traits survive to reproduce in that local environment.

Genetics

How might these desirable traits be passed on from one generation to the next? Some initial answers to this question were first obtained by the Austrian monk Gregor Mendel. Mendel's laws of heredity were based on a series of cross-breeding experiments that he conducted using garden peas. During these experiments he discovered that observable traits could skip generations. He proposed that traits were passed on via invisible internal "factors." In addition, he postulated how these factors were inherited. Other scientists later identified these factors, by then known as *genes.* Mendel's laws are the foundation of "classical" genetics.

Biochemistry

During the latter half of the 19th century other scientists were working on the biochemistry of plant, animal, and human cells. Many of the chemical reactions inside cells, as well as the cell constituents themselves, were recognized: these included lipids, carbohydrates, nucleic acids, and most of the amino acids that serve as the building blocks of proteins. For example, Fischer correctly proposed that the chemical link in protein was established with a peptide bond between each adjacent amino acid. (See Figure 9.2.)

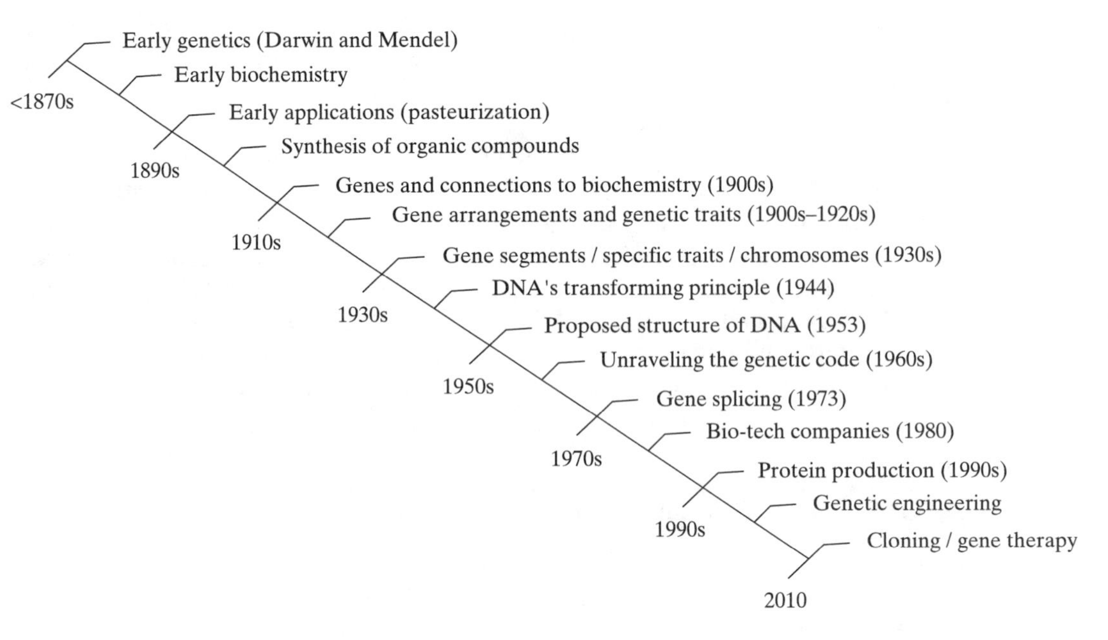

Figure 9.2 Milestones in biotechnology.

9.4.2 Discovering the Function and Structure of DNA

During the 20th century, an interplay between genetics and the biochemistry of cells became clear. By the 1930s, the research of Morgan, McClintock, and other scientists made it clear that "genes were not some kind of theoretical entity" (to quote Darnell et al., 1986) but were related to a biological, genetic material inside cells. Then, in 1944, Avery, MacLeod, and McCarty showed that deoxyribonucleic acid (DNA) was a transforming substance, or principle, that could genetically alter bacteria. This work of Avery and colleagues in 1944 was augmented by another set of experiments by Hershey and Chase in 1952. It was shown that a bacterial virus could infect another host bacterium, *only if DNA* entered the host's cells. In summary, by the early 1950s, the integration of 100 years of research pointed clearly to the role of DNA as the carrier of genetic information.[2] Scientists also knew from biochemical testing that the DNA molecule consisted of *phosphate, deoxyribose,* and four compounds called *bases: adenine, cytosine, guanine,* and *thymine.* But they did not know, overall, how these six units were integrated together to form DNA.

Thus a fundamental breakthrough in molecular biology and genetics came in April 1953. James Watson and Francis Crick, working at Cambridge University in England, deduced a structure for DNA that was consistent with all the existing observations and chemical analyses. They proposed a double helical structure for

[2]It is noted, as a curiosity, that around 1870, the Swiss biologist Friedrich Miescher isolated deoxyribonucleic acid (DNA) by adding hydrochloric acid to human pus cells. He called the gray precipitate he obtained nuclein but was unaware of its significance in heredity.

DNA. As an oversimplification, this structure resembles a "twisted rope ladder." Subsequently, other experiments confirmed their proposed structure.

A few years later, Crick and other scientists described the central dogma of molecular biology—how the DNA molecule makes proteins via a messenger molecule called RNA. It was proposed that a protein's amino acid sequence was encoded in the sequence of the DNA molecule. There followed a period of intense research, by many research teams, culminating in 1966 when they "cracked the genetic code" for the 20 amino acids involved in protein production.

9.4.3 The First Gene Splicing Experiments

In 1973, Herbert Boyer of the University of California at San Francisco and Stanley Cohen of Stanford University were the first scientists to cut the DNA and recombine it. Their first experiments cut and joined (recombined) pieces of the same type of bacterial DNA. But subsequently, they joined different types of DNA with a circular minichromosome (a plasmid). They then introduced this recombinant DNA into bacteria. Clones of the new DNA were then generated as the bacterial host divided and multiplied. In 1976, the modern biotechnology industry was born when Genentech became the first company devoted entirely to genetic engineering and the manufacture of products and processes incorporating biotechnology. In 1980, Genentech became the first biotech company to go public.

9.5 A BIOSCIENCE REVIEW

9.5.1 Cells

Despite the enormous variation in nature, all living things on earth have one thing in common: they are all made up of *cells.* Even a single-celled organism such as yeast is as much alive as us. At the same time, one can think of a transition in size from an atom, to a molecule, to a single-celled organism like yeast, to a tissue or an organ made up of thousands of cells, and then to a complete animal. One cell is surrounded by a membrane, and within this is a jellylike fluid, *cytoplasm.* Cells reproduce by dividing so that one cell is turned into two. Each one of the new cells gets a copy of the original genetic information.

DNA is a key constituent of cells and is the genetic material in all organisms. In any particular organism, the precise arrangements of the DNA's chemical links are used to store and then direct genetic information, not unlike a computer's databases and programs. This genetic program controls almost everything the cell does. For example, the information stored in genes is used to generate proteins (see Section 9.5.2).

Genes control the inheritance of traits from the parents to the offspring. Thus the specific pieces of DNA that make up the many genes in humans pass on specific traits such as eye color from one generation to the next. The interesting point is that all DNA looks much alike no matter which plant, animal, or person it comes from. However, the genetic information that is carried in the DNA makes all living things different and indeed unique.

Most cells are of the *eukaryotic* type, which means they have a nucleus that contains the DNA. Some single-celled organisms do not have a nucleus: these are called *prokaryotic* cells and were the focus of much attention in the 1950s and 1960s when molecular biologists were studying DNA and the processes involved in gene expression. *Escherichia coli* (*E. coli*) is a single bacterium of this type and has been much used in research. Even though such prokaryotes have no nucleus, they still contain DNA. Thus, for the most part, gene expression is very similar in both prokaryotes and eukaryotes.

9.5.2 The Role of Proteins in Life Processes

The cells of all living organisms contain proteins. Some proteins play a structural role, giving shape and substance to cell walls and membranes, while others control a series of chemical reactions. Cells make thousands of different proteins to carry out such functions, and the proteins produced are determined by the DNA in the cells' nuclei.

The role of the DNA in the production of proteins is to "store" information as genes and then, via the messenger RNA, allow the creation of proteins that are needed for energy and life.

This process is sometimes referred to as the central dogma of molecular biology. The key concept is that the flow of genetic information is unidirectional[3]: from DNA to protein, via a messenger molecule called RNA (Figure 9.3).

Proteins are complex structures composed of chains of amino acids. The amino acids are made up of carbon, hydrogen, oxygen, and nitrogen. Some amino acids contain sulfur (Figure 9.4). These may be linked in different combinations; since the chains may be formed in many ways, using anything from only 10 to over 100 specific amino acid links, the number of different possible proteins is large. For example, the human body can make about 60,000 to 80,000 proteins out of 20 amino acids.

9.5.3 The Role of Enzymes in Life Processes

Each step of a metabolic process is controlled by a protein molecule called an *enzyme*. Enzymes are particular types of proteins that speed up chemical reactions. Thus they serve as catalysts during many of the biological processes described in this chapter that occur within cells. For example, an enzyme called RNA polymerase catalyzes the transcription of DNA into mRNA.

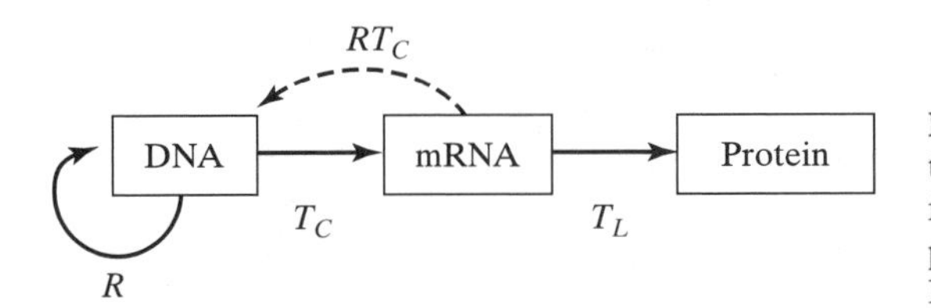

Figure 9.3 The "central dogma" states that the information flow is unidirectional from DNA to mRNA to protein. The processes of transcription, translation, and DNA replication follow this rule.

[3]It should be noted that some RNA viruses carry out reverse transcription, producing a DNA copy of their viral RNA genome.

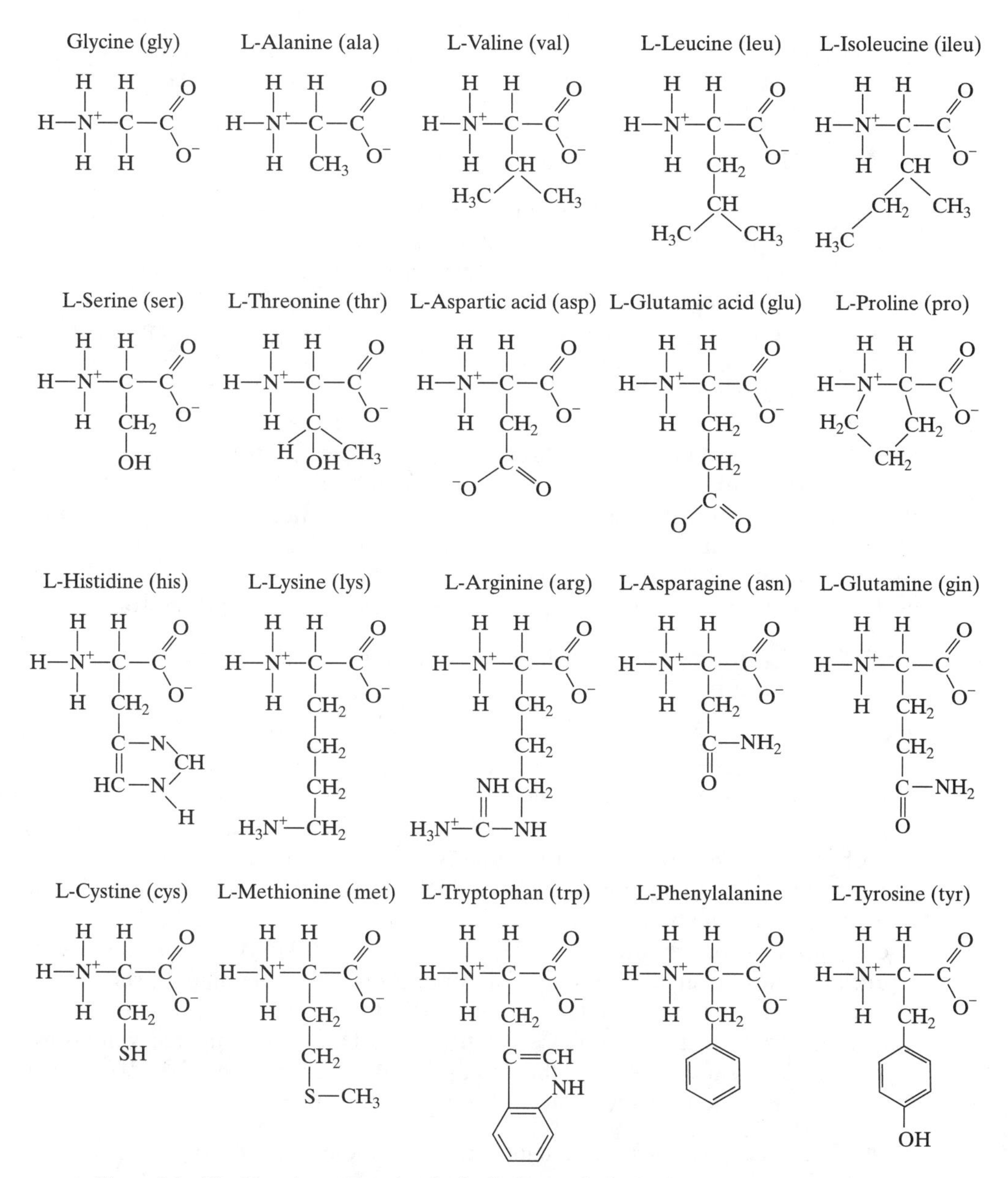

Figure 9.4 The 20 amino acid molecules in the human body. Both names and their common abbreviations are shown above each simplified molecular structure.

In addition, enzymes function in many areas of the human body. For example, amylase in salivary glands breaks down the carbohydrates in bread and pasta into simpler chemicals; pepsin secreted from the stomach lining breaks down the proteins in fish and meat dishes; and lipase secreted by the small intestine breaks down the fats in butter and cheese.

Additionally, enzymes can be produced synthetically and are used in everyday situations—for example, in laundry detergents to break down proteins that may have been spilled on clothing.

9.5.4 The Genome and Chromosomes

The genome is the complete genetic information or total DNA content of an organism. It is packaged into chromosomes, which are genetic structures consisting of tightly coiled threads of deoxyribonucleic acid (DNA) and associated proteins. The genome functions as a blueprint for making an organism. It contains all instructions for building cell structures and directing life processes throughout the lifetime of the organism. Different organisms have different numbers of chromosomes per nucleus. Humans have 46 chromosomes arranged in 22 pairs plus XX or XY—except in the sperm or the egg, which has 22 chromosomes plus X or Y. The nuclei of most human cells contain two sets of chromosomes, each set contributed by one parent. Chromosomes can be observed under a microscope; when stained with certain dyes, they reveal a distinctive pattern of light and dark bands. In karyotype analysis, observable differences in size and banding patterns distinguish the chromosomes from one another. It is also possible to detect major chromosomal abnormalities such as that of Down's syndrome, in which there is an extra copy of chromosome 21.

9.5.5 DNA (DeoxyriboNucleic Acid) as the Carrier of Genetic Information

The science and engineering of deoxyribonucleic acid (DNA) are central to most biotech industrial processes. Why is DNA so special? The answer is that DNA is the genetic material for probably all living organisms.[4] (It should be pointed out that some viruses use RNA rather than DNA as their genetic material.) Specifically, DNA carries the genetic information that is vital for the cells in a person's body to grow, function, and divide normally. Thus, DNA has a universal role in defining and regulating life on the planet, from simple bacteria to complex organisms like us. It is also remarkable for other reasons, including the fact that DNA can replicate itself in a precise fashion over the lifetime of an organism. DNA directs the production of proteins necessary for all cellular functions including its own synthesis.

Earlier in the text, the double-helical structure of DNA was compared with a twisted rope ladder. To further understand the structure of DNA, imagine two strands of "beads" that have been twisted together. This double-helical structure is made up as follows:

- *Each "bead"* is a 5-carbon, sugar/phosphate molecule called a *nucleotide.* Each nucleotide contains a base shown on the right side of Figure 9.5. There are four possible bases: two purines (adenine [A] or guanine [G]), or one of two pyrimidines (thymine [T] or cytosine [C]) (Figures 9.6 and 9.7).

[4] D. Mascarenhas (1999), in his lectures, makes an informal but pedagogically helpful analogy between the genetic information stored in DNA and the notes and music stored on a standard cassette tape. Each track of the cassette tape contains the information for one song, just like one length of DNA stores the information for one gene. Between each track (or gene) there is a break before the next song (or gene) starts. Also, with the naked eye, nothing can be seen on the cassette tape—but the music is there ready to be expressed. Similarly, in the DNA strand, the genetic information is there, ready to be expressed.

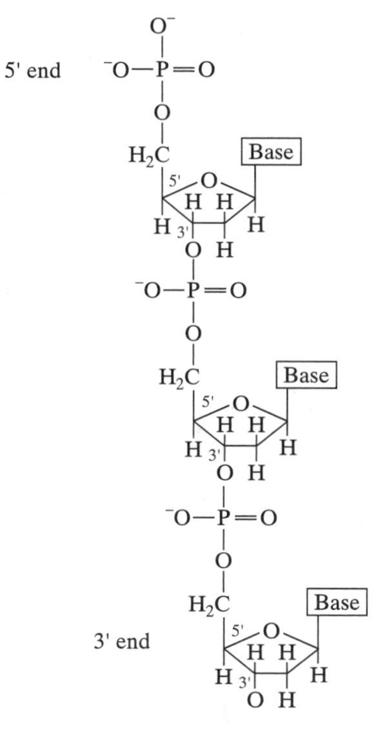

Figure 9.5 Schematic representation of the nucleotide that makes up DNA.

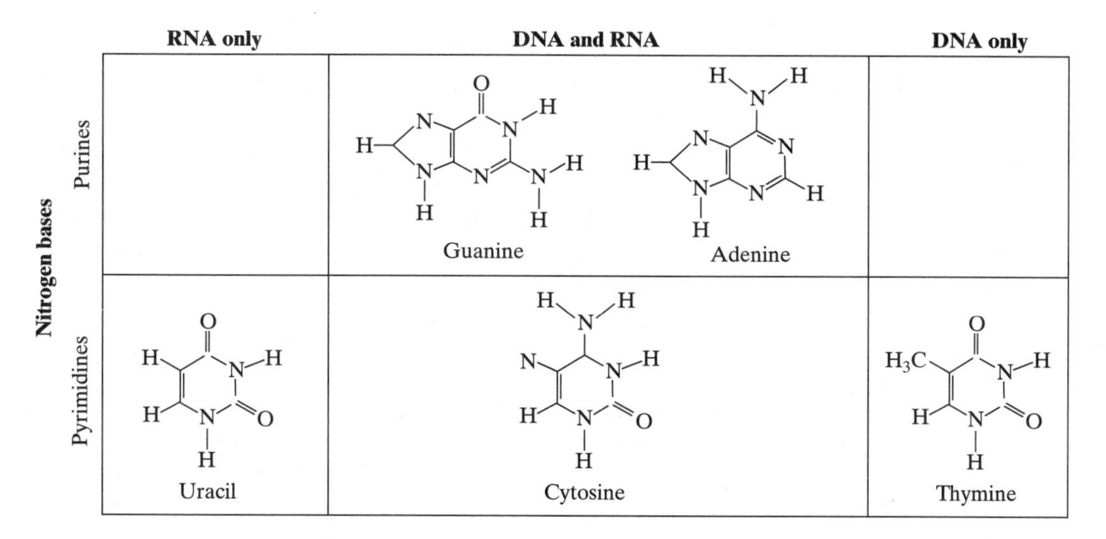

Figure 9.6 The five nitrogenous bases of DNA, and RNA—see Section 9.7.2.

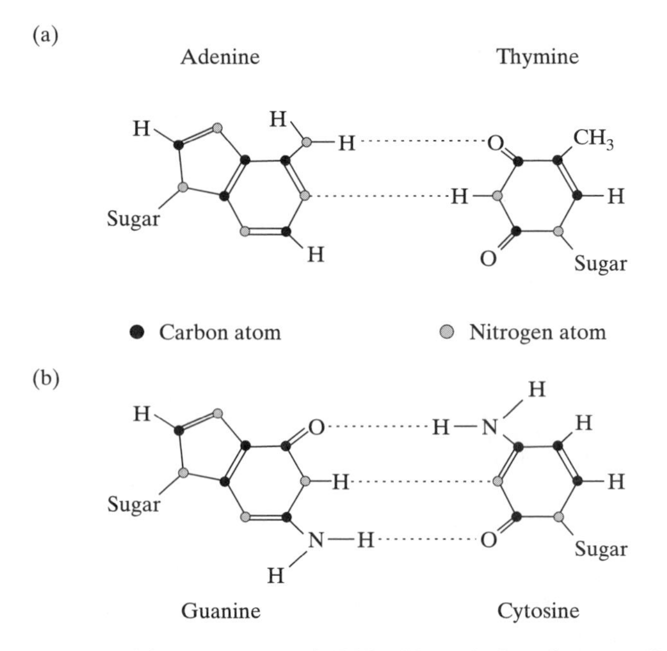

Figure 9.7 Base-pairing arrangements in DNA. (From *An Introduction to Genetic Engineering* by Desmond S. T. Nicholl, © 1994. Reprinted with the permission of Cambridge University Press.)

- *To create the links within one strand,* the nucleotides are attached to each other by a phosphodiester bond between the phosphate group at the 5′ carbon atom of one nucleotide and the 3′ carbon atom of the next nucleotide. These bonds are shown on the left of Figure 9.5. This creates a directionality of 5′ → 3′ to the string of nucleotides.
- *To create the links between the two strands,* the bases on one strand are loosely attached with hydrogen bonds to the bases on the opposite strand (Figure 9.7). Base pairing follows a fixed rule: A always pairs with T; G always pairs with C. Thus, a sequence of ATGG . . . on one strand means the other strand will have a complementary sequence of TACC . . .

5′ .... - ATGGCTACCAAGGTA .... - 3′
3′ .... - TACCGATGGTTCCAT .... - 5′

The genome can be described in terms of the number of these base pairs. DNA in a simple bacterial cell such as *E. coli* is made up of about 4×10^6 base pairs: all of these are shown together in Figure 9.8. By contrast, a human genome is about 3 billion base pairs.

9.5.6 DNA Replication and Its Relationship to Cell Division

Before a cell in an organism divides, the DNA replicates, as illustrated in Figures 9.9, 9.10, and 9.11. As shown in Figure 9.9, the double-stranded DNA molecule unwinds into two single strands separating the A-T and C-G base pairs. Next, with

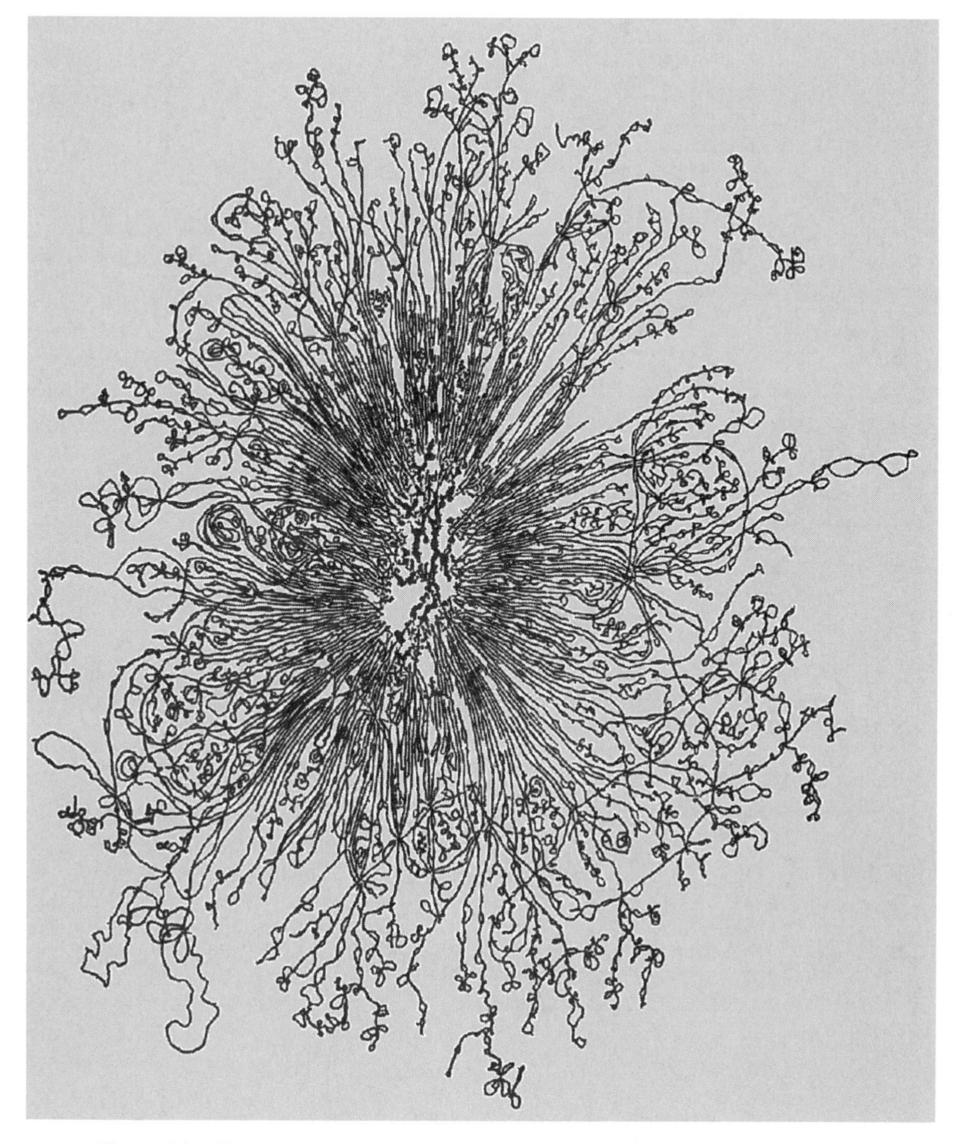

Figure 9.8 Electron micrograph of an *Escherichia coli* (*E. coli*) DNA molecule.

the help of certain enzymes, each strand picks up the bases of free nucleotides in the cell. This is shown schematically in Figure 9.10. Once again, the base-pairing rules apply in this pickup process. In this way, each new double helix becomes a duplicate of the original one because, as shown in both Figure 9.10 and Figure 9.11, the original strands act as the template, specifying which bases are to be added to the growing strands. Thus when the cell proceeds to divide, each of the new

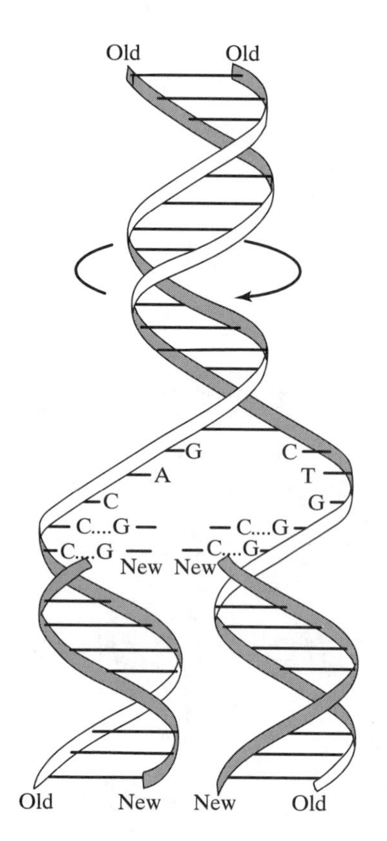

Figure 9.9 Untwisting of linear DNA strands during replication. The strands untwist by rotating about the axis of the unreplicated DNA double helix.

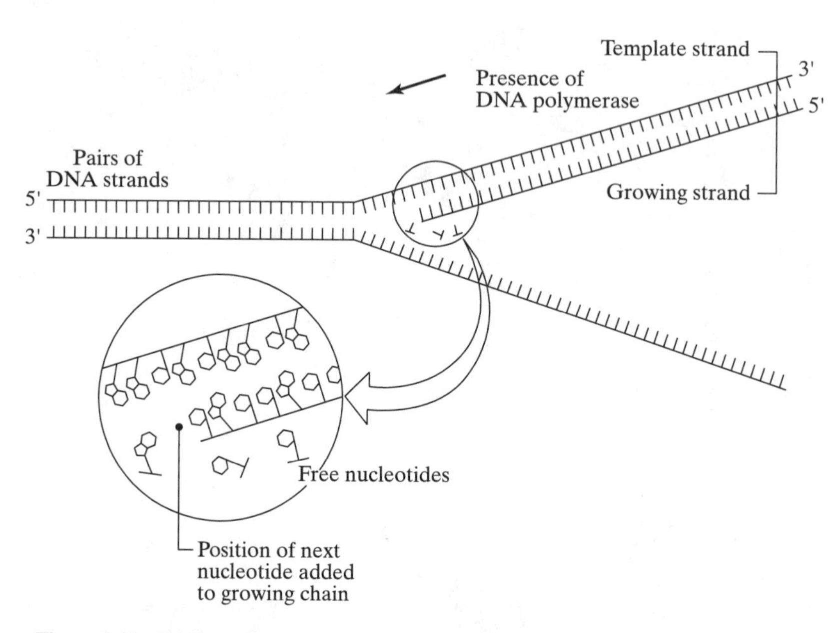

Figure 9.10 DNA replication involves the addition (bottom left) of free nucleotides. The bases are seen in this diagram as the hexagon shapes, and the sugar/phosphate bonds that become the strands are the attached "bar lines." DNA polymerase and other proteins catalyze the process.

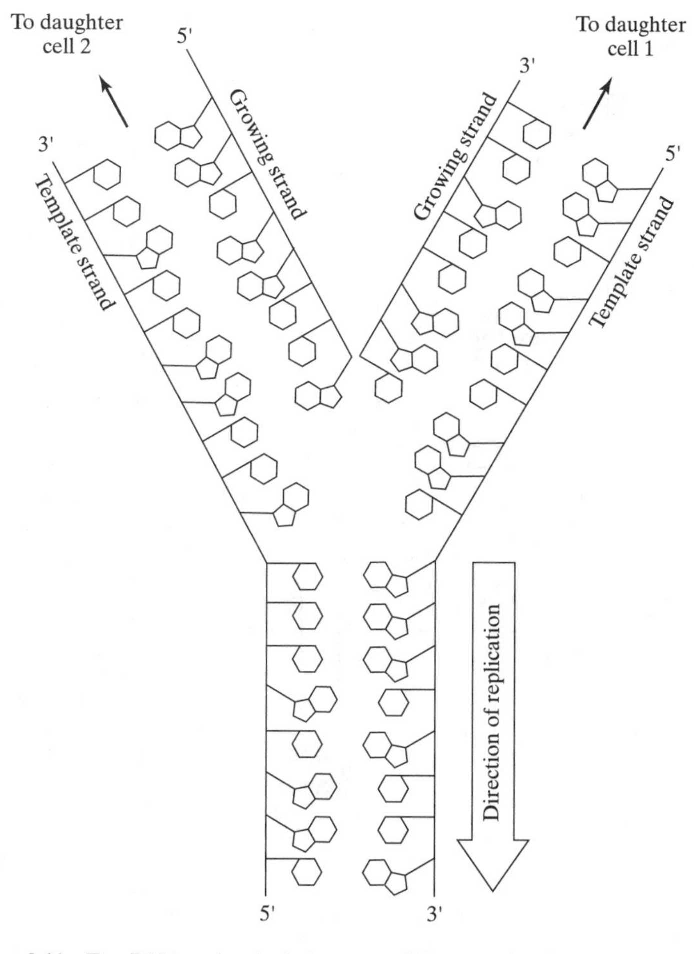

Figure 9.11 Two DNA molecules being created from one, leading to two daughter cells on the upper right and left of the figure.

daughter cells receives a set of DNA molecules identical to that of the original cell—as indicated in Figure 9.11.

9.6 BIOPROCESSES

9.6.1 Gene Expression: Connection between Genes, DNA, RNA, and Proteins

Genes are the basic unit of heredity. Each gene is a specific sequence of DNA nucleotides that carries the information to direct synthesis of a protein. The process by which the genetic information encoded on DNA flows into the cell's "protein factories" is a complex set of biochemical reactions (Figure 9.12).

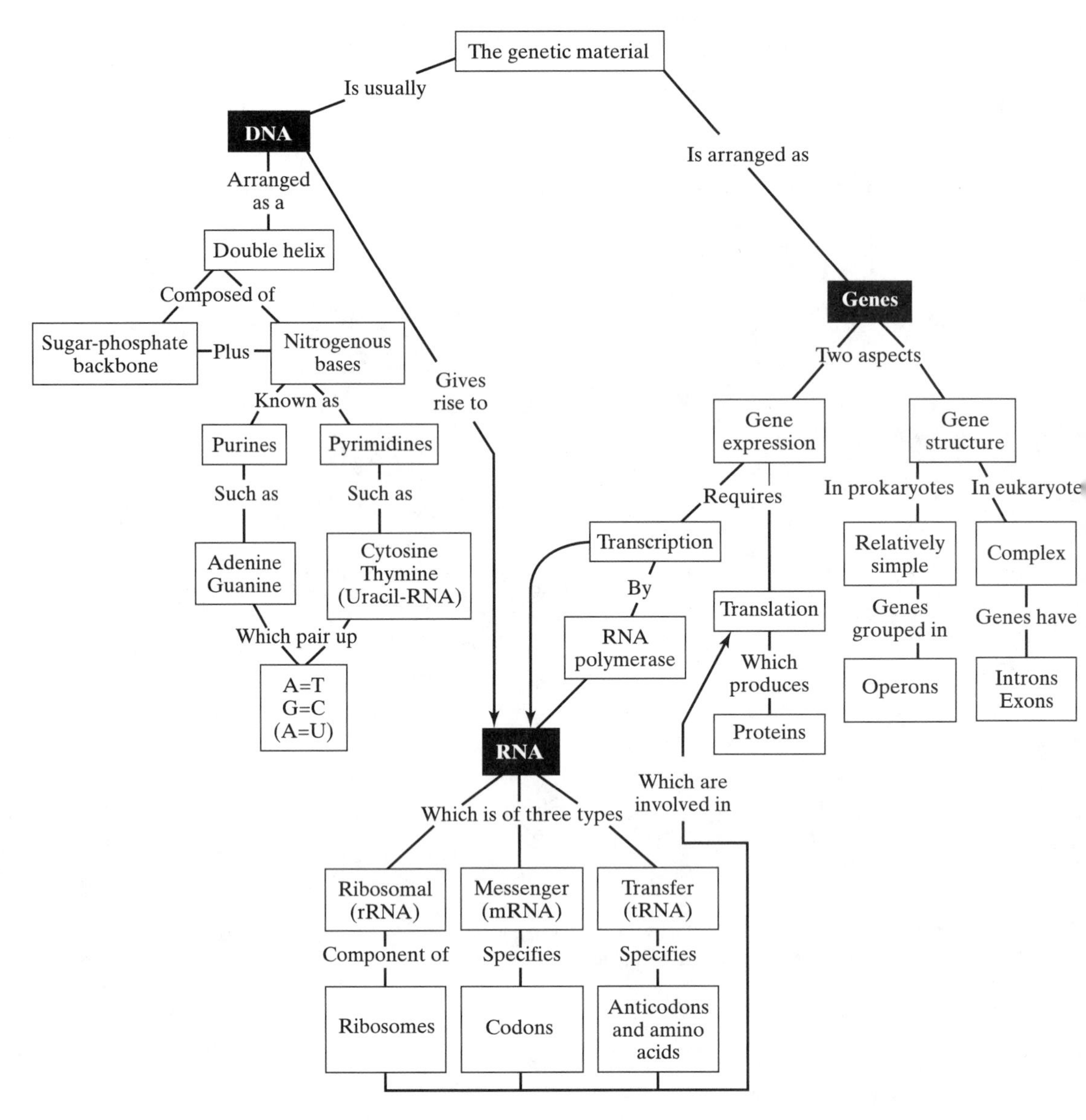

Figure 9.12 Summary of gene expression. (From *An Introduction to Genetic Engineering* by Desmond S. T. Nicholl, © 1994. Reprinted with the permission of Cambridge University Press.)

9.6.2 RNA (RiboNucleic Acid)

A key substance in gene expression is *ribonucleic acid* (RNA). The synthesis of functional proteins out of information encoded in DNA requires several types of RNA. These include ribosomal RNA (rRNA), messenger RNA (mRNA), and transfer RNA (tRNA). Gene expression also requires several proteins, for example, RNA polymerases.

RNA resembles DNA, but there are some major differences:

- RNA is a chainlike molecule but single stranded.
- RNA also has four bases, but RNA contains the base *uracil* (U) instead of the base thymine (T) that is used in DNA.
- The sugar in RNA is ribose, whereas the sugar in DNA is deoxyribose. Both are 5-carbon sugars (pentoses), but the DNA sugar lacks an oxygen atom at the 2′ carbon atom. The RNA sugar is thus called ribose, while the DNA sugar is called deoxyribose—that is, a ribose without oxygen at the 2′ carbon atom. The abbreviated terms RNA and DNA are derived from these descriptions of the sugar molecules.

9.6.3 Transcription

The first step in gene expression is a process known as transcription, in which a DNA molecule carrying the genetic information is used as a template to synthesize an mRNA molecule.[5] Transcription is similar to DNA replication in that a single strand of DNA is used as the template for the synthesis of a complementary nucleic acid sequence.

In Figure 9.13a, the shaded ellipse represents the enzyme called *RNA polymerase.* As implied by the arrow, the polymerase moves along the DNA. Over short regions, the polymerase temporarily unwinds the two DNA strands as it moves along the genetic information. RNA polymerase also plays a role in linking the proper ribonucleotides of mRNA to each other, in order to form the mRNA chain. The DNA acts as the template to specify which is the correct ribonucleotide to add. This proceeds according to the base-pairing rules except that uracil (U) in the RNA replaces thymine (T) in the DNA. Note that the newly formed mRNA strand has the same sequence as the nontemplate DNA strand. For any given gene, only one of the DNA strands is used as the template for mRNA synthesis.

9.6.4 Promoters

How does this whole process get triggered? The answer is that in addition to these base-pairing rules, there are other important regulatory sequences and orderings specified in the DNA strands (see Nicholl, 1994; Okamura, 1998). Thus "upstream" from the specific gene being expressed on the top right of Figure 9.13a, there are DNA sequences to which the RNA polymerase can bind. These sites for starting transcription—that is, which can temporarily bind the RNA polymerase to the DNA strands—are known as *promoters.* These promoter sites can also be seen as clever switches. Thus, when certain food products enter our system, the switches of the promoter sites are activated, thereby setting off the desirable chain reaction consisting of (DNA → mRNA → protein synthesis → needs of the cell → needs of the body). Speaking colloquially, this is why one generation of parents should encourage the next generation to eat a good breakfast. For example, a group of genes known as the *lac operon* in the *E. coli* bacteria in our intestines codes for the enzyme that can

[5]Transcription also refers to the synthesis of rRNA and tRNA, but it is the mRNA that is translated into proteins.

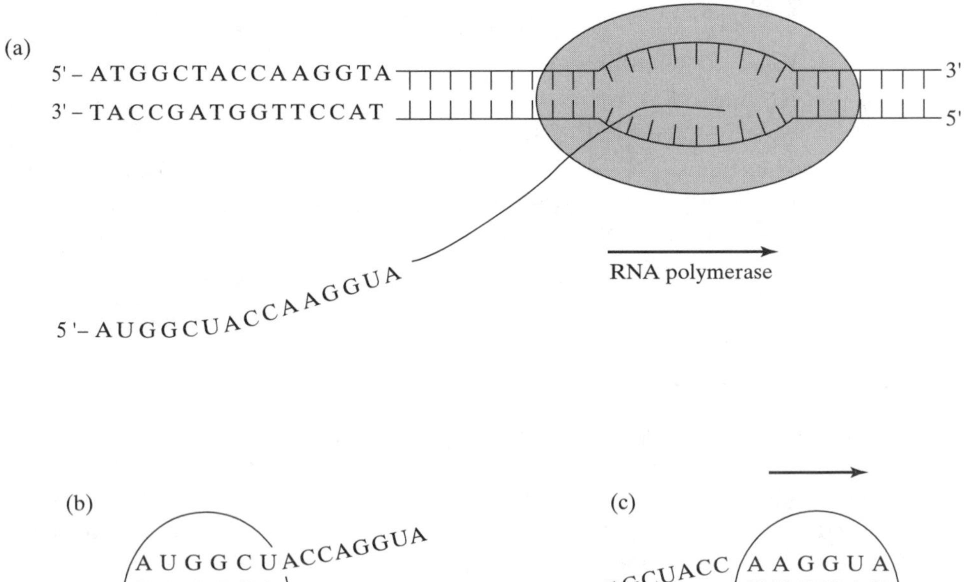

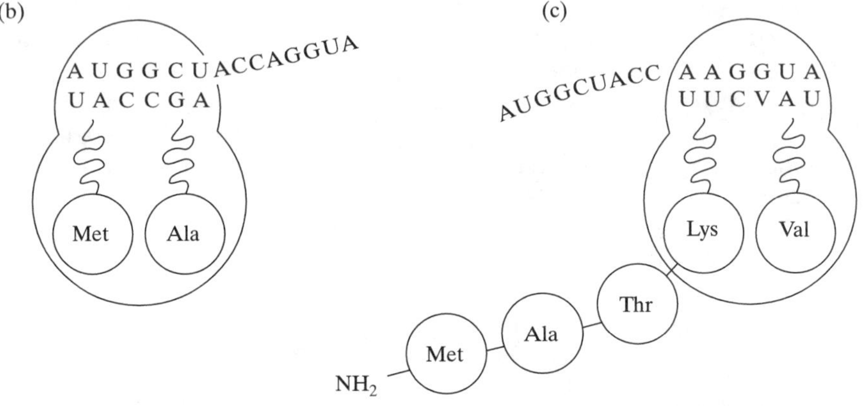

Figure 9.13 Transcription in the upper sketch and translation in the lower two sketches. In (a) the DNA strands are temporarily separated by RNA polymerase, so that the mRNA can be formed during gene expression. (From *An Introduction to Genetic Engineering* by Desmond S. T. Nicholl, © 1994. Reprinted with the permission of Cambridge University Press.)

break down lactose, or milk sugar. However, the "upstream" operator, or promoter, for this group of genes is a site that can only get triggered in the presence of lactose. The *E. coli* bacteria and lactose thus begin a complex chain reaction for breaking down other food in the intestine.

9.6.5 Translation or Protein Synthesis

Translation is the process by which proteins are synthesized using the information carried by mRNA, the product of transcription. Recall that proteins are made up of amino acids. The specific nucleotide sequence of the mRNA determines which amino acids are to be linked to create a particular protein. A combination of three nucleotides (e.g., ACG, GGG, CAG) is called a *codon* and specifies an amino acid. Figure 9.14 describes the correspondence between the codons and the amino acids

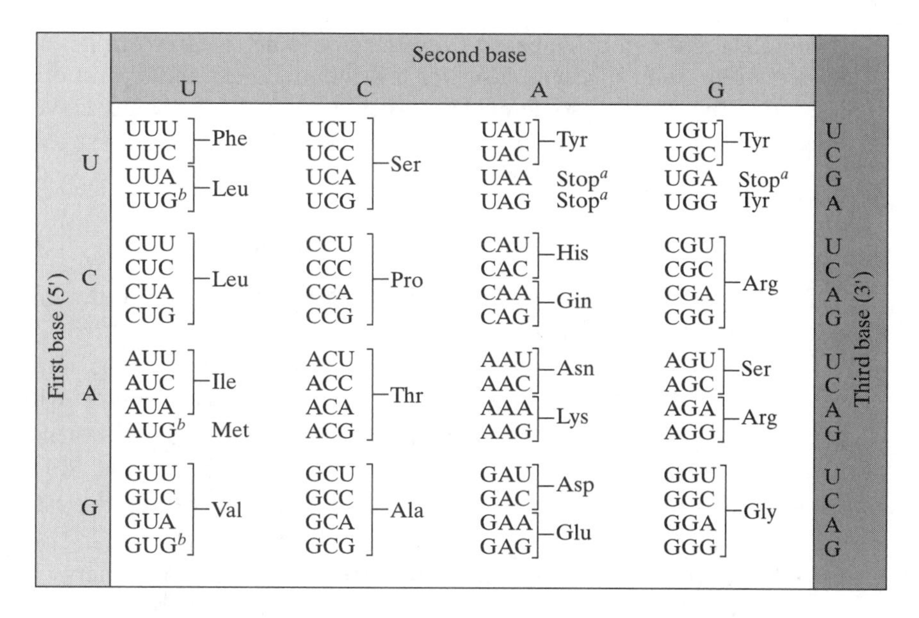

First base (5')	Second base: U	Second base: C	Second base: A	Second base: G	Third base (3')
U	UUU Phe	UCU Ser	UAU Tyr	UGU Tyr	U
U	UUC Phe	UCC Ser	UAC Tyr	UGC Tyr	C
U	UUA Leu	UCA Ser	UAA Stop[a]	UGA Stop[a]	G
U	UUG[b] Leu	UCG Ser	UAG Stop[a]	UGG Tyr	A
C	CUU Leu	CCU Pro	CAU His	CGU Arg	U
C	CUC Leu	CCC Pro	CAC His	CGC Arg	C
C	CUA Leu	CCA Pro	CAA Gin	CGA Arg	A
C	CUG Leu	CCG Pro	CAG Gin	CGG Arg	G
A	AUU Ile	ACU Thr	AAU Asn	AGU Ser	U
A	AUC Ile	ACC Thr	AAC Asn	AGC Ser	C
A	AUA Ile	ACA Thr	AAA Lys	AGA Arg	A
A	AUG[b] Met	ACG Thr	AAG Lys	AGG Arg	G
G	GUU Val	GCU Ala	GAU Asp	GGU Gly	U
G	GUC Val	GCC Ala	GAC Asp	GGC Gly	C
G	GUA Val	GCA Ala	GAA Glu	GGA Gly	A
G	GUG[b] Val	GCG Ala	GAG Glu	GGG Gly	G

Figure 9.14 The "genetic code": nucleotide sequences for amino acids. A total of 64 three-letter codons can be formed from the four letters (A, C, G, and T) of the DNA nucleotide bases. However, there exist only 20 amino acids, not 64. This means that there is more than one codon for most of the amino acids: for example, Valine (Val.) is shown in four entries at the bottom left of the table. Three of the codons (UAA, UAG, and UGA) signal "stop" and the end of an mRNA chain.

that they specify. The process by which the mRNA sequence is "translated" into an amino acid sequence is described in the following paragraph.

Translation requires ribosomes, which are composed of ribosomal RNA (rRNA) and ribosomal proteins. Ribosomes are the protein factories of cells and are depicted in Figures 9.13b and 9.13c as the indented elliptical shapes.[6] The ribosome moves along the mRNA chain, and successive amino acids are linked into a growing protein chain, as shown in Figure 9.13c. This process requires another type of RNA—transfer RNA, or tRNA. There is at least one type of tRNA for each amino acid. With the rules of base pairing coming into play again, the mRNA's codon sequence is recognized by tRNA molecules that are (1) carrying an anticodon of complementary ribonucleotides and (2) carrying the amino acid that is specified by the codon. Thus, as the ribosome moves along the mRNA, tRNAs bring the appropriate amino acids to be added to the growing protein chain. This process continues until the last codon, the "stop" codon, is reached. As shown in Figure 9.14, this last codon can be UAA, UAG, or UGA, which do not specify an amino acid but rather signal that the protein

[6]Nicholl (1994) describes the ribosome as a "complex structure that essentially acts as a *'jig'*—[e.g., see Chapter 7, Figures 7.17 to 7.19]—which holds the mRNA in place so that the correct amino acid can be added to the growing chain."

is complete and can be released from the ribosome to carry out its function. Some proteins remain in the cytoplasm for cell maintenance and operations. Others are needed in the nucleus; some proteins, such as digestive enzymes, leave the cells for other functions.

9.6.6 Summary

The following analogy is somewhat exaggerated but:

- The genetic information carried on DNA is analogous to the designer at the top of Figure 4.17 in Chapter 4.
- The mRNA and transcription are analogous to the process planning steps that convert design information to the specific details of manufacturing.
- Ribosomes are the sites where proteins are translated or synthesized, and thus analogous to fabrication machinery.
- The proteins are the manufactured product, which then do the work of the cells.
- These proteins can be seen as the "workers," who bring energy back into the overall system. The created proteins and enzymes subsequently enable other processes, such as "triggering" the promoters, which then make copies of the genes. The whole process, summarized here, thus becomes self-replicating.

With the review of both biosciences in Section 9.5 and bioprocesses in Section 9.6, some aspects of biotechnology and specifically genetic engineering and manufacturing can now be considered in more detail.

9.7 GENETIC ENGINEERING I: OVERVIEW

9.7.1 Motivation and Goals

The most common type of genetic engineering in biotech uses *recombinant DNA technologies* to transfer genetic information from one organism to another for a potentially useful purpose. For example, in agriculture, there is now the possibility of new breeds of plants and animals. In medicine, there is also help for people with genetic diseases. Genes have now been identified that are involved in Huntington's disease, cystic fibrosis, sickle-cell anemia, retinoblastoma, and Alzheimer's disease.

Also in medicine, the manufacturing of useful proteins through genetic engineering is a basic goal of many biotech firms. There is growing interest in the manufacture of cytokines (such as interferon), which are important for immune system function.

Recombinant DNA technologies can also be used to produce insulin. Diabetic people fail to produce sufficient quantities of the protein insulin and hence are unable to control their sugar metabolism. A daily injection of insulin can regulate their system without interfering directly with the other bodily chemical reactions. In the past, the insulin could be obtained only by expensive extractions from a hog pancreas. Now, genetic engineering has enabled bacteria to produce a plentiful supply, plus help for patients who were allergic to hog insulin. Injecting natural or synthetic proteins (such as insulin) into an organism can induce temporary changes in the organism's protein content and function. Permanent changes can also be obtained in an organism, if it is possible to manipulate protein synthesis inside its cells.

9.7.2 The Essential Steps

Figure 9.15, by Nicholl (1994), provides a "snapshot" of manufacturing by gene cloning.

- *Step 1:* The original DNA of an organism, plant, or animal is first cut into fragments.
- *Step 2:* It is then joined to a carrier, called a vehicle or a vector.
- *Step 3:* This recombined DNA is then reintroduced into the cell of a host organism. With the proper manufacturing controls, clones grow.

Gene splicing is one of the most routine and fundamental steps in gene cloning and is a mainstay of biotech research activities. It is essentially a cut-and-paste process similar to film splicing, except instead of film stock, pieces of DNA are used, as shown in Figure 9.16.

9.7.3 Step 1: Cutting DNA Using Restriction Enzymes

In the first step of Figure 9.15, DNA is cut into fragments at precisely defined nucleotide sites by *restriction enzymes*. This process is shown in Figure 9.17.

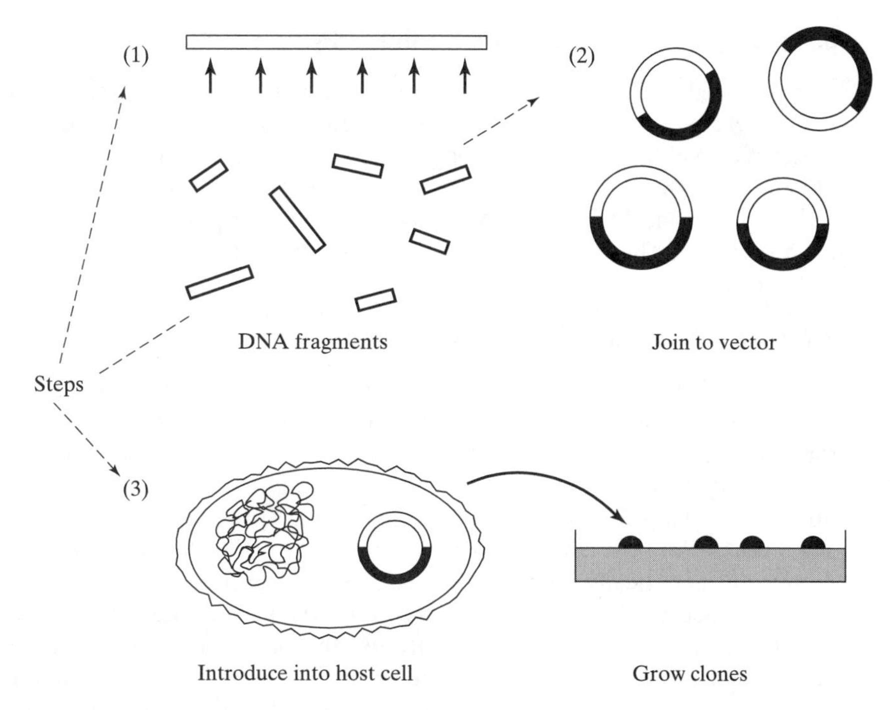

Figure 9.15 Schematic diagrams of gene cloning. (From *An Introduction to Genetic Engineering* by Desmond S. T. Nicholl, © 1994. Reprinted with the permission of Cambridge University Press.)

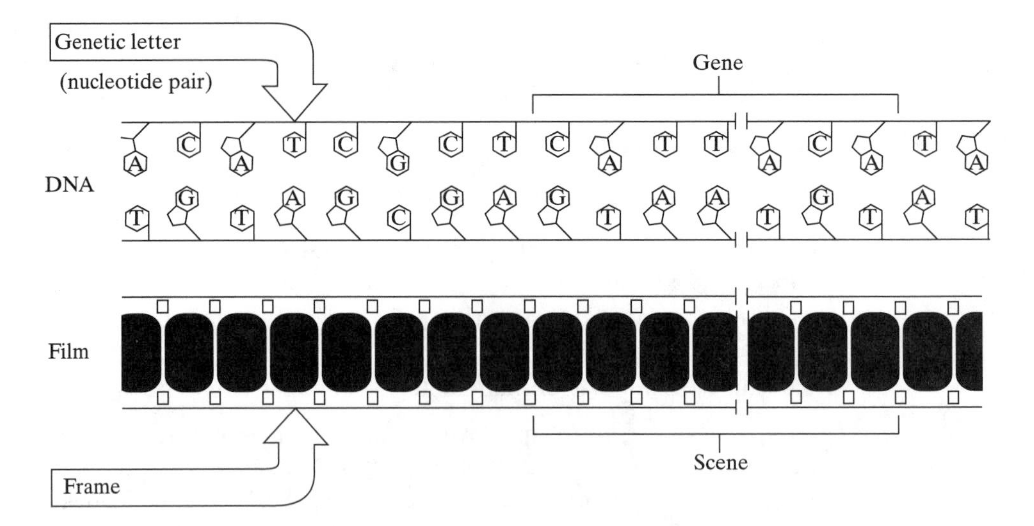

Figure 9.16 Movie-film splicing as an analogy for gene splicing. Only a short section of a gene is represented, and in fact many nucleotide pairs and movie-film frames would be contained in the two "bracketed" areas on the right of the figure (from *Understanding DNA and Gene Cloning* by Drlica, Copyright © 1992. Reprinted by permission of John Wiley & Sons, Inc.).

9.7.4 Step 2: Joining DNA Using the Enzyme DNA Ligase

Once isolated, a DNA fragment is ready to be joined to another DNA molecule. This requires the presence of another enzyme called *DNA ligase.* This joining process is a key aspect of recombinant DNA technology: *the DNA of one organism can be linked to the DNA of a completely different organism.* This is shown in step 2 of Figure 9.15 and in Figure 9.18.

This second DNA molecule is usually a circular *plasmid* or a *phage*—short for bacteriophage. These are shown in Figures 9.19 and 9.20. The plasmid or phage is called a *vector.*

9.7.5 Step 3: Vectors, Hosts, and Cloning

The plasmid or phage vector is used to carry the recombinant DNA into a *host* cell where the genetic material can be propagated. Host cells are most commonly bacteria or yeast, single-celled organisms that can exhibit phenomenal growth rates. They are therefore an important tool in cloning. In a manufacturing context, these hosts can be viewed as the "transfer line for mass production."

When the host cells with the recombinant DNA divide, they produce a large number of genetically identical clones, as shown in part 3 of Figure 9.15. The most common bacterial host is *Escherichia coli. E. coli* is in our intestines right now, as well as in those of most animals. (Out of interest, this bacterium gained notoriety in recent years when several people were killed or sickened by undercooked hamburgers that had an excess of *E. coli.* However, this form of *E. coli* is different from laboratory strains, which are not pathogenic.) *E. coli* is the bacteria of choice for gene cloning because it has been so well studied over the last few decades that it is now characterized and its functions well understood. *E. coli* cells—at high magnification—are

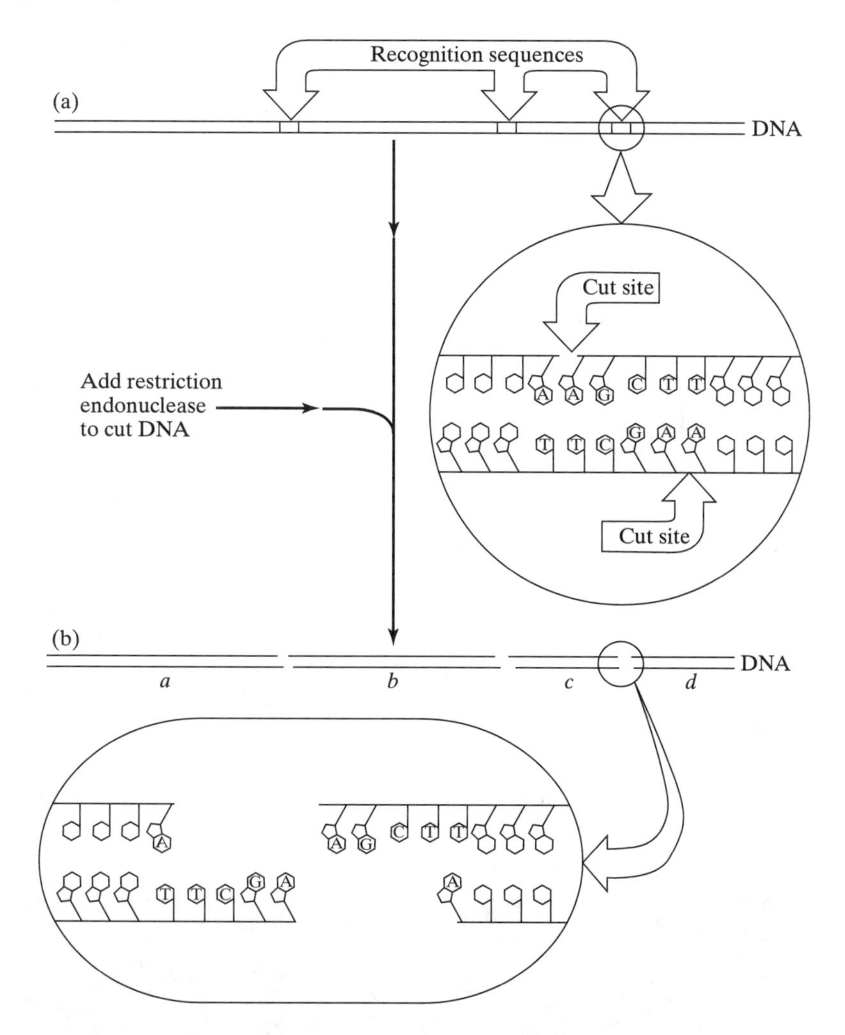

Figure 9.17 Cutting of DNA into short "staggered" pieces. When a restriction enzyme is added to the DNA, it binds to the DNA and cuts it. Note there are three cut sites in the top diagram. This converts the DNA molecule into four shorter molecules: a, b, c, and d. Each has "sticky ends" that can subsequently form base pairs with other DNA in the splicing procedures (from *Understanding DNA and Gene Cloning* by Drlica, Copyright © 1992. Reprinted by permission of John Wiley & Sons, Inc.).

shown in Figure 9.21; growing bacterial colonies—more or less at life size—are shown in Figure 9.22.

9.7.6 Transgenic Plants and Animals

A transgenic plant or animal is one that has been altered to contain a gene from another organism, usually from another species. Genetic manipulation of plants is well established as the science of selective breeding. Direct manipulation of plant genes is a newer but relatively commonplace technique substantially similar to the gene cloning methods for bacteria and yeast.

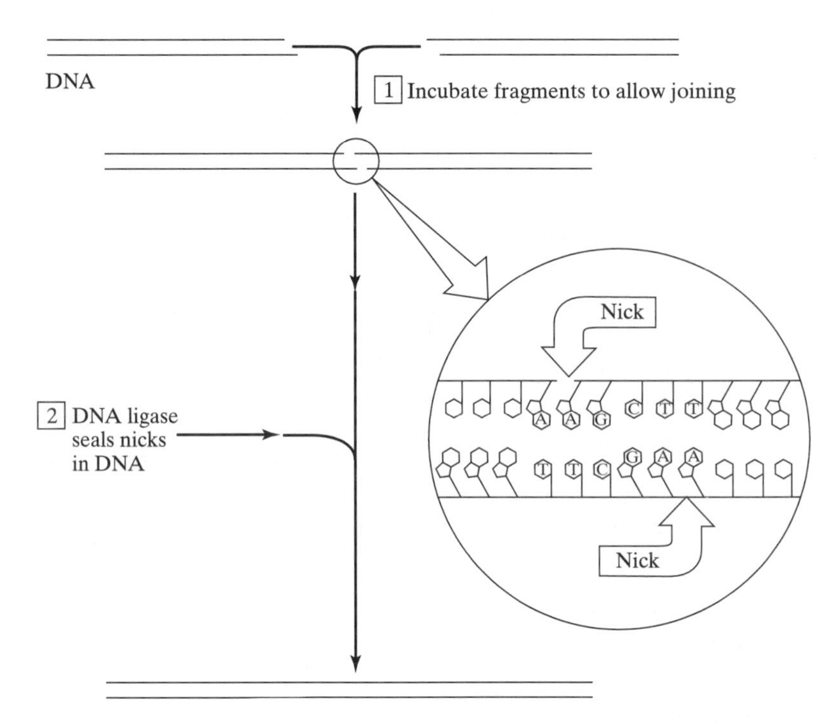

Figure 9.18 Joining two DNA fragments. The complementary ends of DNA facilitate the joining process. "Nicks" are enzymatically sealed by DNA ligase (from *Understanding DNA and Gene Cloning* by Drlica, Copyright © 1992. Reprinted by permission of John Wiley & Sons, Inc.).

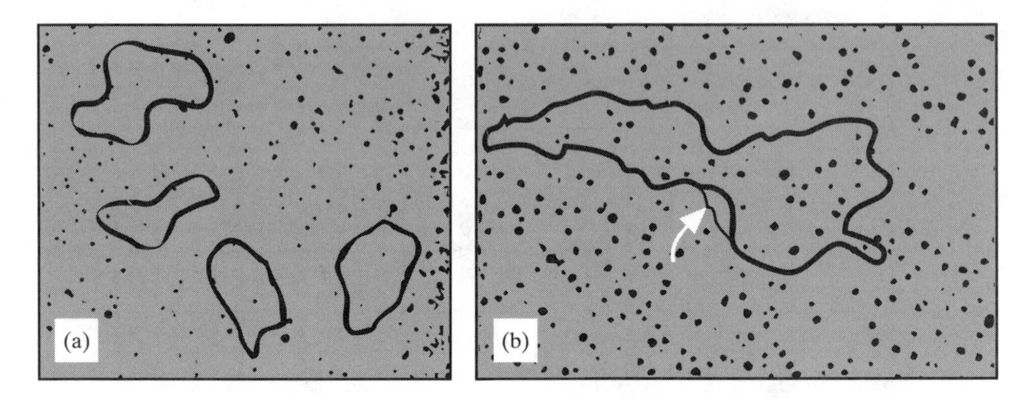

Figure 9.19 An artist's impressions of (a) Circular plasmids and (b) an enlargement of a plasmid showing a short region becoming single-stranded.

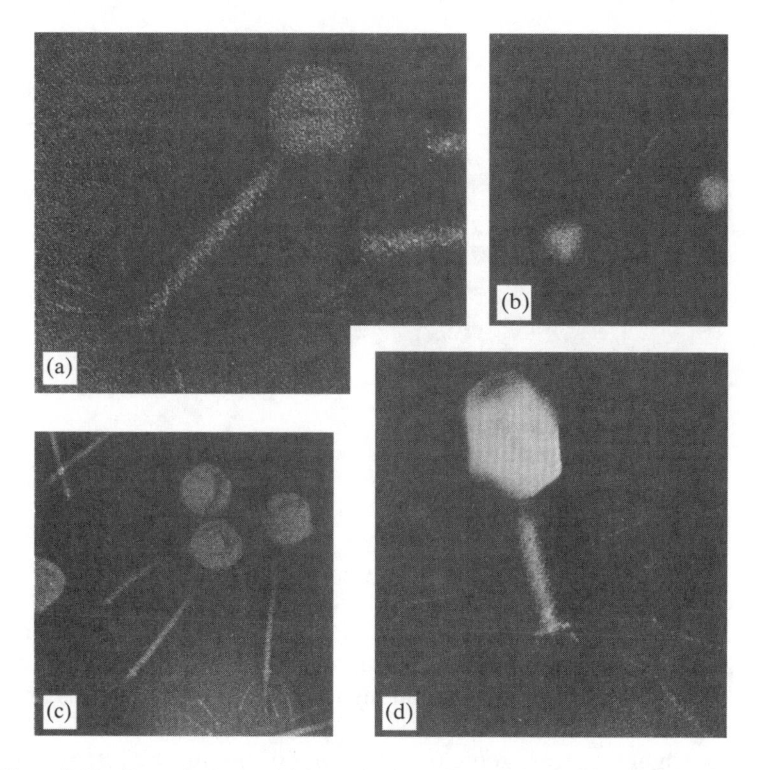

Figure 9.20 Bacteriophages: (a) bacteriophage P2, ×226,000; (b) bacteriophage lambda, ×109,000; (c) bacteriophage T5, ×91,000; (d) bacteriophage T4, ×180,000 (photomicrographs courtesy of Robley Williams, University of California, Berkeley).

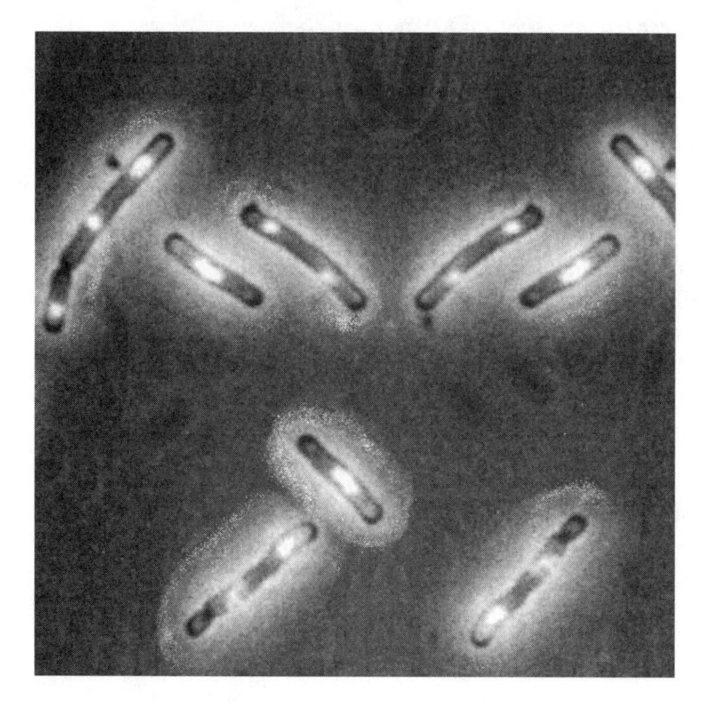

Figure 9.21 Photomicrographs of *E. coli* bacteria.

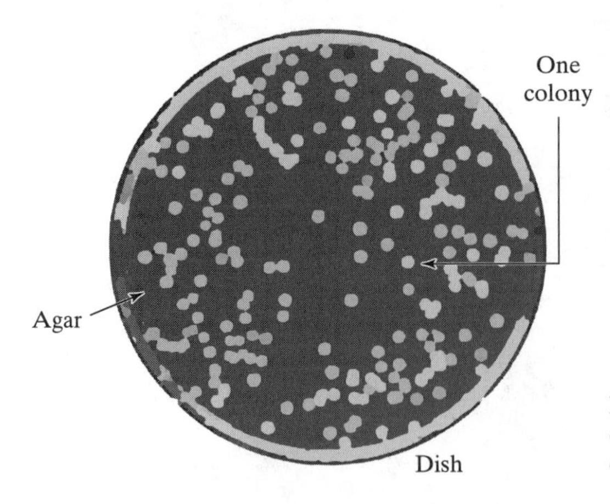

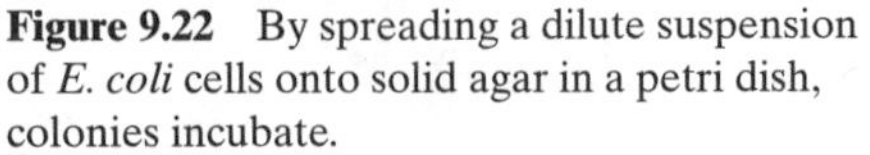
Figure 9.22 By spreading a dilute suspension of *E. coli* cells onto solid agar in a petri dish, colonies incubate.

Transgenic animals including humans have recently been a source of much public discomfort with genetic engineering. Producing a transgenic animal such as the "supermouse" shown in Figure 9.1 is a complex process, requiring a mix of gene cloning and sexual reproduction techniques to ensure the transgenes are found in cells of the animal.

Will biotech create transgenic people? One of the most controversial aspects of genetic engineering is the potential for excising or adding genetic information to correct mutant or missing genes.

9.8 GENETIC ENGINEERING II: A CASE STUDY ON GENE CLONING OF HEMOGLOBIN

9.8.1 Introduction

The previous section gave a summary of the cutting, joining, and cloning procedures in biotechnology. The analogy with movie-film splicing and joining was presented from Drlica (1992). This is a helpful analogy, but the great difference in dimensional scale between DNA and movie film should be emphasized. The diameter of DNA's "twisted rope ladder" is only 2 nanometers, whereas the movie film might typically be 2 centimeters wide. This represents a 10^7 difference in width. For length comparisons, the helical DNA (best seen at the top of Figure 9.9) has a pitch of 3.4 nanometers, spanning 10 base pairs per cycle, whereas one frame of the movie film is again approximately 2 centimeters. The handling of these minute sections of DNA calls for special techniques, including:

- Isolation of DNA from cells
- Preparation of enzymes for cutting, modifying, and joining DNA
- Cloning to create billions of cells
- Radio labeling fragments of DNA to make radioactive probes that find desired genes

9.8.2 A Case Study on the Procedure for Obtaining Clones of Hemoglobin DNA

The techniques listed earlier perhaps come to life more easily if they are described in the context of a particular gene cloning procedure. The goal of this case study is to describe how the genes encoding hemoglobin were cloned for the first time. The diagrams are from Chapter 8 of Drlica's book (1992), and the reader is referred there for a more comprehensive description.

Why is this procedure of interest? Hemoglobin is an important protein in our blood. It is responsible for moving the oxygen in our lungs into body tissue. A number of genetic disorders have been associated with defective hemoglobin, in particular sickle-cell anemia. To study a serious illness of this type, scientists need a supply of hemoglobin genes for a variety of experiments. Gene cloning is not an end in itself but a general technique for generating enough "work material" whose structure can be analyzed and functions understood. For example, it is important to understand how one hemoglobin gene is switched on and another switched off. The studies of cancer and AIDS also involve the need for a bulk supply of cloned genes that can be analyzed and related to disease formation.

The complete layout for the procedure needed to find the hemoglobin genes is shown in Figure 9.23. Once found and isolated, the genes are grown in bulk from bacterial hosts.

9.8.3 Step 1: Obtaining DNA (Top Right of Figure 9.23)

Cells can be opened up to release the DNA and proteins. Subsequently, other procedures can separate the DNA from the proteins. However, many other steps, described in this section, are then needed to separate specific genes, such as those encoding hemoglobin, from the millions of copies of thousands of other types of genes in the DNA. The experiments described in the following sections, used the livers of rabbits to obtain the hemoglobin cells, but most kinds of body tissue could have been broken up to obtain the raw material.

Practical Techniques

Mechanical stirring broke down the cell walls, releasing the DNA and the proteins. Detergents and enzymes were added to the solution to break down the proteins and also to inactivate any nucleases that could have cut the DNA. Phenol was then added, and vigorous stirring moved any remaining protein into the phenol. The mixture was allowed to stand so that a DNA-rich layer could separate out in a test tube, subsequently to be removed by a pipette. Additional centrifuging further purified the DNA, which was syringed out for Step 2.

9.8.4 Step 2: Packaging into Bacteriophages (Center Right of Figure 9.23)

Rabbit DNA and phage DNA are first individually cut by a restriction enzyme. The fragments of rabbit DNA are then mixed with the phage DNA, and the enzyme DNA ligase is added to join the DNA fragments.

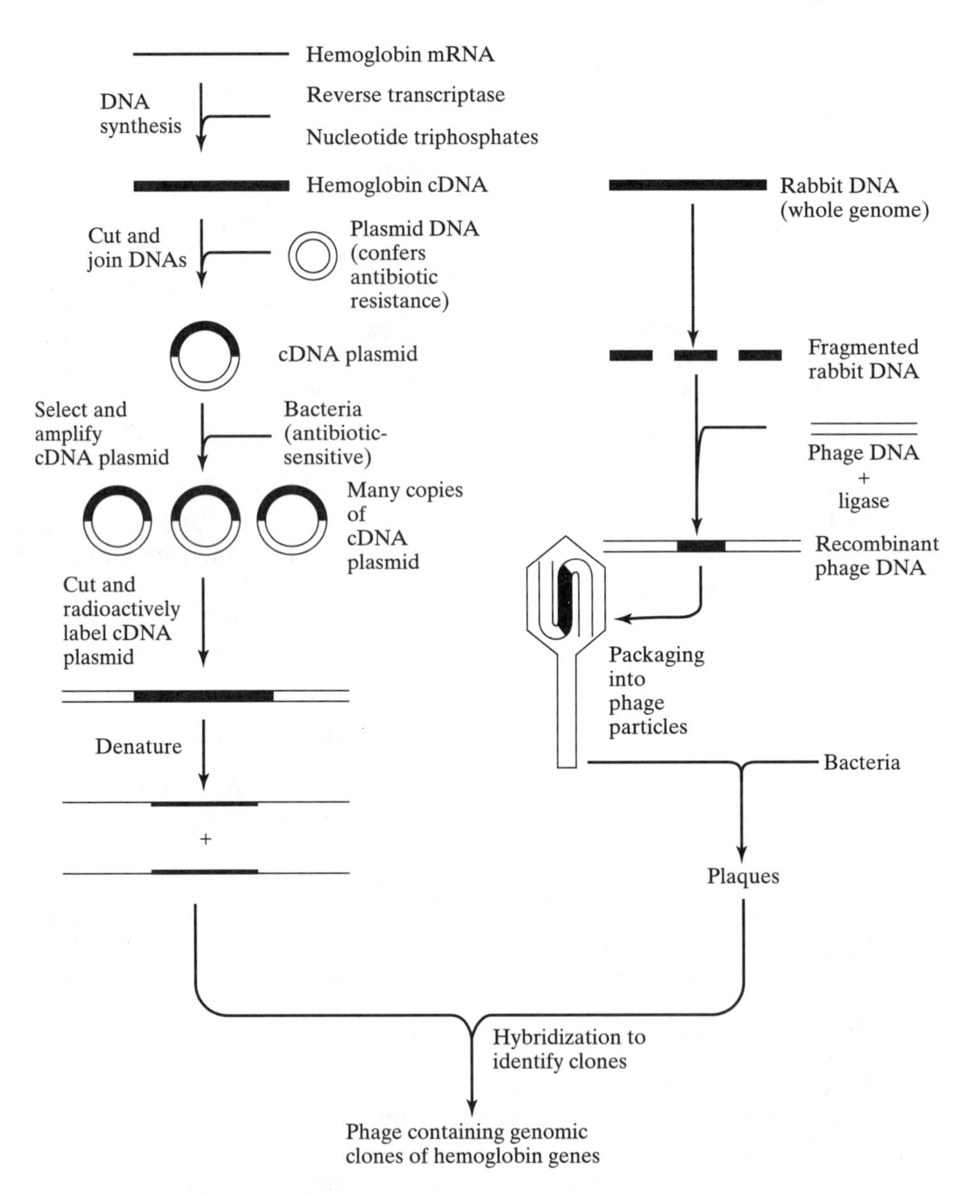

Figure 9.23 Obtaining clones of hemoglobin in bulk for later studies on sickle-cell anemia. On the right side, fragmented rabbit DNA was packaged into the phage vector. On the left side, radioactive probes of cDNA were prepared and made in bulk by splicing into plasmids and growing them in *E. coli.* At the bottom, the phage particles were grown in bulk, and then those containing the desired hemoglobin genes were isolated with the probes using complementary base pairing (from *Understanding DNA and Gene Cloning* by Drlica, Copyright © 1992. Reprinted by permission of John Wiley & Sons, Inc.).

At this point, some of the phages contain rabbit DNA after mixing; the phages are then coated with phage proteins, packaging the DNA inside phage particles, as shown on the right of Figure 9.23. The phages are then used to infect *E. coli* hosts on agar.

Unfortunately, only a very small number of the plaques formed during bacterial growth contain the hemoglobin genes. The challenge is to eventually locate this extremely small number of the desired plaques. Radioactively labeled nucleic acid probes can be used for this purpose. The next two steps are both concerned with the preparation of these probes, and doing so in such a way as to have enough material to test all the plaques. Probe preparation is shown on the left of Figure 9.23.

9.8.5 Step 3: Preparing mRNA as a Template for Probes (Upper Left Side of Figure 9.23)

One potential source for radioactive probes is messenger RNA from the gene being sought. Hemoglobin represents a special case, because the mRNA in red blood cells is mainly hemoglobin mRNA.

Practical Techniques

Red blood cells from a rabbit were centrifuged to obtain a pellet from which the RNA could be isolated from the other cell constituents, with phenol and alcohol treatments and further centrifuging to further purify the RNA. To separate the mRNA from the ribosomal and transfer RNA, the mixture was passed through a glass column filled with cellulose. A feature unique to mRNA molecules is that a sequence of only adenines (A) is added on to the end of their chains. This characteristic can thus be used to separate out the mRNA molecules from the other types of RNA present—tRNA and rRNA. Single-stranded DNA composed of thymidines was attached to the cellulose, and, through base pairing, these attached to adenines on the mRNA only, allowing the other types of RNA to pass through the column, thus separating them from the mRNA. The hemoglobin mRNA was then removed from the column by warming, which broke the A-T base pairs.

9.8.6 Step 4: Increasing the Supply of Probes by Cloning with Plasmids (Center Left Side of Figure 9.23)

The mRNA is then converted into cDNA using nucleotides and reverse transcriptase—an enzyme that can use RNA as a template for DNA synthesis. However, to obtain a larger supply of cDNA, it is first desirable to clone the cDNA into circular plasmids to be replicated in *E. coli,* as shown on the center left side of Figure 9.23.

Practical Techniques

The splicing of cDNA with a plasmid, followed by cloning in *E. coli* hosts, took place as in Step 2. However, it was still necessary to isolate the particular *E. coli* cells that had taken up plasmid DNA into which the rabbit cDNA had been inserted. The procedure that allows this is described in Figure 9.24. The procedure deliberately chose a plasmid with two genes for antibiotic resistance: one for tetracycline resistance (tet R) and the other for ampicillin resistance (amp R). The plasmid was cut within the amp R gene such that those plasmids that contain the cDNA insert no longer have a functional amp R gene. Then, for example, if tetracycline was present, it killed those cells that had not taken up any of the plasmid. Next, the colonies that had been formed on the tetracycline-containing agar plate (Figure 9.24d) were tested to find ones that failed to grow on an ampicillin-containing agar. This distinguished *E. coli* cells containing plasmids joined to rabbit cDNA from those that did not have rabbit DNA attached to them. Figures 9.24e and 9.24f show the two types.

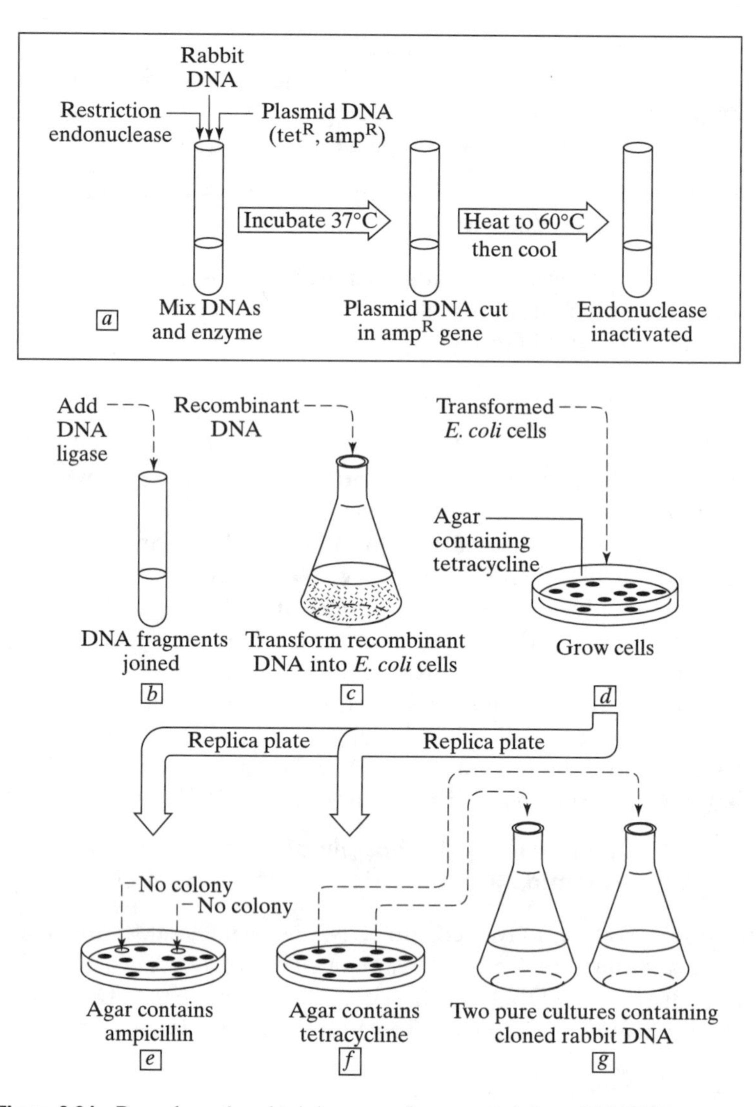

Figure 9.24 Procedures for obtaining pure clones containing rabbit DNA: (a) plasmid DNA with resistance to tetracycline and ampicillin is mixed with rabbit cDNA; (b) DNA ligase is added to join the DNA fragments; (c) *E. coli* cells act as host; (d) plasmid-containing cells are selected by growth on agar containing tetracycline, and cells with plasmids joined to rabbit cDNA are identified by screening on ampicillin-containing agar; (e) and (f) these cells grow only on tetracycline; (g) pure cultures containing cloned genes (from *Understanding DNA and Gene Cloning* by Drlica, Copyright © 1992. Reprinted by permission of John Wiley & Sons, Inc.).

The cDNA could then be isolated, made single-stranded (typically by boiling), and radioactively labeled.

9.8.7 Step 5: Final Screening by Nucleic Acid Hybridization to Isolate Genomic Clones (Bottom of Figure 9.23)

The final step is to screen the colonies grown from the phages containing the original fragmented rabbit DNA and to find which ones have the desired hemoglobin genes (recall this is on the right side of Figure 9.23). The DNA in each plaque is tested to see if it will hybridize with the radioactive hemoglobin cDNA nucleic acid complement to the hemoglobin genes being sought. The technique relies on the principle of complementary base pairing between the cDNA probes and the desired hemoglobin genes in those very few plaques described in Step 2.

Practical Techniques

After the phages were used to grow *E. coli* colonies on an agar plate, a piece of filter paper was placed on the agar and removed. The cells were thus transferred to the paper, which was placed in a dilute solution of sodium hydroxide (lye or caustic soda). The first key feature was that the sodium hydroxide caused the hemoglobin DNA to become single-stranded. The second key feature was that when the single-stranded, radioactive cDNA was next added, complementary base pairs formed *only if* the filter-paper-bound DNA contained the likewise single-stranded hemoglobin DNA gene being sought. The radioactive cDNA bound to the filter paper, identifying the locations of the rabbit hemoglobin gene being sought. The filter paper was next washed to remove any radioactive probes that were not base paired. X-ray film was then used to find bacterial colonies containing the hemoglobin gene of interest. This procedure resulted in a pure culture of *E. coli* in which each cell contained a phage into which the hemoglobin gene of interest had been cloned.

In summary, the cDNA radioactive probes derived from mRNA were used to probe the plaques formed from the bacteria-containing phage, which contained rabbit DNA, and eventually isolate the desired genomic clones of hemoglobin.

9.9 BIOPROCESS ENGINEERING

9.9.1 Bioreactors

A bioreactor is a container or vessel in which a biological reaction occurs. For example, fermentation takes place in a bioreactor: it is a process of growing (incubating) microorganisms on a substrate containing carbon and nitrogen (the "food" for the organisms). A natural example of a bioreactor is a pond, which "manufactures" algae or pond scum.

Bioreactors can range from small bench-top fermentors holding 1 liter to larger 1-million-liter production units. In addition to producing various food products, bioreactors are used to generate industrial chemicals, enzymes, and biofuels. Using microorganisms to generate fuel is of particular interest to researchers in energy-poor countries. In Finland, for example, baker's yeast has been used in a bioelectrochemical device in which the chemical process of substrate oxidation-reduction generates electrical energy.

The biosynthesis process, as well as the resulting relationship between cell growth and bioproduct formation, differs according to the type of bioreactor. Two conventional methods are batch and continuous culture fermentation. Bioconversions can also take place on moist but solid substrates.

A primary task in biotechnology processing is to design, operate, and control bioreactors such that conversion rates and yields are economically feasible. Another challenge is keeping cells and catalysts alive as they are put through various mashing, mixing, heating, and other processes. For this reason, bioprocess engineers must not only be conversant with process development, equipment design, and scale-up but also understand what is needed to keep organisms viable and growing at an optimal rate.

A common type of bioreactor is a mechanically stirred tank, which utilizes a three-phase (gas-solid-liquid) reaction. In this type of device, gas is sparged into the bottom of the vessel, then mixed with the liquid phase of the fermentation process by a mechanical stirrer (Figure 9.25). There are strict constraints in such a process. For example, a steady supply of oxygen gas bubbles for aerobic fermentations is critical. Stirring must be rapid enough to disperse gas bubbles, develop a homogenous liquid, and ensure a solid suspension. Overstirring tends to shear cells, while understirring may asphyxiate them. Another challenge is optimizing heat removal rates; faster fermentation creates a faster rate of heat production. During scale-up, surface-to-volume ratios decrease, reducing the rate of heat removal.

Sterility is another major challenge. Processes must be absolutely aseptic. Elimination of unwanted organisms is required to ensure product quality and to prevent contaminating organisms from displacing the desired production strain. This creates significant design and operation difficulties, particularly in combination with other process requirements. For example, a big problem has been designing high-quality temperature sensors that can stand up to repeated sterilizations.

The molecular reactions inside the bioreactor govern the growth characteristics of cells. Cell growth patterns depend on a variety of factors, including oxygen availability, nutrient supply, pH, temperature, and population density. In a typical batch fermentation, cell growth follows four distinct phases: lag, exponential, stationary, and decline. Little observable growth occurs in the lag phase, during which the cell restructures its biosynthetic mechanisms to take account of the environment. In the exponential phase, the cell grows as quickly as possible given these factors. (In industrial bioprocessing, this phase is typically measured in terms of the time required to double the concentration of cells, known as a biomass.) Exponential growth ceases for a variety of reasons, including nutrient depletion, physical overcrowding, or buildup of by-products of the metabolic process. There follows a stationary phase in which the enzymes that catalyzed rapid growth, along with excess ribosomes, are degraded to create other enzymes or supply fuel for cell maintenance. When internal energy sources are depleted, the cell is unable to carry out basic functions. The result may be cell lysis (breakage) or inviability (inability to reproduce). In the depletion phase, the biomass reduces (Figure 9.26).

Optimizing bioreactor functionality involves a number of discrete process variables, including aeration, agitation, mass and heat transfer, measurement and control, cell metabolism, and product expression and preparation of inoculates. Development of expert systems for real-time monitoring and adjustment of the

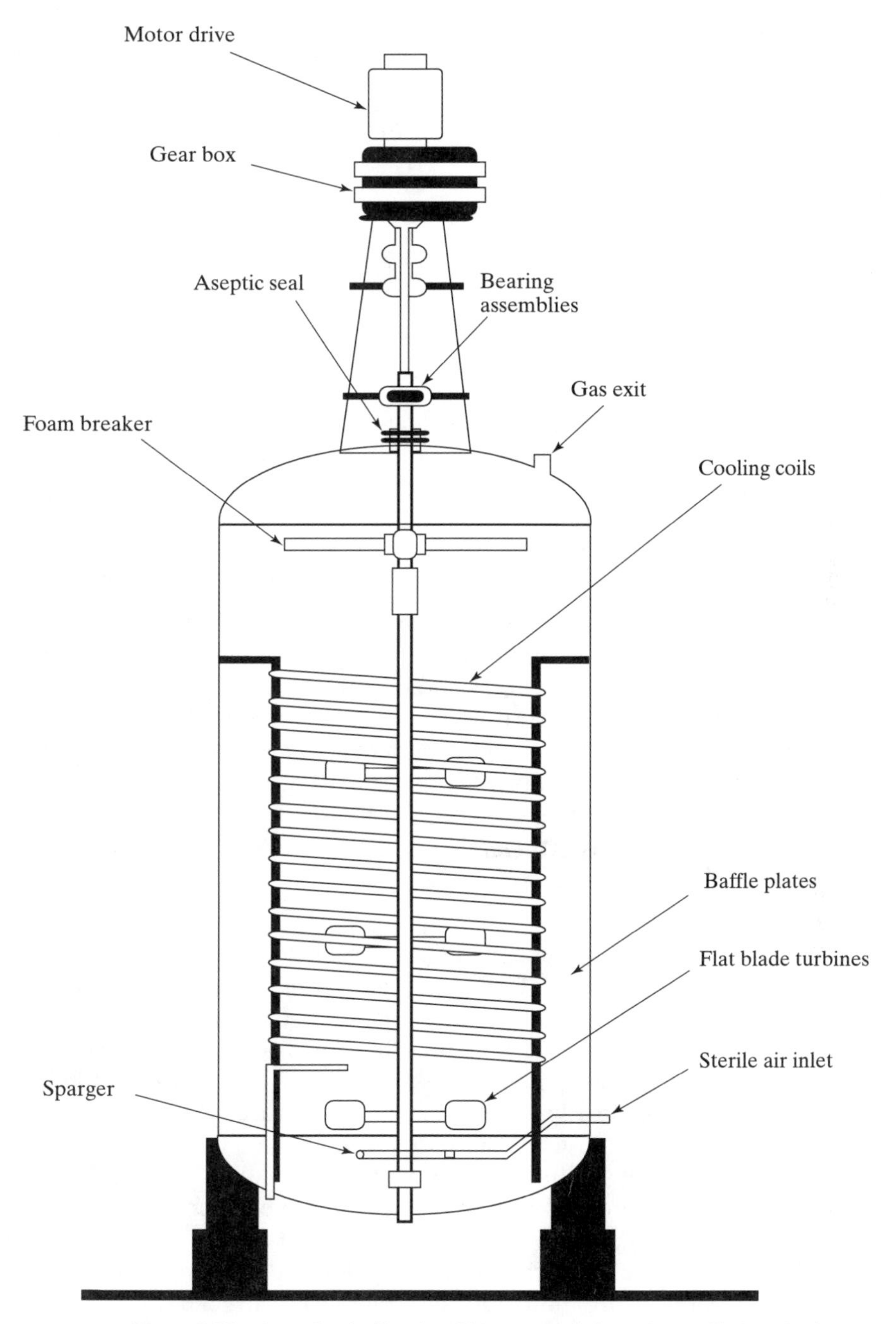

Figure 9.25 A mechanically stirred bioreactor (adapted from Shuler, 1992).

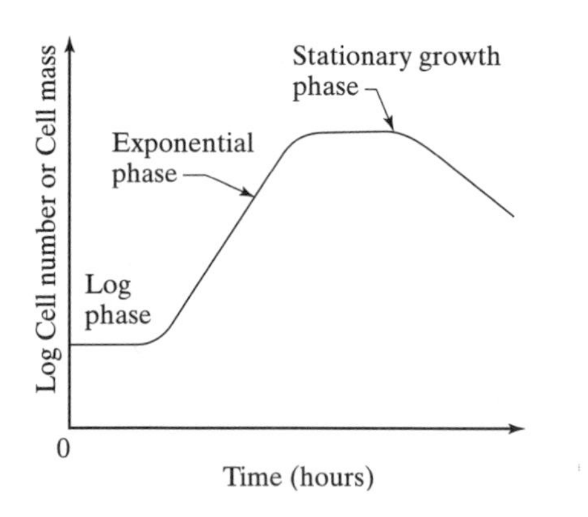

Figure 9.26 Batch fermentation growth curve.

fermentation process is one area of research. One key goal is to be able to correctly identify process problems and correct them online. For a fermentation process, this includes sensor, equipment, and process failure monitoring. Control and batch maintenance are also important candidates for automation.

9.9.2 Postprocessing

The bioreactor phase is the heart of the bioprocess. However, the recovery and purification of a fermentation product are essential to any commercial process. The degree of difficulty in the recovery and purification process depends heavily on the nature of the product. Typical downstream product recovery and enrichment processing includes filtration, crystallization, and drying techniques. Packaging and shipping the product are also important postprocessing activities.

9.10 MANAGEMENT OF TECHNOLOGY

9.10.1 Present Trends

Significant advances have been made in a very short time, and these will continue to foster industrial growth in biotechnology. For example, the Human Genome Project has been under way since 1990, jointly funded by the National Institutes of Health and the Department of Energy. Also in May 1998, a collaboration was announced between The Institute of Genomic Research (TIGR)—a private, nonprofit genetics laboratory—and Perkin-Elmer—the main manufacturer of DNA sequencing instruments. These projects are focusing on sequencing the 3 billion base pairs of human DNA and identifying approximately 60,000 to 80,000 human genes. Naturally enough, these rival projects and enterprises like the Celera Genomics Group, are creating a highly competitive business environment as they race for completion.[7]

The early phases of such research are devoted to mapping each human chromosome as a step toward ultimately determining all the genes in the DNA sequence.

[7]For popular press reviews, see "The Race to Cash in on the Genetic Code," *The New York Times,* August 29, 1999, Section 3. Also, *Science News,* Vol. 154, October 10, 1998, p. 239; and *The New Yorker,* June 12, 2000 p. 66. The Human Genome Project plans to finish sequencing the human genome by 2003 and have a working draft in 2001.

In the process, researchers are developing methods to automate and optimize genetic mapping and sequencing.

Subsequent phases will focus on the development of "molecular medicine" based on early detection of disease, effective preventive medicine, efficient drug development, and, possibly, gene therapy or gene replacement. Research efforts are also under way to sequence the genomes of bacteria, yeast, plants, farm animals, and other organisms. Much of the technology developed during genome research, particularly the automation and optimization routines, will greatly benefit the biotech industry.

9.10.2 Manufacturing

Despite these advances in knowledge, the path from the laboratory to the market is full of obstacles. Biotech research is expensive, time-consuming, and frequently fruitless. The *scale-up* to economically feasible production levels is also a tremendous challenge.

Biotechnology is in many ways related to one of its parent disciplines, chemical engineering. However, it is far more difficult because its raw materials, catalysts, and products are living organisms, which are inherently more fragile and temperamental than petrochemicals and other substances. Stringent product safety requirements, especially for therapeutics, create special problems for commercial production. Equipment and facilities must meet strict safety and quality control standards to ensure product purity. To date, standards for critical processing components (such as valve design and function) are still being established, making it difficult to design and build biotechnology equipment and systems. Lacking manufacturing expertise, many companies are sticking to research and licensing their technologies to biotech firms that have already developed production capabilities. This is similar to the "fabless-IC" model presented in the management of technology section of Chapter 5.

The long approval process and other product development risks make it especially important to shorten the critical path for bringing a new product to market. With the inherent challenges associated with scale-up of industrial bioprocesses, it is important to begin developing synthesis at the bench and pilot plant scales well before clinical trials are concluded (Figure 9.27). This is similar to the general push for concurrent engineering described in several of the earlier chapters.

In addition, there are significant regulatory barriers. New medical compounds must be rigorously tested in numerous clinical trials using strict Food and Drug Administration (FDA) regulations. The approval process typically takes from five to seven years, and the likelihood that a product will fail is high. Agricultural regulation is less rigorous; the federal government recently relaxed its regulations of field testing and marketing of genetically engineered crops.

9.10.3 Investment

Given the potential range and impact of its commercial applications (Figure 9.28), it is not surprising that biotech attracts much attention on Wall Street and with venture capitalists. Throughout the 1980s, hundreds of cash-infused new companies sprouted up. Each vied to beat the others to market with a breakthrough product. Beginning in the 1990s, the well-known pharmaceutical companies also became involved in biotech. In many cases, these larger pharmaceuticals bought, or acquired a major

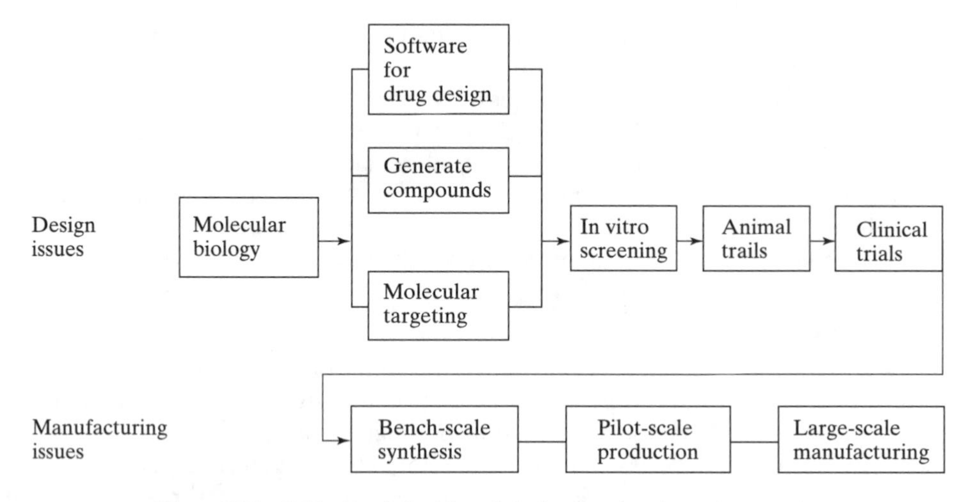

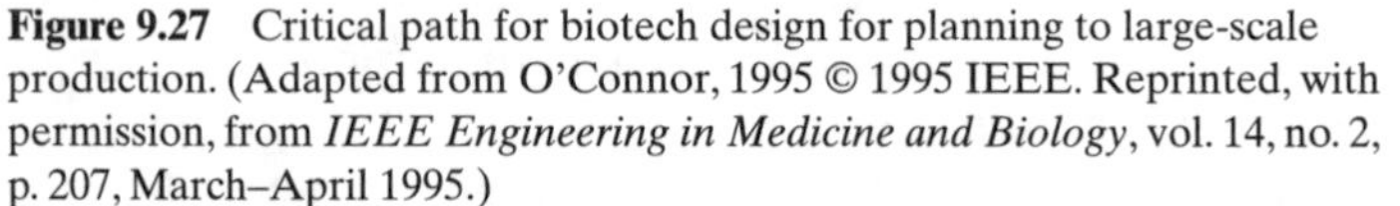

Figure 9.27 Critical path for biotech design for planning to large-scale production. (Adapted from O'Connor, 1995 © 1995 IEEE. Reprinted, with permission, from *IEEE Engineering in Medicine and Biology*, vol. 14, no. 2, p. 207, March–April 1995.)

interest in, the smaller start-up companies. As a result, today's picture is one in which a wide range of company types exists. At one extreme, individuals at universities with *molecular and cell biology* departments continue to start private research-oriented companies. The integrity of such practices is discussed in Kenney (1986). At the other extreme, the large pharmaceuticals have set up production lines for well-established materials. In between, the industry pioneers such as Chiron and Genentech continue with a balance of basic research into new products and the production of well-established products.

9.10.4 The Future

As can be seen in the popular press, biotechnology research creates more ethical debates than the industries reviewed in Chapters 5 through 8. Part of the controversy stems from popular fascination and fears about the potential dangers of "messing with nature." Michael Crichton and Hollywood have helped fuel such concerns with reconstituted dinosaurs wreaking havoc in tropical islands.

However, beyond the fanciful terrors of science fiction, biotech does indeed provoke a host of real ethical as well as practical concerns. Selective breeding was once the only method available to develop desirable plant and animal characteristics over several generations. By contrast, genetic engineering can clone precisely defined species. This may be less threatening for the well-known cloning of sheep. But it is clear from the recent government bans that society feels threatened by the same possibilities for humans. Does society have the right to intervene so forcefully in the evolutionary processes? What are the implications for biodiversity?

Privacy and equity are also a concern. Suppose an insurance company decides to do a DNA test on all its potential clients, and it finds that one of the clients inherited part of some DNA sequence from his or her grandmother that makes the client a likely candidate for a heart attack at 45. Although the cause is clear, a cure is not

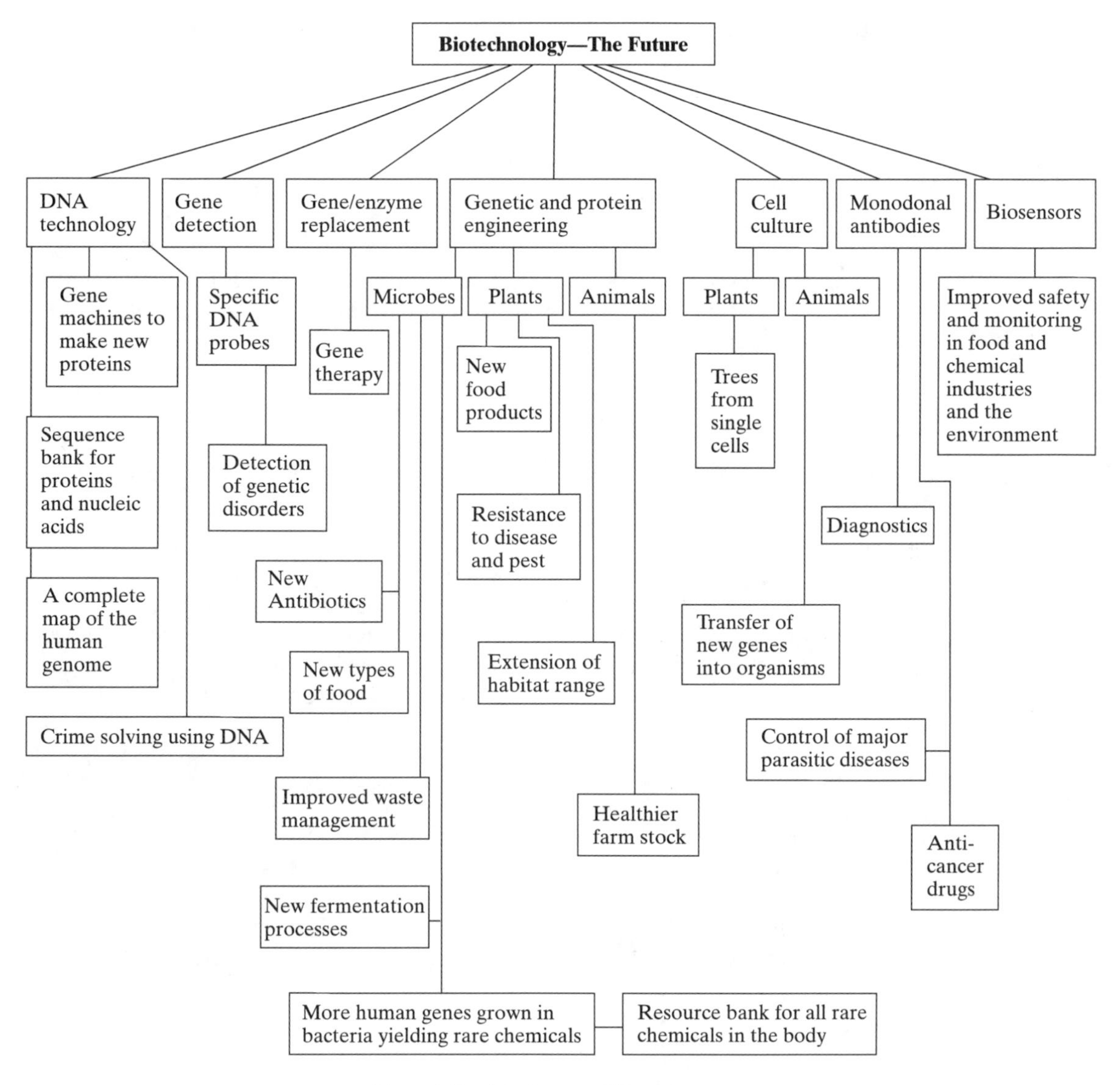

Figure 9.28 The future of biotech.

available yet. As a result, the company refuses to insure the potential client and informs other insurance companies of the risk factor. Is this ethical?

9.10.5 Summary

This brief chapter on biotechnology has been included because the field will experience rapid growth and provide many future career opportunities for people interested in manufacturing engineering. It can be seen that the issues range from interesting scientific principles that are fundamental to life itself, to the engineering of gene splicing, to the operation of bioreactors, and finally to the ethical issues mentioned.

9.11 GLOSSARY

9.11.1 Amino Acids

The building blocks of proteins. There are 20 different amino acids that link together via peptide bonds during the process of protein synthesis on the surface of the ribosome (see Transcription and Translation) according to the genetic information on mRNA.

9.11.2 Bioreactors

Vessels for biological reaction through fermentation or other transformation processes. Interferon, for example, is manufactured in a genetically engineered fermentation process.

9.11.3 Biosensors

Combining biology, IC-design, and IC-microfabrication technologies, biosensors are devices that use a biological element in a sensor. Biosensors work via (a) a biological molecular recognition element and (b) a physical detector such as optical devices, quartz crystals, or electrodes.

9.11.4 Cell

The smallest unit of living matter capable of self-perpetuation.

9.11.5 Central Dogma

The concept in molecular biology in which genetic information passes unidirectionally from DNA to RNA to protein during the processes of transcription and translation.

9.11.6 Chromosome

A subcellular structure consisting of discrete DNA molecules, plus the proteins that organize and compact the DNA.

9.11.7 Codon

A set of three nucleotide bases on the mRNA molecule that code for a specific amino acid. For example, the codon C-G-U represents arginine. There are 64 unique codon combinations. Some amino acids can be specified by more than one codon sequence.

9.11.8 DNA (Deoxyribonucleic Acid)

The genetic material in every organism. It is a long, chainlike molecule, usually formed in two complementary strands in a helical shape.

9.11.9 Enzymes

Enzymes are proteins that catalyze reactions.

9.11.10 Fermentation

A process of growing (incubating) microorganisms in a substrate containing carbon and nitrogen that provide "food" for the organisms.

9.11.11 Gene

The basic unit of heredity, a gene is a sequence of DNA containing the code to construct a protein molecule.

9.11.12 Gene Cloning

One of the most common techniques of genetic engineering, gene cloning is a way to use microorganisms to mass-produce exact replicas of a specific DNA sequence. The cloned genes are often used to synthesize proteins. The basic technique is to construct recombinant DNA molecules, insert the resulting DNA sequences into a carrier molecule or vector, and introduce that vector into a host cell so that it propagates and grows.

9.11.13 Genetic Code

Figure 9.14 shows the 64 possible codons and the amino acids specified by each.

9.11.14 Genetic Engineering

The practice of manipulating genetic information encoded in a DNA fragment to conduct basic research or to generate a medical or scientific product. Selective breeding is one of the oldest examples of genetic engineering. The much newer gene cloning technology is now fairly common practice. Genetic engineering is used in three areas: to aid basic scientific research into the structure and function of genes, to produce proteins for medical and other applications, and to create transgenic plants or animals.

9.11.15 Genome

The total genetic information of an organism.

9.11.16 Host

A cell used to propagate recombinant DNA molecules.

9.11.17 Natural Selection

The process by which a species becomes better adapted to its environment—the mechanism behind the evolution of the species. The process depends on genetic variations produced through sexual reproduction, mutation, or recombinant DNA. Variant forms that are best adapted tend to survive and reproduce, ensuring that their genes will be passed on. Over hundreds and thousands of generations, a species may develop a whole set of features that have enhanced its survival in a particular environment.

9.11.18 Nucleotide

A building block for DNA and RNA. A nucleotide is composed of three parts: a sugar, a phosphate, and a chemical base. In DNA, the bases are adenine (A), guanine (G), thymine (T), and cytosine (C). RNA contains A, G, and C, but has uracil (U) in place of thymine.

9.11.19 Proteins

Proteins are long-chained molecules containing chains of amino acids. Twenty different amino acids are used to make proteins.

9.11.20 RNA (Ribonucleic Acid); mRNA; rRNA; tRNA

A long, chainlike molecule formed of the nucleotides A, G, U, and C. Cells contain different RNA types, including messenger RNA (mRNA), ribosomal RNA (rRNA), and transfer RNA (tRNA). Each type of RNA performs a specific function in protein synthesis (see Transcription and Translation).

9.11.21 Recombinant DNA

A DNA molecule made up of sequences originating from two different DNA molecules. Recombination occurs naturally or through genetic engineering, and is a fundamental source of genetic diversity. The process usually involves the breaking and reuniting of DNA strands to produce the new DNA.

9.11.22 Transcription and Translation

The processes by which genetic information on a DNA molecule is used to synthesize proteins. During transcription, a strand of mRNA is synthesized from a DNA template. During translation, the genetic information on mRNA is read by tRNA on a ribosome (a cell's protein factory) in order to build the chain of amino acids that form a protein.

9.11.23 Vector

A DNA molecule, usually a phage or plasmid, that is used to introduce DNA into a host cell for propagation.

9.12 REFERENCES

Avery, O. T., C. M. MacLeod, and M.McCarty. 1944. *Journal of experimental medicine* 78: 137–158.

Campbell, D. 1998. The application of combinatorial strategies to the identification of antimicrobial agents. Paper presented at the Symposium on Microarrays and Drug Resistance, the American Society of Microbiology. Atlanta, GA.

Darnell, J., H. Lodish, and D. Baltimore. 1986. *Molecular cell biology.* New York: Scientific American Books.

Drlica, K. 1992. *Understanding DNA and gene cloning: A guide for the curious.* New York: John Wiley & Sons.

Economist. 1998. Science and technology. 13 (June): 79–80.

Hershey, A. D., and M. Chase. 1952. *Journal of General Physiology* 36: 39–56.

Mascarenhas, D. 1999. DNA technology in plain English. *Lecture notes from the Symposium on Twenty-Five Years of Biotechnology,* May 13, UC Berkeley Extension.

Nicholl, D. S. T. 1994. *An introduction to genetic engineering.* Cambridge University.

Okamura, S. M. 1998. *Genes required in the **a**/alpha cell type of saccharomyces cerevisiae,* Ph.D. dissertation. University of California, Berkeley.

Palmiter, R. D., R. L. Brinster, R. E. Hammer, M. E. Trumbauer, M. G. Rosenfeld, N. C. Birnberg, and R. M. Evans. 1982. Dramatic growth of mice that develop from eggs microinjected with metallothione in Growth hormone fusion genes. *Nature* 300 (December): 611–615.

Watson, J. D., and F. H. C. Crick. 1953. *Nature* 171: 737–738, 964–967.

Watson, J. D., N. H.Hopkins, J. W. Roberts, J. A. Steitz, and A. M. Weiner. 1987. *Molecular biology of the gene.* Menlo Park, CA: Benjamin Cummings.

9.13 BIBLIOGRAPHY

Bains, W. 1993. *Biotechnology: From A to Z.* Oxford University.

Baker, A. 1994. Engineers further biotechnology's reach. *Design News* 29: 29.

Berger, S. A., W. Goldsmith, and E. R Lewis. 1996. *Introduction to bioengineering.* Oxford: Oxford University Press.

Bruley, D. F. 1995. An emerging discipline: Focus on biotechnology. *IEEE Engineering in Medicine and Biology* 14 (2): 201.

Bud, R. 1993. *The uses of life: A history of biotechnology.* Cambridge University Press.

Economist. 1995. A survey of biotechnology and genetics: A special report. 334 (7903).

Ezzell, C. 1991. Milking engineered "pharm animals." *Science News* 140 (10): 148.

Kenney, M. 1986. *Biotechnology: The university-industrial complex.* New Haven: Yale University Press.

O'Connor, G. M. 1995. From new drug discovery to bioprocess operations: Focus on biotechnology. *IEEE Engineering in Medicine and Biology* 14 (2): 207.

Rosenfield, I., E. Ziff, and B. Van Loon. 1983. *DNA for beginners.* Writers and Readers, U.S.A.

Shuler, M. L. 1992. Bioprocess engineering. In *Encyclopedia of physical science and technology* 2. San Diego, CA: Academic Press.

Timpane, J. 1993. Career paths for MS and BS scientists in pharmaceuticals and biotechnology. *Science* 261 (5125): 197.

Watson, J. D. 1968. *The double helix.* Atheneum Press.

CHAPTER 10

FUTURE ASPECTS OF MANUFACTURING

10.1 RESTATEMENT OF GOALS AND CONTEXT

The goals of this book are to:

- Illustrate *general principles* of manufacturing (Chapters 1 and 2).
- Review some of the main manufacturing techniques needed during the *product development cycle* of a consumer electromechanical product (Chapters 3 through 8).
- Review the *emerging market of biotechnology* manufacturing (Chapter 9).
- Explore *management of technology* and *cross-disciplinary* issues (Chapter 10).

In a teaching environment, it is useful to augment these themes with a semester-long project in which a simple device is designed, manufactured, and analyzed from a business viewpoint. It is also beneficial to visit factories and write case studies on the daily challenges of manufacturing and the future growth of companies. No one would want to visit a medical doctor who had never had any hands-on experience. Likewise, students destined for careers in technology can benefit greatly from seeing the real world of industry early in their careers.

The book is targeted at a class consisting of both engineering and business students. This mix of student interests, combined with a focus on group-oriented case study projects or consulting projects, requires a certain amount of flexibility and compromise on everyone's part.

The book and its related lectures are deliberately a survey of each manufacturing topic and also focus on issues in today's business environment. This has two obvious limitations:

First, as far as depth is concerned, each of the chapters deals with technical material that could, on its own, be expanded into a complete textbook. In addition, all the chapters contain technical material that is somewhat unresolved. Thus the reader is encouraged to follow the many ASME/IEEE-type conferences that take place each year to present the latest research discoveries. Similarly, the management style reviews, found at the beginning or end of most chapters, could use an expanded, rigorous analysis of the type found in the *Harvard Business Review* or *California Management Review.*

Second, there is the problem of keeping the material up to date. Rough drafts of the book were updated each semester for various manufacturing classes at all levels, and between each revision substantial changes kept occurring. For example, in recent years the high-tech community has witnessed the rise of Java and Jini for embedded systems (see Waldo, 1999); the introduction of network computers such as the InfoPad; the increased popularity of handheld computers such as the Palm Pilot; the trend toward the 12-inch silicon wafer with its $2+ billion plant; and the continued growth of the U.S. automobile industry. Regretfully, other companies have experienced a decline. For example, Apple now holds only about 4% of the U.S. market share in personal computers; though, of course, Apple's supporters hope that its new designs and marketing will change the company's performance for the better.

10.2 MANAGEMENT OF TECHNOLOGY

Despite the mentioned limitations, especially the changing nature of the world of technology, there still seems to be the need for a textbook that presents an integrated analysis of the detailed fabrication techniques *and* the management of technology.

Management of technology (MOT) can be approximately defined as the set of activities associated with bringing high-tech products to the marketplace. Specific issues include:

- Identifying who the customer is
- Analyzing investments in technology; for example, with Hewlett-Packard's Return Map
- Launching creative products within a "learning organization" that uses TQM
- Fabricating a prototype and scaling up to highly efficient mass production
- Creating new methods of high-technology marketing
- Exploiting Internet-based, business-to-business capabilities; for example, outsourcing

This last chapter summarizes the specific methodologies and approaches that are useful in this integration of the many facets of a high-tech, globally oriented, commercial enterprise.

10.3 FROM THE PAST TO THE PRESENT

10.3.1 Mass Production and Taylorism

Chapter 1 reviewed the history of manufacturing. It included some details of the industrial revolution (1780–1820), the importance of interchangeable parts (Colt and Whitney), and organized mass production with a division between design and manufacturing (Taylor and Ford). In fact, the commercial concept of the division of labor dates back before the industrial revolution to the beginning of the 18th century, if not earlier. Adam Smith championed it vigorously in his *Inquiry into the Wealth of Nations* (see Plumb, 1965). Frederick Taylor was an even stronger advocate of the division of labor in his *Principles of Scientific Management* (1911).

Oriented to the mass production of mass consumption products, Taylorism created pyramid-shaped hierarchies with sharply defined boundaries between functional areas such as design, manufacturing, and marketing.

Following the tenets of Taylor's scientific management, creative work was distinctly segregated from the manual labor performed on the factory floor itself. Shop floor personnel were called "hired hands" and were actively discouraged from exerting control over or offering input into the decision-making process. Information tended to flow down vertical channels in the corporate structure.

In summary, management was strictly top-down, with managers setting everything from high-level corporate objectives to procedural methods at the bottom rung.

10.3.2 Today's Customized Production

Taylor's structures and practices are poorly suited to the new competitive environment in which product cycles are short, markets are fragmented, and quality and speed are of the essence. For example, Curry and Kenney (1999) capture this new "playing field" in their recent article "Beating the Clock: Corporate Responses to Rapid Changes in the PC Industry." Section 6.5 of this book reviews these dynamics in more detail. And of course the popular press frequently picks up the theme. A recent article in *Forbes* magazine—"Warehouses That Fly"—succinctly captures the speed of production in the PC industry. In summary, it emphasizes that the "old" idea that inventory is kept in a big warehouse is dead. For the PC industry in particular, the inventory is flying through the air in the belly of one of FedEx's or DHL's cargo planes or being sorted at the airport hub for next-day delivery (Tanzer, 1999).

Overall, what is needed is a cross-functional, interdisciplinary approach that fosters communication and integration among all parts of an enterprise. A fundamental prerequisite is for a firm to become a "learning organization" that constantly examines its production practices for potential improvements (Cole, 1999). In contrast to the compartmentalized mentality of Taylorism, the learning organization reintegrates manufacturing with planning and creativity. It promotes integrated problem solving among all groups, including shop floor personnel, production staff, design engineers, and managers. The chief benefits of this approach are to enhance learning by doing and develop a cross-disciplinary understanding of product design, production, and marketing.

10.4 FROM THE PRESENT TO THE FUTURE

10.4.1 The New Competitive Environment

Competing in manufacturing today is very different from even a decade ago, let alone 100 years ago. For example:

- Fragmented, customized markets have largely replaced mass markets.
- Markets and competition are global.
- Product life cycles have accelerated significantly. Motorola pagers and cellular phones have a life cycle of 6 to 12 months. The style of Nike shoes changes almost once a month depending on the season.
- Customers are more educated and demand many things at once: low prices, personal service, superior quality and performance, and shorter delivery times.

10.4.2 Generic Technical Solutions

All companies are now aware of the mismatch between rigid Taylorism on the one hand and demands for rapid response, lean production, and fast-cycle product development on the other. Helped by newly formed human resource (HR) departments, such companies strive to develop group problem-solving strategies. There has also been a trend to use business process reengineering (BPR) and shed unnecessary layers of middle management. The following innovative process technologies can then help a company deal with today's competitive pressures in several ways:

- *Design:* Innovative design methods and DFM/DFA can increase a product's performance, lower production and material costs, and open new market opportunities.
- *Rapid prototyping:* CAD/CAM and rapid prototyping technologies can accelerate time-to-market by improving the design/manufacturing/marketing interface.
- *Computer integrated manufacturing (CIM) systems:* Flexible, reconfigurable production systems and equipment can help a company operate profitably even with frequent changes in production volumes and product design.
- *Expert systems and databases:* A firm's "knowledge capital" first and foremost resides with the people in the learning organization. However, their skills and knowledge can go only so far. Computerized methods for knowledge capture and dissemination are vital.
- *Internet-based, business-to-business collaborations:* Two of the fundamental technological changes now adding to the World Wide Web's infrastructure are (a) distributed computing and (b) client-side or browser-side processing. These applications are expanding the capabilities of distributed design, planning, and fabrication environments. Direct business-to-business transactions, minimizing "markups," are improving the efficiency of the supply chain.

10.5 PRINCIPLES OF ORGANIZATIONAL "LAYERING"

The previous section proposed some generic solutions for commercial success. However, the range of products, processes, and services is so huge that there is rarely a "one-of-a-kind, miracle technological solution" that will give one particular company a dramatic lead. And even if one company does discover a great technology that puts it ahead for a while, it is usually not long before the competitors catch up. A prime example of this might be the original Apple desktop. It has been admired and copied by everyone else to the point where it no longer holds any market edge. Great technology alone does not win the day, at least not for very long. There is no single model for success, given the complexity and volatility of markets and technologies.

Thus instead of prescribing a *post-mass-production model,* it is more useful to think in terms of general organization and management principles for an intelligent manufacturing enterprise. The following sections offer general principles that, when *layered* upon each other, should be equally applicable to companies dedicated to mature products needing continuous improvement as well as to companies developing cutting edge products (see Figure 10.1).

The primary organizational recommendation is to give simultaneous attention to the following layers:

- *I, quality assurance (or TQM) in a learning organization:* constantly improve efficiency in existing operations.
- *II, time-to-market:* introduce methods and technologies that bring products to the market first.
- *III, aesthetics in design:* introduce radical product innovations that captivate consumers.

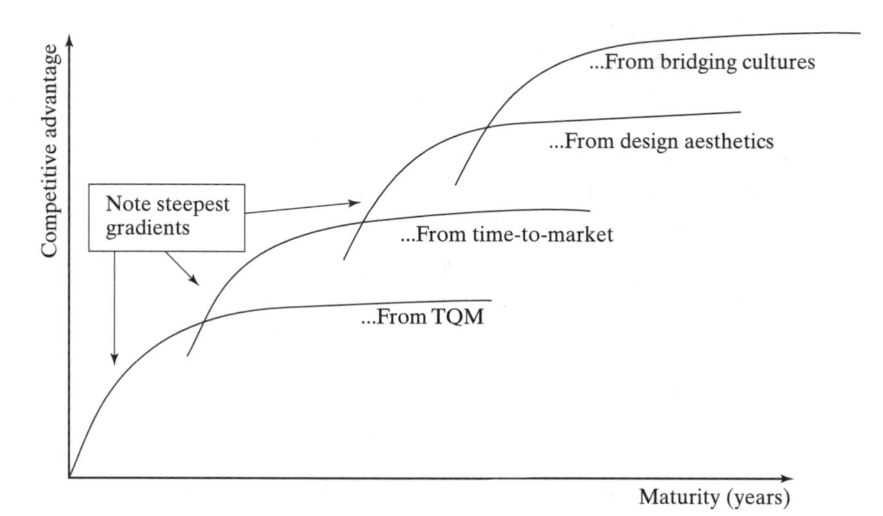

Figure 10.1 A proposed principle of organizational layering as a way to maintain the growth of the firm: this schematic emphasizes that the greatest business success arises from being on the early gradients of these new trends.

- *IV, cross-disciplinary product development:* effectively position the company for future products that synergistically exploit the overlap between mechanical, electrical, biotechnical, and other disciplines.

Consider Figure 10.1: First, it may be reasonably proposed that TQM[1] was first fully exploited by Toyota, Honda, and the Japanese DRAM manufacturers in the 1980s (see Leachman and Hodges, 1996; Cole, 1999; Spear and Bowen, 1999). As shown on the left side of the figure, they obtained the most benefit from being the first to champion the TQM movement. After a while, all other manufacturers realized the genuine importance of TQM. Once this happens, the playing field becomes level in the figure. A competitive advantage will now no longer come from TQM; rather, it is a prerequisite for survival.

Second, it may be further proposed that companies such as Intel and Motorola were the first to exploit time-to-market to the maximum during the early to mid-1990s. Now all companies are aware of the importance of this concept, and the curve is flattening out.

Third, it can be reasonably proposed that companies such as Nike, Ford, and Motorola are some of today's champions of design aesthetics in their market sectors. These seem to be the first companies that understand that customers and consumers are now looking beyond TQM and time-to-market to the more subtle expressions of "flair" or "edge" in product design.

Fourth, it will be argued in Section 10.9 that the next wave of technical products will integrate multiple disciplines, especially exploiting biotechnology. Today's organizations must be alert to these trends and scenarios—perhaps by creating new R&D teams or precompetitive "skunk works projects."

10.6 LAYER I: THE LEARNING ORGANIZATION

Cumulative, broad-based learning accelerates subsequent rounds of problem solving, with the result that overall technical competence and responsiveness increase with each generation of product. Speaking colloquially, if a company wants to be a leader in the manufacture of the $(n+1)^{\text{th}}$ generation product, it almost certainly has to be a leader in the $(n)^{\text{th}}$ generation product. This was an essential finding in Cohen and Zysman's research, reported on in their book *Manufacturing Matters* (1987). They argued strongly, with supporting evidence, that the United States should not give up the manufacturing aspects of product development and merely become

[1]Total quality management (TQM) is being used here because it is a very familiar term. However, the actual phrase TQM has fallen out of favor somewhat with practitioners in the quality assurance (QA) movement, as described by Cole (1999). Cole's colleague Dr. Kano in Tokyo notes that TQM should not be a series of banners strung up around the factory; indeed the presence of such banners usually means there is a quality problem! The genuine goals of TQM or QA are to create a corporate culture that (a) supports and rewards continuous improvement, (b) ensures that continuous innovation is respected and accepted as a necessary condition for sustained corporate success, and (c) sets performance standards and pay raises based on quality achievement.

a service industry to the world.[2] Even though companies might subcontract specialized manufacturing functions (such as rapid prototyping by SLA), it is still crucial to be working as an *integrated team* with the subsuppliers.

Such integrated teams consisting of the original product designers and selected subsuppliers should have the following characteristics:

- Cross-functional teams must bear shared responsibilities.
- All team members must be held accountable for the team's performance.
- Managers must empower individuals and facilitate teams.
- All functional units must cooperate to achieve a common goal.
- Conflicts should be resolved at the level where the best information and knowledge exist to address the problem.

Observers emphasize that the openness of an intranet brings out even more need for honest sharing of ideas inside a learning organization. But how far and to what level of sharing a firm must go *outside* its boundaries to all suppliers is still something that business-to-business relationships are struggling to understand. This section includes a recent quote from the *Economist* (Symonds, 1999):

> In the past the rules of business were simple: beat the competition into submission, squeeze your suppliers, and keep your customers ignorant the better to gouge them. At least everybody knew where they stood. The new technology makes an unprecedented degree of collaboration possible, but nobody can predict how far that will reach outside the boundaries of individual firms.

Berners-Lee (1997) and Shapiro and Varian (1999) describe this broader collaboration as establishing a collective *information universe* for an enterprise or alliance. It is analogous to a legal system where individuals act according to their own drives but in the context of a macroscopic understanding of the bigger picture and its morals. Extrapolated to the firm, or an alliance of firms, the intention would be to establish an information universe where all members would instinctively resonate with global performance criteria.

The university environment that might create the "lifelong learning" opportunities for members of such a learning organization is described in a forward-looking article by Lee and Messerschmitt (1999). One key idea is to return to the "Oxbridge" model of small, village-like learning communities led by a resident tutor. These might exist inside a company but have broader ties to a nearby major university, where special events occur and from which professors can link to these distributed "learning villages."

[2]During the 1992 presidential campaign, one of then-President George Bush's advisers announced that "it doesn't matter if the United States is making computer chips or potato chips." Luckily, this attitude does not seem to have prevailed into the next century.

10.7 LAYER II: COMPRESSING TIME-TO-MARKET

For many years, companies focused on product design and treated manufacturing as a constant factor or as a separate *ex post* concern. This has now changed. In all industries, the simultaneous design of *product and process* has been shown to be a "best practice" (Black, 1991; Leachman and Hodges, 1996).

Outsourcing of manufacturing to specialists increases, rather than decreases, the need for simultaneous design of product and process. As recommended in Koenig (1997), Borrus and Zysman (1997), Handfield and associates (1999), and Cole (1999), all subsuppliers and final customers must be brought into the concurrent design and continuous improvement loop. Employees are encouraged to interact with their peers outside of the firm. The goals are to stay on top of manufacturing developments in their field and to pursue distance learning/continuing education opportunities in manufacturing processes that have been outsourced. Even though the physical manufacturing process may be outsourced, the internal designers must still keep pace with the knowledge in that field.

In summary, for all firms, the manufacturing process is always a part of a larger production chain extending from subsuppliers through the firm to its customers. When technologies and products change rapidly, arm's-length relationships among companies can make it difficult to respond quickly. A firm's suppliers and customers are an important source of ideas and problem-solving capabilities. If, for example, a particular design modification changes the specifications for one or more components, it is useful to coordinate the proposed change with that component's supplier in order to bring in the supplier's ideas and alert it to changes it may need to make to its own processes. Vice versa, the potential of new manufacturing methods must be put in the hands of component designers so that they can exploit new procedures and ideas.

The following infrastructural tools have been shown to significantly compress time-to-market and to foster bidirectional communications between design and manufacturing:

- Concurrent engineering (Chapter 3)
- Rapid prototyping (Chapter 4)
- Computer integrated manufacturing (Chapters 5 through 8)
- Expert systems and database management (Chapter 8)
- Internet-based collaborations (Chapter 4)

As a first example, this explains the sustained success of companies such as Intel, in which each successive generation of ICs gets produced faster than the one before.

As a second example, the design and manufacture of the InfoPad, described in Chapter 6, were accelerated by the use of DUCADE (Wang et al., 1996). This software environment for electromechanical products performed electrical CAD and mechanical CAD concurrently.

10.7.1 Maintaining a System Perspective between QA and Time-to-Market

As a note concerning "the big picture," there will always be an inherent trade-off between total quality assurance and time-to-market (see Cole, 1991, 1999). This is not a new phenomenon. In Chapter 1, it was mentioned that Eli Whitney was first criticized by his customers for slow delivery. Later he was congratulated for the quality and repairability of his guns. But along the way there must have been some tense negotiations!

In today's era of shrinking product cycles in many high-tech markets, the rewards of being first to market are very high in terms of market share and profit. Furthermore, the latter provides the source for funds for the next generation of technology. This often occurs when a producer such as Intel or Microsoft can make orders of magnitude of improvement over earlier products and versions of the same product. If these performance improvements are highly valued by customers—such as high-speed computing for certain customers—some quality problems will be overlooked. Increasingly, this implicit bargain is institutionalized in beta testing.

The most dramatic example is Microsoft 2000, which had 500,000 prerelease customers participating in its beta testing. This broader view of customer awareness shows that if the right bargain is struck between supplier and consumer, then the best possible product can be delivered at the right time.

10.8 LAYER III: AESTHETICS IN DESIGN

Engineers and technologists tend to be a little disparaging toward discussions that involve art and aesthetics. But a question worth pondering is: If miniaturized electronics are destined to be a part of our everyday life, much like clothing and housing, why can't technology be softened to suit the human desire for comfort, elegance, and fine design?

One major way to maintain a competitive advantage over the next few years will be to acknowledge the importance of artistry and design aesthetics in consumer products. This may seem a rash prediction; however, it is supported by an examination of the following three companies that have pulled ahead of their respective competition by devoting more attention to the artistic aspects of common products:

- Ford has reintroduced some of the excitement seen in its older designs to the new car business. Perhaps the Mustang is a good example. The seminal American sports car has returned to the sculpted look of the 1960s rather than the functional, boxy look of the early 1980s. Ford also seems to have hit the exact needs of today's consumer markets with its Explorer.
- Motorola and Nokia have continued to miniaturize and stylize the cellular phone. The Nokia exchangeable face plate in different jazzy colors is aimed at the teenage market of "on-the-move-but-let's-keep-in-touch." For maturer

fashion victims, the Motorola StarTAC can conveniently be worn under a Georgio Armani suit and not ruin the line. Even with jeans, Motorola products aim to be worn with style and not just provide communication ability. The StarTAC's size and elegance appeal to the fashion sensibilities of Wall Street investors and Silicon Valley computer programmers alike.

- Nike and, more recently, Hilfiger, continue to entice a huge number of people to buy \$100+ running shoes because their designs have an "edge" that stands out. "Edge" does not get measured by one obvious factor. It is a combination of shape, material, color, and feel, backed up by effective advertising and sports-hero endorsement. Nevertheless it is a property that teenagers sense immediately. At the time of this writing, it seems that Nike and Hilfiger have tapped into it and Levi's has lost it. But again, things change quickly.

These are intuitive issues that are best discussed informally in the classroom. For further reading, a charming monograph by Jim Adams called *Conceptual Blockbusting* (1974) is a good place to start. In addition, most large cities have a museum of modern art where inspirations for the shape of future products can often be found.

10.9 LAYER IV: BRIDGING CULTURES TO CREATE LEADING EDGE PRODUCTS

Future products will be cross-disciplinary and involve synergy between mechanical, electrical, biotechnical, and other disciplines. The following discussion will show that this confluence of different technologies creates *a spiral of increasing capability where all technologies drive each other to higher achievements.* This trend will certainly continue to be a central aspect of 21st century manufacturing.

The reader is first invited to study Taniguchi's Table 10.1 grouped under the headings of (m) mechanical, (e) electrical, and (o) optical.

- *Normal manufacturing* delivers the precision needed for (m) automobile manufacturing, (e) switches, and (o) camera bodies.
- *Precision manufacturing* delivers the precision needed for (m) bearings and gears, (e) electrical relays, and (o) optical connectors.
- *Ultraprecision manufacturing* delivers the precision needs for (m) ultraprecision x-y tables, (e) VLSI manufacturing support, and (o) lenses, diffraction gratings, and video discs.

The data emphasize that the precision at any level has been more easily achieved as the last few decades have gone by. The greatest benefit has probably come from CNC control, where the axes of factory-floor machines have been driven by servo-mechanisms consisting of appropriate transducers, servomotors, and amplifiers with increasing sophistication of control (Bollinger and Duffie, 1988). This closed loop control of the machinery motions has probably had the biggest impact on the improvements in precision and accuracy over the last 50

TABLE 10.1 Products Manufactured with Different Levels of Precision (Courtesy of Taniguchi, 1994).

		Examples of Precision Manufactured Products		
	Tolerance band	Mechanical	Electronic	Optical
	200 μm	Normal domestic appliances, automotive fittings, etc.	General purpose electrical parts, e.g., switches, motors, and connectors	Camera, telescope, binocular bodies
Normal manufacturing	50 μm	General purpose mechanical parts for typewriters, engines, etc.	Transistors, diodes, magnetic heads for tape recorders	Camera shutters, lens holders for cameras and microscopes
	5 μm	Mechanical watch parts, machine tool bearings, gears, ballscrews, rotary compressor parts	Electrical relays, condensers, silicon wafers, TV color masks	Lenses, prism, optical fiber and connectors (multimode)
Precision manufacturing	0.5 μm	Ball and roller bearings, precision drawn wire, hydraulic servo-valves, aerostatic gyro bearings	Magnetic scales, CCD, quartz oscillators, magnetic memory bubbles, magnetron, IC line width, thin film pressure transducers, thermal printer heads, thin film head discs	Precision lenses, optical scales, IC exposure masks (photo, X-ray), laser mirrors, X-ray mirrors, elastic deflection mirrors, monomode optical fiber and connectors
	0.05 μm	Gauge blocks, diamond indentor top radius, microtome cutting edge radius, ultraprecision *X-Y* tables	IC memories, electronic video discs, LSI	Optical flats, precision Fresnel lenses, optical diffraction gratings, optical video discs
Ultraprecision manufacturing	0.005 μm		VLSI, super-lattice thin films	Ultraprecision diffraction gratings

Notes:
CCD—charge couple device
IC—integrated circuit
LSI—large scale integration
VLSI—very large scale integration

years (Figure 10.2). Important advances in machine tool stiffness have also occurred. Advances in this field have especially been the focus of the research work by Tlusty and colleagues (1999).

It is valuable to compare Figure 10.2 with Figure 10.3. In semiconductor manufacturing, the minimum line widths in today's semiconductor logic devices are typically 0.25 to 0.35 micron wide. These line widths have been decreasing rapidly since the

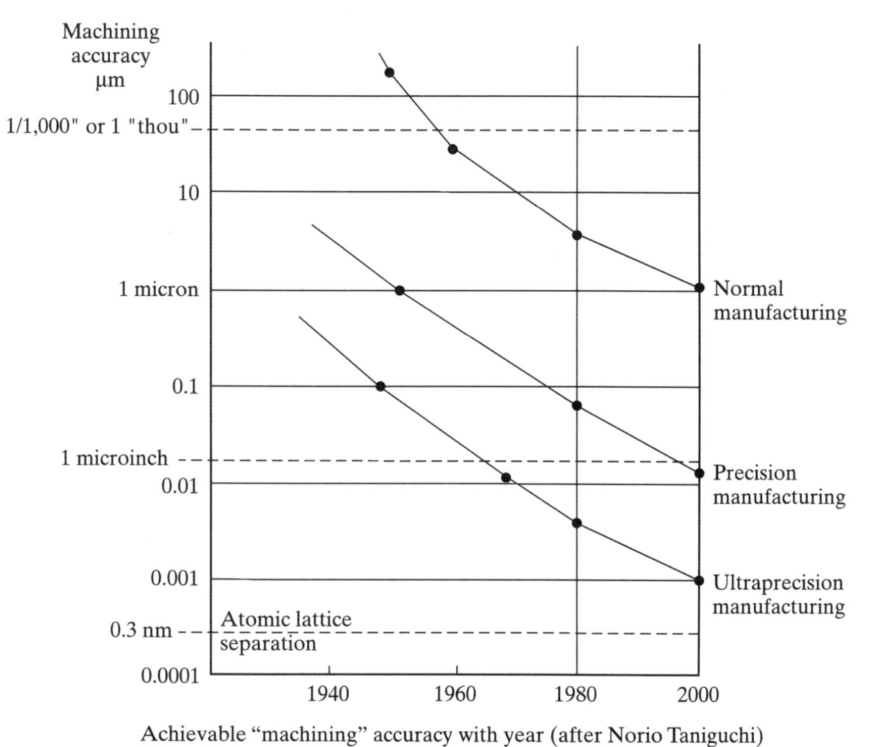

Figure 10.2 Variations over time in machining accuracy.

introduction of the integrated circuit around 1960. A large number of technological improvements in VLSI design, lithography techniques, deposition methods, and clean room practices have maintained the size reduction shown in Figure 10.3 over time.

The semiconductor industry is concerned that today's optical lithography techniques are not accurate enough to maintain the trend in Figure 10.3. Chapter 5 shows a diagram of the projection printing technique used during lithography. The UV light source is focused through a series of lenses. Any distortions in these lenses might cause aberrations in the lighting paths. Furthermore, when the minimum feature size is comparable with the wavelength of the light used in the exposure system, some diffraction of the UV rays limits the attainable resolution.[3] The dilemma being faced is clear: designers are demanding smaller transistors and circuits, but UV lithography is reaching its limits.

The natural limit of UV-lithography semiconductor manufacturing today is generally cited to be line widths of 0.13 to 0.18 micron (see Madden and Moore, 1998). This has prompted major research programs in advanced lithography, sponsored by alliances of semiconductor manufacturing companies (see Chapter 5). Intel, Lucent, and IBM each have their own alliance, each with its own preferred solution

[3]For reference: 0.35 down to 0.25 micron lines make use of UV systems with wavelengths of 365 down to 248 nanometers (deep UV). Line widths of 0.18 down to 0.13 require a 193-nanometer laser.

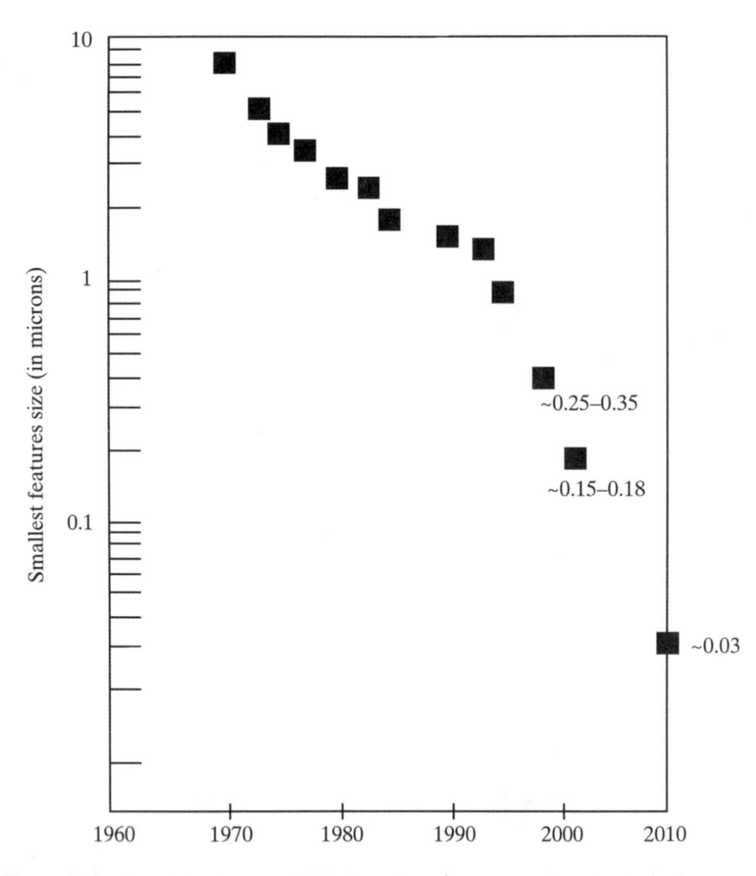

Figure 10.3 Trends in the precision of semiconductor transistor logic devices. Today typical values are 0.25 to 0.35 micron falling to 0.13 to 0.18 micron as the book goes to press. Research projects are aiming for below 0.1 micron and possibly 0.03 micron by the year 2010. More information on such research is given in the Semiconductor Association Roadmap (see SIA Semiconductor Industry Association **<http://www.semichips.org>**).

to the lithography challenge. One example is the alliance between Intel, three national laboratories, and semiconductor equipment suppliers (Peterson, 1997). Using extreme ultraviolet (EUV) lithography and magnetically levitated stages, the project has the goal of achieving line widths below 0.1 micron, perhaps eventually reaching 0.03 micron. While such technologies are not expected to be commercially available soon, they represent examples of how the trend in Figure 10.3 can continue to be satisfied.

In general, as might be expected, cost increases with desired accuracy and precision. For IC wafer fabrication, the ion implantation devices cost $1 to $2 million. Step-and-repeat lithography systems are several million dollars. The equipment for all aspects of IC manufacturing is extraordinarily expensive, leading to the projected costs of $2.5 billion fabs for the 300-mm wafers and 0.13 to 0.18 micron feature sizes. The fabrication of such machines in turn demands highly accurate machine tools and metrology equipment. Thus, while a standard 3-axis CNC milling machine might cost only $60,000 to $150,000 depending on size and performance, the machine tools for the ultraprecision machining of products listed at the bottom of Table 10.1 could cost an order of magnitude more, requiring air-conditioned rooms and frequent calibration by skilled technicians.

As mentioned in Chapter 2, Ayres and Miller (1983) provides the succinct definition of computer integrated manufacturing (CIM) as "the confluence of the supply elements (such as new computer technologies) and the demand elements (the consumer requirements of flexibility, quality, and variety)."

Many examples of this confluence are shown in Table 10.1. Improvements in one technology can be the *suppliers* to the *demands* of another complementary technology. In particular, the pressing demands of the semiconductor industry for narrower line widths spur all sorts of innovations in the machining of magnetically levitated tables (see Trumper et al., 1996), precision lenses, optical scales, and diffraction gratings shown on the bottom right of Table 10.1. Likewise, the complement is true: improved microprocessors have created vastly more precise factory-floor robots and machine tools.

In summary, the precision mechanical equipment allows the precision VLSI and optical equipment to be made, which in turn allows the mechanical equipment to be better controlled and even more precise. This is a spiral of increasing capability where all technologies drive each other to higher achievements.

How might this spiral be extended to a broader set of disciplines, especially biotechnology? Predicting the future is a dangerous game, especially in a textbook, but some synergies might include the following topics:

- The use of computers to decode the human genome and participate in biotechnology is a safe bet for an important area of synergy, where the "needs" of genetic discovery make "demands" on the computer techniques.
- Another safe bet for predicting important synergies is the creation of biosensors for monitoring and diagnosis. Such sensors combine biology, IC design, and IC microfabrication technologies, with a biological element inside a sensor. Biosensors work via (1) a biological molecular recognition element and (2) physical detectors such as optical devices, quartz crystals, and electrodes.
- Specifically, biosensors may well find their most successful applications in the synergy between silicon-based chips and molecular devices. Such a device embedded in the skin could monitor chromosome and cell health. Then, if small deleterious changes were detected, the sensor could essentially prompt the wearer to go to a doctor for some kind of health booster or medicine. To some extent, the wristwatch-like devices that contain insulin and an epidermal patch for penetration through the skin are examples of devices that are a synergy between mechanical design, electronics, and biotech. In principle, this synergy is an extension of wearable computing to biological implants and monitoring devices.

10.10 CONCLUSIONS TO THE LAYERING PRINCIPLE

1. In any company, *sustained growth* will depend on the day-to-day implementation of quality assurance, time-to-market, design aesthetics, and an awareness of new cross-disciplinary opportunities.
2. In idle moments, everyone dreams that he or she can invent and develop a fantastically new product and become "filthy rich." Nevertheless, the story behind most of today's new products shows many ups and downs over a long period before the product becomes an apparent "overnight success." The Palm Pilot is such a story. And even when a product is a clear market leader, such as Apple's original iconic desktop for the Macintosh, there is no guarantee that the product can stay ahead without attention to all the issues listed here.
3. The principle of layering has thus been advocated in this book so that today's students, destined to be the technology managers of the future, do not graduate believing that a "one-of-a-kind miracle solution" will lead to fame and fortune.

10.11 REFERENCES

Adams, J. L. 1974. *Conceptual blockbusting: A guide to better ideas.* San Francisco and London: Freeman.

Ayres, R. U., and S. M. Miller. 1983. *Robotics: Applications and social implications.* Cambridge, MA: Ballinger Press.

Berners-Lee, T. 1997. World-wide computer. *Communications of the ACM* 40 (2): 57–58.

Black, J. T. 1991. *The design of a factory with a future.* New York: McGraw-Hill.

Bollinger, J. G., and N. A. Duffie. 1988. *Computer control of machines and processes.* Reading, MA: Addison Wesley.

Borrus, M., and J. Zysman. 1997. Globalization with borders: The rise of Wintelism as the future of industrial competition. *Industry and Innovation* 4 (2). Also see Wintelism and the changing terms global competition: Prototype of the future. Work in progress from Berkeley Roundtable on International Economy (BRIE).

Cohen, S., and J. Zysman. 1987. *Manufacturing matters: The myth of the post industrial economy.* New York: Basic Books.

Cole, R. E. 1991. The quality revolution. *Production and Operations Management* 1 (1): 118–120.

Cole, R. E. 1999. *Managing quality fads: How American business learned to play the quality game.* New York and Oxford: Oxford University Press.

Curry, J., and M. Kenney. 1999. Beating the clock: Corporate responses to rapid changes in the PC industry. *California Management Review* 42 (1): 8–36.

Handfield, R. B., G. L. Ragatz, K. J. Petersen, and R. M. Monczka. 1999. Involving suppliers in new product development. *California Management Review* 42 (1): 59–82.

Koenig, D. T. 1997. Introducing new products. *Mechanical Engineering Magazine,* August, 70–72.

Leachman, R. C., and D. A. Hodges. 1996. Benchmarking semiconductor manufacturing. *IEEE Transactions on Semiconductor Manufacturing* 9 (2): 158–169.

Madden, A. P., and G. Moore. 1998. The lawgiver: An interview with Gordon Moore. *Red Herring Magazine,* April, 64–69.

Peterson, I. 1997. Fine lines for chips. *Science News* 152 (November 8): 302–303.

Plumb, J. H. 1965. *England in the eighteenth century.* Middlesex, U.K.: Penguin Books.

Rosenberg, N. 1967. *Perspectives on technology.* U.K.: Cambridge, England: Cambridge University Press.

Shapiro, C., and H. R. Varian. 1999. *Information rules.* Boston: Harvard Business School.

Spear, S., and H. K. Bowen. 1999. Decoding the DNA of the Toyota production system. *Harvard Business Review,* September/October, 97–106.

Symonds, M. 1999. The Net imperative. *Economist,* 26 June.

Taniguchi, N. 1994. Precision in manufacturing. *Precision Engineering* 16 (1): 5–12.

Tanzer, A. Warehouses that fly. *Forbes* October 18, 120–124.

Taylor, F. W. 1911. *Principles of scientific management.* New York: Harper & Bros.

Tlusty, G. 1999. *Manufacturing processes and equipment.* Upper Saddle River, NJ: Prentice Hall.

Trumper, D. L., W. Kim, and M. E. Williams. 1996. Design and analysis framework for linear permanent-magnet machines. *IEEE Transactions on Industry Applications* 32 (2): 371–379.

Waldo, J. 1999. The Jini architecture for network-centric computing. *Communication of the ACM* 42 (7): 76–82.

Wang, F-C., B. Richards, and P. K. Wright. 1996. A multidisciplinary concurrent design environment for consumer electronic product design. *Journal of Concurrent Engineering: Research and Applications* 4 (4) : 347–359.

10.12 BIBLIOGRAPHY

Bessant, J. 1991. *Managing advanced manufacturing technology: The challenge of the fifth wave.* Manchester, U.K.: NCC Blackwell.

Betz, F. 1993. *Strategic technology management.* New York: McGraw-Hill.

Busby, J. S. 1992. *The value of advanced manufacturing technology: How to assess the worth of computers in industry.* Oxford, U.K.: Butterworth-Heinemann.

Chacko, G. K. 1988. *Technology management: Applications to corporate markets and military missions.* New York: Praeger.

Compton, W. D. 1997. *Engineering management.* Upper Saddle River, NJ: Prentice Hall.

Dussauge, P., D. Hart, and B. Ramanantsoa. 1987. *Strategic technology management.* Chichester, U.K.: John Wiley & Sons.

Edosomwan, J. A., ed. 1989. *People and product management in manufacturing.* Amsterdam: Elsevier.

Edosomwan, J. A. 1990. *Integrating innovation and technology management.* New York: John Wiley & Sons.

Gattiker, U. E. 1990. *Technology management in organizations.* Newbury Park, CA: Sage Publications.

Gattiker, U. E., and L. Larwood, eds. 1998. *Managing technological development: Strategic and human resource issues.* Berlin: Walter deGruyter.

Gaynor, G. H. 1991. *Achieving the competitive edge through integrated technology management.* New York: McGraw-Hill.

Gerelle, E. G. R., and J. Stark. 1988. *Integrated manufacturing: Strategy, planning, and implementation.* New York: McGraw-Hill.

Lee, E. A., and D. G. Messerschmitt. 1999. A highest education in the year 2049. *Proceedings of the IEEE* 87 (9): 1685–1691.

Martin, M. J. C. 1994. *Managing innovation and entrepreneurship in technology-based firms.* New York: John Wiley & Sons.

Monger, R. F. 1988. *Mastering technology: A management framework for getting results.* New York: Macmillan.

Parsaei, H. R., and A. Mital, eds. 1992. *Economics of advanced manufacturing systems.* London: Chapman & Hall.

Parsaei, H. R., W. G. Sullivan, and T. R. Hanley, eds. 1992. *Economic and financial justification of advanced manufacturing technologies.* Amsterdam: Elsevier Science.

Paterson, M. L., and S. Lightman. 1993. *Accelerating innovation: Improving the process of product development.* New York: Van Nostrand Reinhold.

Rubenstein, A. H. 1989. *Managing technology in the decentralized firm.* New York: John Wiley & Sons.

Shapiro, H. J., and T. Cosenza. 1987. *Reviving industry in America: Japanese influences on manufacturing and the service sector.* Cambridge, MA: Ballinger.

Souder, W. E. 1987. *Managing new product innovations.* Lexington, MA: D.C. Heath and Company, Lexington Books.

Susman, G. I., ed. 1992. *Integrating design and manufacturing for competitive advantage.* New York: Oxford University Press.

Suzaki, K. 1987. *The new manufacturing challenge: Techniques for continuous improvement.* New York: Macmillan.

Szakonyi, R., ed. 1988. *Managing new product technology.* New York: American Management Association.

Szakonyi, R., ed. 1992. *Technology management: Case studies in innovation.* Boston: Auerbach.

Warner, M., W. Wobbe, and P. Brodner, eds. 1990. *New technology and manufacturing management: Strategic choices for flexible production systems.* Chichester, U.K.: John Wiley & Sons.

APPENDIX: A "WORKBOOK" OF IDEAS FOR PROJECTS, TOURS, AND BUSINESS PLANS

A.1 WHO WANTS TO BE AN ENTREPRENEUR?

A.1.1 Essential Attitudes: The Creative and Strategic Side

The key characteristics of a successful entrepreneur include:

- An ability to read market trends and consumers' wants, needs, or desires
- A blend of creativity in both product design and business operation
- The willingness to take financial risks while remaining emotionally balanced
- A passion for success combined with an overwhelming drive to succeed
- The desire to "change the world" rather than—in a blatant sense—"get rich"

A.1.2 Essential Attitudes: The Mundane Side

Also, there are mundane, day-to-day activities that any entrepreneur should consider:

- Mission statements
- Retreats that build communication and integrity while instilling a sense of urgency to satisfy the mission statement
- Performance parameters that are clear to all personnel
- Display boards to keep the organization focused on the mission and sales record
- Daily meetings as a "learning organization" to track deadlines
- Interactions between subproject groups via time lines and formal PERT charts
- Market scanning methods to track competitors
- The ability to circulate and share ideas without criticism
- Rewards for the know-how and problem-solving ability of people, acknowledging that no amount of expensive equipment and software can substitute for creativity
- Integrating knowledge on "downstream" manufacturing (internal and outsourced)
- Openness to outside ideas and emerging technologies

This combination of creative leaps and the mundane issues is the key to initial success and long-term growth.

A.2 PROJECTS ON PROTOTYPING AND BUSINESS

This appendix is a "workbook" for exploring and practicing entrepreneurship within an engineering context. A semester-long project on CAD, prototyping, and integrated manufacturing is discussed first. Engineers get excited when they are challenged to build "wild gizmos." Quite simply it is fun, and it involves both left-brain analysis and right-brain creativity. However, it is more instructive to include a more businesslike analysis of the potential market for the devices being built. These projects give a flavor of how to be entrepreneurial in a small start-up company. Also in large companies, people might well start off in careers with an engineering job title, but after only a few years, all sorts of management positions are likely to open up. Recall that a structured approach to product development was presented in the preface and in Chapter 2. The "clock face" diagram is intended as the "glue" that holds together the wide variety of processes presented in Chapters 3 through 10. For this reason it is reproduced again here as Figure A.1. Another view of concurrent

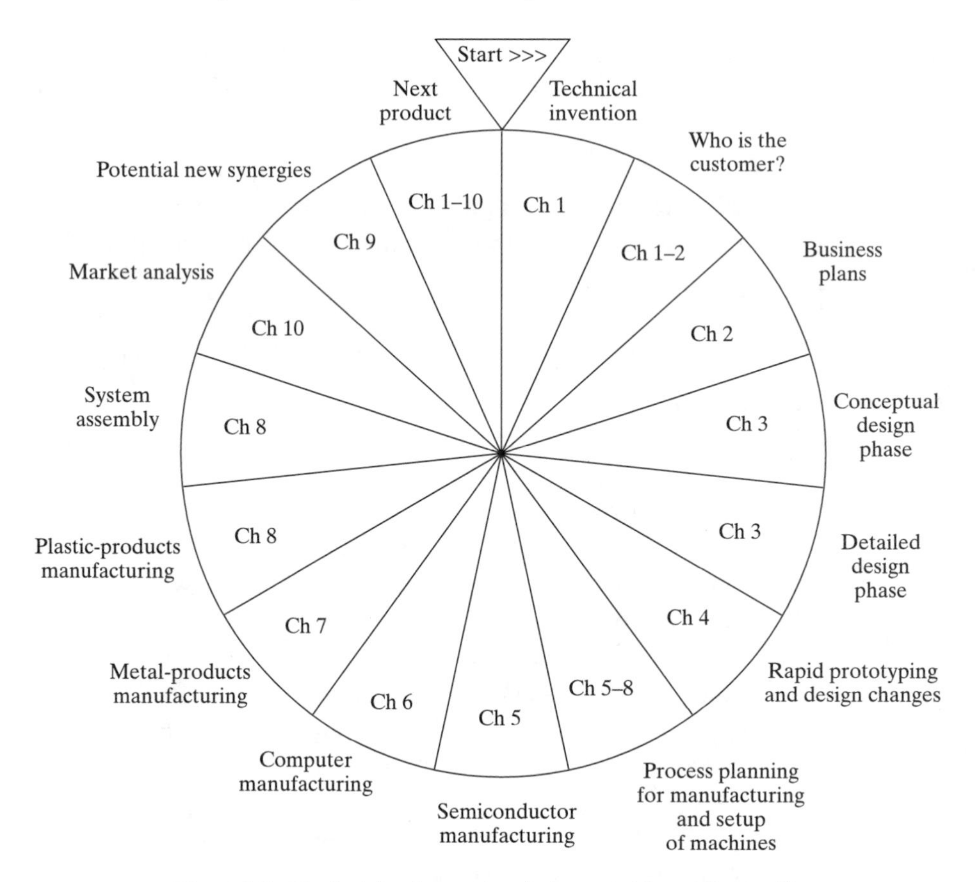

Figure A.1 Product development cycle (repeated from Chapter 2).

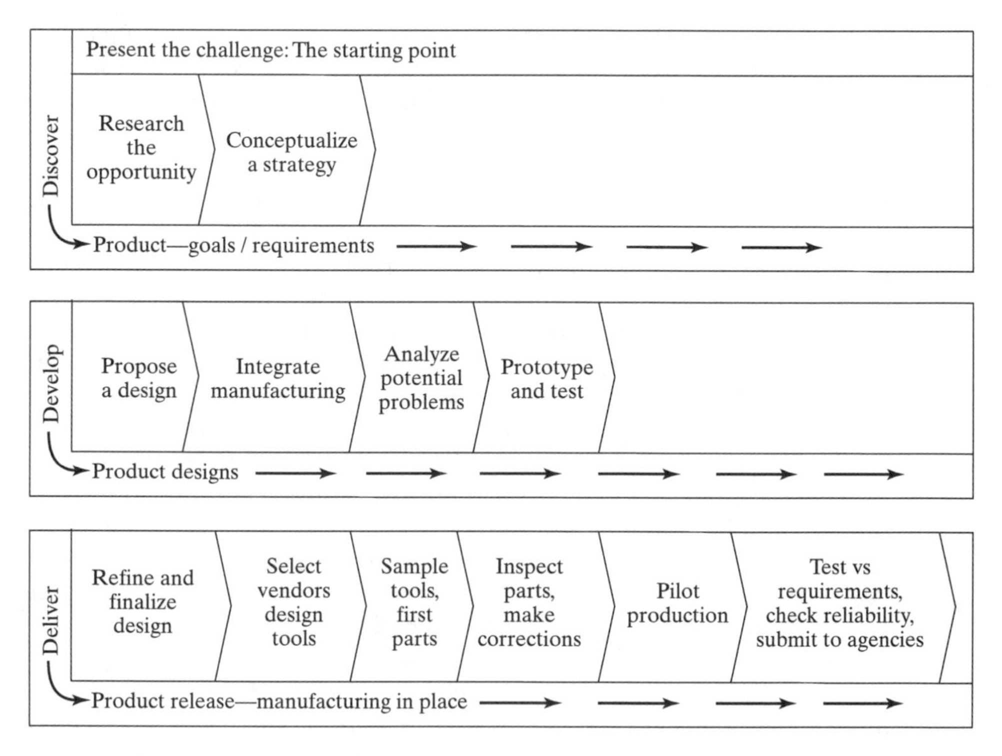

Figure A.2 The GE (1999) model of "discover, develop, and deliver" for a new product.

engineering and manufacturing is from the GE handbook (1999); it shows a linear overlapping time line. These also mention the outsourcing steps to subsuppliers (Figure A.2).

A.3 PROJECT STEPS AND MAKING PROGRESS[1]

These project-based case studies have progressed over a number of years. As might be expected, many mistakes, or at least errors in judgment, have been made along the way. Therefore, some of the factors that seem to make this semester-long project work the best are shared in the following section.

A.3.1 Forming the Design Team

Working in groups of four to six: Modern products are hybrid combinations of all engineering disciplines plus computer science. In the ideal situation, the groups also contain a mix of engineers and businesspeople. At the same time, however, there is a natural tendency to team up with friends or people from the same research lab. These

[1]These generic steps can be used to guide a semester-long class project. They also approximate the steps needed in industry to produce a design, a prototype, and a business plan. The business plan will be much more extensive in real life.

associations are encouraged because some very creative concepts that are well executed seem to come out of groups that have a natural flow or "chemistry." Whatever the group makeup, there is a benefit from understanding the sociology of developing group dynamics and dealing with the unstructured product development environments that seem to typify today's businesses.

Choosing a project with "edge appeal": Perhaps the most impressive observation over the years has been that group members blossom under challenge. At the beginning of the semester, there might be some anxiety about being stretched to build and control a radio-based, electromechanical product with an unusual application. However, by the end of the semester, remarkable working prototypes get built. Ultimately, it is best to build a device that exhibits that undefinable modern quality called *edge.* Such products get students to think creatively during design and to act energetically during the manufacturing operations. The final device can also be shown off to a friend or a potential employer with pride of ownership.

Providing a modest budget: A fixed budget allows the groups to go out and buy supplies at a local model shop or electronics store. The maximum that has been used is $500 per group for projects involving radios and cameras. An amount as low as $100 per group works fine for projects with less electronics. A fixed budget has the benefit of constraining the scope of the project and is of course a lesson for life.

Crossing the chasm (see Moore, 1995): To understand the critical aspects of the product development process, it is best to focus on the beginning of the *modified market adoption curve* shown in Figure 2.3. Predicting the first market niche for a product helps to sharpen the design intent and the manufactured complexity.

A.3.2 Conceptual Design

Creating a conceptual design: The groups develop best by producing, in the first few weeks of the semester, a conceptual design on 22 inch $\times$ 28 inch artist's paper using pencil sketches. This serves as the front end of the project and its associated business venture. It might seem that the pencil-and-paper sketch is old-fashioned at first, but it is the only way to be really creative and to get buy-in from all group members.

A.3.3 Detail Design

Using any preferred CAD environment: In the middle part of the semester, the groups focus in the best by executing a *detailed design* and presenting a preliminary, though brief, business plan. This business plan can be a short version of the one shown in Section A.4. Shortly after this deadline, the CAD drawings should be turned into ".STL" files and sent to the local SFF machines. Alternatively, preparations for milling and assembly should be made.

A.3.4 Prototyping

Fabricating a real device: Becoming competent in some basic SFF and other fabrication practices seems to be a key feature in the success of the course.

A.3.5 Trade Show

Explaining the device to others: At the end of the semester, the groups enjoy presenting posters and demos in a "trade show environment," where the complete class and visitors can mill around and ask questions. It also gives practice in delivering the "25-word punchy explanation" of what the device does and how it might be sold.

A.3.6 Business Plan

Analyzing mass-production methods and the potential market: In the end, the longer term market response is the acid test. Therefore, it is instructive to write a modest business plan, not as detailed as in a standard MBA program but a plan that is reasonable from a mass-production price point. A key aspect is trying to estimate the batch run needed to amortize the costs of plastic injection dies and special chips or printed circuit boards. Further details are in the next main section.

A.4 OUTLINE OF A SHORT BUSINESS PLAN

A.4.1 Cover Page

- Name of the product and group members
- *Mission statement* (a succinct statement less than 25 words)
- A scenario of how the product will be used
- The "head bowling pin," or first market niche (Moore, 1995)
- How much it will cost when sold at Radio Shack/Sharper Image (10 words)
- How much this means for the final cost immediately after manufacturing (10 words) (Typically it is at least a 1 to 4 ratio between manufacture and retail.)

A.4.2 Additional 10 Pages

1. Description of the product and how it works (two pages).
2. Intended market: who will use it, and where will it be used (two pages)?
3. How much has the product cost so far to get it to the prototype (two pages)?
   a. Material cost =
   b. Prototype cost =
   c. Person-hours of work assuming 80K annual salaries × 4 or 5 in group =
   d. Overhead costs assuming a 1,000-square-foot office space at rent of $2.50 per square foot, electricity, phone, and so forth.
   e. Three NT workstations, networking, CAD license, printer, and other peripherals
4. How much will it cost for first-year operations (two pages)?
   a. Start-up costs and advertising
   b. Equipment, legal fees, accounting services, patents, and the like
   c. Payroll
5. How will the company work (two pages)?
   a. Will it need inexpensive overseas manufacturing?
   b. What is the structure of the company?

c. Who are the principals?
d. What are the markets?
e. Who is the competition? Do they have a "barrier to entry"? Will they steal the idea?
f. How long is the development time line?

A.4.3 Worst, Likely, and Best Case Scenarios (Three Tables + Text)

A key aspect of a business plan is the expected return over a five-year period. The information shown in Table 2.4 should be considered. Magrab (1997) provides additional information for a variety of scenarios—particularly those involving R&D overruns. A useful exercise is to construct three tables of this kind, with a rationalization for each: worst case scenario, most likely scenario, and best-case scenario.

A bank lender or venture capitalist will look at this information immediately after he or she understands the general mission statement for the product and the pedigree of the company founders.

A.5 PROJECT SELECTION

The preceding discussions emphasize that in today's engineering careers, a hybrid set of skills that include mechanical, electrical, materials, and computer science is most valuable. Even traditional industries are impacted by embedded computer control. Given the fact that a large number of engineers and businesspeople are destined for this hybrid engineering environment, the projects in recent years have focused on the consumer electronics market.

Project-based case studies on (a) mouse-input devices for simple virtual reality systems, (b) telepresence devices, (c) miniature radios, and (d) GPS devices are given on the next few pages. These seem to have been successful topics and could be extended in other classes. It is worth expanding on some logistics here:

- In some years, after "putting the word out," another research team on campus has been identified that would like to have this class develop prototypes for them. In these situations, the external research team is usually very willing to supply some prepackaged internal electronics for the prototypes. It is also useful to invite this outside research team into the class to pose as a "client" for whom the devices are being built. From there on, additional funds enable the fabrication of a professional-looking prototype that can be marketed.
- In other years, the starting point has been less well defined, and the student groups have built their own electronics. These more open-ended case studies often needed a larger budget of around $500 per group. And in these situations it is preferred to have a mix of engineering types in each group, with each one inclusive of some electrical engineers.
- In other years, a distinctly more conservative approach has been taken, and the students have been provided with off-the-shelf consumer devices that they "deconstruct" and repackage with enhancements into new devices.

What could be other project-based case studies in the future? Bearing in mind that the groups seem to blossom when challenged by products with "edge," it's always worth choosing something that is just emerging in research laboratories that might soon go mainstream and be configured as a consumer item.

In a 1995 article, Poppel and Toole proposed that the following topics are on the "bleeding edge of technology": ATM, desktop 3-D graphics, nanotechnology, object-oriented database management systems, personal digital assistants (PDAs), voice recognition, and virtual reality. At the time of this writing, several years later, these are certainly topics worthy of consideration.

A.6 PROJECT 1: ENHANCED MOUSE-INPUT DEVICES

This first project-based case study resulted in some prototypes such as Figure 3.6. These were enhanced mouse-input devices for virtual reality software, video games, and other applications. To begin, a small motion sensor, mounted on a printed circuit board, was supplied to each group. These prefabricated packages were supplied from an industrially sponsored research team in the electrical engineering department. The product possibilities that were first suggested but not limited to were: 3-D mouse inputs, joystick inputs, and head mounted inputs. The industries that were first targeted included video game companies, engineering training companies, and animation, motion capture companies.

As the class progressed during the semester, and the various groups thought about the real marketing needs, other devices began to emerge. These used the same basic technology of the sensor mounted on the small board. The emerging devices included:

A swing-measuring golf club: This used an embedded sensor package mounted near the handle of the club that could measure the directions of the user's swing.

An acceleration/speed monitoring device: This was a dashboard mounted device that could measure the acceleration of a drag-strip car, for which the students did market research at a local raceway.

A wrist mounted direction controller: This was a control device for a wheelchair.

All the above devices worked reasonably well at the concept level. The two devices discussed next represent the two extremes from that year's class. The first one more or less worked, but the response time was slow. And in general, the rest of the class doubted that a market could be established for the product. On the other hand the second project was very successful, and several more prototypes were fabricated by a local machinist after the class ended. These days, many people dream of being a rich entrepreneur. Thus, it is worth emphasizing that one of the students from the electrical engineering research team graduated and is now one of the main partners of a company that focuses on the last of the projects: the machine vibration measurement device <**http://www.xbow.com**>.

Intelligent dancing shoes (a ludicrous concept): In this device, a sensor was embedded in dance shoes that would twitch the dancer's feet correctly for the desired steps in ballroom dancing. The sensors were connected to a program that was preprogrammed to the correct foot steps.

A vibration sensor for machinery (a sane concept): This was a simple black box containing the sensor package that could be attached to machine tools and other industrial machines to measure in-use vibrations.

A.7 PROJECT 2: BLIMP-CAMS, CART-CAMS, AND TELEPRESENCE DEVICES

In this second project-based case study, products for the emerging market of telepresence devices were developed.[2] The possible products described in this section are just emerging in the marketplace. Thus they are shown in the left corner of Figure 2.3. Prototype versions of a blimp-cam can be seen at the following URL: **http://vive.cs.berkeley.edu/blimp.** These zeppelin-like devices are now being used in museums and in art galleries. They can be remotely controlled from a Website to navigate around an exhibit and enjoy a full 360-degree view of an art object or display.

The class groups were each given a budget of $500 and asked to extend the basic theme of remote presence or telepresence to other environments. The groups were challenged by saying that any variation of "X-cam" that could be demonstrated, that was interesting, and that had market appeal was encouraged. The devices included:

- *Cart-cams:* These were able to climb stairs and navigate around a real estate agent's show home. The perceived end use was that potential clients could navigate themselves around the house from a remote Web browser and see the room layout.
- *A rail-cam:* This device was designed to be installed upside down on the ceiling of a retail store. Here, the intended use was to allow the remotely located store owner to view the way in which the merchandising displays were set up to best impress the buying public.
- *The Internet doorbell:* In this prototype, students built a fully functioning "intelligent doorbell" that was linked into a remote Website. The scenario was as follows: the UPS truck comes to deliver a package, and the owner is at work. Instead of missing the delivery, the doorbell activates the owner's Website and he or she talks over the Internet to the delivery person. Instructions are given on where to leave the package, and a signature authorization is given.
- *Videoconferencing systems:* New display screens were created that would rotate and focus on the person speaking rather than a general wide-angle view.
- *A train-cam that was mounted on the front of a model train:* The camera output was viewable by any Web browser, and the view from the model train was enjoyed by a remote user. This project was fun to watch. As the train went in

[2]Professor John Canny (2000) and his students were key motivators of that semester's project.

and out of tunnels and through the model train stations, the images were quite captivating. It seems that model train builders and enthusiasts are very interested in sharing each other's layouts and designs over the Web.

- *A smart-cart:* This was a modified shopping cart that could be used in a supermarket to log the products being selected, calculate the running total, and prepare the credit card information in advance.
- *A laser mouse:* This device could be used for conference room demonstrations of material on a Website. Today, the Website demonstrator usually has to sit at a terminal with an attached mouse. This newer device put the operator up in the podium at the display board while still providing the dual ability to operate the Web.

A.8 PROJECT 3: MINIATURE RADIOS FOR CONSUMER ELECTRONICS

In these project-based case studies, the emerging market of miniaturized radios was used as the basis for prototyping new products.[3]

In the first year of this project, the class groups were given a budget of $500 and asked to respond to a consumer world with "radios everywhere." Groups were organized to contain both electrical and mechanical engineers. They were asked to use off-the-shelf radio components to build devices. Examples included:

- *A night escort system:* This was a pagerlike personal security system. The group designed it for initial use on a large university campus. The general scenario was that, especially after evening lectures, a student could check out and rent the small device, which would be worn on the way to a darkened parking structure. In the event of a problem, it would transmit a signal to a receiver in the police station. The local position of the transmitter could be approximately located on a large campus because many of them now have existing network infrastructures. For example, the presence of the Metricom Richochet RF communication network could allow the night escort to be integrated with the existing receiving units that are placed about 250 meters apart. The transmitted signal to them could allow the position of the wearer to be analyzed and approximately located. This group created a lightweight package using stereolithography, which could be attached to a belt or key chain. A simple button or pull-pin actuation mechanism was designed for easy operation. While initially designed for a campus, the extension to commercial parking lots, malls, and hotels is clearly feasible. Such devices are now entering the market.
- *A multimedia personal protection and monitoring biofeedback helmet:* This device was designed and fabricated to be worn underneath, and yet integrated into, a firefighter's protective helmet. The group members enjoyed deconstructing one of the off-the-shelf devices from Radio Shack. They then fabricated

[3]Professor Robert Brodersen, Professor Jan Rabaey, Brian Richards, Susan Mellers, and their students at the Berkeley Wireless Research Center were key motivators of these projects.

their own helmet casing that incorporated the two-way radio, a video camera, and environment monitoring capabilities (such as temperature, smoke, and light detection; in a more advanced unit it might also be possible to measure the wearer's pulse and body reactions). The group envisioned that the individual firefighters would be in much better communication with their captain. Also, the captains and the central coordinators of a fire would be able to follow each firefighter's precise location and local fire conditions, thereby orchestrating events more effectively.

- *A wireless system for keeping track of personal effects:* This project responded well to the idea that miniature radios will be incredibly cheap and embedded in countless everyday devices. Given this assumption, the group assumed that people will be able to execute simple communications with those everyday items that are easily misplaced: car keys, wallets, and purses being the obvious candidates. The group thus provided a simple pagerlike device that could be worn on a belt, with prearranged communication channels that would be a wireless connection to the whereabouts of these everyday items. Similar devices could also be used by parents to keep track of young children, especially in crowded environments such as department stores, ballparks, and ski resorts.
- *Outdoor radio collar device:* This product was a two-way radio device that was attached to the user's jacket collar. It allowed clear communication without restricting movement or activity. The unit's operating range of up to 2 miles was expected to be sufficient for most hiking, skiing, or climbing activities. The device was also designed for total hands-free use by attaching an ear button speaker and an in-line voice activated microphone. The unit was designed to easily attach to the collar of any conventional jacket. Eventually it was imagined that a fashionable jacket might be designed with an integrated unit sewn into the collar lining.
- *A wireless communication device based on the Infopad:* This group, consisting of a mix of electrical engineers and mechanical engineers, built a more local version of the Infopad. Their "in-touch" communicator worked in a local corporate or academic environment. It offered the same features as a cellular phone combined with access to the Internet. Thus it was possible to keep up with one's e-mail and messages during the course of a day when meetings and activities would be on the go rather than at a fixed desk location. Communications were digital and used the public radio bandwidth. The group also included a video display that worked to a limited degree in the time available. In the longer term this would be used to display campus maps, schedules, and announcements.

In a second year on the same topic, each group was provided with a Motorola Talkabout Radio. This updated version of a "walkie-talkie" has become a popular consumer product. For example, family members can keep in touch with each other—over a 2-mile radius—while bike riding or shopping.

Groups were asked to pull the radio apart and reuse the "insides" as the basis for their own consumer product. This made the assignment somewhat easier than the previous devices, where the groups had to organize much of their own circuitry. Actually, several groups still preferred more of a challenge and ignored the assigned radio in favor of their own devices.

- *One-way radio receiver:* This product was designed as a one-way radio receiver for sports teams, instructional groups, or military organizations. It deliberately provided communication on a one-way basis to allow for quick instruction and direction without interference. It had a range of approximately 2 miles, was small in size ($3 \times 2 \times 0.75$ inches), and was very light in weight. The product featured adjustable channels, so that even with multiple instructors in an area, such as at a ski school, each instructor could operate on a separate channel. It was powered by two AA batteries with an operational life of over 30 hours. Two brackets located on each side of the receiver made it possible to attach the product on an arm, around the neck, or in a belt. Stereolithography (SLA) was used for the packaging.
- *Warehouse telecommunication device:* This product was designed as an easy-to-use communication device that could be integrated with a standard industrial back brace. The product was aimed at people working in a lumber yard, factory, or parcel delivery service. The unit required only one finger to operate and also had the potential for a hands-free model in the future. All components were contained within two stereolithography housings that were intended to be worn on the waist and shoulder strap of a lumbar support belt. This setup eliminated the need for having a broadcast/intercom system in a facility, which many workers find annoying and distracting.
- *Hands-free, waterproof, two-way communication device:* This device was designed to allow active communication between a student and the instructor during waterskiing. The system was therefore designed to be water-resistant. It included a radio, a voice-activated headset, a voice-activated microphone, and an ear-bud speaker attached to an ear hook. The unit came with a flotation case, so that if it was dropped into open water, the system would float. Fused deposition modeling (FDM) was used for the packaging.
- *Child monitoring product:* This provided two-way communication between a parent and a child. The communication device was embedded in a stereolithography package in the lining of a teddy bear, with access to the unit through a zipper located in the back of the bear. It was designed in such a way that all the user functions were facing the back, so that the parent could change the frequency, channel, and codes without actually taking the stereolithography package out of the lining. Voice activation was an important feature of this product.
- *Portable video and audio conferencing device for automotive assembly lines:* This product was a portable video and audio conferencing device for assessing live-action situations on an automotive assembly line. It allowed real-time interaction between people from any distance. It was a wireless product utilizing the Internet and off-the-shelf hardware and software. At the "trade show" the following scenario was demonstrated: A production engineer in a factory setting wears the headset, which is connected to his or her notebook

computer. The headset with the small video camera and microphone allows video input to the notebook computer and also audio input/output. The notebook computer holds the software to connect to the Internet and the videoconferencing software. Actually, the camera was detachable and could be held inside a machine tool to look at a particular problem. An engineer at the home site or headquarters could then receive video and audio over his or her computer and carry out a dialogue on how to fix assembly problems at the factory in another country. A small stereolithography package was built for the camera attachment and the wires into a standard headset. A similar product is at <**www.xybernaut.com**>.

A.9 PROJECT 4: GPS-BASED CONSUMER PRODUCTS

In these projects, the emerging market of global positioning systems (GPS) was used as the basis for prototyping new products. As in previous years, each group was also provided with a Motorola Talkabout Radio. However, there was no obligation to stay with this format, and other units were developed. For example, one group was interested in developing MP3 devices, and another was interested in "energy scavenging" to make such portable units less reliant on standard batteries.

- *The "Voyager" was an integrated bicycle computer and global positioning system (GPS) unit:* It mounted on standard bicycle handlebars via two custom brackets. An ergonomically shaped thermoplastic housing contained the two separate circuit boards and liquid crystal display (LCD) screens. The shape of the housing minimized interference with the rider's hands and allowed information to be read easily. The bicycle computer circuit board continually read the wheel speed via a magnetic sensor that was mounted on the bicycle fork. The wheel speed data were converted into instantaneous velocity by the bicycle computer. Based on this data, the LCD could display instantaneous speed, average speed, riding time, and total elapsed time. The bicycle computer also computed elapsed trip distance and continuous mileage. The actual display was chosen by the user, based on a pair of buttons on the face of the device. Concurrently, the GPS circuit board read data from the GPS satellite constellation. Based on these data, the LCD displayed a variety of information, including position (latitude, longitude, and elevation), average speed, bearing, and time of day. The GPS also stored way-point and route data. By means of an embedded memory chip, the combined unit stored and displayed trip data to a PC via an output port. The CAD files for the product are shown in Figures A.3 and A.4.
- *The "Inlande" was targeted at land surveyors:* It integrated a GPS unit with an organizer. Coordinates of a location were established using the GPS and stored in the organizer. The large amount of memory of the organizer allowed for storage of tens of thousands of data points, which could be uploaded to a computer to plot them on a map. The unit came equipped with a belt clip to carry it around easily, and it was rugged enough to handle any terrain and climate.

Figure A.3 Solid model of the Voyager bicycle GPS system.

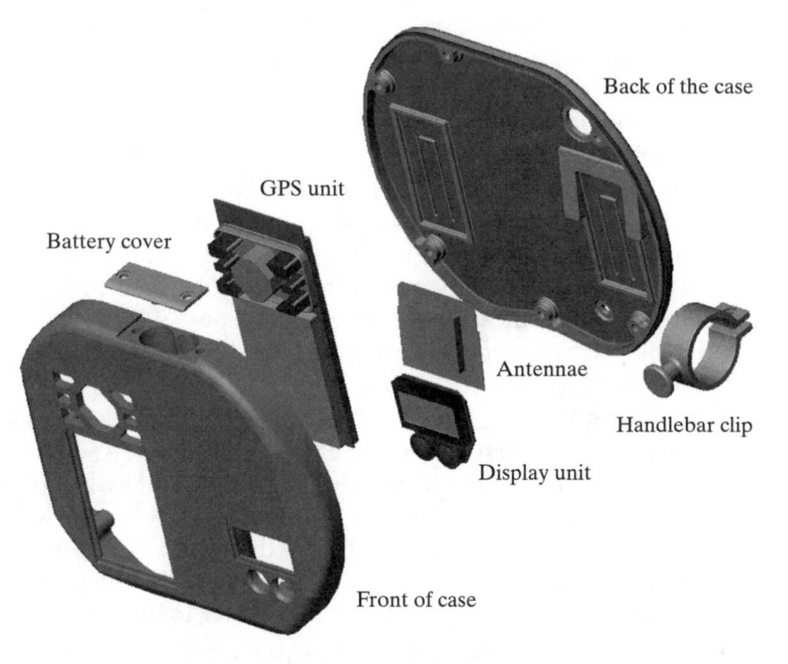

Figure A.4 Exploded view of solid model.

- *The "Child Locator" was a personal device that could detect the location of its bearer and transmit this information to a central station:* The intended use was for a large crowded public area such as an amusement park, where parents could locate a lost child by referring themselves to a base station. The system made use of existing GPS and radio transmitting technology and included two main units: the first unit was the portable module, and the second was a central base station that received the position. Individual devices continuously emitted their coordinates with a unique channel and code.

- *The "MP3 Music Player" was designed for consumers who wish to use their computers to play music, but do not want to be tied to a monitor or keyboard:* It was designed to control and display real-time song information in conjunction with Nullsoft's Winamp MP3 player. The product consisted of a display module with an integrated infrared receiver and software to connect to the Winamp. The display module provided song information and player status on a high-contrast liquid crystal display. When used in conjunction with a remote control, the IR receiver allowed the user to switch songs, load play lists, and modify Winamp settings such as equalizer positions and volume. The display module communicated over serial ports. This feature gave a user freedom to mount the unit anywhere the cables could be routed, typically up to 40 feet.
- *The "Solarcator" combined the GPS with the battery-extending capability of solar power:* The front of the unit was a liquid crystal display with a keypad for user input. The back of the unit held a solar panel that charged an internal set of rechargeable alkalines. When placed in direct sunlight, the panel charged the internal batteries. Using nonrechargeable alkaline batteries, the unit could run for approximately 15 hours. With the solar panel and rechargeable alkalines, the battery life could be theoretically extended by a factor of 30 to 450 hours in optimal conditions.
- *The "nTouch" was a wireless networking module for the Handspring Visor PDA—"Palm Pilot" clone—based on the industry standard IEEE 802.11 communication protocol:* The module allowed the Visor to communicate at speeds of 2 Mbps within a 100-meter range of a hub. Enabled Visors could also communicate directly with other Visors, or any 802.11 device such as Apple's iBook. Communication took place over unregulated radio frequencies, thus there were no connect or usage fees.
- *The "GPS Xtreme 2K" (GX2K) was a GPS module that also interfaced with the Handspring PDA:* In this product the Handspring Visor utilized a standardized expansion slot and the Springboard development board to interface and send geolocation information to the PDA platform. Software was written enabling the PDA to poll the GPS unit, display the geolocation information on the PDA display, and perform other miscellaneous control functions. The GPS Xtreme 2K contained its own power supply in order to process GPS signals.
- *The "NAVTalk" combined two-way radio and the GPS in a single, lightweight, easy-to-use device:* It was developed for outdoor use by hikers and campers. However, firefighters and some military units could also be potentially large customers. Also, it was perceived as a safer solution for rescue missions because of its reliable GPS signal reception and short-range radio.
- *The "Pulse" was an integrated heart monitor and walkie-talkie:* It was designed for athletes in training and elderly people needing health monitoring. The unit was capable of monitoring an individual's heart rate. The heart rate was measured using a pulse sensor clip that could be attached to either an ear lobe or a

finger. It could then transmit a radio signal to another walkie-talkie when an abnormal biorhythm was detected outside the user-defined operating range.

Technology projections indicate that within the next few years it will be entirely feasible to use CMOS technology to fully install a radio on a chip (see Rudell et al., 1997; Chien, 2000). Devices that truly emulate Dick Tracy's famous wristwatch-size, two-way radio will then be consumer items. It will be interesting to pursue this opportunity in future projects.

A.10 CONSULTING PROJECTS[4]

The projects described in the last four sections were largely successful at the concept level; namely, in the final "trade show," the electromechanical package functioned satisfactorily. In fact, they usually functioned in a rather fragile way, and the functionality did not last much beyond the demo. And here and there a few "kluges" were needed. Despite these deficiencies, the concept was well demonstrated.

It is of course unreasonable to expect a highly robust system ready for market testing. A reasonable goal is a device that more or less works and shows basic functionality, backed up with a modest business plan. This plan should include how the product would be developed over a more substantial period of time and with more budget.

Another method of running the course is the "consulting project," where teams of three or four students work together through the semester as "consultants" for a local production facility. The consulting project allows student teams to get hands-on experience with working production systems. It functions well in a class where the fabrication of a full prototype or model is not as convenient—perhaps because of limited staff support or laboratory facilities.

Each team develops a report on its client, summarizing the product, materials, and current production methods. Teams then identify and quantify a specific problem related to quality or throughput, measure and analyze relevant data, research alternative production methods involving automation, and propose solutions with cost analysis.

The client facility can be anything from a cement mixing plant to a high-volume coffee house. The facility must be involved in the production at volumes where issues of production are important.

Each team gives presentations on the progress of its consulting project during the semester. These are in the form of timed 10-minute Powerpoint (or Internet) presentations given by the team to the class in a computer lab. This allows the instructor and the class to provide feedback at each step. A 10-page report is required after the final oral presentation.

In Phase I, the team finds an appropriate facility and contact person. Finding a friendly and enthusiastic contact person makes the project much more enjoyable. It is important to clarify that this is an educational project and that there are no guarantees

[4]Professor Kenneth Goldberg has contributed the written material in the next few pages.

about the results. Often companies are concerned about confidentiality of their production methods and levels. One option is to keep the company identity confidential in all presentations (see examples 11 and 12 in the following section). Also, teams can offer to scale production numbers or to suppress confidential details, but the company must agree with the group to present enough information to describe the project. The facility should be close enough to permit at least three site visits by the team. Also, Phase I includes a report on how team responsibilities will be divided.

In Phase II, the team prepares a presentation on the client, history, product, and the current methods of production. Photographs and plant layout plans with flowcharts are very helpful. Teams should also use the Internet and the library to research similar production facilities and give statistics about their clients' market and industry in general.

In Phase III, the team identifies and quantifies a specific research problem related to quality or throughput. Clearly it is impossible to analyze everything in the facility. Often this is related to current methods that are labor intensive and could be improved with automation. The objective could be to reduce product variance, reduce waste, reduce production cost, improve throughput, modify product design, and so forth. See the examples in the following section. Where appropriate, teams should describe existing equipment models and get performance characteristics from the vendors. The key is to identify specific numerical metrics to measure and compare performance and show how these relate to cost and profit. Teams can gather data from the facility using direct measurements (e.g., with stopwatches or rulers) or data provided by the client. Teams can analyze these data statistically and present results graphically. Computer simulation can be useful during Phases III and IV.

In Phase IV, groups should use the Internet and/or phone book to find alternative solutions, usually involving automation (or new models of existing machinery). One very useful resource for finding industrial vendors is the online Thomas Register. Students can then quantify predicted improvements and provide a cost analysis, taking into account factors such as labor, increases in market share, decreased liability risks, and the like. This final report is presented to the class and to the client, whose feedback is included in the written report.

A.10.1 Examples from Recent Consulting Projects

Examples from recent consulting projects are as follows:

1. Flour delivery methods for Diamond Team Noodle Company
2. Facility layout analysis for Berkeley Farms ice cream plant
3. Automation of molding and decorating for San Francisco Chocolate Co.
4. Statistical analysis of cocoa bean weights for WWW Chocolate Co.
5. Simulation and cycle time analysis for Hauser Window Shade Company
6. Analysis of meter refurbishing process for PG&E
7. Optimal choice of stereolithography build orientation for rapid prototyping
8. Scheduling of grinding and polishing pads for Komag Magnetic Disk Co.
9. Analysis of automated packaging machine for Procter & Gamble
10. Efficiency analysis of rod breakdown machines for the Saudi Cable Co.

11. Workstation assembly sequence design for a network computer company
12. Automated packaging analysis for pharmaceutical products
13. Bottle conveyor failure analysis for Pyramid Brewing Co.
14. Waste wax analysis for J and S Candles
15. Solar wafer load time analysis for SunPower, Inc.
16. Mail-order processing analysis for Peet's Coffee Co.
17. Torque data analysis for GM-Toyota NUMMI auto plant

A.11 OVERVIEW OF POSSIBLE FACTORY TOURS

Factory tours may be organized to reflect the technical chapters of the book:

CAD/CAM
Rapid prototyping and SFF
Semiconductor manufacturing
Computer manufacturing
Metal-products manufacturing
Plastic-products manufacturing
Biotechnology manufacturing

These can be analyzed in two-week segments. The first lecture of a segment can deal with the essential technical and economic factors of a technology. In the subsequent lectures it is most instructive to visit a factory and listen to one of its senior managers review the history, future, and economic competitiveness of the company. During the second week, information about the company can be gathered concerning the technical capabilities and economic strengths or weaknesses. The last lectures can consolidate the technical aspects of the field.

A.12 RATIONALE

The topic of manufacturing covers a very broad field, much broader than the seven basic areas listed in the preceding section. The rationale for choosing those particular industries is now briefly reviewed.

CAD and rapid prototyping represent the front end of most manufacturing concerns and launch the new product, or a redesigned old product, into the development cycle. Larger companies may well have their own in-house rapid prototyping facilities. On the other hand, many of these new processes are best outsourced to a specialized bureau. Specialized technicians in these smaller service-oriented companies can usually "tweak" the rapid prototyping machines to get the best accuracies out of them. A visit to such a company is an excellent experience. The list of Websites—for example, <**www.metalcast.com**>—given at the end of Chaper 4 provides the contact information for a number of rapid prototyping services both in the United States and abroad.

Semiconductor manufacturing (and semiconductor equipment manufacturing) remains a major economy in the United States, even though many of the new "fabs" are in Asia or restricted to U.S. states that have provided the leading IC manufacturers with handsome tax benefits, well-funded infrastructures, cheap natural resources, and an adequate stream of well-trained engineers to run the plants. A visit to any kind of fab is an impressive demonstration of precision manufacturing.

Computer manufacturing covers the whole endeavor from chips to board assembly to case assembly. Ideally a class can visit some kind of local production line to see the detailed steps needed to assemble one of today's consumer electronic or computer components. These production lines might well include layout areas, detailed assembly, reliability testing, and quality assurance methods. Many companies that today provide the outsourcing for the main computer companies have also established design for manufacturability teams and integrated quality assurance programs (see, for example, <**www.solectron.com**>).

Metal-products and plastic-products manufacturing are the next areas of interest. It is especially instructive for students to visit machine shops that specialize in products such as high-accuracy molds for plastic injection molding, ultraprecise stepping equipment for lithography, disc-drive assembly systems, and biotechnology equipment. Many rapid prototyping shops and some machine shops also carry out plastic injection molding. A tour can often include several processes at once.

Biotechnology is an important area from a manufacturing perspective. However, it is new enough that most university courses have not yet included some of the basics in the engineering curriculum. This book attempts to rectify this omission with the material in Chapter 9. However, a visit to some kind of biotech firm is even better. Biotech companies tend to be clustered around the San Francisco Bay Area, Boston, and Baltimore. To get a sense of biotech manufacturing, students in other parts of the country may be obliged to visit their local microbrewery. From a class perspective, an interesting focus can be the scale-up from successful "test tube" research to efficient flexible production systems. Sensors and control-system software are the topics that need to be more widely introduced into this industry.

A.13 FACTORY-TOUR CASE STUDY WRITE-UP

For the assignments that follow the factory visits, the same project groups should prepare a 3,000- to 5,000-word case study (not a "case" in the traditional business school use of the term). It is reasonable to complete the written assignment two weeks after a visit. The suggested format for the case studies is given in the next section. The tone of the material should be of the type seen in consulting reports; that is, formal and factual. One can be "direct," but one should always be polite. As the semester proceeds, it is a good idea to photocopy the best case studies and circulate them to the rest of the class. Also, the best reports can be passed on to the companies.

About 90% of the needed information comes from the tour and the talk, and a little information and background can be obtained from the Internet. It seems to take a well-organized group about four hours to draft a report and another hour to edit and clean it up.

As a final note of encouragement, it can be emphasized that the format for the case study is exactly the kind of preinterview research a person might do when looking for a job. The use of the template to gather data on the company can formulate interesting questions. This creates a great impression and helps in landing the position.

On another spin, if a person feels like being an investor in the stock market, the template also asks the critical questions one might need answered before investing hard-earned cash.

A.14 SUGGESTED FORMAT AND CONTENT FOR THE FACTORY-TOUR CASE STUDIES

A.14.1 Vital Statistics of the Company

Ideally these are listed in a spreadsheet-type "box" on the first page of the study:

- Name and location
- Type of business
- Typical client
- Major clients
- Major rivals or competitors
- Public or private company
- Gross revenues and profits
- Internet URL

A.14.2 Size and Scope

The material to cover in this area can be written out in bullets or prose and might cover the following topics:

- Annual gross sales
- Approximate profits
- Gross margin of the company
- Potential economies of scale
- Possible cash flow problems in this business
- The pricing structure
- Number of people employed
- Balance between engineers and other staff
- Recent hiring rate
- Projected hiring rate

A.14.3 Market Analysis

This material will be more descriptive than the preceding data-oriented aspects:

- Major product in the context of the clients listed earlier.
- How do they react to the competition (the prisoner's dilemma)?
- Do they build "to order" or "for inventory"? If the latter, do they have good inventory control?
- What is a typical delivery time (faster or slower than rivals)?
- History of product development
- Future of market or projected growth (flat versus high)
- Comments about distribution, packaging, image

A.14.4 What Is Unique about This Industry or Company in Comparison with Other Industries?

A company near the top right of Figure 2.2 fabricates an "old" product and, to be competitive, must focus on quality, incremental improvements, and finding the correct market niche. By contrast, manufacturers of PDAs and new computer gizmos are in the bottom left corner and need to focus on market identification. Some companies, like the subcontracting firms that specialize in printed circuit board assembly, really have no product of their own but provide a DFM and board assembly techniques: they must therefore focus on customer satisfaction and securing large quantities of subcomponents at an excellent discount. Biotech firms might focus more on patents and FDA approvals. Each sector and each company has specific issues that affect competitive advantage. It is useful to surf the Web and make comparisons with rival companies.

A.14.5 Engineering Analysis

This section can briefly describe the following:

- Major and/or unique skills
- Equipment value
- Equipment age
- Needed updates to equipment or skills
- Rework, scrap, and environmental considerations

A.14.6 Why Has This Company Been Successful in the Past?

This could cover a wide range of topics. Answers might be:

- Good relationship with consumer
- Good market niche (If so, what is its size?)
- High-quality engineering
- Good advertising
- Geographical positioning

- Staying ahead of the competition
- Willingness to invest in new equipment
- Market diversification
- Management style (e.g., horizontal or hierarchical?)

A.14.7 What Is Needed to Stay Successful in the Future?

Again, this might cover a wide range of possible answers:

- New engineering developments (with examples)
- New markets
- Improvements in quality assurance (e.g., TQM)
- New marketing techniques
- New divisions in other parts of the country
- New equipment or techniques
- Possible diversification
- New personnel strategies
- The dangers or opportunities facing the company

A.14.8 An In-Depth Recommendation

In the last section of 1,000 words or so, it is interesting to imagine that the group has been given a realistically large budget to improve productivity in one division of the company.

Recommendations can be in any area that suits the group strength: business, engineering, or policy. It is important to consider the return on investment (ROI) and other perspectives.

A.14.9 Overview

Based on the rest of the case study:

What is the company-identified core competency?
Who are the product champions?
What does your group think the core competency is?
What are the corporate values of the company?
What is the vision of the CEO?
What are the problems facing the company?
What are the generic solutions?

A.15 REFERENCES

Bowen, H. K., K. B. Clark, C. A. Holloway, and S. C. Wheelwright. 1994. *The perpetual enterprise machine.* Oxford and New York: Oxford University Press.

Canny, J. F. 2000 *Telepresence devices at UC Berkeley.* See **http://vive.cs.berkeley.edu/blimp.**

Chien, G. 2000. Low-noise local oscillator design techniques using a DLL-based frequency multiplier for wireless applications. Ph.D. thesis. University of California, Berkeley.

Dillon, P., and D. Roth. 1998. The next small thing: The right way to make mistakes. *Fast Company,* no. 15: 98–110 with the extract on page 108 (publisher address: 77 North Washington Street, Boston, MA 02114-1927).

GE. 1999. *The GE Engineering Thermoplastics Design guide.*

Magrab, E. B. 1997. *Integrated product and process design and development.* Boca Raton and New York: CRC Press.

Moore, G. A. 1995. *Inside the tornado.* New York: Harper Business. (also see *Crossing the Chasm,* an earlier book by the same author.)

Morone, J. G. 1993. *Winning in high-tech markets.* Boston: Harvard Business School Press.

Poppel, H. L., and M. Toole. 1995. The bleeding edge of information technology. *Red Herring,* 82–86.

Rudell, J., J. J. Ou, T. Cho, G. Chien, F. Brianti, J. Weldon, and P. R. Gray. 1997. A 1.9 GHz wideband IF double conversion CMOS integrated receiver for cordless telephone application. In *Proceedings of the International Solid State Circuits Conference.* San Francisco, CA.

A.16 BIBLIOGRAPHY

Allen, B. 1991. Choosing R&D projects: An informational approach. *AEA Papers and Proceedings* 81 (2): 257–261.

Norman, D. A. 1990. *The design of everyday things.* New York: Doubleday. 1990

Norman, D. A. 1998. *The invisible computer.* Cambridge, MA: MIT Press.

A.17 URLS OF INTEREST

Some interesting URLs on wearable computers are a resource for future projects:

www.microopticalcorp.com
www.5050LTD.com
www.InfoCharms.com
www.symbol.com
http://www.cs.cmu/edu/afs/cs/project/vuman
www.xybernaut.com

A.18 CASE STUDY: THE "PALM PILOT"

The June/July 1998 edition of *Fast Company* included an extensive review of Palm Computing's development of the popular *Palm Pilot.* The following extract is part of that longer article by Dillon and Roth (1998).

A.18.1 The Right Way to Make Mistakes

What's the secret to breakthrough entrepreneurial success? If the 15-year saga of the Palm Pilot is any guide, it's not perfect execution. That's too much to ask of any new company. Rather, it's making mistakes, learning from them faster than the competition, and marshaling resources to stay in the game until you're in a position to win.

Here are three lessons to take away from the Palm Pilot's success.

A.18.1.1 Sometimes the Wrong Product Is the Best Teacher

Palm Computing's first product, *Zoomer,* was a failure. But it was a necessary failure. From it, Palm learned what consumers really wanted from PDAs. With Zoomer, concedes founder Jeff Hawkins, "We produced a flawed product." The second time around, he says, "I wanted no excuses for failure. Every step of the way, we investigated the relationship of every decision we made to the product we were trying to deliver."

A.18.1.2 If You Want a Bit of Margin for Error, Build a Lean Organization

One of the miracles of the Palm Pilot was that Hawkins and his colleagues[5] did so much with so little. Apple spent an estimated $500 million in designing, building, and marketing the Newton. Microsoft has invested $250 million in Windows CE, the key to its entry into the handheld market. Palm Computing needed just 28 people and $3 million in development spending to make a prototype of the Pilot. Even today, amid runaway growth, the company employs fewer than 300 people.

Credit Donna Dubinsky's less-is-more leadership style. "Early on," she says, "Palm Computing's controller was a woman who kept a close eye on the cash flow. And our motto was 'Cash is Queen.' If we had always been on the verge of running out of money, we would not have been able to negotiate the best possible deal for the company."

A.18.1.3 There Is No "I" in Team

It's easy to trace Palm's success to the vision of its founder. But the journey from brainstorm to breakthrough required a committed team. "When I look at my own contribution to the success of the Pilot," says Ed Colligan, one of the early developers, "I can point to one major thing, and that's helping to build the best team in the industry. I can take some credit for that—and for trying to make Palm Computing a fun, exciting, energetic place to work. The team, rather than a good product idea, is what really makes things happen."

[5]See <www.handspring.com> for their new company.

Index